Photomask Fabrication Technology

Photomask Fabrication Technology

Benjamin G. Eynon, Jr.
KLA-Tencor Corporation

Banqiu Wu
Photronics, Inc.

McGraw-Hill
New York Chicago San Francisco Lisbon London Madrid
Mexico City Milan New Delhi San Juan Seoul
Singapore Sydney Toronto

The McGraw-Hill Companies

Library of Congress Cataloging-in-Publication Data

Eynon, Benjamin G.
Photomask fabrication technology / Benjamin G. Eynon, Jr., Banqiu Wu.
p. cm.
Includes bibliographical references and index.
ISBN 0-07-144563-3 (alk. paper)
1. Integrated circuits—Masks. 2. Masks (Electronics). 3. Microlithography. I. Wu, Banqiu. II. Title.

TK7872.M3E99 2005
621.3815'31—dc22

2005047886

1 2 3 4 5 6 7 8 9 0 DOC/DOC 0 1 0 9 8 7 6 5

ISBN 0-07-144563-3

The sponsoring editor for this book was Stephen S. Chapman and the production supervisor was Richard Ruzycka. It was set in Century Schoolbook by International Typesetting and Composition. The art director for the cover was Handel Low.

Printed and bound by RR Donnelley.

This book is printed on recycled, acid-free paper containing a minimum of 50% recycled, de-inked fiber.

Contents

Contributors

Paul DePesa *Synopsys, Inc., Mountain View, California* (Chap. 2)

Florence O. Eschbach *Intel Corporation, Santa Clara, California* (Chap. 6)

Benjamin G. Eynon, Jr. *KLA-Tencor Corporation, San Jose, California* (Chap. 1)

Emily Fisch Gallagher *IBM Corporation, Essex Junction, Vermont* (Chap. 8)

Mark Gesley *Skyler Technology, Inc., Walnut Creek, California* (Chap. 3)

Derek Kay *Synopsys, Inc., Mountain View, California* (Chap. 2)

Michael Lercel *IBM Corporation, Essex Junction, Vermont* (Chap. 8)

Ted Liang *Intel Corporation, Santa Clara, California* (Chap. 6)

Dhirendra P. Mathur *Photronics, Inc., Brookfield Connecticut* (Chap. 8)

Gary Meyers *Synopsys, Inc., Mountain View, California* (Chap. 2)

Richard M. Silver *NIST, Gaithersburg, Maryland* (Chap. 5)

Robert J. Socha *ASML, Santa Clara, California* (Chap. 7)

Alan R. Stivers *Intel Corporation, Santa Clara, California* (Chap. 6)

Edita Tejnil *Intel Corporation, Santa Clara, California* (Chap. 6)

András E. Vladár *NIST, Gaithersburg, Maryland* (Chap. 5)

Banqiu Wu *Photronics, Inc., Austin, Texas* (Chap. 4)

Preface

The quality and functional capabilities of nearly all of today's electronic devices and the production lines that produce them are dependent on key internal semiconductor components. These components, called integrated circuits (ICs), are manufactured through an optical reduction process known as photolithography; the pattern transfers of which are made possible only by master templates called *photomasks*. The ability to provide IC manufacturers with sets of perfect photomasks for each device they produce is the determining factor in keeping pace with the electronics industry's quest to shrink ICs and other related products ever smaller.

Photomasks are defect-free optical templates that serve as printing masters for the fabrication of all ICs. These "masks" as they are sometimes called, are made from highly-polished slices of fused silica (glass) coated on one side with a thin absorber layer of light-blocking metal or metal oxynitride. The pattern for one layer of a microchip is etched into the absorber layer such that a high-precision optical exposure system can transfer that pattern onto a silicon wafer or other semiconducting substrate. In mainstream photomask production, chromium serves as the absorber for both binary intensity masks (BIMs) and alternating aperture phase-shift masks (AAPSMs). Molybdenum silicon oxynitride serves as the absorber for embedded attenuated phase-shift masks (EAPSMs). Other exotic materials and applications are being investigated even now in hopes of finding commercial application in niche market products such as flat panel displays, thin film heads, printed circuit boards, optical switches for fiber-optic cable telecommunication, microelectronic mechanical systems (MEMS), and many other emerging nanotechnology applications, or for meeting future mainstream IC device specification requirements (i.e., those found in everyday products such as computers, calculators, cameras, and automobiles).

Since Texas Instruments invented the integrated circuit in 1959, photomask fabrication has become an integral part of semiconductor wafer production. Starting out as simple 1X-magnification optical stencils, photomask fabrication technology has continued to advance with the needs of high-end semiconductor wafer production. As long as the IC features required were larger than the wafer exposure tool's source wavelength, photomask making was relatively straightforward. However, when IC feature sizes matched or fell below the exposure tool's source wavelength, photomask fabrication moved squarely into the middle of the critical path of future IC business success. Very strict mask critical dimension (CD) and feature placement specifications, intensive capital equipment investment, unique raw materials and applications, and special expertise requirements for photomask fabrication technologists resulted in the formation of an independent photomask industry. In the 1990s, this photomask industry was symbolized by a group of photomask manufacturers, electronic design automation (EDA) companies, photomask substrate and pellicle raw material suppliers, equipment manufacturers, and end users. Also growing at this time were international professional societies, annual symposia, and peer-reviewed professional journals.

Global photomask production is a moderate-scale industry by annual revenue measures. By comparison, gross revenue of the photomask industry (roughly 1.5–2.0 billion U.S. dollars) falls well below 5 percent of the size of the overall semiconductor industry.

Annual photomask-specific symposia began in 1981, and presently, technical publication counts approach 1000 each year. Accumulated publications on photomask fabrication are in the several thousands. This book serves as a reference manual for those currently in the photomask industry, a training resource for newly hired photomask employees, and a textbook for university coursework in microlithography and nanotechnology.

The objective of this book is to complete a comprehensive reference book on photomask technology, including scientific theory, fabrication processing methods, quality control parameters, and an overview of the industry's technological future.

The contents of this book are designed and written for photomask process engineers, photomask equipment manufacturing engineers, photomask raw material production engineers, wafer fabrication engineers, professors and researchers, university students, and technical management.

This text serves as a handbook describing the science and technology of industrial photomask production. It includes fundamental principles, industrial production flows, technological evolution and development, literature reviews, present state-of-the-art technologies, and predictions for the photomask industry's future.

The contents consist of an introduction to the photomask industry, data preparation and pattern formation, the pattern transfer process, photomask metrology, defect reduction and control, resolution enhancement techniques, and future photomask technology trends.

The book contents (chapters) and styles were designed and proposed by the principal authors—Benjamin Eynon and Banqiu Wu. The individual chapters were written by the principal authors and industry expert chapter contributors.

Benjamin G. Eynon, Jr.

Banqiu Wu

Photomask Fabrication Technology

Chapter

1

Introduction

Benjamin G. Eynon, Jr.

1.1 Overview

While absolutely essential to the manufacture of nearly every electronic product made today, such as the semiconductors that go into computers, cellular phones, and automobiles, the photomask's role in the proliferation of semiconductor technology is rarely recognized. Many books have been written on the subject of semiconductor processing theory and techniques, but very few works have been published solely on photomask fabrication.

This book serves as an overview of the processes and technology associated with producing the various types of photomasks used in the fabrication of semiconductor devices and other components that use a photolithographic patterning process. It is intended to serve as a text for university engineering students and a source of reference for both photomask and semiconductor lithography technical professionals.

The contents of this book are organized in chronological order according to the photomask fabrication process (see Fig. 1.1) with special topics following. After the introductory chapter, the data preparation step is explained in detail in Chap. 2 as it pertains to providing the photomask writing, inspection, and metrology tools—the necessary pattern information required to produce and certify the exact device layer intended. Chapter 3 describes the processes that take place once the "prepped" data are sent into the cleanroom where the photomasks will be fabricated. The data files are loaded into the writing tools so that the resist film on the mask blank can be exposed by an electron beam writer or a laser writer, then postexposure baked (PEB), and then developed using solvent or aqueous chemistry. Chapter 4 then describes the pattern transfer step whereby the features printed in the resist are transferred into the absorber layer(s) by means of a wet acid or dry plasma etch. After chemically stripping the resist layer, the photomask has essentially been built. The remaining steps are for either the validation of pattern specifications, cleaning and finishing for the intended application, or certification of compliance to specifications for customer shipment.

The photomask metrology steps are next to be completed to ensure that all physical and optical properties are within specification tolerances. These steps include *critical dimension* (CD) measurement, pattern placement (registration) measurement, and possibly some special

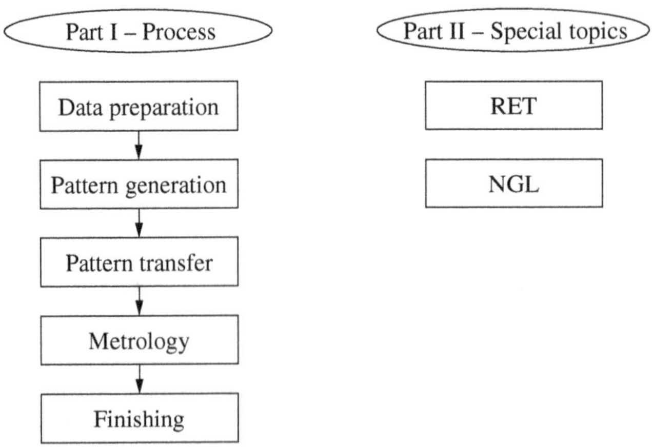

Figure 1.1 Content flowchart.

physical and optical characteristic measurements depending on the intended application. Chapter 5 describes these metrology steps and the theory behind them in detail. On completion of metrology certification, the photomask undergoes a finishing process as presented in Chap. 6. In this chapter, the inspection, repair, cleaning, and pellicle mounting steps are defined and discussed. With the completion of the steps detailed in Chap. 6, the photomask is complete and ready for use in the lithographic process for which it was fabricated.

The remaining two chapters cover special technology topics relevant to the photomask industry. Chapter 7 describes the various *resolution enhancement techniques* (RETs) in practice today. The principles of *optical proximity correction* (OPC) and phase-shifting mask (PSM) theory are described in detail. Chapter 8 then presents the technologies and issues of the future of photomask fabrication. In this chapter, both the fledgling strategies and those still very much in the research and development stages are described with respect to their principles of operation and the obstacles each faces to gaining full commercial adoption.

1.1.1 History

Prior to the 1960s, electronic devices were manufactured by soldering discrete components together to form the desired electrical circuits and, ultimately, the desired electronic components or products. The result was the creation of a working device, albeit primitive by today's standards, regarding number of functions, physical size, amount of power consumed, quantity of heat generated, ergonomics, and degree of control via user interface, to name a few categories. During this time period, the ubiquitous vacuum tube had also recently been replaced by the transistor, the adoption of which resulted in a huge reduction in electronic device size and heat load. Also, governmental and consumer populations were demonstrating their desire for even better, smaller, faster devices by funding the research and then buying the resulting products at an ever-growing rate. The semiconductor industry had begun.

Even though transistors were a vast improvement over vacuum tubes, electrical devices still comprised discrete components, until the invention of the integrated circuit in 1959. With this invention, the need for a multistep microlithographic process that had the ability to transfer an entire circuit pattern onto a semiconductor substrate was evident. Leveraging the ability to design and build all components (such as resistors, capacitors, and transistors) on a single piece of semiconductor material (such as silicon and gallium arsenide) using common process steps allowed for the manufacture of much smaller and less expensive devices. So, shortly after semiconductor device physicists came up with clever ways to ion-dope silicon and to connect and insulate *p* and *n* regions to form unique electrical components, another group of individuals created a process for producing high-quality optical patterns used to transfer circuit pattern layers onto these semiconductor materials. Without these master circuit layer patterns, called *photomasks*, no microchip could be built using a microlithographic process, and the electronics industry as we know it would not exist.

In the beginning, photomasks were not much more than optical stencils used to transfer patterns for single layers of semiconductor devices onto the semiconductor substrate material. Today, the resolution and image placement specifications required of photomasks have essentially turned them into removable transmissive elements in the wafer exposure tool's optical path. This is an enormous change from 40 years ago, and has spawned new fields of study such as *design for manufacturability* (DFM), new technologies such as complex resolution enhancement techniques, and countless feature CD uniformity and linearity improvement programs. To make matters worse, the photomask defect tolerance is zero at printable severity since it serves as the master from which all device patterns are printed in the lithographic process. A single defect on what is called a single-die photomask could result in printing nothing but zero-yielding devices (i.e., "garbage"!). So, not only must all of the pattern features be of the correct size, position, transmission, and phase, but there must also be absolutely no printable contamination or flaws in the pattern plane either, such as dirt particles, stains, smudges, and scratches. Optical element specifications and the requirement for zero printable defects make building photomasks very different from building the actual electrical devices.

Back in the early 1960s, all companies wishing to produce semiconductor devices (later called microchips) needed to start from scratch. Managers in each manufacturing company assigned a small group of people to create a process line to produce its very first photomasks. The term "mask shop" was coined as these small, separate groups within each larger semiconductor manufacturing company were leveraged to supply the master patterns for all *integrated circuits* (ICs) being built. To establish these processes, early mask shop pioneers were challenged to select the most suitable substrate and absorber materials and then select or even build the requisite equipment that could produce these photomasks. Following the stencil analogy, the absorber material needed to block exposure radiation while the substrate material needed to allow 100 percent of it to pass. More discussion on these materials will follow later in this chapter.

At this early stage, photomask manufacturing was a manual process performed by a skilled few. The masking material, called Rubylith, was the principle design medium used to transfer the circuit pattern to the photomask itself. Rubylith was made up of two thin sheets of plastic adhered to each other. The base layer was a stiff, scratch-resistant plastic that was transparent to the green light used in the subsequent photomask exposure process. The absorber layer was a red-colored plastic that absorbed green

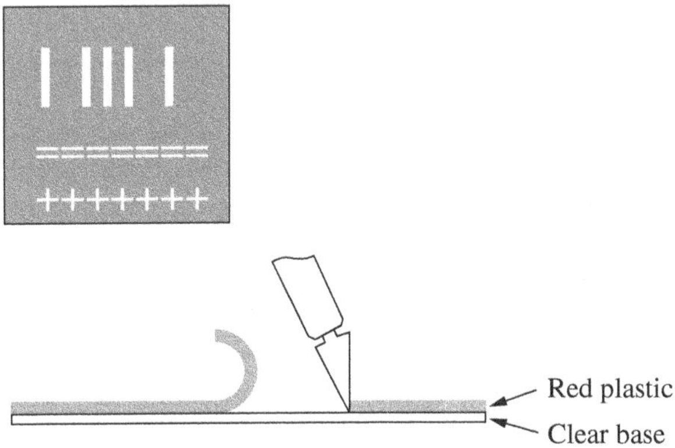

Figure 1.2 RubyLith.

light, rendering it opaque under exposure. So the first "advanced" design work station consisted of a light table, masking tape to hold the Rubylith in place, a hobby knife, and a straight edge! Designs were trace cut by hand using the hobby knife and straight edge, and then the red plastic was peeled off leaving the red-on-clear circuit pattern (see Fig. 1.2).

The Rubylith pattern was then reduced and replicated via a photoemulsion coating (sensitive to green light) on a thin piece of glass. This piece of glass was called a *reticle* and was an intermediate component toward making the final photomask.The remaining steps were accomplished by reducing and replicating the high-magnification image from the reticle to an emulsion layer on another glass plate. After photoemulsion development, the final product was the photomask (see Fig. 1.3). The

Figure 1.3 Basic 1X photomask.

reduction ratio was such that whatever scale the design was cut originally was reduced via intermediate process steps to the desired semiconductor device size (1X).

Once the photomask was built, the Rubylith and reticles were no longer needed technically. They were stored in case of irreparable photomask damage so another copy could be made. This damage did come occasionally from careless blunders of course, where the photomask was inadvertently scratched or even broken into pieces, but it also occurred from normal use on high-running IC devices. The reason was that early photomasks were used in a contact printing process. The mask was placed in a contact printing exposure tool, photoemulsion side down, and then pressed against the photoresist coating of a semiconductor wafer for exposure. After exposure, the photomask was lifted off the wafer, ready for another wafer exposure. Since the photoemulsion material was not very durable for this type of process, multiple copies of the photomask had to be made from the reticle to avoid excessive yield degradation from wafer to wafer.

Making multiple photomasks proved costly, so stronger materials were sought out to replace the photoemulsion layer. Ultimately, chromium was selected for its durability to physical contact and for its optical density to exposure radiation.

To further reduce costs, direct writing of the photomask via electron beam exposure tools came into existence in the late 1970s. Laser tools were also introduced in the 1980s as an alternate photomask writing method to e-beams. By writing the photomask in one step, all the other costly intermediate steps were eliminated. The need to store RubyLith and glass reticles was replaced with electronic file backups, and photomask fabrication cycle time decreased due to the fewer number of process steps.

1.2 Materials

In principle, the fundamental material components of a finished photomask comprise a dimensionally stable, optically homogeneous and transparent substrate, and a layer of thin opaque or phase-shifting absorber. Figure 1.4 shows the very basic configuration as used in a wafer exposure tool. The exposure energy must pass through the photomask from the backside, then through the patterned absorber, and finally through the protective pellicle membrane. The patterned image is then reduced through the exposure tool's lens system where it finally exposes the photosensitive resist coating on the wafer.

1.2.1 Substrate

Even though a part of the semiconductor industry, the photomask is an optical component and therefore has some unique specifications

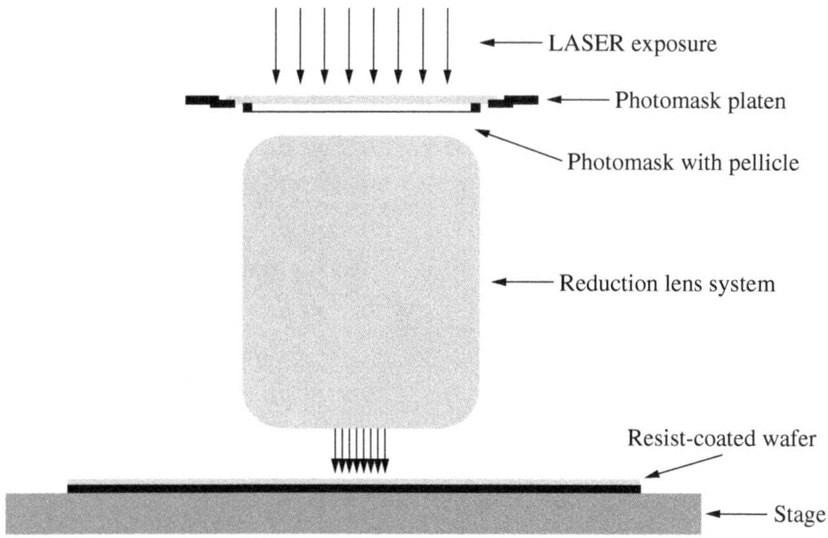

Figure 1.4 Wafer scanner block diagram.

compared to its electrical wafer counterparts. Photomask substrate requirements include both physical and optical properties and must be tightly controlled since the photomask serves as the master copy for all subsequent pattern exposures. There is very little room for error regarding these physical and optical qualities. Optically, the material must be transparent at the wafer exposure tool's wavelength and must have minimal reflectivity as well. Many different glass and crystalline materials have been used over the years, such as soda lime glass, white crown, quartz, and various grades of fused silica. Today, the standard photomask substrate material used for advanced lithographic applications is amorphous fused silica. This material possesses high transmittance and low reflectance at the actinic (commercial exposure) wavelength, in addition to possessing superior physical properties.

The fused silica manufacturing process varies across the supplier base, but in general, it is deposited at high temperature into a mold. There can be one or more annealing steps where the temperature is ramped and cooled to "heal" impurities, but the end result is a large block of high-purity glass called a boule. Proprietary manufacturing processes make it such that the optical property homogeneity is maximized, and optical aberrations are minimized. The boule shape is typically a tall cylinder whose diameter is somewhat larger than the diagonal axis of the desired photomask blank. The cylinder is sliced into disks and then reshaped into a square by trimming four straight edges. Optical defects

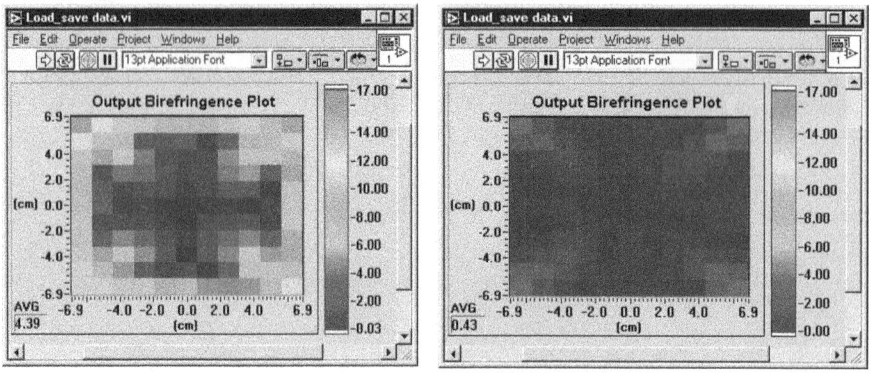

(*a*) (*b*)

Figure 1.5 (*a*) Poor birefringence substrate; (*b*) superior birefringence substrate.

in the boule reside around the circumference due to stresses induced from the mold. Knowing this, the glass substrate manufacturer builds the mold large enough so that the defects do not encroach into the photomask blank area needed for actual printing. Figure 1.5 shows an optical stress map of two photomask substrates manufactured under different processes. The substrate in Fig. 1.5*a* was manufactured using a nonideal process, while the process of Fig. 1.5*b* was much more suited to producing substrates with lower stress. Photomask pattern CD and placement errors are proportional to the stress in the glass. While minor according to specifications at the time of this publication, these errors may prove more significant as semiconductor designs continue to shrink, and optical errors consume more of the wafer lithography error budget.

Initially, all edges and surfaces are cut slightly larger than the final desired size to allow for the removal of material during the rough and fine polishing processes that bring the final dimensions within specified tolerances. Again, since a photomask is an optical product, care must be taken to ensure that the slices are perfectly straight and parallel to avoid errors such as the formation of a wedge or localized concave/convex lens areas (see Fig. 1.6*a* to *c*). The polishing process cannot correct a wedge (slicing) problem, but is meant to improve surface roughness and overall flatness.

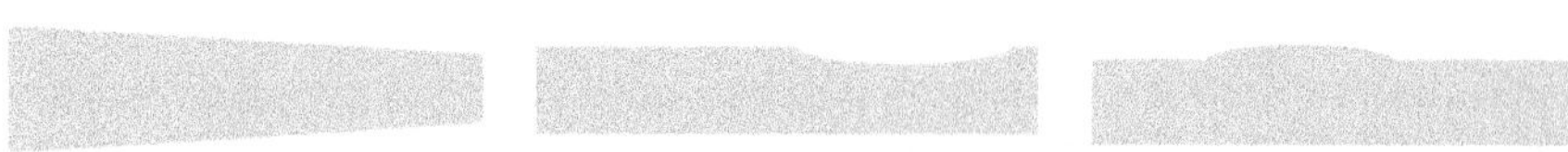

(*a*) (*b*) (*c*)

Figure 1.6 (*a*) Wedge error; (*b*) concave error; (*c*) convex error.

Physical dimensions and tolerances are specified and controlled for use in wafer exposure systems and for handling by human beings. Factors considered and standardized include:

1. Side length fixturing (*x*, *y*, squareness)
2. Focal plane fixturing (thickness, flatness, surface roughness, wedge)
3. Edge beveling for human safety
4. Final weight (human and robotic handling)

Optical substrate tolerances are also considered, which include:

1. Percent transmission
2. Percent reflection

Other factors not specified but that come as a result of appropriate photomask substrate use include:

1. Stresses due to fixturing (clamping, vacuum)
2. Forces due to mounting orientation (gravity)
3. Localized stresses due to the attachment of a pellicle
4. Factors due to movement during exposure (momentum, inertia)
5. Thermal expansion and contraction due to irradiation
6. Crystalline compaction due to long-term irradiation
7. Creation of nucleation sites for progressive defect formation
8. Overlay matching to other photomasks (device layers) over time

Substrate material must be able to transfer the pattern image with near perfect accuracy, so it is also required that no material properties cause any image distortion of any kind and that no flaws exist in the material such as particles, microcracks, or voids.

In the earliest days of photomask making, any specifications that existed for dimensions and tolerances varied from company to company. The exposure tool industry has since collaborated to set these standards and has updated them as technology requirements dictated. Photomask shapes and thicknesses have gone through many iterations over time for many different reasons and applications. They have been used as noncritical components in high reduction magnification situations, and have also served as critical optical components in lower magnification situations where the features to be printed were smaller than the wavelength used. More details on the subwavelength regime and its effect on the *mask error enhancement factor* (MEEF) are covered in Chap. 7.

When the photomask was used as a proximity printing master in the 1980s, or even as a 5X reduction element in the 1990s, requirements for flatness and rigidity were fairly loose by today's standards. In these cases, the photomask substrate could be large but relatively thin (7.0 in. × 7.0 in. × 0.15 in., for example). When the exposure tool industry moved to a 4X reduction scenario at the same time when the wafer pattern features required to be printed were near the actinic wavelength size, the photomask substrate needed to maintain its shape without sagging. Sagging errors caused by the mounting arrangement in typical wafer exposure tools manifest themselves as pattern position errors on the wafer that can lead to unacceptable yield loss. By increasing the photomask substrate thickness to 0.25 in. in this example, the mask substrate became more rigid, meaning that there was less sag and therefore less pattern position error.

Commercial photomask sizes over the last 20 years have ranged from 4 to 7 in. square sizes and have also taken on some nonsquare shapes. There were also some cases of smaller than 4-in. and greater than 7-in. photomasks as well during early development or other specific research applications. In addition to square-shaped substrates, nonsquare shapes have included a 7.25-in. round shape that was used in some 1X projection printers, and one company manufactured a stepper that required the photomask maker to cut a 5.0 in. × 5.0 in. × 0.090 in. substrate into a 3.0 in. × 5.0 in. shape. Beveling was also performed on the newly cut sharp edge for safety.

Without prior knowledge, one would assume that the smaller substrates were used in the past for simpler technology and the largest substrates were used for the newest technology. As it turns out, the largest commercial (7.0 in. × 7.0 in.) substrates used to date were used for older 1X projection printing applications. Physical specifications change as reduction factor and device layer pattern requirements change. However, the reduction factor has gone up from 1X in the 1960s and 1970s to 5X in the 1980s then back down to 4X in the 1990s and 2000s. Of course, 1X magnification today would be the most difficult-to-achieve specification since whatever is on the mask actually would print on the wafer. In other words, a pattern feature size or placement error would not benefit from a reduction factor, so it would print at the same severity as was on the photomask. Back when a 1X reduction factor (i.e., no reduction) was used, device features were not challenging optical system resolution limits, so success was achieved for design objectives in that era. The photomasks could also be larger then because they were oriented vertically in many wafer exposure tools. The forces of gravity do not make the photomask sag in this orientation, so pattern integrity was maintained without needing to thicken the substrate for pattern stability.

When designs did eventually push the limits of 1X lithographic capability, a decision was made to increase the magnification factor across the industry to 5X on the newly developed wafer stepper. At the same time, the photomask size was standardized at 5.0 in. × 5.0 in. × 0.09 in. Any size or placement error on the photomask was also reduced by the magnification factor. This was very beneficial to the photomask maker as many of the errors seen on the photomask were unresolvable after the image was reduced in size by five times. Specifications on the equipment needed to build 5X photomasks could also be relaxed. The period of time that 5X wafer exposure tools were used was called the "mask maker's vacation." The only downside to making the change from 1X to 5X was that chip designers would have to restrict their maximum design size to 5X compared to the space available on a photomask under 1X wafer lithography conditions.

When device design sizes became too large to fit on the 5-in. photomask substrate, and the need for a more rigid substrate was evident, the industry again made a move to accommodate. Wafer exposure tool companies standardized on a new magnification and a larger photomask substrate size. The magnification went back from 5X down one step to 4X, and the substrate size went up in both side length and thickness to 6.0 in. × 6.0 in. × 0.25 in. These two factors, in addition to the ever-tightening pattern fidelity specifications ended the "vacation" enjoyed by the mask industry. Pattern errors were now reduced by only 4X instead of 5X—a 20 percent loss in margin. To add to that, the area covered by the patterns increased on the new larger substrate, which forced the requirement to maintain all pattern specifications over this larger area as well.

To be complete, mention must be made of the lithography industry's serious consideration of moving to an even larger substrate standard. A significant amount of investment and effort in the late 1990s explored the possibility of adopting a 9.0 in. × 9.0 in. × 0.4 in. substrate. Besides the increase in the tool cost of ownership, many other infrastructure factors that seemed to drive the need for the 9-in. substrate were systematically dismissed over time such that the need was mitigated—at least for the time being.

1.2.2 Absorber

Up to this point, the discussion has centered primarily on the photomask substrate. Assuming that a photomask blank manufacturer can procure high-quality glass, both from a physical and optical standpoint, the next step is to coat the substrate with the optical masking material called the absorber. The two major absorber categories are called *binary* and *phase shifting*. The purpose of a binary absorber film is to serve as a simple stencil. All opaque areas are meant to completely block exposure tool illumination, and all clear areas are meant to pass all illumination. The phase-shifting film is more complicated and will be covered in more detail in Chap. 7. It is enough to say here that the phase-shifting absorber allows some illumination to leak through in order to provide destructive interference where unwanted radiation would have added constructively if the mask were built with a binary absorber.

Binary. Early photomask pioneers used different types of opaque absorber materials to either paint or tape the first primitive patterns onto small thin pieces of glass, which were used as contact print templates for semiconductor devices. As the industry evolved, chromium (or chrome) was selected as the standard *binary intensity mask* (BIM) absorber material for its durability and optical properties. The chrome absorber layer went through several developmental morphologies where thickness and reflectivity were adjusted until it reached a form in the 1980s that held fairly constant through the end of the century. As depicted in Fig. 1.7, the chrome absorber is actually a stacked film comprised of roughly 20 percent chrome oxide *antireflective* (AR) layer on top of the remaining 80 percent pure chromium. The industry standard thickness became about 1050 Å, which meant that the stack component thicknesses were roughly 800 Å and 250 Å for the chrome and chrome-oxide AR films, respectively.

The sputtering process is used to apply the chrome layer to the glass substrate. The technological aspects of this process are well covered in other texts, so a thorough treatment is not necessary. Compared to the wafer process, a few unique issues are encountered when sputtering photomask

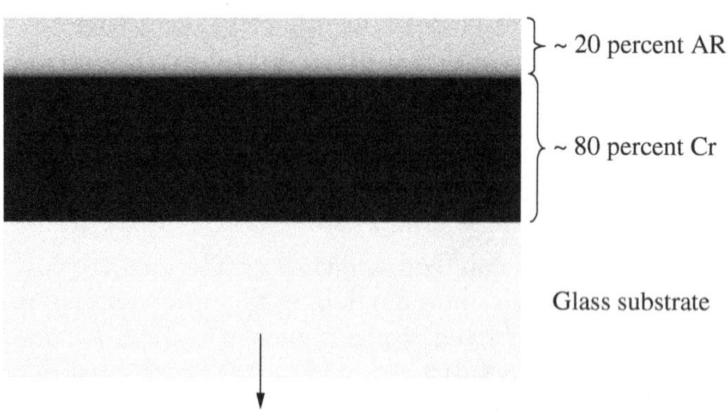

Figure 1.7 Partial cross section of a typical chrome photomask blank.

substrates. The first is that the area to be covered is square and not circular. Deposition must be uniform out to the corners of a photomask—the most difficult region in which to maintain thickness uniformity. Special physical baffling and other process optimizations have been made over the years to provide continuous improvement to the chrome absorber layer thickness and optical property uniformity.

The *optical density* (OD) of the chrome absorber was just above the specified 3.0 and the reflectivity was around 10 percent down to the 365-nm (i-line) lithography tool exposure wavelength. Optical density was dictated by chrome layer + antireflective layer thickness, while reflectivity was controlled primarily by modifying the stoichiometry and thickness of the antireflective layer on the chrome surface. For older generations (365 and 436 nm), the Cr reflectivity remained fairly constant at about 12 percent. It was not until the industry moved to wavelengths below 365 nm that photomask reflectivity started to increase substantially. It is now about 20 percent reflective at the 193-nm exposure wavelength.

It is true that at the 193-nm exposure wavelength, chrome OD falls as thickness is reduced. One might make a suggestion here to keep the chrome absorber thickness high to avoid the risk of falling below OD requirements and specifications, but thinner chrome is desired for current and future cases where the pattern feature aspect ratio is increasing. As device pattern features shrink, unless accompanied by a corresponding reduction in absorber thickness, the aspect ratio of the final etched feature will increase correspondingly. At some point, the aspect ratio will become too great, and the feature will be impossible to resolve by either e-beam or laser patterning, or by the subsequent pattern transfer processes. In the early 2000s, it was determined that the minimum practical OD for a binary absorber could be reduced from 3.0 to around 2.5 without adverse effects. Thanks to this assertion, "thin chrome" emerged as a new product in support of extremely small feature resolutions such as those found in optical proximity correction applications.

Phase shifting. In an effort to extend the lives of expensive wafer exposure tools, resolution enhancements came in the form of a phase-shift mask (PSM). The term *phase shift* applies to the incident radiation in a wafer exposure tool relative to the radiation that has undergone a change in phase due to either the lack of, or addition of the photomask material. An unaltered phase area where an incident wave travels is called a "0° phase" region, and an area of a photomask that has been π-phase shifted is called a "180° phase" region. There are two major categories of PSMs—*alternating aperture phase-shift masks* (AAPSM) and *embedded attenuated phase-shift masks* (EAPSM). Nicknames and pseudoacronyms for these two resolution enhancement technique types are many, and some of them are listed below:

AAPSM	EAPSM
APSM	EPSM
Hard shifter	Soft shifter
Strong shifter	Weak shifter
Quartz etched	Moly, MoSi

From a material's standpoint, the alternating aperture class of PSM is built from a standard *chrome-on-glass* (COG), or binary photomask blank. No special materials are used since the action of phase shifting is accomplished by removing portions of the fused silica substrate by means of a wet or dry etch. So in the AAPSM case, phase shifting the incident exposure radiation is accomplished by *removing* material (glass) in strategically selected areas in order to enhance feature resolution.

Conversely, EAPSM masks rely on the properties of an *added* film to accomplish the desired phase shift. These films allow a small percentage of the incident exposure radiation to "leak" through so as to enhance the resolution capabilities of the exposure tool beyond what was possible with binary photomasks. The actual transmission percentage of films varied during development phases in the i-line (365-nm) stepper era, but became standardized at 6.0 percent with its widespread adoption during the KrF (248-nm) scanner era. In all cases, the phase-shifting film needed to be designed such that it inverted the phase of the radiation it passed 180° relative to the radiation traveling through the transparent substrate nearby. More on the principles of phase shift can be found in Chap. 7.

The film of choice for both the KrF and ArF (193-nm) eras is *molybdenum silicide oxynitride* (MoSiON). Overall stoichiometry of MoSiON EAPSM films varies by the blank supplier and needs to be changed globally if the film is deposited in a homogeneous manner, or individually if a bilayer approach is used. The requirement, regardless of film composition, is to achieve the 6.0 percent transmission target while simultaneously causing a 180° phase shift in the leaked or transmitted radiation relative to the incident wave.

Experimental EAPSM absorber materials have been created to provide transmissions much greater than 6.0 percent while still maintaining 180° phase-shifting properties. Such high transmission films have not been used commercially yet, but may become relevant as a means by which to improve process margins on certain types of pattern features (e.g., hole patterns). These high transmission films may help find yet another lower-cost alternative to rapid implementation of more *expensive next generation lithography* (NGL) technologies (see Chap. 8).

Regardless of the absorber film's transmission percentage, the material scientist must consider not only the optical properties of the film at actinic wavelength, but also at the wavelengths of all tools used in the manufacturing process. For example, an EAPSM absorber material must also have sufficient contrast at the inspection tool's wavelength to result in the detection of any pattern defects requiring repair. While a film may be usable under actinic exposure conditions, the absorber film must not be too transmissive to be inspected under standard manufacturing process conditions.

Because the phase-shifting film is partially transmissive, a layer of more opaque chrome + AR must be deposited on top for subsequent border application purposes (see Fig. 1.8). Wafer exposure tool sensing devices and wafer fab automation barcode readers require high contrast images for proper performance. The wafer exposure tool's adjustable masking blades also have enough placement error to require some sort of opaque placement error zone on which to land during setup. The photomask maker is then required to provide a *chrome-on-MoSi* border on all EAPSMs to achieve a high contrast (chrome-on-glass) pattern in the borders.

In certain pattern types and with most high transmission EAPSM absorber films, some chrome must be left over the MoSiON absorber to ensure that too much radiation leakage through the absorber does not occur. Photomasks that utilize this strategy are called *tritone* masks for the three transmission levels present— fully transmissive glass regions, partially transmissive absorber regions, and nontransmissive chrome regions. The chrome features can be found in the scribe area also where they are used to facilitate more accurate metrology readings for certain applications.

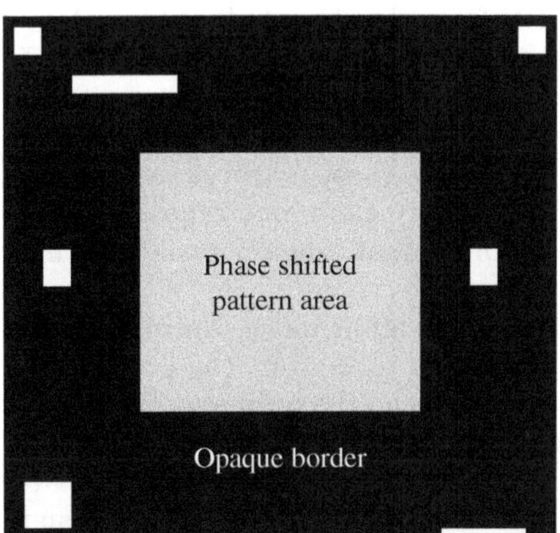

Figure 1.8 General EAPSM layout.

1.2.3 Resist

As is the case with wafer fabrication, a photosensitive yet etch-resistant coating is patterned and later stripped once the pattern is transferred to the underlying film. It is important to mention resist as a photomask material because photomask blanks are shipped from the blank supplier to the photomask manufacturer precoated with resist. Anyone familiar with wafer fabrication knows that bare wafers are purchased as exactly that—bare. For single-exposure photomasks (i.e., binary), no other resist coating is required of the photomask manufacturer to completely fabricate the mask. So, for the better part of three decades, resist coatings were provided solely by photomask blank suppliers. It was not until phaseshift masks entered mainstream manufacturing that a second (third or fourth) resist recoating was required of the photomask maker. A comparison of the process flows of BIM, EAPSM, and AAPSM will be discussed later in this chapter.

In contrast to optical-only processing as in wafer lithography, photomask lithography can be accomplished by means of either an optical or electron beam process. In the 1960s and early 1970s, photomasks were processed using primitive optical methods since these exposure systems were lower cost, easier to manufacture, faster, and had sufficient capability relative to e-beam systems. It was not until later in the 1970s that e-beam exposure made its debut in commercial mask making. Photomask blank manufacturers then needed to provide two different resists to photomask manufacturers depending on the lithography strategy employed. Obviously, the photoresist was coated for optical processing and the e-beam resist was coated for e-beam processing, but in both cases, the photomask manufacturer could simply load the blank onto the exposure tool right out of the box.

As stated above, photomask manufacturers were finally required to coat the resist as part of standard PSM flows. In almost all cases to date, photomask exposure tools used for what is called the "second write" step in PSM processing have been the optical laser type since maximizing feature resolution was not the main goal. As pattern feature sizes continue to shrink and specifications on those features tighten, an e-beam exposure may be required to satisfy customer requirements. The photomask maker will need to implement e-beam resist coating capability at that time.

Another degree of freedom the device designer has to optimize lithographic performance in the mask shop is to lay out the pattern in reverse tone. Positive-working resists have been the mainstay for nearly 20 years until recently when negative-working e-beam resists reappeared in commercial photomask production. At the time of this writing, blank suppliers provide both positive and negative e-beam resist coatings on either binary or EAPSM blanks. Also, all resist coatings made by photomask manufacturers for PSM processing (second write step) are the positive optical type.

Specifications on photomask resist quality are inferred by the photomask maker based on monitoring the dose required to produce the desired CD size on the mask after resist development. Resist categories quoted by blank manufacturers include both thickness average and uniformity, but do not include any guarantees for photospeed ranges. Because of the difficulty maintaining not only the resist thickness and photospeed uniformity in the corners, but also the chrome + AR thickness and optical property uniformity and the substrate optical quality uniformity, photomask blank manufacturers specify two general grade categories. The high-grade category guarantees that all specifications will be met all the way out to the corners, while the lower-grade category guarantees that specifications will be met only in a centered circular-shaped region where the corners are ignored.

1.2.4 Pellicle

The word "pellicle" comes from the Latin and refers to a thin transparent film. While added to the finished photomask on completion, the pellicle serves to keep foreign particles from falling onto the photomask pattern surface where they could become killer defects on the wafer (also called "repeaters"). The critical requirement of the pellicle is that it performs this function while also serving to faithfully transmit the pattern image without negative optical impact. Figure 1.9*a* depicts the standard top view and cross section of a finished 5-in. binary photomask

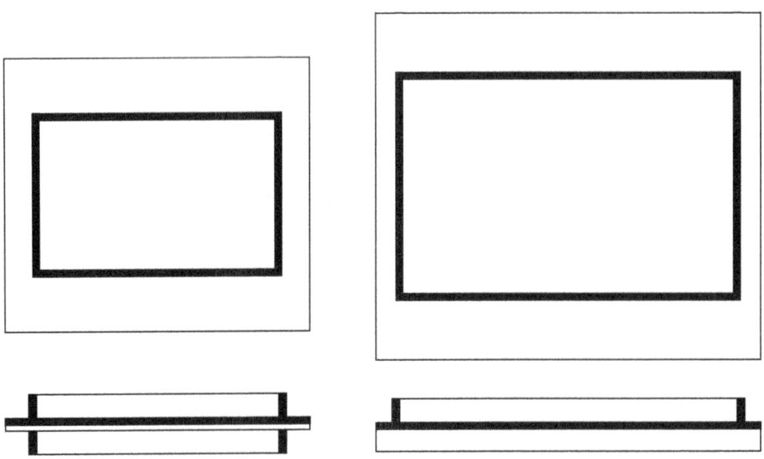

(*a*) (*b*)

Figure 1.9 (*a*) 5-in. mask with two pellicles; (*b*) 6-in. mask with single pellicle.

with both backside and frontside pellicles mounted. Figure 1.9*b* shows the standard 6-in. configurations. Note that only one pellicle is used in the 6 in. case. An easy explanation for this integrates the notions of depth of focus relative to the pattern plane with the wafer exposure tool wavelength era in which each type is used. A mask size of 5 in. was used in the g-line and i-line eras where exposure tool focal depths were larger as were typical airborne defects, thanks to lesser filtration. For this reason, a pellicle needed to be attached to both sides of the 5 in. × 5 in. × 0.09 in. mask to ensure that any airborne defect that fell on either side of the mask would remain out of focus during wafer exposure, thereby not printing.

On the other hand, 6-in. masks became mainstream during the higher NA i-line, KrF, and ArF eras where the depth of focus became much shallower. Added to that the thickness of the 6 in. × 6 in. mask grew to 0.25 in. for dimensional stability. A side benefit of this change was that now any defect landing on the backside of the photomask was exactly 0.25 in. away from the pattern's focal plane, and particles were filtered in the wafer fab environment to much smaller levels. The result was that only the pattern side of a 6 in. × 6 in. × 0.25 in. photomask needed to have a pellicle attached for defect protection.

The materials that comprise pellicles vary somewhat from supplier to supplier, but the principal components are:

1. Rigid aluminum frame
2. Black-anodization (frame coating)
3. Inner frame wall adhesive
4. Frame adhesive
5. Transparent membrane
6. Membrane adhesive

The frame takes on many shapes depending on the available exposure field and fixturing limitations of the wafer exposure tool in which the photomask will ultimately be used. The exposure field limitation dictates the X-Y extents of the maximum usable pattern area. All device patterns must be designed to fit within this rectilinear shape on the photomask. The largest field size available on a commercial wafer exposure tool is 104 mm × 132 mm, but there are many older systems still in service that have smaller field sizes and are of different rectilinear shapes. The fixturing apparatus that transfers the photomask or secures it during wafer exposure also has constrained the shapes of pellicle frames throughout history. Frames were built in square, rectangular, and octagonal shapes depending on the location of loader platen clamps and various end effector designs.

The membrane materials employed are dependent on the wafer exposure wavelength and are organic in composition. Nitrocellulose material is used for g-line and i-line applications, and fluoropolymer material is used for KrF and ArF. These membranes are extremely thin (typically less than 100 μm) and therefore are quite fragile. Extreme care must be exercised during handling and storage so as not to scuff or tear the membrane. If the membrane does become damaged, the entire pellicle (frame and all) must be replaced since there is no practical way to replace just the membrane material while maintaining absolute cleanliness on the pattern surface.

Adhesives are used for multiple applications. The membrane is adhered to the upper frame surface, and the frame itself is adhered to the photomask surface. To eliminate the possibility of particle intrusion, the adhesives form an air-tight seal at the membrane-frame and frame-photomask interfaces. Adhesives have been refined over the years in response to adverse conditions during shipment and use. The first was outgassing that occurred over time causing chemical haze to form on the surface of the photomask. New formulations were developed that eliminated this problem for the most part during the g-line and i-line eras. The second issue was that the frame adhesive needed to be "weak" enough such that a technician with reasonable strength and dexterity could use a special prying tool to remove the pellicle from the photomask surface for reworking purposes.

One interesting aspect of the frame that has evolved over the years is the addition of a vent hole (also called a pressure relief valve, or PRV). Before the vent hole was added, photomasks shipped via air would experience pellicle inflation during high altitude (low pressure). The membrane would "puff up," sometimes to the point of making contact with the box lid in which it was stored. This contact would produce a scuff that rendered the mask useless before it ever made it into the wafer fabrication cleanroom. To alleviate this problem, a small hole was drilled into the frame allowing the pressure to equalize between the encapsulated air under the pellicle membrane and the ambient air. Since the statistical chance that a small particle could in fact be sucked through the hole during pressure equalization is nonzero, holes can either be covered with a filter material or be cut as a tortuous path through the frame, or both.

1.3 Patterns

1.3.1 Device features

The primary reason photomasks exist is to provide the pattern content for the transfer of one layer of a semiconductor device. In the early days of the industry, patterns were exact copies of what was to be printed on the wafer. These patterns were laid out as right-angled rectilinear features in what became known as "Manhattan" patterns (named for the straight city streets of Manhattan, New York). All photomask writing equipment have been built to handle Manhattan patterns as base capability.

Angled-line patterns were also employed due to either the layout efficiency or the nature of the device being fabricated. Difficulties exist, however, when the photomask writer does not have the capability to draw edges at angles other than 0°, 90°, 180°, or 270°, or only has the capability to draw certain standard angles (e.g., 45°, 135°, 225°, and 315°). The writing tool will break a non-standard-angle feature up into many smaller building blocks to create the smoothest line edge it can. There is an unavoidable line-edge roughness resulting that may or may not be detrimental to the wafer lithographer, depending on the resolution capability of his or her process (see Fig. 1.10).

The final major category of basic pattern features is the hole pattern. Also referred to as "contact" or "via," these patterns are worth mentioning separately because holes behave differently than line/space patterns. Recipes and other process parameters for contact and via layers in the photomask fabrication process are often times different from those for line/space layers due to several factors including optical effects during exposure, fluid dynamic effects during development, and loading effects during plasma etching. More thorough discussions on each of these topics will be presented in subsequent chapters.

1.3.2 Optical proximity correction

In the mid-1990s, feature sizes began approaching the resolution limits of the exposure systems, so pattern fidelity errors became problematic for device designers. Feature corners rounded, narrow line ends pulled back, and those features packed closely together (dense features) printed

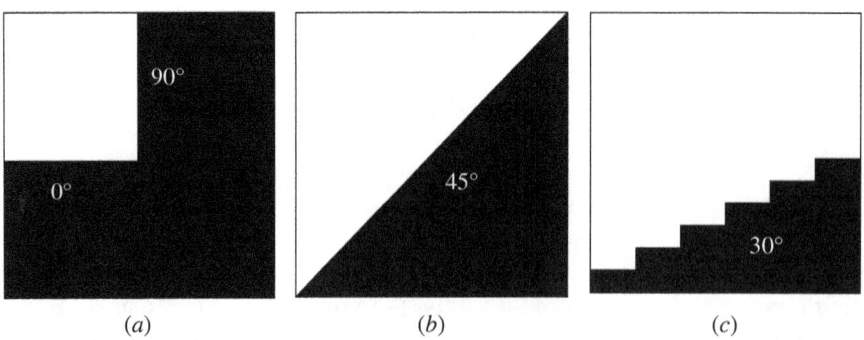

Figure 1.10 (*a*) Standard "Manhattan"; (*b*) standard 45° angle; (*c*) nonstandard angle.

differently than those standing alone (isolated features) on the wafer. A way to improve the fidelity of corners and line ends during wafer lithography was to alter the photomask feature patterns in the areas where the errors occurred. The term optical proximity correction (OPC) was coined as a means of modifying design data only in the proximity where corrections were needed so as to produce the correct optical image.

OPC design alterations were originally classified as either mild, medium, or aggressive, depending on the degree of added complexity to the original design (see Fig. 1.11*a*).

Mild OPC usually meant that only line edges were adjusted on certain features and not reshaped in any way (see Fig. 1.11*b*). This "decoration" was usually to account for simple dense versus isolated line printing differences. Moderate OPC entailed changes to feature edges in the form of jogs, hammerheads, and serifs (see Fig. 1.11*c*). While still part of the originally designed features, the final shapes were altered to such a degree that it sometimes made it difficult to even recognize what the original pattern shape looked like. Aggressive OPC usually meant that extra features, such as scatter bars and scatter slots, were added to the design. These extra features were below the resolution limit of the exposure tool, which meant that they were extremely small and challenging for the mask maker to fabricate well. To complicate matters, these *subresolution assist features* (SRAFs) usually were used in the presence of both mild and moderate OPC as well (see Fig. 1.11*d*).

Chapter 7 describes in detail the theory and functionality of OPC features, but one can see that the impact to a photomask maker was substantial as standard Manhattan-type designs were suddenly decorated with more detailed features as seen in the moderate OPC case. The impact was felt in the cleanroom where minimum feature sizes of

Figure 1.11 (*a*) No OPC; (*b*) mild OPC (size adjustment); (*c*) moderate OPC (hammerheads + serifs); (*d*) aggressive OPC (scatter bars + serifs).

these jogs, serifs, and hammerheads dropped over 2X from the main feature size, and it was also felt in the computer room where these newly decorated patterns resulted in much larger data file sizes, which forced upgrade purchases for computer CPU speed and disk space. If the drop in wafer exposure tool magnification from 5X to 4X was not enough of a challenge for the mask makers in the 1990s, the second of the "one-two punches" to end the infamous mask makers' vacation was the introduction of OPC to device designs.

An interesting point to make here is that mask makers, for the most part, have not taken ownership of decorating patterns themselves, but rather accept what is delivered to them by their customers (device designers). Most mask makers now control the degree to which designs are altered by providing design rules dictated by the fabrication process or computer capabilities, but there was much suffering in the attempts to produce what ended up being closer to wafer (1X) feature sizes compared to 5X or even 4X larger sizes... almost overnight!

1.3.3 Complementary (BIM + PSM)

In one interesting phase-shift application, two photomasks are used for completing essentially one wafer exposure. In this case called complementary PSM (or cPSM), the design is split into two patterns and two photomasks are built—one BIM and one AAPSM. Designers need to cleverly position opaque chrome patterns on the BIM so as to complement the phase-shifted areas on the AAPSM, as well as account for the overlay error between the two masks. The cPSM strategy has been used more widely in the semiconductor field, but has also been employed for the fabrication of other products where extremely high resolution patterning is required.

1.3.4 Dummy fill

All of the features in the pattern area of the photomask discussed thus far are used to either define or refine the shapes of the features to be printed on the wafer during exposure. However, there is another pattern category that can be called on to aid the *chemical-mechanical polishing* (CMP) steps in wafer processing. These patterns are generally called dummy fill and are usually regularly spaced contacts in an array. While a part of the pattern area, the metrology and defect specifications can be much more relaxed since dummy fill features serve no electrical purpose. They exist to ensure that the pressure exerted by the grinding action of a CMP tool is evenly distributed across the device. If dummy fill patterns are employed, the result is a much more planar wafer surface providing significant depth-of-focus (DOF) benefit for subsequent layers in the fabrication of the device. There is also a benefit to populating wide open spaces in the design with dummy pattern features in that it ensures that loading effects remain uniform across the field during plasma etching steps.

1.3.5 Fiducial patterns

The device pattern is usually centered on the photomask surface. The primary purpose for building the mask in the first place is to provide the pattern for a single device layer, but many other types of patterns on the photomask outside the perimeter of the device pattern are required to make the photomask completely functional in a manufacturing setting. The term "fiducial" was adopted by the wafer lithography tool industry and means *a pattern outside of the array*. Some of these outside patterns are used to assist photomask fabrication, some assist wafer fabrication, some provide unique identification, some provide historical record, and still others identify whose product the photomask and the final device is.

Mask metrology marks and regions. As in most manufacturing processes, there are certain points along the way where a measurement is taken, or inspection is performed in order to ensure that the product in build is worthy of continuing on to the next step in the process. Photomask fabrication is no different. CD metrology features, graduated resolution targets, pattern placement (registration) metrology features, phase and transmission metrology targets, and overlay metrology targets among others are used for this purpose. Since the device area is designed to perform a specific electrical function, all nonfunctional photomask metrology features must be placed outside the device pattern area. This region is commonly called the scribe line or kerf area and is usually included with the device pattern file at the time of order placement. Mask makers will often add patterns outside of even the scribe line if there is no room or no permission given to add any extra features to the customer-provided scribe pattern. The specific features used depend on the metrology tool requirements and have changed dramatically over the years.

Exposure tool alignment marks. Other features are placed far outside the device pattern area to provide alignment targets for wafer exposure tools. These patterns also vary in appearance, but serve to provide both coarse and fine alignment feedback to the stepper or scanner tool's photomask alignment system. There is no device pattern comprehension on the part of the exposure tool, so all alignment must be performed by means of fiducials. Once alignment is complete, the photomask's device layer pattern is presumed to be in perfect alignment with the patterns already printed on the wafer, and exposure commences.

Barcodes. To aid the cleanroom automation effort, large barcodes are written and etched into the absorber simultaneously with the creation of the intricate device patterns. The purpose of the barcode is to provide the wafer fab's photomask management system with an error-free method of cataloging all photomasks quickly and easily. In some fabs, the number of active photomasks can reach the tens of thousands! Large amounts of time would be wasted without an automatic means by which to record and track the status and whereabouts of each individual photomask.

Titles. Human-readable titles are also written into the absorber in the same way as the barcode. These titles typically serve to label the mask with the writing tool used to create patterns, the date and time of exposure (writing), and sometimes the customer company's name. Since there is no machine-based function for titling (only satisfies human requirements), titles are usually located near the extreme edges of the photomask itself, where some amount of lithographic fidelity degradation can be tolerated.

Logos. Logos are usually only placed by merchant mask companies as a kind of advertisement, or permanent record of who the manufacturer was. These logos are not too interesting nowadays, but when the industry was less controlled, designers would embed microscopic logos into unused areas of the pattern region. While not common anymore, mostly because device "real estate" is so costly that none is wasted, a careful eye may still see one of these special logos from time to time! Figure 1.12

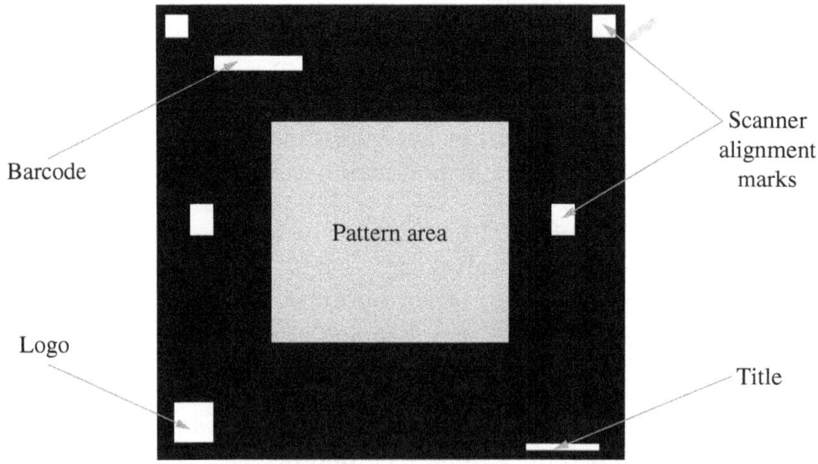

Figure 1.12 Example of fiducial pattern locations.

shows an example of the many categories of fiducial patterns on photomasks made today.

1.4 Fabrication Flows

1.4.1 Binary intensity mask

The least complex photomask process today is that for the binary intensity mask, also called the chrome-on-glass mask. Figure 1.13 shows the general process flow for a contemporary BIM. In this process, there is only one absorber (chrome) and only one pattern transfer step from the resist to the chrome layer. Therefore, only one write operation is required. Also, since the chrome absorber has no phase-shifting properties, only CD and registration metrology is required. Pattern types can include both Manhattan and OPC types, so the inspection and defect repair tools need to be capable of handling all types of features in either 0 or 100 percent transmission categories. Standard wet cleaning chemistries also do not adversely affect the chrome or AR layer, so no special cleaning processes are required.

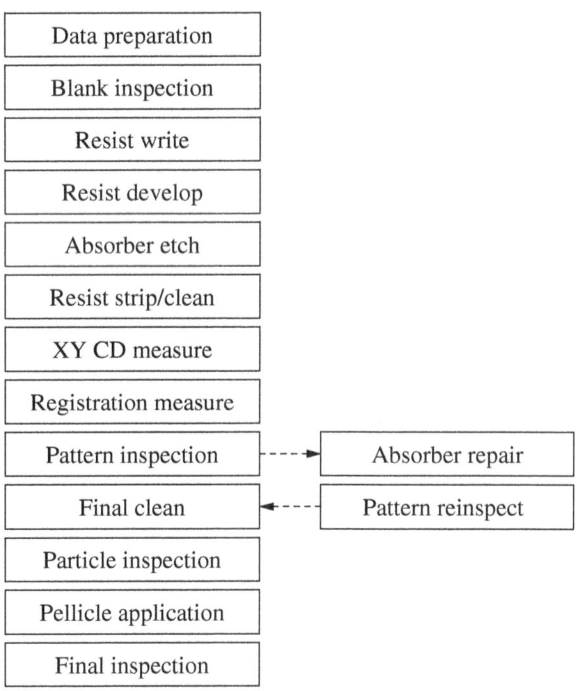

Figure 1.13 Basic BIM flow.

1.4.2 Embedded attenuated phase-shift mask

The embedded attenuated phase shift mask process utilizes two writing and pattern transfer steps as seen in Fig. 1.14. This is primarily because there are two absorber layers that have very different functions. The purpose of the first writing step is to transfer all pattern data into the top chrome layer. Once this is accomplished, the chrome layer is used as the masking layer for the underlying phase-shifting layer (usually MoSiON). At this point, the mask must be repatterned and etched such that the chrome is removed over specific areas of interest in the circuit pattern region. The need for repatterning forces the mask making community to add the resist coating step to its flow. Once the mask is recoated, it is then exposed with the second write pattern such that subsequent etching removes the chrome over the circuit pattern, thus allowing the phase-shift layer to fulfill its purpose.

The chrome that remains after the second pattern transfer etch can serve multiple purposes, depending on location. One common use is in the border area where scanner alignment targets and barcodes are found. For accurate sensor detection, simple opaque-and-clear target features are better than PSM-and-clear materials due to the lack of

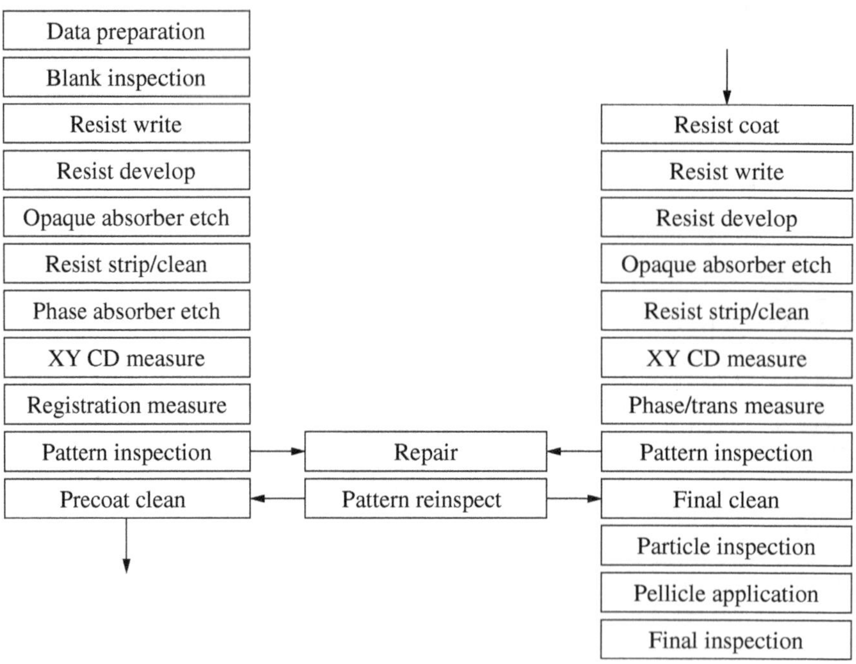

Figure 1.14 Basic EAPSM flow.

unwanted optical effects. The full chrome-on-MoSiON stack is needed to provide a high-contrast opaque/clear transition for the scanner's alignment sensors and barcode readers to "see." Also in the border, the chrome is used around the perimeter of the device area as a permanent frame for wafer exposure. The scanner's masking blades now need not be extremely accurate in and of themselves since the chrome edge serves the chip size determiner.

As mentioned before, chrome is also left over certain areas in the device area when constructive interference causes exposure in unwanted areas of the wafer. These chrome patterns are also called tritone features. The major difference between border chrome patterns and tritone features is that the chrome-to-MoSiON pattern overlay requirements are much tighter for tritone features than they are for border patterns.

The remaining pattern is the MoSiON device pattern on the clear glass substrate. There are extra metrology steps for the PSM layer that measure the actual phase-shift and actinic transmission through the material, as well as the overlay error between the chrome and MoSiON layers. Inspection and repair processes needed to be adapted to handle the MoSiON material. Inspection tools accommodate the higher transmission and phase-altering layer, and repair tools are at least made to remove unwanted MoSiON as well. There is still no repair deposition material that truly matches the 180° phase and 6 percent transmission characteristics of the MoSiON film simultaneously, much less for various high transmission films.

Not shown in the flow is a rework process in the cases where a pellicle becomes damaged or where contamination appears in a critical pattern area, forcing the need for pellicle replacement. In the BIM case, the photomask can withstand a great many cleaning processes due to the durability of the chrome layer. The MoSiON layer is much less durable to the ammonium hydroxide rinse chemistry commonly used in photomask cleaning and therefore does not have the long life expectancy regarding pellicle reworks as seen with BIMs.

1.4.3 Alternating aperture phase-shift mask

Alternating aperture PSMs come in many varieties, but all share the same strategy wherein portions of the glass substrate are etched away to a depth that provides a 180° phase shift relative to nearby unetched clear regions still at 0° phase. The substrate used in all cases is a standard chrome-on-glass blank, but special glass etch (commonly called quartz etch) steps are added to achieve localized destructive interference during wafer exposure. To accomplish the etching of the glass in one clear region while leaving the remaining clear areas untouched, a subsequent writing step(s) is required following the initial Cr patterning steps (see Fig. 1.15). The following sections describe the major AAPSM categories and associated process differences.

Dry etch only. The simplest AAPSM process flow is that of the *dry etch only* process. Assuming the mask has already been processed through the chrome patterning steps, it is cleaned and recoated with a resist in preparation for the second writing step. During the second write, only those areas within the pattern that require glass etching are exposed and developed. Once the resist is patterned, the mask is placed into a plasma etch tool to etch the glass to a depth that produces a 180° phase shift. Subsequent metrology is performed to ensure that the glass trench depth is correct and then the mask is finished in much the same way as a BIM would be.

There are risks associated with the use of this process. The first is that if one overetches the trench depth, the mask will not perform correctly in use and is flatly rejected. Second, if portions of the pattern are blocked during glass etching, removing the glass during repair can be very difficult if defects are in contacts or even more complex OPC pattern shapes. These defects must indeed be removed as they now present unwanted phase transitions that ultimately print on the wafer. Finally,

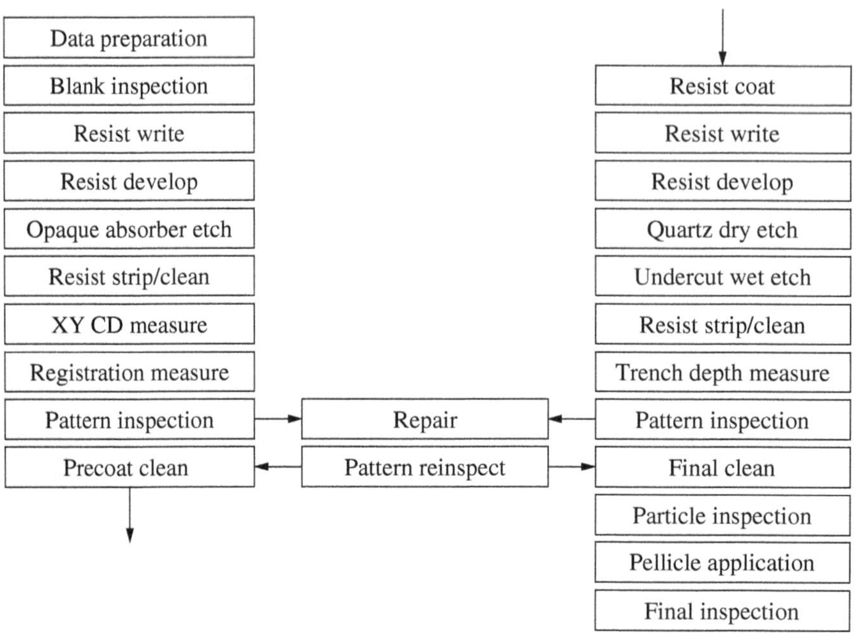

Figure 1.15 Basic AAPSM flow.

unless the glass etch results in perfect 90° trench sidewalls, some radiation will be "stolen" by the sidewalls during exposure, resulting in less intensity through the clear area than was desired initially. Figure 1.16 shows how some radiation is scattered as a result of the presence of a sidewall that is shallower than 90°.

Voted dry. To alleviate all of the problems above, a multi-lithography-step process can be employed. Usually, this process comprises three "voted" glass etches resulting in stepped areas of 60°, 120°, and 180° in the phase-shifted trench. The term voted here means that the etched areas are laid out in such a way that they leave exposed glass edges or steps. The first advantage to using a voted triple-etch process is that the overall trench depth can be controlled better if measured just prior to the final glass etch step. Assuming that the sum of the first two etches did not produce the desired trench depth up to that point, the third etch can be custom calculated taking any errors from the first two etch steps into account.

The second advantage to voting the etch process has to do with defect mitigation. This is best discussed by example. As mentioned above, blocked glass etch areas that occur in a single etch process have a high probability of printing since they exhibit a 0 to 180° phase transition. While fairly straightforward to repair in isolated device regions, blocked etch defects occur randomly and at some point may fall on very complex pattern features that are beyond any repair tool's capability to fix. However, if the voted-etch process were employed instead and the same complex pattern area was blocked during any of the three quartz etches, the resulting defect may not print—at least not catastrophically (see Fig. 1.17). The reason is that between each etch step, the photomask is stripped, cleaned, recoated, reexposed, and redeveloped. In all this processing, the etch defect would obviously be removed. If a new

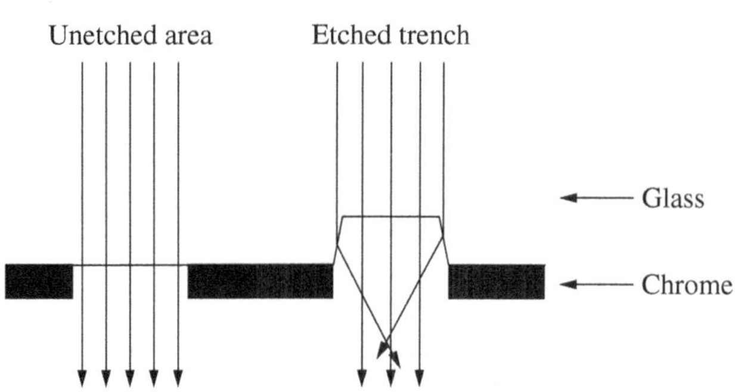

Figure 1.16 Intensity decrease in etched trench.

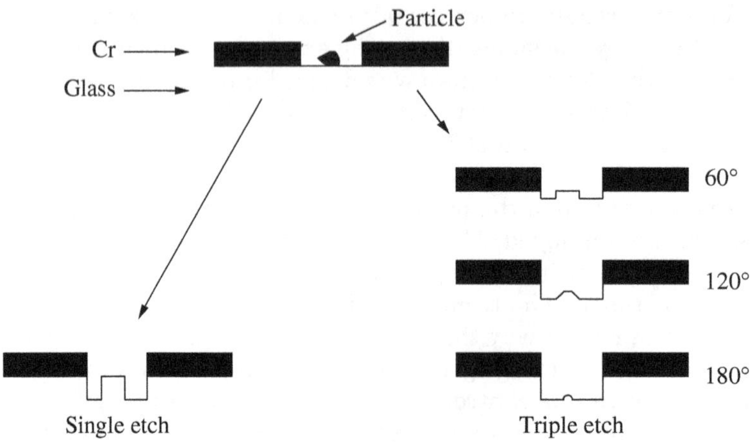

Figure 1.17 Phase defect mitigation using multiple glass etch steps.

etch-blocking defect does fall into the pattern again, the statistical chances of it falling on the exact location of a prior defect, thereby moving the printability risk higher, is near zero.

There are some significant disadvantages to utilizing the voting process in a manufacturing process. The first is that more design layout "real estate" is required. Instead of a single trench region having a vertical sidewall, the voting process requires that the trench edges become "terraced" away from the absorber edge. In effect, the edge is spread out over some nonzero distance, which requires more room to accomplish. In many high-density designs, feature packing density is such that it cannot accommodate the extra lateral distance required.

Aside from physical layout constraints, the voted etch process is also subject to lower process yield due to the increased number of process steps. As with any manufacturing process, the more process steps, the lower the yield thanks to handling defects, process drift, machine error, and of course, human error. All these errors mean that there is either rework or scrap and hence additional cost to manufacture each photomask. The utilization of the voted etch process should only be used in the manufacture of AAPSMs when the repair process is not fully capable. A way to reduce the real estate issue is to merely reetch the same area three times, but lower yield and higher cost issues still remain.

Dry + wet undercut. The AAPSM process that seems to embody the best of all worlds is the *dry + wet undercut* process. Leveraging the best of the dry only process, the fabrication of the dry + wet undercut AAPSM only requires a total of two write steps. Fewer corresponding process steps are employed as well, so there are fewer chances for handling-related defects and all related process errors.

While not truly voting the placements of the quartz etch regions, the etch process leverages the multietch process and is broken into two steps. The first step is a fairly anisotropic (directional) dry etch and is stopped short of the full depth that would produce a 180° phase shift. Instead of continuing with a subsequent dry etch, an acid etch is used to complete the process. The isotropic action of the acid etches the fused silica substrate both vertically and laterally. One advantage of this isotropic etch is that the chrome absorber is etched laterally at the same time the remaining trench depth is etched. The result is that the intensity through the etched trench is the same as the unetched trench since there are no sidewalls present to steal some of the light traveling through. The dry to wet etch depth ratio needs to be set such that the undercut is not so great as to make the overhanging chrome too susceptible to damage during subsequent processing and cleaning (see Fig. 1.18).

Another advantage to having a complementary isotropic wet etch following the dry etch is that the effect of any quartz bump defects can be reduced to the point where they no longer print on the wafer. A particle that blocks the quartz etch during the dry step leaves a potential printing defect if the dry step accounts for more than 60° of the 180° phase shift. When subjected to the wet etch, these defects at least partially dissolve in both the vertical and lateral directions making the

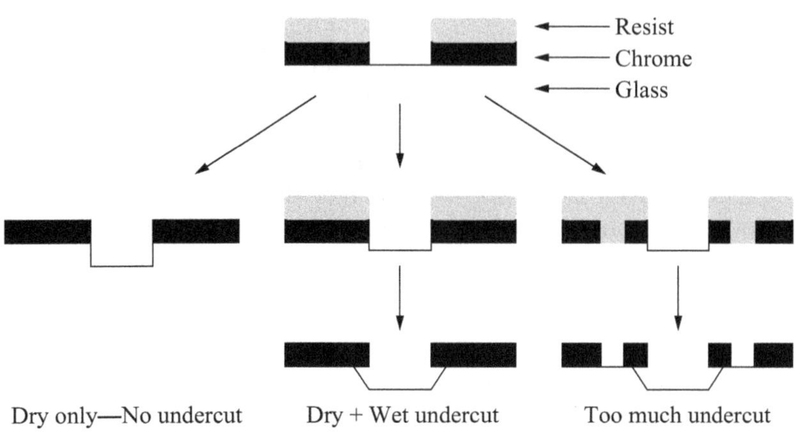

Figure 1.18 Etch strategies.

defects shorter (having less of a phase-shift effect), and smaller laterally, hopefully falling below the lithography process's resolution limits.

If there is a disadvantage to adding a wet etch step, it is that the acid used is hydrofluoric acid (HF). Anyone who has had HazCom training, or who has worked around semiconductor processing facilities is familiar with this dangerous acid. The photomask industry had managed to avoid the use of HF for over 40 years, claiming to be a much safer place to work within the high-tech industry because of it, until this process gained popularity. The good news is that all of the HF storage, delivery, and disposal learning in the semiconductor industry has been leveraged for the introduction of hydrofluoric acid into photomask fabrication facilities.

1.5 End Products

The need for photomasks was born out of the semiconductor industry in which each photomask contained the pattern information for a single layer of a device. The end product produced by these photomasks was a microchip. Of course these microchips were then utilized by the entire electronics industry for computational, timing, data storage, and many other applications. These basic functions have now been applied to a host of products in both commercial and military applications. Basically, almost everything today that plugs into a wall or runs off battery power contains microchip technology, built by photoimaging the mask patterns onto silicon wafers. One can see then the importance of the role of the photomask in the history and continuance of the semiconductor industry itself.

Throughout the semiconductor industry's evolution, devices were invented or eventually shrunk to sizes beyond the capabilities of the macroprocessing techniques employed, and the photolithographic process was leveraged as the means by which to fabricate them. Wave guides, flat panel displays, optical gratings used in the photonics industry, thin-film heads for computer disk drives, *microelectromechanical systems* (MEMS), and now even biotechnological applications utilize photomasks for any photoimaging processes required in their fabrication. As long as optics is the means by which devices are patterned and mass produced, photomasks will have a place in their fabrication.

1.6 Future Issues

The future of the photomask industry can be made or broken by a number of issues. Regarding technology, the ability to manage the physical shapes and ultimately, the electrical properties of the circuit pattern features from the design layout all the way to the wafer itself will separate the successful from the unsuccessful companies. The photomask is an integral part of this formula in that pattern shapes intended by designers need to be faithfully and consistently replicated in the photomask. As a matter of fact, once a wafer lithography process is established on a certain photomask pattern capability *process of record* (POR), consistency from then forward can take higher precedence over trying to implement major improvements to the photomask process. Of major concern is the consistency in corner rounding, line end shortening, and overall fidelity of OPC features. Once designers have a repeatable lithography process that gives them the ability to accurately predict the shapes and sizes of critical circuit components, any changes to this predictability could mean large yield loss and months of work to reestablish a new baseline.

What this means to the mask maker is that the installation of a new e-beam writer, plasma etcher, repair system, or a new resist process could radically change the ability to more accurately represent the pattern features in the design. In the captive or foundry photomask fabrication facility setting, the ability to communicate such changes is fairly simple since all groups work under the same banner so to speak. In the merchant mask making situation, the communication is not as straightforward and requires an overt effort on both the mask makers' and wafer lithographers' parts to ensure that critical information is passed in a timely manner.

Technological modifications that do stand to take place in the near future include absorber and resist material changes, scanner wavelength and magnification changes, and RETs that are unique to each device manufacturer. The chrome absorber may need to be changed in order to facilitate better CD and phase uniformity. Resists will also continue to change as new formulations enable better resolution. Scanner wavelengths, while extending their useful lives thanks to immersion lithography, still stand to drop one or two more times in the future. Below 193 nm, photomask substrate materials need to change from standard fused silica to either calcium fluoride or fluorine-doped fused silica. Of course, all the accompanying dimension metrology, phase/transmission performance, as well as defect mitigation improvements need to scale with resolution improvements.

There is (always) some talk about changing the demagnification factor from 4X to something higher. While this may give the mask maker a little more leeway on metrology, it may also drive the need to increase the mask substrate size to something larger than 6 in. × 6 in. × 0.25 in. The original proposal that almost took hold at the end of the 1990s was to move to the 9-in. format. Although the change never took hold then, a magnification change will most likely force the revisitation of this topic to discuss size changes to 7 or 9 in. again. And, as the demands on resolution dictate specific RETs for the type of product being built, the photomask industry will need to incorporate the associated processes and metrology required to produce these high-precision optical products.

One of the biggest hurdles in attempting to incorporate the aforementioned technological advancements is the lack of infrastructure within the photomask industry. On the equipment side, there are only a few suppliers for certain key processes. Many of these companies are small and do not have the resources to make major changes without some kind of outside financial assistance. The photomask fabrication facilities have also traditionally lagged behind the wafer fabrication facilities in *statistical process control* (SPC), floor control software, and defect control sophistication. While there has been steady improvement over the years, improving the trajectory to intercept wafer fab capabilities will become more and more critical as photomask specifications continue to tighten.

The photomask industry is also relatively small compared to the semiconductor industry, less than 5 percent by revenue numbers. Therefore the critical mass, or the ability to effect change, is limited by the amount of investment that can be allocated toward research and development, major infrastructure changes, or even just buying the occasional piece of metrology equipment to handle a new type of RET technology. To complicate matters, the rising cost of doing business has forced consolidation over the years to where there are now very few merchants, foundries, and captive facilities capable of fabricating photomasks at the most advanced technology nodes. With fewer customers to sell to, photomask equipment suppliers are also strained in their ability to recoup their R&D investment. It is becoming harder to maintain the current business paradigm in light of the technology, infrastructure, and business pressures mounted against the photomask industry. The future will undoubtedly bring many changes as the photomask becomes treated less like a commodity product and more like an integral element in the wafer lithographic optical system.

Chapter

2

Data Preparation

Paul DePesa, Derek Kay, and Gary Meyers

2.1 Introduction

"In the beginning there was Rubylith, and it was good."[1] Yes, 40 years ago the transition from a designer's schematic to a set of photomasks involved strips of plastic on a light table.* With good eyes and a sharp knife, a design schematic was made real. The first *mask data preparation* (MDP) was a "layout" of red Mylar.

Mounted above the table was a reduction camera. The negative from this camera, which reduced the size of the polygons in the Mylar layout by a factor of 10, was used to make a 2-in. mask. This mask was contact printed onto wafers. As a result of this "contact" these masks didn't last very long, and wafer runs (a hundred was an average lot then) required that masks be made over and over again, because after about 10 contact prints the induced defects made the mask unusable. This led to the unfortunate view of the mask as a commodity. Only recently has this perception begun to change. A mask is now recognized as a distinct and equally important element in the wafer lithography process.

This chapter covers mask data preparation as it is today. It describes a general flow that may be used as a model to guide you, and covers the basic steps leading to a high quality mask set.

Mask-writing tools require input data that are made up of trapezoids.† In fact, one can still hear people refer to a mask layer's size as its "trap count." The objective of the MDP step is to convert the polygons of the design database to trapezoids in a machine specific format, including making sure that the vertices of all the trapezoids are on a uniform grid. The smaller this grid is, the more accurate the mask. However, this also makes MDP processes longer, and possibly mask write times longer. This breaking up of the design polygons into more primitive trapezoids is referred to as *fracturing*.

There are different mask writing tools, and most of these tools have their own formats. Moreover, mask inspection tools and mask measuring tools require independent files. These are usually derived from the fractured mask data, but can be processed from the original design data.

More data need to be written on the mask than just the design patterns. The "lanes" between the patterns on the mask are used to saw the resultant wafers into individual chips. Wafer fabs normally use this saw-lane area for test pattern placements to monitor the wafer processing, and so these test patterns need to be created and written on the mask as well. They are destroyed, of course, during the sawing process. Additionally, alignment patterns are required to align a pattern relative to the previous wafer-imaging step. These are placed on the periphery of the mask to allow the stepper to remove translational and rotational

*A large translucent table lit from beneath.

†Some machine formats accept rectangles only.

displacement. Barcodes and titles are also written to aid in tracking and identifying the mask.

2.2 MDP Flow

Fracturing is the link between *integrated circuit* (IC) design and mask making. After design completion, which is marked by a *layout versus schematic* (LVS) run followed by a successful *design rule check* (DRC), data need to be created for use in manufacturing the mask set. These data include the full complement of mask inspection and metrology files required to ensure successful wafer production. Note that the total number of masks for a state-of-the-art process is now above 40, and the data associated with this task are now in the hundreds of gigabytes. Figure 2.1 shows a basic flow of the data creation steps from IC design

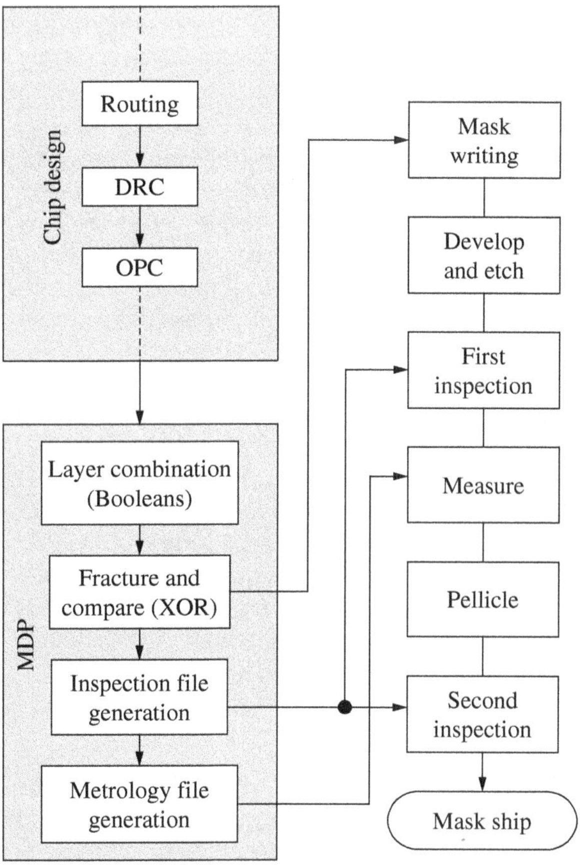

Figure 2.1 A simplified data preparation and mask manufacturing flow.

through to mask ship. In some cases, these steps may not always be separable and are merged into a single step. Input data to MDP are expressed in a *layout interchange format*, a format that allows the design database to be exported to MDP tools. The output data are machine specific, e.g., MEBES and JEOL. Because capabilities differ from machine to machine, and the specifications may differ for different layers, a set of masks is almost never written entirely on one tool. Hence, different formats may need to be generated for different mask layers.

2.2.1 Layout interchange formats

A computer-aided design (CAD) environment is normally used to create the layout of an integrated circuit. The output of the CAD tool is the layout of an IC and is represented in a format that describes the layout in polygons.

In the mid-1970s, Calma Company sold a CAD system known as *Graphic Design System II* (GDS2). This system exported data in a format known as Calma Stream. Today it is the industry standard for IC layout interchange, and it has come to be known as GDS2 Stream, or just GDS2. The GDS2 format is a digital (binary) file format.[2]

A significant feature of the GDS2 format is its ability to use hierarchy in the layout description. Instead of describing each individual polygon, it is possible in GDS2 to describe a structure made up of polygons and *refer* to it in another structure. This reference may be single references, as shown in Fig. 2.2, or an array reference. An array reference is simple statement that gives the structure name and the spacing in X and Y, along with the number of repetitions. The syntax may differ among software tools, but the savings in file size are enormous.

In 2002, a new layout interchange format for ICs was proposed by Semiconductor Equipment and Materials International, Inc. (SEMI) under the name Open Artwork System Interchange Standard (OASIS)[3] Its purpose is to replace the GDS2 stream format, and its objective is to make the interchange file more compact, as well as to overcome other shortcomings of the GDS2 format. The process of switching from GDS2 to OASIS is still underway.

Apart from these two formats, a number of other formats are sometimes used in dedicated applications.* The vast majority of design layouts today, however, are still interchanged from one package to another in GDS2.

2.2.2 Boolean operations

Not every mask layer needs to be "drawn." Some mask layers can be "derived" with a logical operation between drawn layers. More advanced

*CIF, Applicon, DXF are examples.

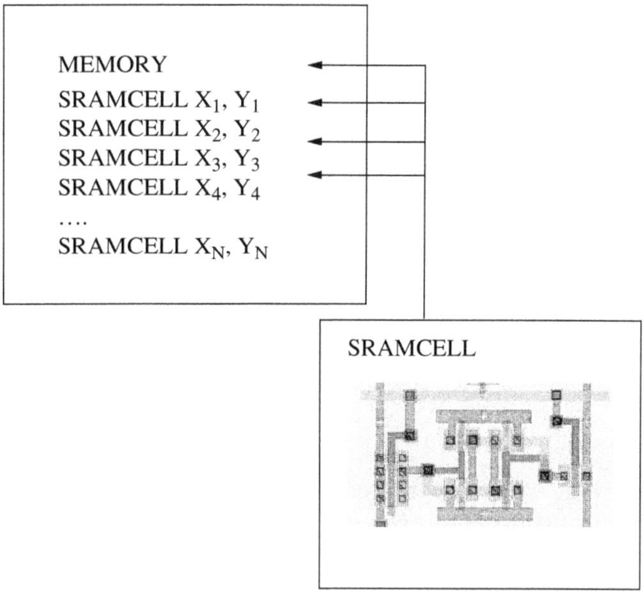

Figure 2.2 The GDS2 format contains cell definitions and cell placements.

processes, with built-in process options such as nonvolatile memory blocks, tend to include more derived layers.

The operations of layer derivation come from a set of rules first formalized by George Boole in the 1850s.[4] Operators such as AND, OR, MINUS, and XOR between layers, and selections based on criteria such as feature area or perimeter, make most derivations. A sizing operation, a negative or positive change in the dimensions of each polygon, may also be part of the derivation formula.

Sometimes layer operations such as these may generate small notches, spikes, small lines, or small spaces. These small features may be just artifacts of the boolean operation and have no relevant function in the design. Such features are poorly resolved during the mask-making process, and the mask may be rejected at the inspection step. Eliminating these small features during MDP can significantly reduce mask inspection issues. These artifacts may be removed by applying a small upsize and subsequent downsize of the same amount to the data. The order is important, as an MDP tool should not let figures that are joined together come apart in a negative sizing operation. A common term for these operations is "gap removal" as shown in Fig. 2.3. The amount of up-size and down-size used in such an operation needs to be chosen carefully to avoid damaging the data. Nonorthogonal edges are the most problematic areas, and need to be checked carefully.

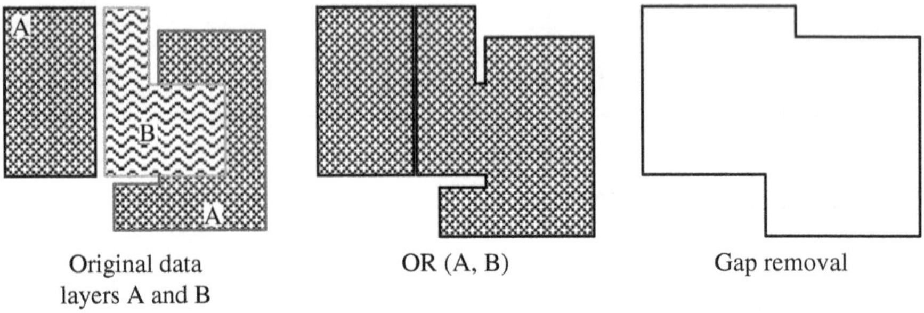

Figure 2.3 A logical OR between layer A and layer B generates small gaps.

2.2.3 OPC and PSM

During data preparation, some layers require correction for optical effects. Of the many resolution enhancement techniques developed to correct these problems, the following are the most common:

- Optical proximity correction (OPC)
- Subresolution assist features (SRAF)
- Attenuated phase-shifting mask (AttPSM)
- Alternating aperature phase-shifting mask (AltPSM)

Refer to Chap. 7 for details on *resolution enhancement techniques* (RETs). Here we only consider the impact of these techniques on the MDP flow.

The OPC step converts layout data elements with few vertices to elements with many vertices and additional line segments. Figure 2.4 clearly illustrates this.

The addition of these segments and vertices results in increased data volume. This increased volume will result in longer processing times.[5] RET also tends to eliminate hierarchy, again with a resultant increase in processing time.

Overall, careful attention must be paid to the processing steps being implemented, as the effect may be damaging to the intended enhancements. Further, these layers with reduced hierarchy and increased vertices will require more compute resources, i.e., more time, and this will have an impact on scheduling.

2.2.4 Mask writer file creation

After RET is applied, and possibly verified by another DRC run, the interchange file is created. This step is still referred to as *tape out*,

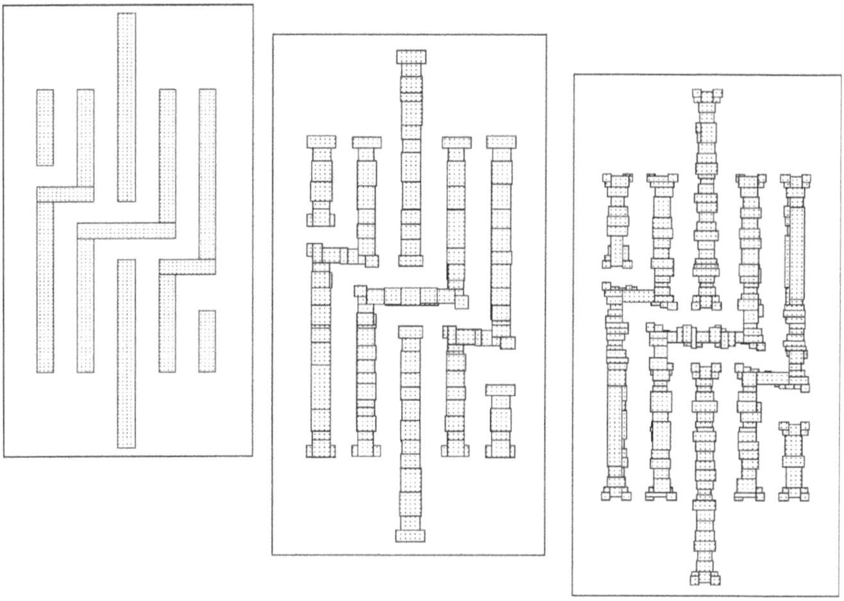

Figure 2.4 Increasing OPC aggressiveness—the image on the left is the original design.

although tapes are rarely used today to transfer data. The design data must now be formatted for reading by the mask-writing tools and the inspection tools. Formats are far from standardized, and each writing tool requires an optimized format specifically designed for it. As mentioned earlier, these formats require trapezoids.

In addition, other data manipulations may be necessary:

- Scaling (up or down)
- Sizing (positive or negative)
- Rotation (90° and 45° increments)
- Mirror (about the 90° and 45° axes)
- Tone reversal

Figure 2.5 shows common examples of these operations.

Scaling is used to adjust for the reduction factor in the lens of the wafer exposure system. Most steppers today use a lens with 4X or 5X reduction. Therefore, the original design data require a 4X or 5X scale increase before writing the mask. There are two common reasons data sizing might be required—(1) to compensate for mask processing and (2) to compensate for wafer processing. Most MDP tools allow sizing operations to be carried out during boolean operations. A pattern mirror

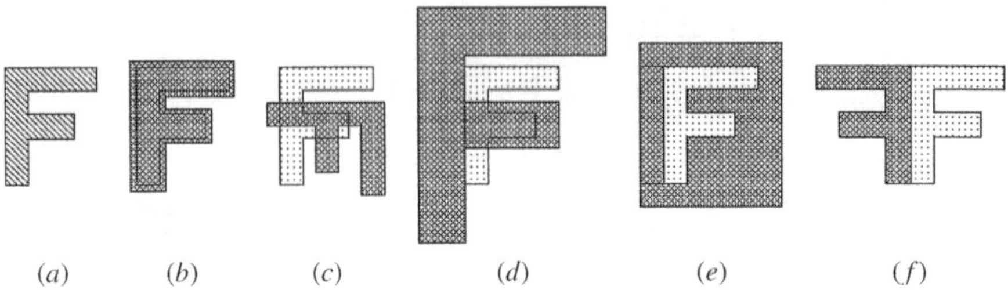

Figure 2.5 (*a*) Original data; (*b*) sizing; (*c*) rotation; (*d*) scaling; (*e*) tone reversal; (*f*) mirroring.

may be required because some write tools are unable to do this themselves. Any of these operations may also be used in the derivation process described previously.

2.2.5 Metrology and inspection file creation

One of the last steps in MDP is the generation of metrology and inspection tool files. In the last decade, there has been a vast increase in the number of measurements required on each mask. Over 100 measurements are commonplace for high-end masks, and a requirement for over 1000 measurements would not be surprising. With such a high number of measurements it has become inefficient, if not impossible, to locate all measurement sites just by looking through the microscope on the metrology tool. Consequently, the ability of an MDP tool to generate these files with some level of automation is mandatory.

There are two types of measurements performed on masks:

- Dimension measurements commonly known as *critical dimensions* (CDs)
  - These are width and space measurements.
- Registration measurements
  - These are placement accuracy measurements.

An MDP tool should make it possible to locate these features manually or automatically, by means of a search operation. The output is a tool-specific setup file that can be loaded into the measurement tool along with the mask, once it is processed. A measurement setup file usually consists of a text file in which coordinates and measurement instructions are written in a proprietary language. These setup files may also contain images used for positional alignment.

The procedure is similar for an inspection tool. The file created, however, is a binary file that the inspection tool uses to graphically compare with the mask. This is called die-to-database inspection. Inspection tools can also compare a die on the mask to an identical die also on the mask. This is the die-to-die inspection mode. Please refer to Chaps. 5 and 6 for further details.

2.2.6 Passing of mask order information

Once the mask data are generated, they are supplemented with additional order information and instructions, and passed to the group responsible for manufacturing the mask. The supplementary information includes such information as shipping data, the pattern file names, and quality specifications. A standard, SEMI-P10, has been developed for mask order information. Its electronic order format contains well-defined keywords, which must be listed in a specific order. Note that the P10 format does not contain strict layout information, just basic instructions for the placement of patterns on the mask. A mask *job deck* must be created for the write tool (refer to Chap. 3).

The fixed format of SEMI-P10 order files facilitates their use for automated order handling, and large mask shops today are set up to receive customer orders in this format.

2.2.7 MDP ownership

Although MDP is performed in steps, the responsibility for performing these steps is variously assigned to groups within a corporate organization. Indeed, some of these steps may be performed in parallel. In some companies, a single group is responsible only for the OPC step, while in other cases a single group is responsible for the entire flow depicted in Fig. 2.1. The following considerations should be taken into account when assigning tasks to different groups:

- Data integrity
- Mask data quality
- Cycle time
- Mask cost and writing time
- Flexibility with respect to the writing tool

When masks are manufactured by an external company, the decision on where to place the "hand-over" point is critical. In summary, it is important to consider the entire MDP flow, and look at all the factors listed in this section (Sec. 2.2.7), when deciding where "MDP ownership" lies.

2.3 Pattern Fidelity and Quality of Fractured Data

The way in which mask data are prepared will have an impact on pattern fidelity and hence on the quality of the mask. Mask quality has a major impact on the final wafer image and therefore on overall chip performance. Some major mask qualities that are affected by pattern fidelity include:

- Line-width control
- Pattern placement accuracy

This section discusses MDP issues that have an impact on pattern fidelity.

2.3.1 Grid snapping

One of the most important MDP parameters is the grid address unit and subsequent coordinate rounding to the chosen grid (grid snapping). Machine data formats require figures to be placed on a machine grid (dictated by hardware characteristics) that are typically coarser than the original design grid. Thus grid snapping can create a deviation from the originally designed line width, which in turn causes a CD error even before the mask is written. Grid snapping can happen in any number of the steps in MDP.

Design data ultimately consist of polygon and path definitions. When a polygon's coordinates are all on a certain grid, grid snapping is avoided by using this grid as the final fracturing grid. A *path* definition consists of a set of centerline coordinates plus a path width. Avoiding grid snapping in such a structure requires not only that the centerline coordinates be on grid but also the width of the path needs to be a multiple of the double grid as well (or the half-width a multiple of the grid). The perimeter of such a structure *can* be on this particular grid; however, if there is a nonorthogonal section in a path, this perimeter is *never* exactly on this grid in the direct vicinity of the oblique section (Fig. 2.6). In other words, when a nonorthogonal path is defined, grid snapping is inevitable along the nonorthogonal edges. Additionally, when a small sizing of the data is applied, nonorthogonal edges require special care, as these edges can lead to gross off-grid situations.

When a sloped line crosses several scan field boundaries, most formats require that at the boundaries these lines be on the fracture grid. Grid snapping here can cause a straight line to form into separate lines joined at the scan field boundaries. If data containing such lines are reinput and refractured several times, these line ends will require special attention.

Grid snapping, in general, either by itself or in combination with sizing constitutes one of the major sources of error during MDP, and needs to be checked and controlled carefully.

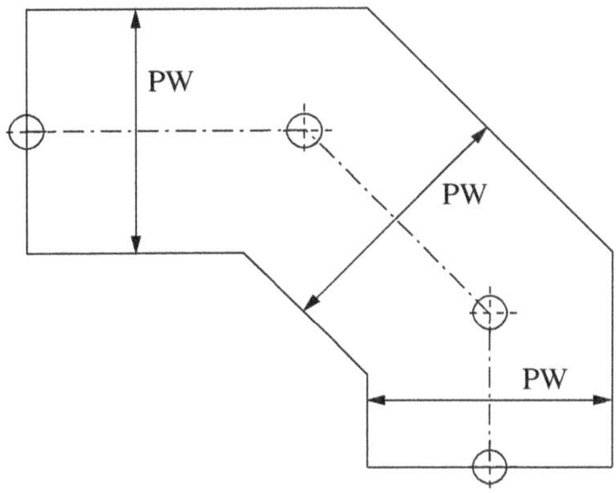

Figure 2.6 A path definition consists of a set of centerline coordinates plus a path width (PW).

Note that a less obvious instance of grid snapping can occur during mask writing if the mask-writing tool uses a fixed address unit, and the mask data are not fractured to its grid or an integral multiple of it.

2.3.2 Data slicing

In elementary, format-specific blocks, input data may be sliced several ways. Slicing differences can lead to differences in the line-width accuracy. If a rectangle is divided into two elements, for example, the line-width error can be different from when the data are not split (Fig. 2.7). The actual difference in terms of line-width error on the mask is a function of many factors. A comprehensive understanding of both the data preparation step *and* the mask-writing tool is

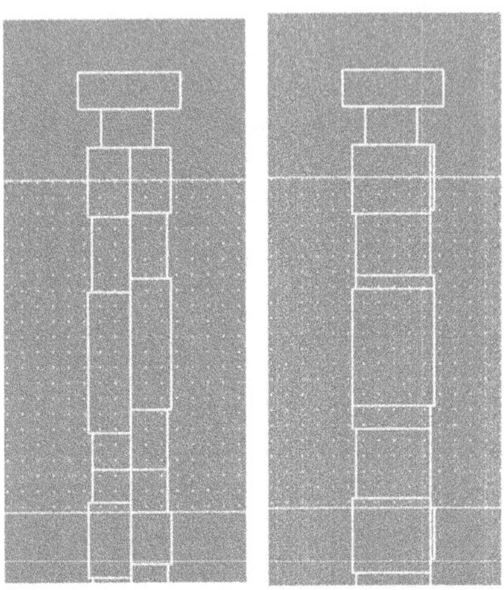

Figure 2.7 Different slicing of input data can yield different line-width errors on the written mask.

required to achieve the best possible MDP and is vital to achieving quality fractured data.

The analysis also needs to consider how slicing should occur to ensure that the original design intent is maintained throughout the manufacturing process. If it is obvious that slicing a certain feature will have an adverse impact on CD control, the slicing should be placed outside the critical areas. Sometimes data are such that exposing the nondata, and using a negative-tone resist, can yield better CD control. That is, if a line is formed by printing the area around the line, the CD control is different compared with printing the line itself. Note that the use of a negative-tone resist requires all data to be reversed in tone, including alignment keys, barcodes, and test patterns.

2.3.3 Scan field stitching

For almost all writing tools, the total exposure area is divided into rectangles, usually referred to as scan fields, blocks, or stripes. The size of the scan fields can vary from a few micrometers to several thousands of micrometers. These scan fields are used to cover the whole exposure area with a combination of stage travel and beam deflection. The stage moves the mask to the beginning of a scan field. Then a combination of beam deflection and stage movement "writes" the content of the scan field. If a data structure is placed exactly on a scan field boundary, this structure could be written in two separate steps (Fig. 2.8). If required, the following measures may be taken to reduce the impact of scan field stitching on pattern fidelity:

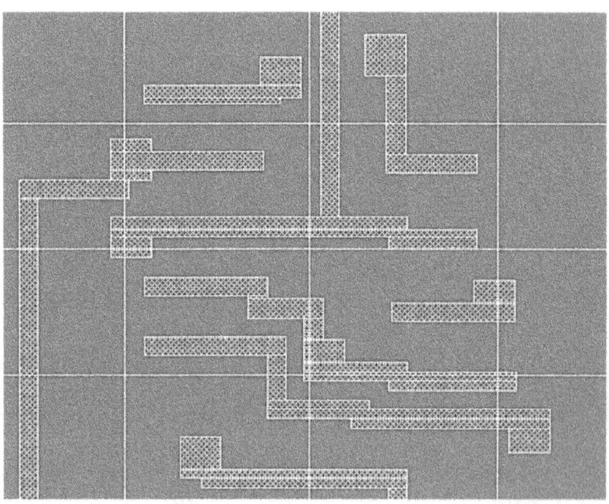

Figure 2.8 The data for mask writer tools is divided into blocks. When these scan fields divide a pattern, the pieces are written in different fields. This can lead to line-width variation.

- When scan fields overlap one another, the software used should avoid slicing of features of a size smaller than or equal to the overlap amount. A feature that is partly in the overlapping zone is then completely written in the adjacent scan field.
- Also, with overlapping scan fields, the features in the overlapping zone can be written with only 50 percent of the exposure dose. When writing the adjacent scan field, this feature is again written with 50 percent of the dose.
- Scan field boundaries may be positioned outside critical areas, making it possible to avoid the impact of scan field stitching altogether. Such an approach is not always practical for mask writing, but when writing small patterns, e.g., direct writing on a wafer, it may be the simplest and most useful technique to overcome potential scan field-stitching problems.

Note that in some writing tool formats, the scan field size is fixed and cannot be altered at all, while other formats permit manipulation in both size and position.

2.3.4 Nonorthogonal edge approximation

As mentioned earlier, nonorthogonal path segments can give rise to grid snapping near the edges. Other deformations of the sloped edges can also occur during the conversion to the mask data formats. Deformations are tool specific, and therefore format specific. Some tools are pixel based and thus approximate the boundaries of nonorthogonal edges (Fig. 2.9). If the pixel size is large, the step may be significant enough to be visible on the mask. Using a smaller pixel size will reduce this problem, although using a smaller pixel size may increase writing time.

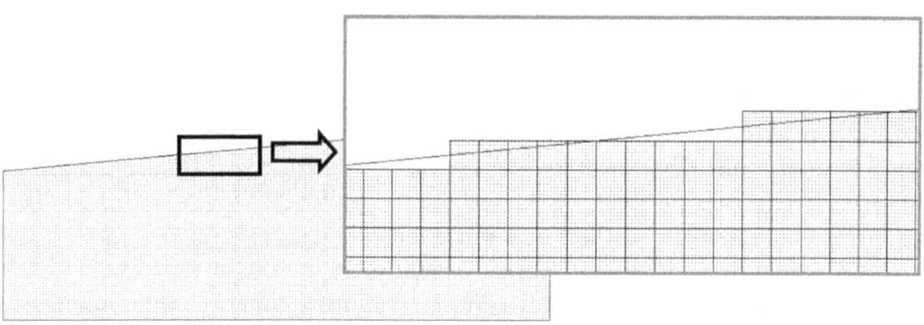

Figure 2.9 Some mask-writing tools approximate sloped lines. The size of the steps is equal to the grid (pixel) size of the data.

Some *variable shape beam* (VSB) writing tools can only print orthogonal and 45° edges. For non-45° edges, an approximation, with a combination of 45° segments and orthogonal segments, follows the sloped edge as closely as possible.

2.3.5 Laser proximity effects

Most mask exposure tools use either a laser or an electron beam. Both of these exposure methods have their limitations and hence an impact on the final pattern fidelity.

One of the main limitations of a laser is the limited optical resolution. The wavelengths of lasers currently being used for mask exposures are in the range of 400 to 200 nm. Features smaller than or nearly equal to these wavelengths are deformed to some extent during mask writing. The most obvious symptom of image deformation is corner rounding, which is similar in effect to the corner rounding experienced when printing wafers.

In a technique similar to OPC, mask data may be altered before writing by the application of serifs to the corners to improve fidelity. For mask inspection, normal, uncorrected data are used.

2.3.6 Electron-beam proximity effects

When a beam of electrons is used for mask writing, a much higher resolution is possible. However, the high acceleration voltages used to create the electron beam charge the electrons with a high kinetic energy and cause them to scatter throughout the atomic levels of the resist and underlying materials. This phenomenon is known as electron-beam proximity effect.[6] The tools that employ correction techniques for this effect are known as *Electron-Beam Proximity Effect Correction* (EBPEC) software.

The most common way of correcting the data for EBPEC is to modulate the actual writing dose. Areas with a high pattern density are given a relatively low dose, while more isolated and smaller features are given a relatively high dose. Alternatively, the edges of features can be moved in small increments.

EBPEC is a CPU-intensive operation. Each individual trapezoid is divided up into small squares, and an individual dose multiplier, or "tag," is assigned. The increase in output file size can be substantial. Because of the relatively long range of these effects, large amounts of data must be considered. When a job deck is used to combine patterns for mask writing, it is possible that a considered range, or neighborhood, will encompass different MDP files. When applying correction on one file, the presence of other pattern data must be taken into account. The simplest way to deal with interacting files is to merge them into one, although this can lead to even larger file sizes. An approach to reduce the processing time is to exploit the hierarchy in the design data.[7–10] Methods have been developed to use the hierarchy despite the fact that electrons are scattered across cell borders.

2.3.7 Loading effect correction

To meet today's tight CD specification, all systematic errors are under continuous investigation. CD control errors neglected in the past are now being investigated and modeled to find ways for correction.

Dry etching has been introduced in mask making for improved pattern fidelity and process control. One of the phenomena observed with dry etching is the etch rate dependency on the amount of etchable surface exposed to the etching plasma. The more the material that needs to be etched, the more the etching plasma will be depleted, and lower will be the etch rate. This is known as *loading effect.*[11–14]

The result of this loading effect is a variation in line width as a function of pattern density. The effects are comparable to the electron-beam proximity effects just discussed, but with a much larger range of influence, although the amount of correction is typically less than 20 nm.

The whole procedure requires that a test mask be written and measured, and the resultant error fed back into the model that is used to correct the data. At this time, almost all available tools move data edges to achieve acceptable results; however, there is some ongoing research into using a variable dose tag, as in EBPEC. This would make it possible to correct for both effects in one processing step.

2.4 Job Decks

Chip patterns are often placed on a mask several times, either as separate instances or in an array. Especially when the number of these repetitions is large, or the data volume of the patterns is large, it pays to use some sort of language.

2.4.1 Common parameters

Job deck formats differ with each writing tool. As depicted in Fig. 2.10, a job deck file is a table of instructions with one or more references to pattern files, combined with placement information expressed as either a single coordinate or an array of coordinates referencing those files.

In this simple example, an MEBES job deck, the "ROWS" commands tell the tool where to place the pattern. The coordinate convention is that the lower left corner of the mask is 0,0. The "SLICE" code at the beginning tells the size of the mask, in this case 17, which equals a 6-in. mask. The units of the command itself are in micrometers.

```
SLICE EDIT,17
GROUP 1,2
SCALE 0.1
OPTION  AA=0.25
CHIP N1,
$ (1,   BIRD$$$-$$-50)
ROWS 24200/62700,6,6000
ROWS 32200/44700,6,6000
ROWS 32200/86700,4,6000
ROWS 120200/44700,11,6000
ROWS 128200/62700,6,6000
CHIP N2,
$ (2,   BIRDREV-$$-50)
ROWS 32200/80700
ROWS 60200/77700
ROWS 80200/38700
ROWS 80200,2,8000/56700,5,6000
ROWS 80200/116700
ROWS 112200/74700
END
```

Figure 2.10 This is an example of an MEBES job deck.

A job deck can also contain other instructions for the tool, including mirroring or scaling, correction factors, column setup parameters, and possibly field, or subfield information. An asterisk denotes a comment.

One oddity of an MEBES job deck is that the coordinates given by the ROWS command, and separated by the forward slash (/), are in Y, X order. Although tool suppliers will supply buyers with appropriate manuals and definitions, MDP software tools are built to check proper syntax.

2.4.2 Optimizing

An ill-conceived job deck can have an adverse impact on the write time and pattern fidelity. This is because these tools have a direct relationship with the *sequence* of commands in the job deck. A thorough knowledge of the writing tool used to write the mask is required to set up the best possible job deck.

One group of writing tools merges the different patterns in the job deck. This local and flattening process is done during the actual writing of the mask, to continue taking advantage of a more efficient data representation.

Since resolution can differ per pattern file in a job deck, some grid snapping may occur during the actual writing of a job deck. Some machines have a fixed spot size so that, even when writing a single pattern, snapping can occur. For masks with very tight specifications, this snapping can take up a significant part of the total error budget.

Again, let us here emphasize that a thorough knowledge of specific tool job decks is critical.

2.5 MDP Processing Time

The latest technology nodes have seen a steep increase in the MDP complexity. The factors contributing to this increase include:

- Increased number of masks
- Increased complexity of layer combinations (booleans)
- Increased data volume caused by:
  - Resolution enhancement techniques
  - Feature size reduction
  - Exposure correction factors
  - Process correction factors
- Different advantages of each writing tool

The estimated increase in data volume, as shown in Fig. 2.11, results in an enormous increase in the MDP turn time, with the total conversion time from DRC to ready-to-write data now exceeding the total time needed to process the mask. Simultaneously, the requirement for the total turn time of a mask set is getting shorter. This section discusses techniques and approaches for reducing the total MDP time.

The most obvious way to reduce the MDP runtime, of course, is the use of a faster computer. Other ways of speeding up data processing include using multiple parallel CPUs in a single machine, splitting up the total task over several machines in a cluster, or using a combination of these two approaches.[15,16]

Because the configuration of data in different data processing steps can vary significantly, some steps are better suited for parallel operations than others. The repetition of identical data cells, requiring a single cell to be placed many times during MDP, suggests the use of hierarchical processing. However, it may not always be possible to take advantage of the hierarchy. For example, if otherwise identical cells are in close proximity with other, differing data, they cannot be handled as equivalent cells because proximity effect correction (optical, electron beam, or etch) does a correction as a function of the surroundings. Therefore, there is the tendency to reduce the amount of hierarchy in corrected data.

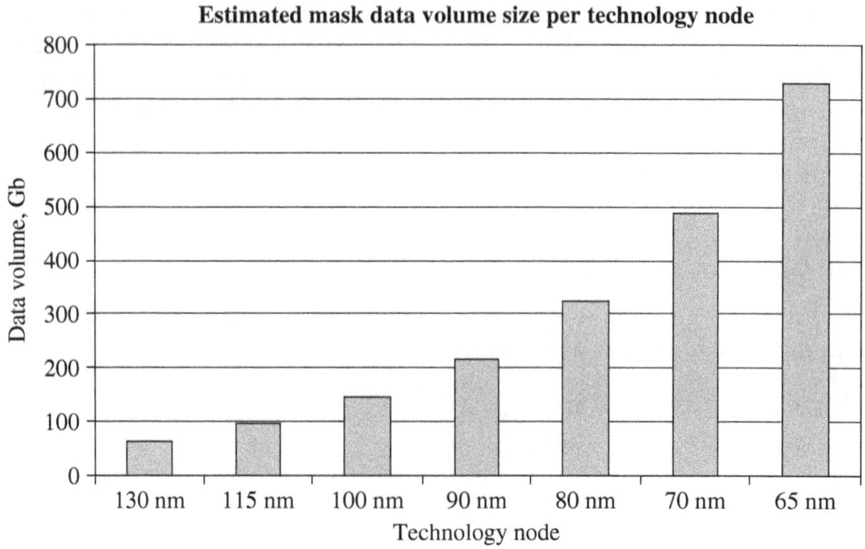

Figure 2.11 The estimated mask data volume increases exponentially as a function of technology node. (Source: ITRS roadmap.)

References

1. http://www.findarticles.com/p/articles/mi_m0EKF/is_6_47/ai_71055571
2. http://3c.nii.org.tw/3c/silicon/SOC/gdsii.pdf
3. http://www.semi.org
4. http://en.wikipedia.org/wiki/George_Boole
5. 2001 International Technology Roadmap for Semiconductors, http://public.itrs.net.
6. Chang, T. H. P., "Proximity effect in electron beam lithography," *J. Vac. Sci. Technol.* 12, pp. 1271–1275, 1975.
7. Eisenmann, H., T. Waas, and H. Hartmann, "PROXECCO—Proximity effect correction by convolution," *J. Vac. Sci. Technol.* B11(6), pp. 2741–2745, 1993.
8. Rosenbusch, A., C. K. Kalus, H. Endo, Y. Kimura, and A. Endo, "On the way to the 1 gigabit: demonstration of e-beam proximity effect correction for mask making," *Proc. SPIE* 3236, 1998.
9. Kalus, C. K., W. Rossl, U. Schnitker, and M. Simecek, "Generic hierarchical engine for mask data preparation," *Proc. SPIE* 4754, 2002.
10. Gesley, M., and M. A. McCord, "100 kV GHOST electron beam proximity correction on tungsten x-ray masks," *J. Vac. Sci. Technol.* B12(6), pp. 3478–3482, 1994.
11. Kwon, H. J., D. S. Min, P. J. Jang, B. S. Chang, B. Y. Choi, K. H. Park, and S. H. Jeong, "Dry etching of Cr layer and its loading effect," in *Photomask and Next-Generation Lithography Mask Technology VIII*, H. Kawahira, ed., *Proc. SPIE* 4409, pp. 382–389, 2001.
12. Lee, J. Y., S. Y. Cho, C. H. Kim, S.W. Lee, S.W. Choi,W. S. Han, and J. M. Sohn, "Analysis of dry etch loading effect in mask fabrication," in *21st Annual BACUS Symposium on Photomask Technology*, G. T. Dao and B. J. Grenon, eds., *Proc. SPIE* 4562, pp. 609–615, 2002.
13. Granik, Y., "Correction for etch proximity: new models and applications," in *Optical Microlithography XIV*, C. J. Progler, ed., *Proc. SPIE* 4346, pp. 98–112, 2001.

14. Kwon, H. J., D. S. Min, P. J. Jang, B. S. Chang, B. Y. Choi, and S. H. Jeong, "Loading effect parameters at dry etcher system and their analysis at mask-to-mask loading and within-mask loading," in *21st Annual BACUS Symposium on Photomask Technology*, G. T. Dao and B. Grenon, eds., *Proc. SPIE* 4562, pp. 79–87, 2002.
15. Amdahl, G. M., "Validity of single-processor approach to achieving large-scale computing capability," *Proceedings of AFIPS Conference*, Reston, Va., pp. 483–485, 1967.
16. Gustafson, J. L., "Reevaluating Amdahl's law," *CACM* 31(5), pp. 532–533, 1988.

Chapter

3

Pattern Generation

Mark Gesley

3.1 Introduction

This chapter traces the evolution and performance of photomask pattern generators. Aspects of patterning associated with wafer direct write, 1× x-ray, and other applications are addressed only so far as there was technology transfer to printing of reticles and masks for photo-optical lithography. The nature of key design features and their benefits will be analyzed for various system architectures and key component technologies. Hopefully this will provide insight to pattern generator systems and the methods tool designers use to achieve mask pattern quality and throughput.

Mask *pattern generators* (PGs) have required large capital investment and years to hammer out various flaws in the initial designs to meet stringent manufacturing needs. In response designers have developed scalable platforms within the constraints of a given architecture to meet the changing needs of the industry in a timely and cost effective manner. Attempts to develop radically new PG systems have been justified on the basis of eliminating the mask altogether.

Lithography capability combined with *integrated circuit* (IC) design needs to determine key pattern generator print quality requirements. PG systems have enabled various mask and IC process technology innovations. The need to improve yield and reduce integrated circuit manufacturing cost drove PG adoption and continues today.

Both patterning and lithography involve transferal of an object to an image plane. Both applications involve optics, precision mechanical handling of the substrate with laser-controlled interferometry, and a substrate develop-exposure process. Differences in the exposure time and print methods and nature of substrate affect the details of the image transfer. Resist exposure and thermal control of the substrate for mask printing require hours, whereas wafer exposure occurs within minutes with high thermal conductivity. Lithography images large fields, now typically with 4× reduction. Pattern generators scan small fields to ensure high accuracy. Scanning is either in a raster or vector motion. Image demagnification varies with optic.

PG architecture can be categorized as either raster or vector. Though early prototypes of both systems were created to address the needs of reduced cost and improved IC manufacturability and yield, each focused on different aspects of the process. The first raster system was dedicated to improving the mask pattern quality. Early vector scan prototypes placed a greater emphasis on wafer direct write.

The PG is part of a large tool set within the mask shop, which is a module within the IC design to manufacturing process. Data manipulation, job deck creation, expose, clean, develop, etch, metrology, inspection, and repair require calibration and control. Pattern generator and mask inspection equipment are typically the two largest capital equipment costs for the mask shop. IC, *electronic design automation* (EDA), mask makers and tool designers are working to develop application and data interfaces as a common infrastructure for physical design and verification. The design to data model is a global solution to reduce cost and improve time to market. New data models and feedback loops between these processes improve work flow integration. For these reasons it is worth understanding the *computer-aided design* (CAD)–PG data interface in some detail.

The data path is a unique feature which differentiates PGs from lithography. The data path transfers and operates on the CAD-generated design, in lieu of a fixed object, to create an aerial image, whose subsequent development and absorber etch of the latent resist image ensures that the pattern is fixed within the upper region of the mask. Novel programmable mask concepts are being developed, which blur this distinction between fixed and electronically defined objects whose image is cast by the beam delivery subsystem.

3.1.1 Pattern generator history

The patterning and use of masks in photolithography was adopted from the very beginning of the invention of the IC and prediction of the laser in 1958. That same year a new company Geophysics Corporation of America (GCA) was founded and a year later acquired a small manufacturer of comparators called David W. Mann, and began to build on Mann's precision motion capability to tackle the problem of precisely aligning circuit patterns between successive layers of a silicon wafer.[1] The capabilities for sub-100-nm electron beam (EB) lithography were also first proposed by D. Buck and K. Shoulders[2] and R. Feynman[3] and then demonstrated in this era with modified transmission[4] and scanning electron microscopes.[5,6] The future source for laser pattern generators, the blue-green argon ion laser, was invented by W. B. Bridges in 1964,[7] and operated one year later in continuous wave (CW) mode.[8]

The photorepeater was introduced in 1961 by D.W. Mann and metal foil masks were made at Siemens.[9] This machine was a shaped beam, light-optic pattern generator. Siemens first production of masks for transistors and diodes occurred in 1966. A five-stage mask technique developed in 1965 by one of the first European institutes for microelectronics, the Arbeitsstelle für Molekularelektronik Dresden also founded in 1961, provides an example of how early masks were produced.[10]

The artwork original was cut with a scale of about 200:1. A double layer rubilith film was used for the artwork originals, and 2-in. photographic plates were used for the reticles and masks. Using a reduction camera the original was reduced about 20 times, to get the reticle. This reticle was then further reduced about 10 times using a step and repeat projection microscope, to get the master mask whose scale was then 1:1 to the final image. By direct photocopying of the master mask one obtained the working mask. Finally, an aligned direct exposure of the working mask onto a wafer created, layer by layer, the micropattern of the chip array. By 1967 bipolar integrated circuits, NOR gates with eight transistors, were made with four masking levels on 25-mm Si wafers with 1.5-mm chip size and 20-μm critical dimensions (CDs). In the late 1960s some improvements were introduced, particularly the six-barrel semiautomatic photorepeater, which improved the figure width and gave lower tolerances. By 1969 complexity grew to where 36 transistors on a chip 1.6 mm × 1.8 mm would be made with nine masking levels and with 10-μm minimal figure width.

By the early 1970s D.W. Mann reticle pattern generators had a high-pressure Hg arc lamp to illuminate the source and a reduction system with 10× demagnification, which printed 10 flashes per second at ~1 megapixel per second, writing primitive shapes on a photosensitive mask in a step and repeat process with 1-μm resolution. Pattern designs were created in AutoCAD and this file was then converted into binary format, which could be fractured into data read by the pattern generator. The light was filtered and projected onto a shutter, which controlled the exposure dose. Mechanical shutters were used for variable apertures. The set of vertical and horizontal aperture blades could be rotated from 0 to 90° to create rectangular images of varying sizes and orientations to form the pattern design. A focusing column imaged the apertures onto the mask plate. The mask plate was secured onto a fine motion stage, which positioned the plate at each exposure location. The rolling element, screw-driven stage did not move during writing.

However, in the mid-1970s *very large scale integration* (VLSI) device complexity required ~10^6 optical flashes driving print times above 10 h per mask, in some cases to 80 hper mask.[11] Not only was this affecting process flow, but tool reliability was impacted by the lifetime of the mercury arc lamp. While these early optical pattern generators were to be soon superseded for leading edge mask patterning work, they remained useful tools for another 20 years.[12]

Bell labs initiated both e-beam and laser PG development prototypes starting about 1970.[13,14] An informative review of the laser PG developments in this era and later can be found in Ref. 15. The Bell Labs laser PG system, and the one developed by TRE Corp. were never commercialized.[16,17] The latter system was designed for 160 Mpixels/s operation, a data rate not achieved by a commercial PG until the advent of CORE and MEBES-IV in the late 1980s/early 1990s. In retrospect the speed and resolution of that laser tool was adequate for commercial application. This wouldn't be the last time that the complexity of commercializing mask PG tools superseded the promise of the technology.

Manufacturing requirements in the late 1960s and early 1970s were growing for fast cycle time of multiple device types. Two to four 3-in. masks or 10× reticles could be printed in an hour with 1-μm resolution and 0.5 μm range of overlay accuracy using a vector-scan gaussian e-beam PG.[18,19] e-beam systems provided a capability to have a flexible method to generate various patterns reliably at high speeds under computer control.[20] The 1× lithography requirement was addressed by the high-resolution capability of e-beam technology, but was not the prime driver.

By August 1974 ATT had installed the EBES system, an electron beam raster scan tool dedicated to mask patterning. The machine used small field electron optics and compound (electrical and mechanical) scanning. Stage position errors were measured interferometrically and corrected electrically through the deflection system to an accuracy of ±1/64 μm. Within a few months it was determined that this system could meet all conventional mask making requirements. As a consequence, optical mask making equipment was shut down.[13]

The EBES system provided three key technical advances—a continuously moving stage, raster scanning, and relatively low (10 kV) beam voltage. The benefit of a continuously moving stage is reduction of stage overhead time. Raster scanning was developed to achieve high placement accuracy on the mask. Repetitive scanning enabled the use of analog and later digital correction of systematic scan errors.[21]

EBES use of 10 kV was driven by the desire to avoid proximity electron beam correction. Vector scan electron beam systems at that time were developed at 20 to 30 kV. This voltage range was a vestige of customizing scanning electron microscopes as lab PGs. The voltage of 10 kV was low enough to keep the backscattered electron range (~0.6 μm) less than the minimum feature size ~2 μm of contemporaneous IC designs. This was a successful strategy until the mid-1990s when mask features sizes approached and then went below this range even at 10 kV, which was one driver that forced the adoption of proximity correction at all voltages in the 10- to 50-kV range.

Perkin Elmer introduced the Micralign projection mask aligner in 1973. The unity magnification optic eliminated contact printing, which revolutionized photolithography. Defect rates were dramatically lower without mask-wafer contact and yields improved substantially. Masks that previously had to be replaced after as few as 150 contacts now lasted for as long as a year. The masks cost device makers up to $1000 each, but their improved durability, due to the projection aligner, justified the investment.

The EBES technology was licensed by Western Electric Corporation to Etec in late 1975 and about a year later to Varian. Etec got to the market first by adopting a copy-exact strategy staying with 20-MHz operation. By 1977 the first MEBES I was delivered to Fairchild Semiconductor and a 16K RAM was one of the first patterns printed. Siemens began electron beam mask production the same year with MEBES s/n7. Varian attempted to double the EBES deflection field to ±500 μm and blanking rate to 40 MHz (Ee-BES-40-6), which achieved only limited market acceptance. Both Etec and Varian were projecting direct write e-beam was about three years away—a theme to be replayed over the ensuing decades to the present day.

The license included the tool design, use of PBS—a positive-tone resist, and COP—a negative-tone resist, which had poor edge profiles but was useful for machine calibration. The resist sensitivity was sufficient to match the tool architecture, print quality, and throughput requirements.[22]

In 1979 Perkin-Elmer's equipment business exceeded $100 million in sales, making it the first semiconductor equipment supplier to break that barrier. That year, the company acquired Etec, the leading producer of e-beam systems for mask making. Early reviews by groups at IBM, Texas Instruments, Etec, and Varian drew attention to the benefits of electron beam lithography.[11,23–25]

A large number of system prototypes were built during the 1970s and 1980s by large vertically integrated IC manufacturers—ATT, IBM, TI, HP, Toshiba, NEC, Fujitsu, Hitachi, NTT, Thomson-CSF. The focus was to varying degrees, as much on direct write on wafer as mask patterning. Later component technologies and platforms built for direct write were often transferred to the mask pattern generators. NTT's EB52 was a 20-kV gaussian beam system for 0.5-µm direct write and mask fabrication with dual-channel electrostatic deflection, which had an 8-bit *digital to analogue converter* (DAC) and 250 ns settling time.[26] This system evolved to the HB55 variable shaped vector system printing in the 20 to 50 kV range[27] with a technology transfer to Hitachi.[28]

HP built a 300-MHz, raster electron beam machine for wafer direct write and mask patterning.[29] The system introduced a 300-MHz, fixed-spot-size, 20-kV column using a thermal-field emission source.[30] Thomson-CSF developed a 15-kV vector-shaped beam system having 0.1-µm resolution for direct write and mask fabrication.[31,32] NTT built a *variable-shaped beam* (VSB) system (EB57) with large-angle, high-speed electrostatic deflection for 0.5-µm lithography.[33] This group contributed to high resolution and high accuracy electronics components development with emphasis on the relationship between DAC bit weight and settling time, which is a key pair of variables in VSB design. Varian explored a 20-kV, VSB system (VLS-1000).[34]

Hughes Aircraft company had a 10-year e-beam pattern generator development, which included a VSG system EBS-4/5.[35] Hughes and Etec developed the first VSB system with continuously moving stage, which was commercialized for direct write on wafer.[36] Key learnings appeared a decade later in Etec's shaped beam raster (*raster shaped beam*) RSB-mask PG, which included 100 A/cm^2 VSB optics, continuous stage motion, high speed digital control, and deflection electronics to support 100-MHz clock rates, and 13-bit DAC with 30-ns minor field settling.[37] Fujitsu built a 20-kV, VSB (NOWEL) system for 16-Mbit *dynamic random access memory* (DRAM) mask technology.[38]

By 1983 more than 50 raster scan electron beam PGs were in the field.[39] Etec introduced MEBES III, the commercialized version of EBES, which became the market leader. By 1986 MEBES s/n100 was installed at Siemens and the first 1-Mbit production reticles were produced. While e-beams were generally known and used for their resolution capability in advanced mask pilot and lab *direct write* (d/w) applications, these early commercial successes did not primarily rely on this attribute. GCA had introduced the step and repeat system for wafer exposure in 1978.

Step and repeat revolutionized photolithography in the eighties, increasing resolution and enabling linewidth shrinks. But feature sizes were still in the 2-μm range, while the e-beam optics easily could have achieved submicron spot sizes.

Market acceptance was based more on ability to manufacture, service, and support a system with reliable operation, with automation of column setup, system calibration, and data preparation. Early commercial mask makers such as Fairchild, RCA, and Oki chose MEBES for its reliability and speed compared to the mechanical Mann PG, and ease of pattern data entry. Market adoption was bolstered by Etec's ability to manufacture relatively high volumes of tools in 1983 to 1984. This was an example of how commercial manufacturing requirements differ from R&D, where reliability, speed, and manufacturability often trump leading-edge performance. Past history doesn't always result in similar behavior. Leading-edge resolution capability created a counterexample in the case of adoption of OPC/PSM in the late 1990s.

Toshiba was one company that developed a raster scan PG and later converted to a vector system with 50-kV-shaped beam printing.[40] Etec also developed raster, and later the AEBLE vector scan systems for direct write. However, only component technology from the latter was transferred to mask patterning.

The vector scan architecture was adopted by most other electron beam PG tool makers. Shaped beam optics was introduced to improve throughput via multiple pixel exposure.[41] *Vector scan shaped beam* (VSB) systems were developed for increased throughput in pilot manufacturing for direct write.[42,43] IBM built VSB pattern generators for prototyping direct write on wafer.[44] The EL-series were for pilot and quick turn manufacturing.[45] Although these systems were not directed to commercial mask patterning of chrome on glass, they had a broad influence on all vector scan shaped beam design. Similar architectures and platforms remain the basis for present day VSB mask patterning systems commercialized by NuFlare/Toshiba, JEOL, Hitachi, and Leica.

The commercialization of i-line projection step and scan systems and transition from 1× to 5× masks for 0.5-μm lithography[46] had an enormous impact on the mask making industry.[47] MEBES III had been printing 1.0 μm features for 1× projection printing, when mask features sizes abruptly became 5.0-μm for 5× reticles. 5× reticles had to be defect free while their 1× predecessors had defect density requirements, but acceptable defect sizes were 1 μm and repair equipment was adequate to meet the need. Nearly 200 mask pattern generators had been purchased for 1× mask manufacture. With these systems available for 5× reticle production, capability and capacity exceeded demand. Thus began the "dark ages" or "mask maker's vacation" that lasted from the mid-1980s through the early 1990s.[48]

In the mid-1980s laser PGs also became commercially available with the advent of the Atec CORE 2000 system.[49] The multibeam linear brush optics enabled high print speed with a fixed beam size at moderate resolution using optical resists. The original Ateq laser PG business plan initially targeted direct write applications, a common approach to this day for purveyor's of new patterning concepts.[15] Of greater importance was Ateq's strategy of following in the wake of wafer lithography, adapting component technologies for imaging and resist systems. CW sources were well-matched in frequency to the excimer lasers used by wafer steppers. Optical resists such as AZ i-line resists developed for lithography could then be adapted for mask making since the 363.8-nm CORE exposure wavelength was close enough to i-line (365 nm). This helped differentiate the product and reduced the need for incremental process technology research and development.

The mask pattern generator market had matured since the early 1980s, at which time it was entirely supplied by e-beam technology. Mask pattern generator product evolution during 1985 to 1995 is shown in Fig. 3.1. The market penetration of laser technology through 2000 is shown in Fig. 3.2. The requirement for additions to basic capacity was a modest 5 to 6 percent average annual growth rate (essentially the same rate of growth as IC unit production) and the primary source of unit demand is, and remains, for technology replacement. For 15 years, this demand was fulfilled by 15 to 30 system deliveries per year worldwide. Laser PGs were first made commercially available in 1987, at the beginning of a protracted recession in the mask making industry. Despite severely curtailed capital spending, laser PGs quickly captured approximately 25 percent of the available market. In 1996, the first systematic growth in the market in over a decade began. As increased capital funding became available, laser PGs rapidly increased market share to over 50 percent of new systems delivered and sales revenues generated.

From the early to mid-1990s Etec introduced major architecture and platform changes for both e-beam and laser mask pattern generators,

	85	86	87	88	89	90	91	92	93
DRAM generation		4 M		4 M	16 M		64 M		256 M
Design rule		0.7 μm			0.5 μm		0.35 μm		0.25 μm
Vector shaped									
NuFlare/Toshiba		EBM130/140	EX-7		EX-7 M	EX-8	EX-8 M		EX-8D
JEOL		JBX-5DII		JBX-6AIII				JBX-7000 MV	
Hitachi	HL-600				HL-700 M			HL-700 MIII	HL-800D
Leica			EBMF 10.5		EBPG-FE				ZBA31M
IBM				EL3/0.25 μm					
ATT/Lepton/Ultrabeam	EBES4								EBES4 II
Raster e-beam									
AMAT/Etec/PE	MEBES III							MEBES IV	
Raster laser									
AMAT/Etec/Ateq			CORE 2000						CORE-2564
Micronic							LRS-18		

Figure 3.1 Mask pattern generator product evolution 1985–1993.

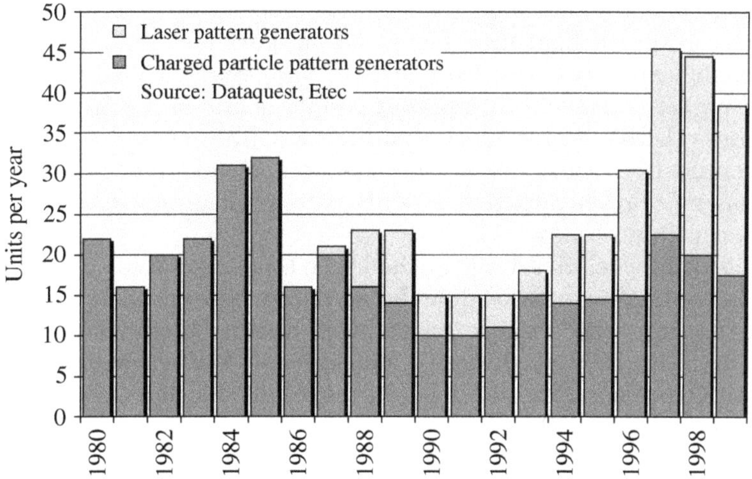

Figure 3.2 Unit delivery history of mask pattern generators. (*Source:* Dataquest, Etec.)

MEBES IV and the ALTA 3000. The former incorporated integrated high brightness *thermal field emission* (TFE) optics and doubled the blanking rate of MEBES III, while the latter increased the number of beams from 8 to 32. These were followed with various platform extensions throughout the 1990s. JEOL, Hitachi, and Toshiba also had active continuous engineering developments to support periodic system and product introductions, see Fig. 3.3.

Reticle enhancement technologies (RETs) including *phase-shift mask* (PSM) and *optical proximity correction* (OPC) began in the mid 1990s and grew into accepted manufacturing technologies. These applications required resolution and tighter process control. The feasibility of improved process latitude using 50-kV e-beam printing had been documented throughout the decade.[50–52] The benefit of moving from 10 to 50 kV e-beam PG systems was the ability to achieve tighter CD control via improved beam resolution, which improved resist sidewall angles.

	95	96	97	98	99	00	01	02	03	04	05
DR[nm] ITRS '03			250		180		130		100	90	80
2 year IC cycle			250		180		130		90		65
MPU gate in resist			200		135		90		53		35
Pilot mask PG intro			180		130		90		65		45
Vector shaped											
NuFlare/Toshiba					EX-11	EBM-3000	EBM-3500	EBM-3500/B	EBM-4000	EBM-4000B	EBM-5000
JEOL			JBX-7000MVII		JBX-9000MV		JBX9000MV II		JEOL-3030		
Hitachi		HL-800M			HL-800M+	HL-900M	HL-950M		HL-7000M		
Leica	WePrint200		ZBA31S	ZBA1H			ZBA-350				
Raster e-beam											
AMAT/Etec/PE	M-4500		M-4500s	M-5000	M-5500	MEBES-X	eXara			RSB	
Raster laser											
AMAT/Etec/Ateq	ALTA-3000		ALTA-3500			ALTA-3700	ALTA-3900	ALTA-4000	ALTA-4300		ALTA-4700
Micronic				LRS 200-25	Omega 6000	Omega 6500	Omega-6600E	Sigma7100		Sigma-248	

Figure 3.3 Mask pattern generator product evolution 1995–2005.

Sensitive chemically amplified resists were needed by 50-kV, VSB systems to avoid mask heating effects and support fast patterning. Negative-tone *chemically amplified resists* (CARs) enabled VSB printing of dense patterns to avoid throughput penalties and minimize line-edge roughness (LER). While 50 kV and high resolution could also be achieved technically with raster e-beam systems, development delays led to increasing market adoption of VSB systems beginning around 2000, which continues today.

The benefits of laser relative to e-beam had been lower capital cost resulting from less complex design, throughput because lasers use parallel beams, lower infrastructure costs because resists were already productized for wafer lithography. In 1999, production of masks in the United States was approximately 30 percent of the global production of 700,000 units, of which about 90 percent, or 180,000 units, were manufactured using laser technology. This production had a value of over $550 million at an aggregate mask *average selling price* (ASP) of $3100. By 1999, 44 percent of the pattern generators active in the United States used laser technology, representing about 65 percent of total capacity. At the same time Japanese mask output was 36 percent of global production—a substantially greater share than the United States. However, only 19 percent of Japanese pattern generators were using laser technology. By 2004 about 75 percent of all masks were produced using laser PGs.

During the 1990s, pilot production mask prices grew 75 percent, including a 50 percent increase in one four-year period alone. The price differential of these masks over peak production masks grew from about 3.5 to nearly 6 times, because the pricing for peak production masks grew by only 20 percent total over the same 10 years. The rate of price growth in peak production masks is less than one-half the rate of value growth in the IC industry as a whole and is attributable to two factors. First, substantial consolidation of the mask industry over the decade produced significant improvements in productivity and second, laser pattern generation was adopted for the production of the majority of these masks.

The demanding OPC requirements needed for subwavelength lithography beginning around the 130-nm node, coupled with a downturn in the market during 2000 to 2003 led to slack capacity demand. Though the complexity and capital cost issues remain, over the years laser expense runs around 20 to 35 percent lower than e-beam, the market has paid the electron beam price premium to have historically adjustable exposure conditions, which enabled early exploration of mask and IC process methods.

Absent a 50-kV raster pilot alternative for 130-nm technology, mask makers invested in solutions for vector pattern preparation and CAR, and accepted the added data preparation and throughput consequences. The demonstrated higher throughput of multibeam laser PGs was not sufficient to overcome the switching costs and risk of requalifying OPC mask designs with OPC printed on higher resolution 50-kV electron beam PGs. Optical resist systems were no longer a distinguishing feature. Resolution became the fundamental factor required by critical layer OPC. The 50-kV electron beam PGs have since dominated the pilot application space and regained a larger segment of production mask making with the installed base of VSB systems handling most mid-critical layers for 130 nm in the low volume market environment post-2000. This latter trend has seen some retrenchment since 2004 with laser technology regaining share in 130-nm node masks used for metal and wiring layers.

The absence of a laser PG capability in the sub-65-nm era would likely alter the pricing of peak production masks. The impact of such a change in production mask prices would not be uniformly felt across the IC industry. Manufacturing costs of very high-volume parts, such as microprocessors and memory, are relatively insensitive to mask pricing. Typically for such parts, a single mask set may be used to print upward of 10,000 wafers, producing several millions of ICs.

Increased mask pricing would primarily affect two IC production segments—the *application specific standard product* (ASSP) producer and the *customer specific IC* (CSIC) producer. ASSP parts are produced in volumes of 100,000 to 1,000,000 units and CSIC parts are produced in very low volumes, as little as a few hundred parts. The price of the first mask set can represent from 5 to 25 percent of the nonrecurring engineering for a new design. Substantial increases in mask pricing would have the effect of reducing the total number of designs that can meet economic criteria for production.

The special importance of this issue arises from the reengineering of the industry, which was accomplished in the 1990s. During this period, the industry shifted from a vertically integrated business model to a decentralized one using significant outsourcing of production. Taiwan and the Asia Pacific region became an important market for equipment. Taiwan's government sponsored ERSO mask shop was established in 1981 with the acquisition of Electromask and their first electron beam MEBES III PG was introduced in 1986. This work led to commercial mask shops being established in 1989 (TMC), 1991 (Innova, TSMC), 1997 (UMC), and 1998 (TC-Toppan, PSMC, and WSMC).[53]

This evolution led to a proliferation of so-called fabless semiconductor companies, which had a focus on higher value products, higher value content of system- and chip-level intellectual property, and major reductions in capital requirements. As a result of this reengineering, CISC and ASSP product types of mid and low volume are most sensitive to mask price. Higher mask prices directly reduce the number of economically viable designs.

Meanwhile VSB mask pattern generators have become the tool set of record for primary mask exposure of leading-edge devices at the 90-nm node. The fate of 50-kV raster e-beam development remains unclear.

NuFlare, JEOL, and Hitachi have built on 30 years of continuous engineering and commitment to their respective technologies.

The pace of laser resolution improvement will determine if this trend can be countered. Advances in lithography components suitable for laser pattern generation have become increasingly difficult to achieve in production. Improving laser resolution using a 198-nm CW laser for incoherent gaussian imaging or adapting a 193- or 157-nm-pulsed excimer for *spatial light modulator* (SLM) is a difficult technical challenge. Since wavelength reduction or increased numerical aperture via immersion lithography is difficult, it is likely the laser raster PGs will continue to pursue the use of advanced gray raster bitmap data manipulations. Both laser PG manufacturers can also be expected to develop various beam modulation and phase-shift illumination techniques to reduce the effective k_1 of their optics.

The laser PGs remain well positioned to handle the growing demand for second-level printing required by PSM, which grew from a research topic in the 1980s to an established mask manufacturing technology in the late 1990s.[54] In this application the ability to provide accurate overlay enabled by light optical imaging, a relaxed set of resolution requirements, and inherent speed advantages make laser PGs a good fit. Electron beam pattern generators have to deal with resist charging, which is compensated by adding conductive components to the resist. While this has been shown to be technically feasible, the approach adds process complexity and is plagued by increase defects.

The cost components associated with vector pattern preparation and yield issues remain. These costs included the engineering, infrastructure, yield loss of pattern preparation, and multiple resist processes. The pattern preparation and resist system infrastructure requires continuous reinvestment, and vector technology is not likely to overcome exponentially declining throughput as feature sizes shrink. However, as long as leading-edge mask pattern quality and resolution remains a driver, high-resolution 50-kV electron beam delivery will remain the preferred technology.

3.1.2 Mask print quality drivers: IC-design and wafer lithography

Mask pattern generation requirements are driven by image quality, throughput and cost of ownership (CoO), mask type, and timing of pilot and production phases of the mask and lithography development cycle. The focus in this section must remain with the specifications. The nature of specific tests is too detailed and mostly proprietary to provide a description here. Suffice it to say, however, that specifications are only of value when connected to a specific test protocol.

Mask complexity is revealed in the growing number of quality parameters. The largest contributor to mask cost and prices is yield, which provides competitive advantage in both cycle time and cost. A principle method for improving mask quality yields is to ensure that the pattern generator prints within these specified parameters in a reliable fashion so that repeated printing is avoided.

Historically, critical dimension, mean to nominal, registration accuracy, and defect size comprised the list. A relatively few constituent specifications such as stripe butting were used to ensure calibration supported tool acceptance and normal quality assurance during operation. These are now augmented with nested to isolated CD, horizontal to vertical CD, linearity, foreshortening, and line-edge roughness.

Another aspect of specifications is that they evolve over time as the manufacturing process increases yield. This evolution is managed in a competitive environment where cost, quality, and cycle time are balanced to meet immediate customer demand. The increased mask complexity and pattern generator print times are significant cost factors in today's operations.

International Technology Roadmap for Semiconductors (ITRS) lithography provides a basis for many mask quality requirements. However, the projected timing of a two year design cycle is studied as it has been the actual pace of leading-edge mask development and tool requirements since the 180-nm node in 1999. Early adopters of advanced mask technology continue to drive this aggressive schedule.[55] The two-year cycle is important as it is used often in current tool development planning, particularly for those organizations interested in serving pilot mask applications. This has become an important barrier to entry for later arrivals to the 130/100-nm technology transition. A new barrier grew from the reluctance of mask makers to requalify their processes on new tool sets for the advanced OPC/PSM designs. The situation was exacerbated with slack demand in the business cycle.

Advanced MPU gates and local interconnect pitch continue to drive mask requirements. 193-nm optical lithography at 4× reduction is the process of record. 193-i immersion lithography now appears to be the preferred future photolithography method.[56] Its resolution capability and feasibility exceeds 157-nm F2 lithography, which has encountered difficulties adapting CaF2 mask blanks, manufacturable pellicles, and complex optics in moderate vacuum as well as a limited scaling to advanced nodes. Table 3.1 shows that 193-i will be stretched to the limit to print 32-nm node gates in a resist with process and optics driving k_1 to the 0.25 limit. The advent of 193-i coupled with a long, complex technology development has pushed the EUV lithography window of opportunity to the 32-nm node. EUV has severe mask handling requirements where blank distortion is extremely critical and will be a precision

TABLE 3.1 Table of Wafer Lithography History and Trends. Pilot Manufacturing Has Been Occurring in Two-Year IC Design Cycles Since 1997

	97	99	01	03	05	07	09
2 year IC cycle	250	180	130	90	65	45	32
MPU gate in resist	200	135	90	53	35	25	18
Source	i-line	KrF	KrF	ArF	ArF	ArF-i	ArF-i
Wavelength λ	365	248	248	193	193	193	193
NA	0.63	0.68	0.75	0.75	0.85	0.85	0.85
Refractive index n	1	1	1	1	1	1.44	1.64
k_1	0.43	0.49	0.39	0.35	0.28	0.28	0.25
$W = k_1\ \lambda/(n \times \text{NA})$	249	179	129	90	64	44	35
Pilot mask PG intro	180	130	90	65	45	32	

mechanical handling driver of the blank and mask throughout the entire expose, develop, inspect, and measure cycle.

Electron projection lithography (EPL) is an advanced lithography technology strategy to address high mask costs by simplifying mask requirements. The ongoing development strives to achieve the wafer patterning requirements at a cost-effective throughput.[57] It remains an open question whether sufficient productivity and cost advantage will exist compared to 193-i.

In any case, it is expected that advanced subwavelength photolithography will require aggressive optical proximity correction, and more complex alternating aperture PSM in spite of the extra manufacturing cost. The *mask error enhancement factor* (MEEF) measures the increase in relative sensitivity to variations in mask critical dimensions. The MEEF was first predicted by early simulations of a phase-shift mask at 5× reduction.[58] These findings were further supported by simulations and experiment, which quantified the effect for 0.35-μm lithography where a 20 percent increase in errors was shown over that expected from the stepper reduction ratio.[59] MEEF is affected by reticle enhancement technologies.[60] Masks can now contribute as much as ~85 percent to in-shot wafer CD uniformities (CDU) for attenuated PSM (APSM) printing of memory devices.[61]

Pilot and production patterning. Figure 3.4 illustrates the insertion profile of equipment technology into the mask industry. In the two year period prior to the beginning of volume production, mask processes are developed using pilot equipment most suitably implemented with flexible electron beam technology. The capacity in this class of equipment peaks at about the time that volume production begins and declines thereafter. Pilot equipment seldom accounts for more than 20 to 30 percent of the total cumulative capacity that will be installed.

The majority of total capacity utilized for a given mask generation is of the peak capability type. If feasible, for cost of ownership reasons, peak

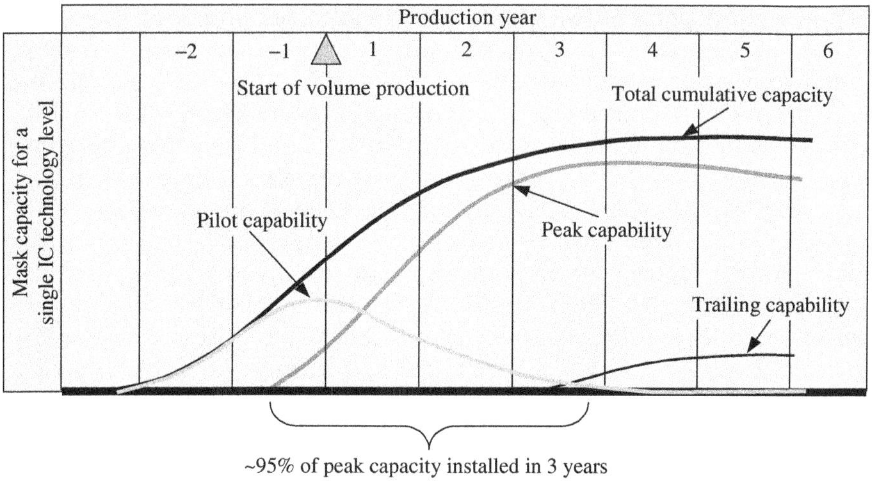

Figure 3.4 Mask technology insertion.

equipment is most suitably implemented in laser technology. The peak class of equipment enters service at the beginning of the volume production ramp of an IC fabrication technology and within three years has reached maximum capacity. Participation in the peak market is very sensitive to introduction timing. A delay of only one year is equivalent to losing one-third of the total market opportunity.

Laser PGs target the production phase of the manufacturing cycle, while early electron beam tool placements support pilot mask process development. Many vendors designate "node *n*/node *n* – 1" to describe their products. This aligns to both the production and pilot phases and further moderation of the terms exist depending on the number of mask layers printed with a given tool. With some systems both applications can be addressed using different operating modes, e.g., by running the machine more slowly and printing finer grids to achieve the pilot requirements.

There are two categories of leading-edge masks. Pilot production masks are those produced during the development of, and first production from, new IC wafer fab technology. Because of the need for user experimentation in process methodology, pilot production masks are nearly all manufactured using electron beam technology with commensurate higher cost contributions for the mask maker. These masks are relatively few in number but command very substantial price premiums because of the role they play in time-to-market competition for the IC manufacturer.

Pilot production orders are generally based on preliminary results and a product roadmap. Delivery of pilot equipment is dependent on minimum useful results and an acceptable improvement plan committing to a series of productivity upgrades to support faster printing at similar lithographic quality specifications. Mask lithography quality is paramount in pilot applications and target times are less than 10 h. Actual practice can be greater depending on the urgency and capability of existing tools.

Peak production masks are those produced as an IC wafer fab technology moves into volume production. These masks are produced in volumes from 15 to 20 times greater than the pilot production masks for the same technology level, which had been produced earlier. These masks cannot obtain time-to-market pricing premiums and are essentially commodity parts. Peak production orders emphasize demonstrated performance with delivery based on the capacity required. Production mask PGs targets are typically in the 3 to 4 h range emphasizing cost of ownership benefits and delivering established manufacturing specifications. For volume production cost of ownership considerations become critical in defining the throughput.

Pilot e-beam product introductions have been on two-year development cycles for the 90- and 65-nm nodes, and this trend is expected to continue through the 45- and into the 32-nm era; see Fig. 3.5. Through the late 1990s laser PGs lagged slightly behind the 2 year cycle,[62] but had extended market share. The pendulum swung in the other direction at the 130-nm node with electron beam systems regaining lost ground. In 2005 the ALTA 4700 is being introduced as a 90/65-nm node capable system, which is consistent with the production requirement timeline shown in Fig. 3.5. At the same pilot electron beam tools must demonstrate 45-nm pilot capability to be selected for use to create leading-edge layers for early lithographic process development.

Table 3.2 provides a view of specifications from various sources as an indication of how leading-edge requirements have tightened over the years. One remarkable aspect is the nearly order of magnitude

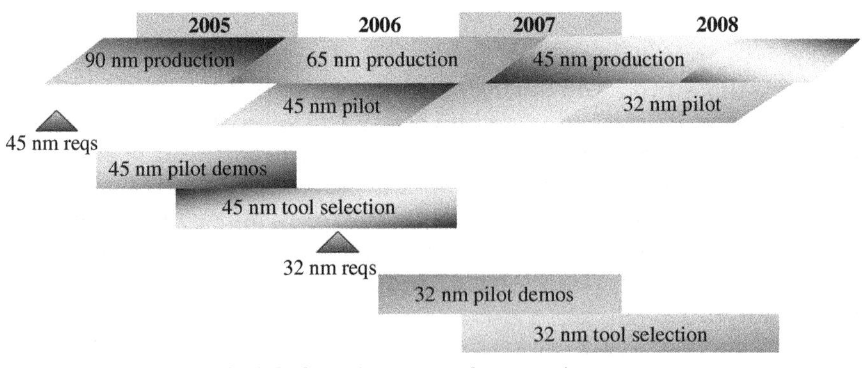

Figure 3.5 Projected schedule for 45/32-nm mask patterning.

TABLE 3.2 Timing of IC Design Cycles and Introduction of Pilot Pattern Generators for Leading-Edge Mask Specifications [nm]

	97	99	01	03	05	07
DR ITRS '03	250	180	130	100	80	65
2 year IC cycle	250	180	130	90	65	45
MPU gate in resist	200	135	90	53	35	25
Pilot mask PG intro	180	130	90	65	45	32
Mask MFS	560	360	200	140	100	72
OPC opaque	280	180	100	70	50	36
Global CDU (3σ)	18	10	7	5	3	2
CD linearity	35	14	12	10	8	6
Position accuracy (3σ)	20	12	10	8	6	4
PSM OL	60	40	30	20	15	10
Address unit	10	1.25	1.25	1	0.5	0.25

improvement over the 10-year span in almost every category. The details of how each value has been measured and nature of specific tool capabilities cannot be covered here. Introduction of OPC has had a nonlinear effect on tool address sizes. Some small address design requirements did not propagate to manufacturing and led to large file sizes, yet pilot tool builders were compelled to support this in the early years. Later these small grids sizes did support the tight CD uniformity needs of subwavelength photolithography. Specifications for positional accuracy have been affected by use of two-point versus multipoint fitting algorithms.[63]

Another aspect that is difficult to capture in a table of specifications is their evolution for a given node over time as the mask transitions from pilot to production. An example of this is revealed by considering the application space, minimum feature size, and CD uniformity as defined by a commercial mask maker interested in production needs compared to values shown in Table 3.2,[64] which are driven by pilot mask making requirements. Production CDU (node) per this reference requires 25 (180), 15 (130), and 12 (90) nm; whereas the pilot needs are 18, 10, and 7 nm, respectively. The lesson is that specifications for a given application space are a moving target. The mask maker must balance the need to qualify masks under conditions that anticipate continuous improvement of the tool, customer needs that may range from early lithographic process learning to low-risk, high-yield manufacturing, and qualification criteria that are dependent on the test methodology imposed at a given moment in time.

Higher reduction ratios and daughter masks. Discussion about increasing the mask-reduction ratio >5× surfaces periodically as a means to ameliorate pattern generator requirements, particularly on critical dimension control. Any purported benefits of this approach have not yet been compelling, at least in part due to the complexity of implementing new reduction optics designs in the wafer scanners.

Daughter masks are another proposal to relieve mask print quality requirements and costs. A two-stage mask process was developed by Texas Instruments to reduce e-beam write time by two orders of magnitude for a 64-Mbit DRAM. The method involved the mix of using both e-beam and optical lithography on a single reticle.[65] The lack of widespread adoption may be a guide to more recent attempts that also must challenge a higher lithography capability or a lower mask cost with compelling switchover attributes to a new mode of production.

The latest Toshiba/Nikon idea is also a return to the photomask repeater, 40 years after its initial introduction with a few twists.[66] A master reticle is patterned at 20× wafer dimensions. The master reticle is photoreduced 5× onto a daughter mask by the photomask repeater—a special purpose stepper. The resulting 4× mask is used in the normal way in a wafer stepper.

The approach requires a reconfigured stepper, which exceeds pattern generator accuracy. The wafer chuck must be replaced by a kinematic mask support. A gradation filter, placed near the conjugate plane of the illuminator, blends the exposures at the master mask boundaries. For small chips <5 mm, a single master mask can be replicated on the daughter mask. For larger chips, multiple masters are required. In the extreme case of a large chip filling the wafer stepper field, 35 master masks would be required. For each case an additional master mask is required for patterning the reticle fine alignment marks.

One challenge is the limited field size on the daughter reticle, which must be stitched together with multiple patterns. This adds a new component to the critical dimension uniformity on the daughter mask. Another question is the patterning throughput of master and daughter masks, which may be an issue at large chip sizes. The data fracture, inspection, repair, and metrology logistics become more complicated for the mask shop, adding to overhead cost. The need for patterning OPC features on the daughter mask would likely also require OPC or PSM on the master.

Part of the promise is that by using a champion i-line stepper with a lens distortion of 10 nm (3σ), Nikon was able to demonstrate 22-nm absolute accuracy over a full 100 mm × 132 mm pattern area. The stepper speed was reduced to achieve this, resulting in a throughput of two daughter masks per hour for a chip using two master masks. The demonstration was aligned to the 150-nm node with 12-nm (3σ) CD uniformity for 600-nm features on the daughter mask.

These results show that a stepper run at very low speed was able to far surpass nominal performance, which is specified over larger fields and at higher throughputs. With further refinements in calibration and correction it is feasible that sub-20-nm absolute placement accuracy could be achieved on daughter masks.

3.2 PG Architectures

3.2.1 System and component analysis

Pattern generator architectures have evolved within a matrix differentiated by print strategy, scanning and source technologies. Each of these choices has had major implications on system and component technology development for the various companies. Choices have also been made that in some cases were initially driven by direct write on wafer application instead of mask patterning. An early driver of the benefit of vector scanning was the observation that for small coverages the exposure time was linear with pattern density. This feature was highlighted in research and prototype direct write applications.

The EBES prototype raster system emphasized pattern placement accuracy, which could be achieved by repetitive scanning. High exposure rate multibeam printing was enabled by CW laser sources and linear brush optics. Laser PGs are exclusively raster based. Viewed as a graphic architecture, rasterization renders vector-CAD input to gray-scale bitmaps and from there to gray scale printing. Raster bitmaps combined with high performance digital electronics and firmware geometry engines permit high bandwidth pixel delivery rates transmission to the beam delivery subsystem in support of the writing strategy.

The various attributes of the raster and vector architectures have been actively debated since their inception in the 1970s, as shown in Table 3.3. Narrowly defined, raster and vector refer to scanning methods. But the related influences on data handling, calibration, and print quality and throughput indicate that these terms should be primarily understood as architectural capabilities.[67]

TABLE 3.3 Pattern Generator Architectures

	Raster	Vector
Laser	Micronic: Omega, Sigmas Etec: CORE, ALTA	None
E-beam	ATT: EBES Etec: MEBES Toshiba-1980 HP-1981	Toshiba- NuFlare JEOL IBM Hitachi Leica Cambridge

One way to distinguish these methods is to consider the channels for high bandwidth data transmission. This information flow is imparted to the beam delivery subsystem at different points for the two architectures. For raster PGs the high bandwidth data transmission impacts the beam via e-beam blanking (time domain) or laser acousto-optic (amplitude) modulation. Vector modulation is driven through the scanning system. High speed systems are created by combining relatively slow modulation rates with parallel exposure of pixels per flash. The combined pixel delivery rate is generally a product of number of pixels per flash times flash rate. This is generally true for all systems with a broad use of the term flash to include gaussian or shaped, or imaged pixels at a given time.

Both raster and vector systems use multipass, offset writing to improve patterning quality by averaging scan field "butting" errors. Consequences of accommodating this feature have different effects within the two architectures. Vector scan multipass printing requires refracturing off-line of the pattern files for each pass. Raster systems handle the requirement within the real time data path.

Vector scan printing defines the tool address grid by the ratio of the deflection field to DAC bit weight. Vector scan placement accuracy can be traded off with exposure rate by adjusting deflection settling times. Raster exposure rates have a quadratic dependence on grid size. The introduction of gray beam printing allowed finer grids to be printed without loss of exposure rate as print pixels could be maintained at relatively large sizes while gray bitmap rasterization enabled fine feature edge placement.[68]

Vector scan print rates are linearly dependent on pattern coverage, hence sparse contact and gate level patterns can be written faster than densely populated patterns. Raster scan exposure rates are independent of pattern coverage.

Laser and electron beam capabilities. The cost of ownership of laser and electron beam pattern generators is dependent on their respective capabilities. Capital cost reductions are achieved with laser PGs in two ways. First, the requirements for electromagnetic shielding and ultralow vacuum are absent, resulting in a simpler design of fewer subsystems, lower manufacturing costs, and lower prices. Second, the facilities used for laser equipment can be less demanding. In some cases, for charged particle equipment, an entire module space may have to be electromagnetically shielded. Both aspects of capital cost reduction produce a moderate CoO benefit. Actual costs vary dramatically for the specific equipment used, the specific type of mask exposed, and other factors relating to the utilization of the equipment.

Laser systems are optimally designed to function in a fixed optical configuration, which results in a fixed exposure rate or throughput. To provide laser systems with significant user adjustments in the form of beam size or spacing is technically difficult and would defeat the benefits arising from the simplicity of laser technology. On the other hand, electron beam systems are designed to exploit their adjustability. Such systems provide user control over a range of conditions of beam size and spacing and of exposure modes. The result is a variable exposure rate or throughput. Typically, laser systems provide behavior that is highly predictable and uniformly robust over a broad range of requirements. These attributes are ideal for volume commercial manufacturing. Electron beam systems typically provide extended capability at the cost of lower throughput. These attributes are suited for pilot, preproduction, and developmental requirements.

Laser PGs can deliver multiple, independently modulated beams through a single optical apparatus to produce exposure parallelism. Throughput is directly proportional to the number of beams used. Similar means of multiple beam exposure has not yet been achieved in a commercial electron beam system. In the applications of economic interest, on average for similarly demanding exposures, laser systems have roughly 2 to 3 times the throughput of charged particle systems. The issue remains that process development of advanced mask technology and resulting risk in switchover cost has lately favored the resolution capabilities of the electron beam systems.

System design constraints. All pattern generator systems have a high-level group of throughput design constraints that must be satisfied. These also define major subsystems requirements. Each design must simultaneously address three groups of exposure rate parameters to minimize overheads. The optics design is affected by all three groups.

The design philosophy is to keep these three sets equivalent and maximize exposure rate without sacrificing lithography quality. If there are inequalities between any of these, then system overheads will exist. The three groups of exposure rate parameters generally hold for vector systems as well. The section on vector print strategy will clarify where the two exposure methods differ in their details of achieving CD control and associated throughput.

The first design parameter set connects dose and resist sensitivity. The second affects data rate to the beam delivery subsystem via flash or pixel delivery rate, for vector and raster systems, respectively. The third imposes constraints on stage velocity and scan field size and frequency characteristics.

To illustrate the interrelations of the system parameters, a raster print strategy is assumed for a hypothetical 50-kV multiple electron beam system that targets the 45-nm node to provide a specific example of how the design could be configured to meet the print quality and throughput requirements. Later we shall explore a similar case for vector scan shaped beam tools.

The purpose of this example is to highlight how system overheads become a factor, how the system might be scaled to 16- to 32-nm print addresses, while achieving subnanometer feature edge placement, and what some of the features and tradeoffs are in a design analysis.

Assume that the beam delivery system comprises 32 beams printing at 50 MHz, exposing an 8 μC/cm^2 resist over a 132 mm × 88 mm mask area (116 cm^2) with a burst data rate of 1.6 Gpixels/s.

A 2 h per phase print limit is imposed to ensure that the total exposure time without overheads remains at 8 h. This has a consequence on the pixel size, which affects critical dimensions through grid snapping effects. Four-phase printing is commonly used to ensure high lithographic quality. A phase is a complete exposure of the mask area in one print sequence.

The exposure rate R, total current I, and resist sensitivity Q are related by

$$R\,[\text{cm}^2/\text{s}] = \frac{I[\text{A}]}{Q\,[\mu\text{C/cm}^2]}$$

The equivalent relation for lasers is stated in terms of joules rather than coulombs. Faster resists reduce the total current or laser power required for a given exposure rate.

Greater exposure rates demand either total current or resist sensitivity be increased. Larger total currents tend to degrade electron beam size due to various space charge effects. For the stated conditions

Exposure time T_{exp} [h]	8
Q_o [μC/cm^2]	8
R [cm^2/s]	0.02
I [nA]	33

The raster pixel delivery rate for a print pixel size P_w [nm] is

$$D\,[\text{pixels/s}] = \frac{R\,[\text{cm}^2/\text{s}]}{P_w^2}$$

P_w [nm]	32
D [Gigapixels/s]	1.6
Number pixels/scan	8192
L_s scan length [μm]	262

In this case the 8K data path memory height keeps the scan length small, which reduces lens distortions and aberrations, while maintaining tolerable stage velocities.

Data path clock frequency [MHz]	50
T_{se} scan exposure time [μs]	164
Retrace time [μs]	20
Scan period [μs]	184
Scan efficiency [%]	89

Scan frequency characteristics are then derived with the first system overhead appearing as a scan inefficiency, which results from having a flyback settling time when the beam is off, and the scan signal has settled within a range correctable by calibration firmware.

One of the benefits of linear-brush multibeam printing is that the scan period is longer than required by single beam scanners. The tradeoff is that linear-brush distortion becomes a lens aberration-distortion component. The scan overhead increases the exposure time from 7.9 to 8.8 h.

The third set of relations connects the stage velocity to two forms of the scan characteristic. The first connects scan length L_s and efficiency to a stage velocity of 5.6 mm/s, to support the given exposure rate

$$R\ [\text{cm}^2/\text{s}] = L_s\ [\mu\text{m}] \times V_s\ [\text{mm/s}] \times \text{scan efficiency}$$

Stage velocities of ~20 mm/s, which would be required for 6.4 Gpixels/s burst data rates, are relatively at low risk. The state of the art would have to be extended to handle stage velocities of 100 mm/s needed to support 10 to 50 Gpixels/s.

Both unidirectional and serpentine stage-velocity profiles plotted in Fig. 3.6 would support 6.4 Gpixels/s with similar stage overheads.

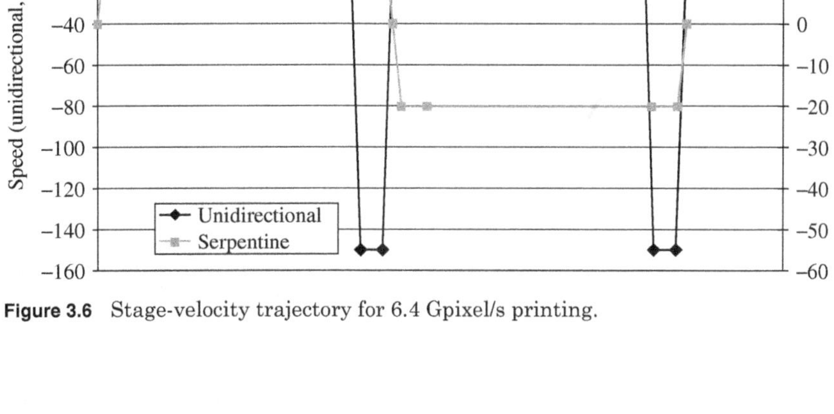

Figure 3.6 Stage-velocity trajectory for 6.4 Gpixel/s printing.

But a unidirectional mode has a simpler set of calibrations. Under these conditions stage accelerations up to 8 mg are projected during non-writing motion and writing speeds of 20 mm/s would be needed to keep stage overhead effects reasonable.

Trajectory and accelerations are other factors, which affect the stage specification, and can add to system overheads. An example of how stage overheads can be derived from a given advanced stage trajectory, velocity, and acceleration profiles is given in Table 3.4. For 6.4 Gpixels/s printing on a 32-nm address, the trajectory-induced accelerations and retraces yield a 72 percent stage efficiency. Using a similar analysis the stage efficiency is 63 percent for the 1.6 Gpixels/s example, in which case the four-pass print time example increases from 8.8 to 14.0 h. The latter value now includes both scan and stage overheads.

Lithography component analysis. A brief rationale is given here for pursuing high beam resolution and low distortion scan fields to achieve nanometer-level CD control. An estimate of stationary random variables effect on CD, ignoring time dependent stochastic processes and writing strategy specific artifacts dependent on scan and stripe orientation, can proceed from a variational method. In practice tool designers build detailed error budgets and use Monte Carlo methods to analyze pattern generator print strategies and provide subsystem tradeoff option to meet mask error budget allocations.[69] Various features can be modeled from gratings, isolated features, and contacts, which are shown to be the most sensitive to focus variations.[70]

Assume the critical dimension $\hat{X}_{\mathrm{CD}}(Q, P_W, d_s)$ comprises dose Q [μC/cm^2], distortion P_W, i.e., positional error sources along the direction orthogonal to the beam image profile, and beam image size including aberrations d_s.

The variation of $\hat{X}_{\mathrm{CD}}$ is

$$\delta \hat{X}_{\mathrm{CD}} = \left(\frac{\partial X_{\mathrm{CD}}}{\partial Q}\right)\partial Q + \left(\frac{\partial X_{\mathrm{CD}}}{\partial P_W}\right)\partial P_W + \left(\frac{\partial X_{\mathrm{CD}}}{\partial d_s}\right)\partial d_s$$

The observable is on the left side of the equation. The term $(\frac{\partial X_{\mathrm{CD}}}{\partial Q})xQ$ describes dose latitude [nm/dose%], ideally measured only in the absence of the latter two terms. A Bossung plot can determine the isofocal condition $(\frac{\partial X_{\mathrm{CD}}}{\partial Q_s}) = 0$, which holds for small perturbations ∂d_s. Exposure away from the isofocal point increases CD variation and also has the consequence that data biasing must be adopted to compensate for the difference between printed and designed feature size.[71]

Measuring CD sensitivity to dose is often complicated by other tool and process effects, see Fig. 3.7. The star marks an experimental measurement using an ALTA 4300, which has a beam size of ~190 nm *full width at half maximum* (FWHM). 1.8 nm per percent was measured

TABLE 3.4 **6.4-Gpixels/s Motion Control and Stage Overheads**

Accel [mg]	8		1		8	8	50		50		50	s/w o/h 0.20	Total cycle per pass	Stage efficiency
Pass-cycle <unidirectional>	Ramp up	Settling into WE window	Print mode	Settling (symmetry)	Ramp down	*y*-move, next stripe	Retrace Ramp up		Retrace constant		Reverse to start			
Scan time [s]	0.26	0.25	5.47	0.00	0.26	0.08	0.31		0.50		0.31		7.63	72%
Velocity [mm/s]		20.1	20.1	20.1					150					
Distance[mm]	0.26	5.0	110.0	0.0	2.6	0.26	23.0		74.3		23.0			
Total travel [mm]	120.2						120.2							
Accel [mg]	8		<1		8		8		<1		8	s/w o/h 0.20	Total cycle per pass	Stage efficiency
Pass cycle <serpentine stage>	Ramp up	Settling into WE window	Print mode-->	Settling (symmetry)	Ramp down	*y*-move, next stripe	Reverse Ramp up	Settling into WE window	Print mod<--	Settling (symmetry) start	Reverse Ramp down			
Scan time [s]	0.26	0.75	5.47	0.75	0.26	0.08	0.26	0.75	5.47	0.75	0.26		7.63	72%
Velocity [mm/s]		20.1	20.1	20.1				20.1	20.1	20.1				
Distance [mm]	2.6	15.1	110.0	15.1	2.6	0.26	2.6	15.1	110.0	15.1	2.6			
Total travel [mm]	145.3						145.3							

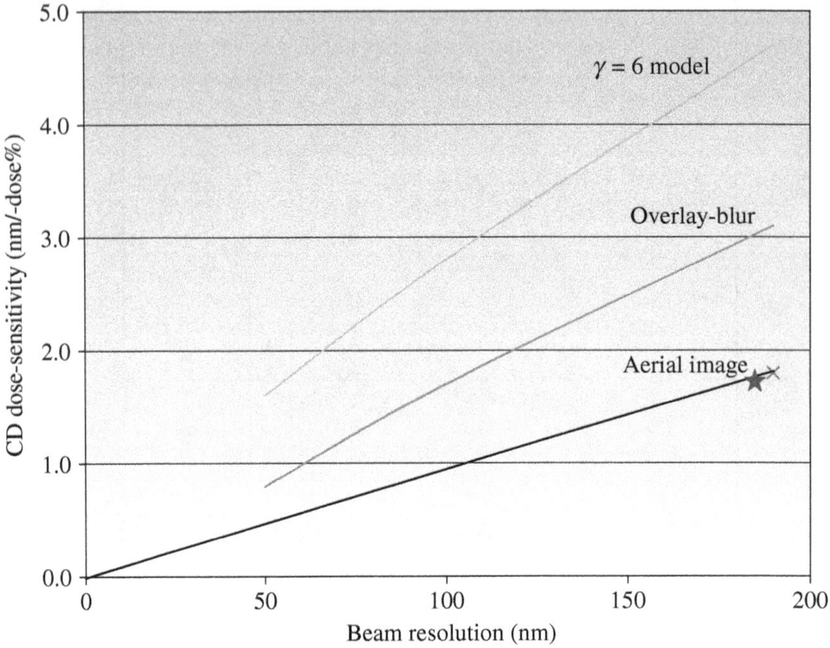

Figure 3.7 Nanometer CD control requires sub-50-nm optics.

using a high contrast resist PEK-130.[64] Completely excluding other tool print artifacts associated with the other two terms is made difficult by tool print methods. The overlay-blur line is taken from experimental measurements on an RSB-Quadra along with a $\gamma = 6$ model.[72] Overlay-blur refers to a nonstandard print method, which reduces nondose related factors by printing at isofocal and minimizing beam position errors. All scan related CD-variance effects could not yet be excluded, which along with process and etch effects might explain why measurements are larger than the aerial image plot.

The general point is that CD variation becomes less sensitive to dose as the optic resolution improves and beam distortion errors are minimized while printing at the isofocal point.

At the isofocal dose the variation reduces to two terms:

$$\delta\hat{X}_{\mathrm{CD}} = \left(\frac{\partial X_{\mathrm{CD}}}{\partial Q}\right)\partial Q + \left(\frac{\partial X_{\mathrm{CD}}}{\partial P_W}\right)\partial P_W$$

The distortion sensitivity term $(\frac{\partial X_{\mathrm{CD}}}{\partial P_W}) = f(d_s / P_W)$ is a function of the *spot-to-grid ratio*, and includes scan or stripe and blanking artifacts.

Reducing the scan length or varying spot to scan-pitch ratio will minimize the distortion-related contribution and in some cases counteract the dose-related CD errors at the isofocal point. Beam placement errors include multibeam brush distortion, scan distortion, retrograde scan artifacts, multiphase overlay, and beam landing errors resulting from nontelecentricity coupled with mask height errors. Beam drift due to thermal and electrical charging for e-beams can be minimized with reregistration during writing. Scan distortion is compensated with various calibration corrections including magnification, rotation, and offset.

Beam blanking modulation also creates position and dose errors. Raster e-beam systems have used pulse-width blanking. Shaped beams are modulated by a shaper driver that deflects the upper shape off the lower one, which also creates image plane motion if the rocking point is not conjugate to the mask plane. Scanned gaussian lasers use amplitude modulation and have velocity of sound effects via the *acousto-optic modulator* (AOM). Spatial light modulation contrast formation is based on diffraction created by small tilt angles of a torsional mirror, which have their own time dependent variations.

All systems have print signatures, which degrade pattern quality. Vector scanning introduces random placement errors within the scan field, which are difficult to calibrate, but are revealed in characteristic spatial frequency domains associated with shot size, subfield, and major field deflections. Raster scan system asymmetries can be revealed in differing feature dimensions in the horizontal and vertical scan directions where printing is essentially within a quasi-one-dimensional field. The x axis has lower distortion where the print window in this axis can be set ~1 μm. Cross-scan calibration errors in a raster system are extremely low along the x axis. The cross-scan raster x axis requires only a small amount of zero-order correction and scan rotation. The y-scan axis has more difficult requirements with respect to calibration of scan length, scan linearity, and settling. The cross-scan (x axis) local CDs sample a longer scan exposure time period and so are more susceptible to noise contributions. Key to improving print quality in these complex systems, many dedicated engineers have also worked on reducing deflection scan noise, ground loops, and electromagnetic interference.

3.2.2 Vector scan printing

Vector scan printing characteristics will now be discussed using system parameters derived from the recent literature. Consider a 132 × 106 mm^2 mask area and set the condition that a 1-h print time per pass with a 3.8 mm/s stage-velocity limit is required. The coverage rate is 0.039 cm^2/s. A fast resist for advanced 50-kV mask patterning is presently at ~5 μC/cm^2. Most VSB optics deliver 20 A/cm^2 and use four-phase printing to achieve high quality. This requires $T = J/Q =$ 50 ns resist exposure time.

Many vector scan systems break the deflection into two or three channels, making use of high resolution, but slower 16- to 20-bit major-field DAC for subfield placement, and faster ~12-bit DAC for minor field deflection inside the subfield.

In some systems, e.g., vector scan gaussian, the slower DACs are used for placing the origin of each primitive shape, and the faster DACs are used for filling in the shape. In addition to deflecting the beam with separate DACs, these DACs drive physically separate deflectors (magnetic or electrostatic).

These dual channel deflection strategies match high accuracy mixed-magnetic and high-speed electrostatic deflection. An example of VSB optics, which adapted this strategy,[43] is shown in Fig. 3.8. The benefit of mixing the final objective lens and magnetic deflection fields is reduced aberrations or distortion.[73] JEOL is notable for having avoided triple channel deflection and in fact transitioned to a single stage electrostatic deflector. Single stage deflectors have fewer problems with matching deflections of the "fast" and "slow" electronics, but sacrifice some speed.

The deflection settling time overhead is a key technical problem that influences accuracy and throughput in vector scan systems. The need for a faster (shaper) blanker driver electronics for shaped or gaussian beams is the analogous raster scan requirement. As a result SELETE has sponsored the development of new transistor technology, in cooperation with JEOL, Hitachi, and NuFlare. The objective is to reduce the settling time to 30 from 60 ns, which is presently the state of the art for 50-kV VSB.[74]

In the present example, let's consider a dual channel deflection strategy that is used with a major field, which is a $1024 \times \mu m^2$. Magnetic major field deflection with beam trajectory calibration can minimize DAC-coil settling time effects. Fast minor field deflections are controlled over 64-μm subfields. In this case there are 1.33×10^4 major fields and 3.42×10^6 minor subfields per mask. Setting the subfield time at 60 ns determines a 110-ns flash cycle time, which includes dwell and settling times. The above situation is proximate to EBM-4000 capability.[75]

Vector scan exposure rate per print phase is

$$\frac{\text{Print time}}{\text{Phase}} = N_{\text{flash}}\ (\text{dwell time} + \text{settling time})$$

where the number of flashes N_{flash} and the flash cycle time is the sum of beam dwell and settling time. Beam dwell time T [s] = Q[C/cm^2]/J[A/cm^2] is a function of resist sensitivity and beam current density. Flash settling time depends on the deflection electronics and is increased when accuracy is to be improved at the expense of throughput.

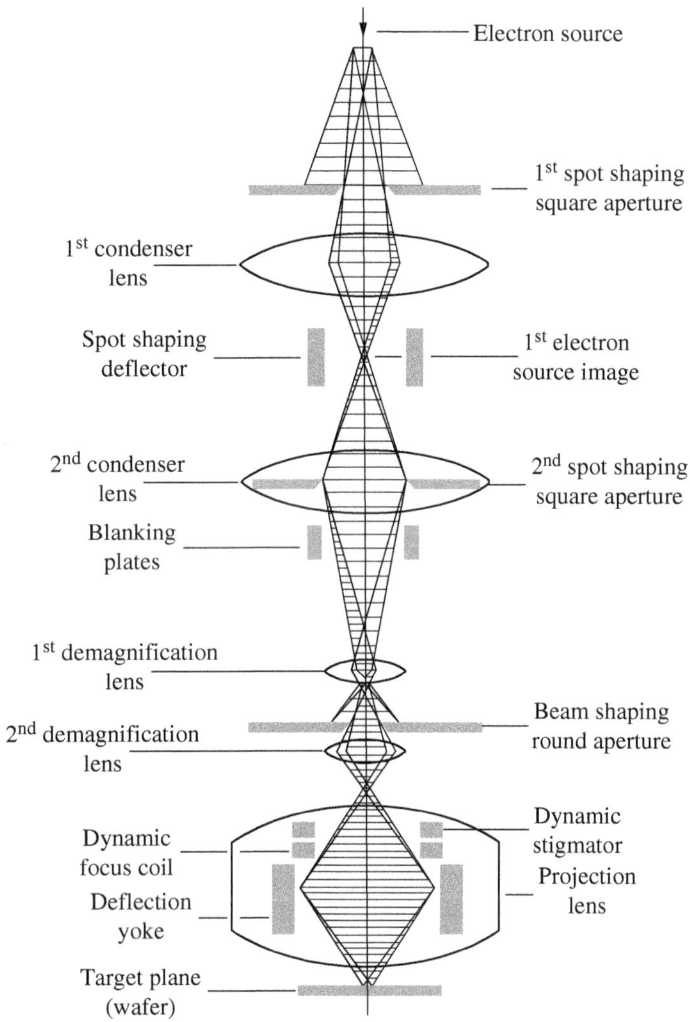

Figure 3.8 VSB optics-linked beam trace.

Projections can then be made based on this capability and device scaling, which will require 10 to 20 percent system speedup per node. Estimated VSB write time performance for various mask technologies is shown in Fig. 3.9.

This scaling includes approximately fourfold increase in flash count per node, but excludes more aggressive estimates that have gone as high as seven times due to OPC. The reason is that by the time the production era occurs, the more radical OPC approaches have often been excluded.

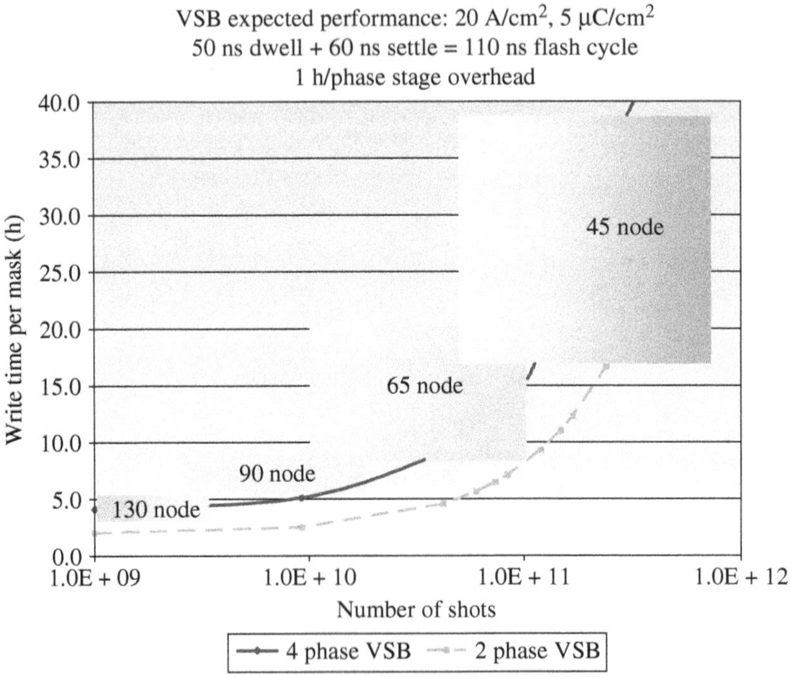

Figure 3.9 Estimated VSB write time performance for various mask technologies.

The strong dependence of VSB throughput on shot number continues to drive development of printing larger groups of shots simultaneously.[76] The approach is an outgrowth of the character projection concept.[44,50,77,78] Various cell repertoires have been studied for application at the 65-nm node, but the benefits and time for implementation indicate that this approach is not yet ready for commercial VSB integration.[79] The promise of cell projection is to reduce the number of total flashes, but this occurs when there are repetitive cells to print. Trends in OPC tend to flatten hierarchy in pattern files. The response is to increase the number of cells, which drives the complexity of calibration and data fracturing. Also, as is well known in direct write, symmetry-breaking patterns have a nonlinear effect on system throughput as the multiple requirements lead to various overheads. Nevertheless effort will continue to be devoted to solving these issues for this promising extension of VSB in spite of already a decade of work attempting to transfer the concept to the stringent demands of mask patterning.

Toshiba/NuFlare. Toshiba has had a long technology development history for direct write development platforms and internal mask making using VSB. Toshiba is one of only a few companies to have developed both raster and vector scan systems over a wide range of beam delivery optics and exposure voltages. Both horizontal and vertical integration within Toshiba drove a focus on system design and product marketing, which targeted the leading-edge pilot application.

The EBM-130/140 was a 15-kV raster gaussian beam reticle writer with continuous stage motion, which was built for 0.5-µm, 1 to 4 Mbyte DRAM.[40]

A major architecture change was made in 1987 with the EX-7 platform adopting vector shaped beam printing[80] and high current density 200 A/cm^2 at 50 kV.[81] The system was targeted for wafer direct write and 1× x-ray masks. The shape-edge blur resolution was about 125 nm for a 1-µm shape. Triangle beams were added to improve angle feature fidelity[82] and along with character projection was added to reduce shot count.[50]

The EX-8 adopted variable shaped beam vector scan with continuously moving stage and was designed to print 500-nm reticles in support of 16 to 256 Mbyte DRAM lithography.[83] A year later the system performance had been evaluated from 15 to 40 kV.[84] The prototype EX-8 specifications were given along with an indication that the platform extension would include 50 kV.[85] Other improvements included active magnetic field cancellation and new calibration methods for deflection and substrate height.[86]

External commercialization began in 1999 and continues under the NuFlare Inc. spin-off. The EX-11[87] became the platform for EBM-3000 mask PGs.[88] Prior to commercialization a beta version (EBM-210VS) was tested by SELETE[89] and subsequently 17-nm CD linearity data test of proximity effect correction showed it could meet the reported 20-nm specification.[90] Evolution of NuFlare/Toshiba specifications with 65-nm projections is shown in Table 3.5.

The EBM-3000 competed with fast laser writers by using a, then relatively fast, 16 µC/cm^2 CAR and short 100-ns beam settling time. Lithographic quality was emphasized as a pilot e-beam using relatively slow ZEP7000 resist and increasing the beam settling time. The EBM-3000 optics resolution was 90 to 120 nm.[91]

A third-order dose modulation proximity correction algorithm was implemented using a two-channel design to handle position and dose with a coarse pixel map for efficient backscatter correction. Resist heating was avoided by adopting four-pass writing and reducing the beam current density from 40 to 20 A/cm^2.[87]

The EBM-3500 system targeted 150/130-nm production.[92] It evolved from the EBM-3000 with new magnetic shielding, improved reticle thermal control, faster electronics with noise reduction, and a multiprocessor pattern data receiver. The EBM-3500B was designed to be 10 percent faster than the EBM-3500 by adding a triangular shape to the shot repertoire. The pattern could then be fractured into fewer shapes.

TABLE 3.5 Evolution of NuFlare/Toshiba Capabilities with 65-nm Projections

System		EBM-5000	EBM-4000	EBM-3500B	EBM-3500	EBM-3000
Target node		65/45	100/90	130	150/130	180/150
Min feature size [nm]	Contacts	360	400	480	600	750
	Gratings	180	280	320	400	500
	OPC	100	120	150	150	150
<x-y> CD (range)		2	3	3	5	5
CD in scan field (range/2)		3	5	7	10	13
CD uniformity (3σ)		4	7	10	13	18
CD linearity (range)		10–15	10	10	15	20
Pattern density (range)		7	10	10	12	13
Corner rounding (radius)		80	100	140	150	150
LER (3σ)		5	7	10	15	20
Design address		1.0	1.0	1.25	1.25	1.25
Position accuracy (3σ)		8–10	12	12	15	20
Overlay (3σ)		6	8	10	12	15
PSMreg(mean ± range/2)		25	30	30	50	75

The EBM-4000 was designed for 90-nm production per 65-nm pilot.[75,93] System enhancements include improved electron optics resolution, electronics noise reduction, and 65-ns deflection amplifier settling time compared to 120 ns for the EBM-3500. It prints with four-pass stripe offset and representative figure *proximity effect correction* (PEC) is used. Writing strategy involves unidirectional stage motion during exposure with three-stage deflection—1.024-mm main, 64-μm subfield, and shot components. Subfields are offset among each pass to reduce shot butting effects. One, two, four, or eight passes can be chosen. Print time is 30 to 40 percent faster than EBM-3500B, which results, at least in part, from rectangle and triangle sorting per subfield that accounts for 10 to 20 percent print time reduction.

Although EBM-4000 optics did not significantly improve resolution continuous engineering on the platform reduced noise and improved subfield settling accuracy. The published data also reveal how subfield deflection settling accuracy affects CD data, see Table 3.6. As expected uncorrected systematic position error components are highly correlated ~75 percent with global field size. Undivided single shot local CD data for horizontal and vertical lines average to 2.3 nm. Assuming independence of the residual error, subfield deflection contributes 1.9 to the 3.0-nm total averaged local CD.

A second check on the subfield error component can be obtained from the 4.2-nm overall CD uniformity, which includes field boundaries and uses 500-nm grating. The global CD uniformity excluding field boundaries

TABLE 3.6 EBM-4000 Data and Analysis of Local CD Contributions

Data	(3σ) [nm]	
CD		
<Single shot>	2.3	Data (2.5, 2.0; horiz, vert)
RSS (<Local>-single)	1.9	Subfield deflection error due to four pass
<Local>	3.0	Data (3.7, 3.4; 3.1, 2.3) (horiz, vert; line, space) no field boundary included 500 nm line/space grating <average> x, y 60 values
Global	3.8	No field boundary included 132-mm square region
RSS (Overall - global)	1.8	Field stitching component
Overall CDU	4.2	Includes field boundary 500 nm line/space grating
Position		
Registration	8.6 x 6.9 y	Image placement (IP) error 35 × 35 grid, 139 mm^2
Overlay	4.0 x 3.8 y	Residual random IP component 3 masks; 13 × 13 grid, 132 mm^2
RSS (Reg-OL)	7.6 x 5.8 y	Systematic IP error component
Small-field overlay	3.2 x 2.7 y	Residual random IP component 3 masks; 5 × 5 grid, 44 mm^2

NOTE: The field stitching component is approximately equal to subfield deflection error.

was 3.8 nm. The residual 1.8-nm field stitching component is highly correlated to the derived 1.9-nm subfield settling effect.

Recently the EBM-4000 specifications have been modified with an EBM-4500 version. CD uniformity was tightened from 7 to 5 nm and CD linearity for 280-nm features relaxed from 10 to 15 nm.

JEOL. JEOL has a long history of developing VSB technology, direct write development platforms, and commercializing mask pattern generators. As with most other vector scan system developers, component, platform, and direct write development enabled subsequent mask pattern generators. JEOL has additional strengths in the horizontal integration of tool technologies across product lines, of which only one is mask pattern generators. While not often known as a technology innovator, the continuous engineering of systems over long periods of time continues to win JEOL a market share in merchant and captive mask shops.

The JEOL pattern generator platform architecture is now based on vector scan shaped beam, 50 kV, step and repeat stage motion. This latter feature results in somewhat different throughputs and print quality artifacts compared to other VSB competitors who adopt write-on-the-fly stage motion. The product marketing is generally aligned with the IC manufacturing, though they also have a history of positioning those tools for insertion at leading-edge pilot phase of the development cycle. The JEOL specifications were shown in Table 3.7.

An early system was a vector scan gaussian beam machine JBX-5DII, which tested 25-50kV, a LaB6 source, and in-lens octupole deflection.[94] A 25-kV thermal field emission gun was later installed.[95] These components were later transferred to the JBX-6AIII mask PG.[96]

The JBX-7000MV introduced a 20 kV, 2 A/cm^2 maximum current density, shaped beam vector scan step and repeat and two-phase offset printing for 5 × mask making of 64-Mbit DRAM at 0.3- to 0.4-μm ground rules.[97] Tool grid size was cut in half to 25 nm and a 10 μC/cm^2 resist could be exposed with 0.7 A/cm^2 current density, which required 7 μs of dwell time per flash for two-phase printing. A six-point substrate clamping design reduced maximum out of plane distortion from 4.7 to 3.2 μm.

The JBX-7000MVII was developed to print 4-in. reticles for 256-Mbyte DRAM class devices.[98] The system could create a map of distortion values for the deflection so that patterns could be mapped more precisely onto optically generated features. The system handled up to 7-in. plates with a laser stage measurement unit of 0.6 nm (λ/1024). Improved thermal control supported 30-nm (3σ) overlay accuracy and 40-nm placement accuracy.

TABLE 3.7 JEOL System Performance Capabilities

		JBX-3030 MV	JBX-9000 MVII	JBX-9000 MV	JBX-7000 MVII	JBX-7000 MV
Node [nm]		90/65	130/100	180/150	250	350/250
Min Feature size [nm]	Contact	240				
	Gratings		100	100	200/300	300/400
	OPC					
<*x-y*> CD (range)			±4			
CD within scan field (range)			8	20		
CD uniformity (3σ)		8	20			40
CD linearity (range)		15	15			
Residual proximity error range				20		
Design address			2			
Position accuracy (range)		±12	±20	±30	±40	±50
Overlay (3σ)		±12	±20	±30	±40	±50

As part of the Association of Super-Advanced Electronic Technologies (ASETs) electron optics development during 1995 to 1997 JEOL developed a 50-kV capability. The transition to 50 kV was made with a shorter electron optical path[99] to improve resolution and a higher single stage 20-bit deflection to deliver 20-nm CD uniformity with 30-nm positional accuracy.[100,101] This prototype 180/150-nm step and repeat VSB led to the JBX-9000MV, which also included improved calibrations and dynamic corrections of astigmatism and field curvature.[102]

The JBX-9000MV II was designed for the 130- to 100-nm nodes and improved calibrations of dynamic distortion and aberrations to achieve 8-nm local CD control. Print strategy continued with a variable shape, vector scan, multipass step and write using a 50-kV LaB6 10 A/cm^2 optic, which could expose the same resist and retain 7-μs flash exposure time of the JBX-7000MV series. The system has a two-pass exposure mode, with multifracturing to accommodate the representative figure method of PEC. Beam metrology, mark detection, electronic noise and thermal control, and fogging correction improved CD uniformity from 20 to 8 nm.[103]

The JBX-3030MV has been developed for the 90-nm technology node and is an upgrade of the 9000MVII platform. The JEOL 3030MV two-stage deflection optics is shown in Fig. 3.10. A triangle shape was added along with a high speed electrostatic beam deflection unit, a high accuracy PEC system, and a glass-in glass-out material handling unit. The system achieves a global CD accuracy of 8 nm (3σ). Placement accuracy capability is reported to be ±8 nm, though the specification states ±12 nm.

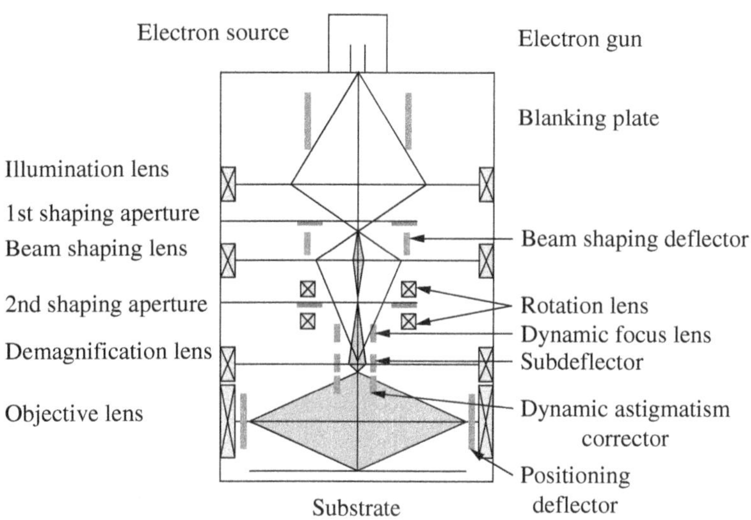

Figure 3.10 JEOL 3030MV two-stage deflection optics (positioning and subdeflector)

Exposure speedup is stated as being three times faster than the previous model.[104,105] The 3× factor is associated with a reduction in the beam settling time by improving the DAC amplifier, which appears to be the major contributor to overall system throughput based on the published data. The system throughput will likely be affected on the order of ~10 percent, but no throughput analysis is yet available. A high alignment-accuracy electron beam mask writing process for PSM has been developed using alignment and height marks, which has a capability for 16-nm (3σ) overlay.[106] A full process for second-level PSM overlay has also been reported to meet 20-nm (3σ) requirements.[107]

Hitachi. Hitachi developed VSB technology, direct write development platforms, and commercialized mask pattern generators. Vertical integration with other manufacturing segments has been a strength, along with involvement in direct write, which has provided technology and development platforms. Mask pattern generator introductions have typically been directed to peak rather than pilot mask production in Japan and Taiwan.

The HL-600 system was developed for 0.5-μm direct write on wafer and masks.[108] The 30-kV VSB optic used dual-channel magnetic and electrostatic deflection[28] with 100-ns settling time and was based on the NTT/Hitachi VSB prototype.[27] The Hitachi system performance capabilities are shown in Table 3.8.

The transition from 20 to 50 kV along with a continuous writing method occurred with the HL-800D.[109] The HL-800M performance demonstrated <40 nm positional accuracy and 40-nm CD linearity,[110] which later was improved to18 nm.[111]

TABLE 3.8 Hitachi System Performance Capabilities

Node [nm]		HL-7000M 90-65	HL-950M 130-100	HL-900M 180/130	HL-800M 250/180	HL-600 500
Min Feature size [nm]	Contacts					
	Gratings		100		500	
	OPC					
<x-y> CD (range)		2				
CD within scan field (3σ)		6	10	10	20	
CD uniformity (3σ)		8	10	20	30	100
CD linearity (range)		10	15	20	40	
Pattern density (3σ)					18	
Position accuracy (3σ)		20	20	30	40	100
Overlay (3σ) 15		15		30	100	
Design address		1.25	2.5	10		25

Improved print quality for the HL-900M was achieved by double shielding the column objective lens, reduced mask-laser mirror distortion, and tighter thermal control.[112]

The HL-950M was designed to print 130/100-nm masks and was based on the HL-900M platform. System improvements include address size being reduced four times to 2.5 nm, fractured pattern data for sub-subfield offset printing, and improved second-order proximity correction.[113]

The HL-950M was rebranded as the HL-7000M under a company reorganization and positioned at 90-nm node.[114] This tool had the same architecture and similar specifications to HL-950M, but shrunk the field size from 2 to 1 mm so the internal tool address grid went from 2.5 to 1.25 nm to improve lithographic quality. Cell projection was introduced to the mask PG to reduce shot count and improve exposure rate. Two-pass exposures with offset sub-subfields helped reduce intrafield shot placement artifacts. Hierarchical data fracturing and improved 100 Mbytes/s disk input/output (I/O) reduced system overheads. Pattern dependent fogging effect corrections were also described. Further improvements in CD linearity have been achieved by introducing a new method for shape size and gain determination, which avoid fluctuations in direct shape size measurement. These improvements have improved CD linearity and local CD control to 5 nm.[115]

IBM. The influence of IBM on mask making has been extensive through pattern generator technology development, research on physical effects such as proximity correction, support of commercial equipment vendors, and in-house mask making expertise. IBM pattern generator development was internally focused and directed primarily to direct write on wafer, though various systems were applied to mask making over the years. The development of a manufacturing electron beam lithography system for direct write on wafer was begun in late 1967.[116] A long-term investment in semiconductor equipment technology development led to innovations adopted within the mask PG equipment industry.

Vector scan gaussian (VSG) beam printing was directed primarily to wafer direct write at TI,[19] HP,[30] Philips,[117] and IBM, where it was also used for 1×, x-ray mask patterning.[118] High resolution vector scan gaussian beam lithography with automated 0.1-μm registration[119] and maximum 4-mm deflection, 14-bit deflection resolution, and 0.5-μm addressability[120] was used for advanced device research.[121] The system was also applied to chrome mask fabrication demonstrating improved pattern definition and fast turn times.[20] IBM showed feasibility of printing 0.5 μm features, which were transferred into chrome absorber masks about five years before this requirement was met in a manufacturing environment by MEBES III.[24]

The development of LaB6 as a stable, bright thermionic electron was widely adopted for electron beam pattern generators by the mid-1980s and remains in use by many vector scan shaped beam systems.[122] Electron source technology progressed further with the development of stable, high brightness Zr/O/W <100> Schottky/thermal field emission sources.[123] A 25-kV TFE electron optics with highest reported effective brightness 9×10^7 A/cm^{2}sr at the substrate was then built for a 100-nm nanolithography vector scan gaussian beam system. This system also achieved the highest current density for beams in the 10 to 30 nm range.[124]

Pioneering work applied finite element analysis to modeling advanced focus and deflection optics for a series of vector and raster scan systems, both shaped and gaussian beams.[73,125] This systematic electron optic modeling of the original moving objective lens concept for large field scanning of extended-object shapes with low distortion and aberrations[126] enabled variable axis immersion lens implementation to advanced e-beam lithography systems.[43,127]

Fixed-[41] and variable-shaped[43] electron optics were conceived to improve vector scan exposure rates by increasing the number of pixels exposed in parallel. These innovative concepts were integrated with the vector scan shaped beam EL-series over a period of almost three decades. The first EL-1 tool was installed in 1973 with the specific purpose of manufacturing LSI logic chips by direct wafer exposure with the ability to print 1.0-μm features over 5-mm fields.[128] The optics comprised a scanned 25 kV, 50 A/cm^2, 2.5 μm fixed shaped beam projecting 25 pixels per flash. To increase exposure rate and pattern repertoire variable shaping was introduced to provide 100 parallel pixel exposure capability. The EL-2 system featured variable shaped, dual channel deflection system and automated major field calibration.[116,129] Outside the field of microlithography variable shaped imaging had been proposed for printing wallpaper.[130]

Character projection was invented to further reduce flash count,[44] adopted for direct write on wafer[50,131–134] but not widely applied in mask patterning because the benefit of reduced flash rate was outweighed by nonrepetitive, OPC-decorated patterns. The method also requires greater calibration complexity and hence loss of pattern quality.

The first EL-3 VSB systems were for 1-μm wafer direct write manufacturing[135] with subsequent extensions to 0.5-μm[136] and then 250-nm lithography.[137] EL-3+ was primarily designed for 1× x-ray masks and increased the voltage from 25 to 50 kV, added a variable axis immersion lens (VAIL) and refined deflection field distortion calibrations (LEARN).[138]

The EL4 was originally operated at 75 kV with emphasis on x-ray mask printing to exploit the benefit of printing 100-nm minimum features.[139] A high, 60 A/cm^2 current density, feature specific to the EL-VSB series was maintained. The electron optics combined a larger number of lenses with a unique three-stage deflection for optimum speed.

The final VAIL lens provided minimized off-axis aberrations (or maximum field coverage) as well as telecentric beam positioning. This feature was required for high throughput wafer direct write. Tool grid size was 3.25 nm and placement accuracy 36 nm (3σ) A shifted focus to chrome-on-glass printing was revealed with a 50-kV VSB error budget targeting the 180-nm node for the IBM EL4+/EL5 series.[140]

Leica. Leica offers variable shaped and gaussian beam vector scan systems, both types for mask making and direct write applications. Leica's metrology leadership and exposure to the research direct write market provide some horizontal integration for its mask pattern generator business. Leica has inherited and acquired technology from several companies including Thomson CSF, Philips (ASM), Cambridge Instruments (1988), and Zeiss-Jena (which became Jenoptik in 1991 and Leica in 1996). Zeiss-Jena built variable shaped beam systems, used for mask making and direct write in the eastern block during the Soviet era. Philips EBM-F and EBPG-4 series were dedicated to R&D applications.

Shaped beam[141] and cell projection were adopted on the WePrint 200 and ZBA 31/32 tools.[142] This VSB direct write feature has not made its way into the mask patterning repertoire. The earlier Leica Jena ZBA 31/32 handled plates up to 7 in. and wafers up to 8 in. The 31 was for masks and the 32 was for wafer direct write. The ZBA31M used a 20 kV LaB6 optic with 100-nm resolution.[143] The ZBA writes on the fly and delivers 20 A/cm^2.

Leica offers the SB351 variable shaped beam mask writer for 90-nm node masks.[144] This system is also used for direct write, and has demonstrated full 65-nm node performance for 1× patterns on silicon.[145] More recently, it demonstrated some performance features at 1× for the 45-nm node.[146] The ceramic stage accommodates fully populated substrates up to 300 mm in size, and has demonstrated writing on 9 in. × 9 in. masks.[147] The SB351 system architecture supports multipass exposure and handles proximity, local heating, and fogging effects with the 90-nm node targeted.[148] As for all Leica ZBA/SB tools, the SB350MW adopts vector scan exposure strategy, a continuously moving stage, and the variable shaped beam architecture. The maximum shot area is 2 μm × 2 μm, and rectangular shapes with 0° and 45° orientations as well as triangles can be exposed in a single shot. The resolution of the pattern design grid was upgraded to 1 nm. The pattern data are fractured into stripes and subfields for exposure, the maximum stripe size is 1.0 mm, and the maximum subfield size is 50 μm. Both the SB350 VSB and VB6 gaussian beam system have been used to demonstrate 1× templates for nanoimprint lithography.[149]

Performance specifications for the SB350 and SB351 shaped beam mask writers are summarized in Table 3.9.

TABLE 3.9 Performance Specifications for the Leica Variable Shaped E-beam Mask Writers

	SB350 MW	SB351 MW
Year of introduction	2002	2004
Stage motion range (x and y)	300 mm	310 mm
Maximum mask size	230 × 230 × 9 mm	230 × 230 × 9 mm
Beam voltage	50 kV	50 kV
Composite placement (3σ)	15 nm	12 nm
Overlay, mask to mean, (3σ)	12 nm	10 nm
OPC minimum feature size	150 nm	100 nm
Critical dimension uniformity (3σ)	10 nm	8 nm
Minimum address grid increment	1.0 nm	1.0 nm
Maximum number of dose levels	256	256

EBES4: Hybrid vector/raster e-beam. The EBES4 was a hybrid vector-raster scan system with fixed gaussian spot, triple-channel deflection, and 230-MHz data path clock rate.[150] EBES4 column (left) and hybrid vector-raster printing (right) are shown in Fig. 3.11. The 20-kV column delivered 1600 A/cm^2 with an effective brightness at the substrate of 5×10^7 A/cm^2/sr.[151] The fixed 0.125-μm spot optic used a single beam crossover at the center of a high-speed, double-deflection, traveling wave beam blanker.[152]

Pattern data were segmented into stripes 256 μm high. Magnetic deflection with 5-μs settling time covered the stripe height. The stripes were fractured into 32-μm cells containing macrofigures, which were addressed with a 100-ns settling electrostatic vectored deflection. The macrofigures

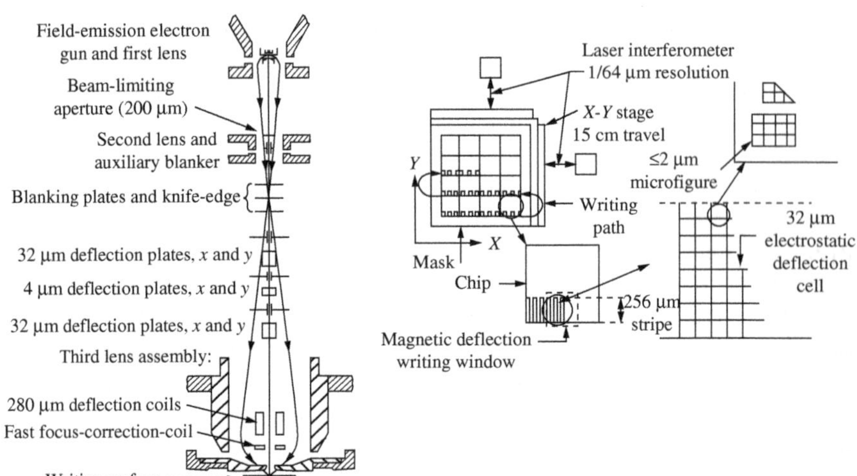

Figure 3.11 EBES4 column (left) and hybrid vector-raster printing (right).

were then split into microfigures with dimensions up to 2 μm . The elemental microfigures were written in a raster scan fashion using terminated transmission line electrostatic deflection having a 1-ns settling time in conjunction with the 500-MHz bandwidth figure generator.

Lepton and then Ultrabeam licensed the gaussian beam,[153] continuous stage motion, EBES4 technology from ATT. The pattern generator operated at up to 500 MHz, and the high overall throughput allowed production of a 16-Mbit DRAM mask in 30 min.[154] Ultrabeam V2000 ran at 20 kV and with 300 to 500 MHz data rates. The mask writer was reported to have a spot size of 0.12 μm, uniformity to 50 nm (3 s), stitching error of 40 nm, and repeatability (overlay accuracy) of 30 nm over a 6-in. reticle.

3.2.3 Raster scan systems

Raster scan printing has evolved over three decades and its evolution through the 1980s through the mid-1990s is shown in Fig. 3.12. Originally developed at ATT, the architecture remains in use at Etec and Micronic. Raster printing scans the beam orthogonal to stage motion. Prior to 1988 feature exposures were binary with the beam either on or off for a fixed amount of time. The resulting exposure was a series of gaussian beams all having the same nominal intensity.

Subsequent writing strategy innovations were introduced to minimize printing artifacts related to scan and dose errors. Virtual addressing (VA) was introduced to halve the input address, while maintaining constant exposure rate per phase.[155] Comparison of virtual address and multipass gray printing is shown in Fig. 3.13.

The concept of multiple-pass exposure (offset scan voting) was developed to reduce stripe butting CD errors at scan boundaries.[156] One approach is to offset each phase by one-fourth of the scan length in the y axis and offset in the x axis by a quarter of the array size to average over butting and distortion errors within the scan.

Interstitial grids produced by multiphase printing can double address resolution, see Fig. 3.14. In this case pixels can be divided into two sets and written in separate phases to cover the tiled grid. Various forms of interstitial grid tiling are possible, e.g., with four-phase, offset printing

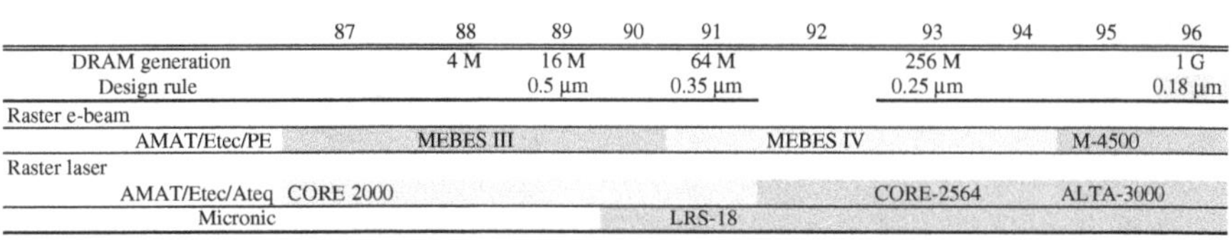

Figure 3.12 Raster PGs in the 1980s through the mid-1990s.

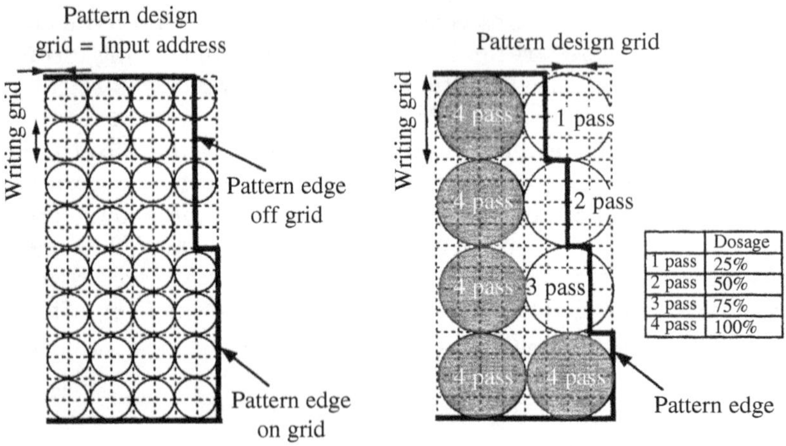

Figure 3.13 Comparison of virtual address and multipass gray printing: (*a*) virtual addressing writing method; (*b*) MPG writing method.

each phase can be offset by half a writing pixel. This forms an interstitial grid one-half the writing pixel size.

Gray beam printing modifies the exposure intensity. Gray-level rasterization of features yields fine feature edge placement without degrading throughput. Finer grids can be achieved with offset printing techniques by alternating the dose map on various print phases, increasing the number of gray levels, or writing with various tiling methods.[157] Laser systems achieve gray levels via intensity modulation of the incident light. Electron beam systems have used pulse width time modulation to

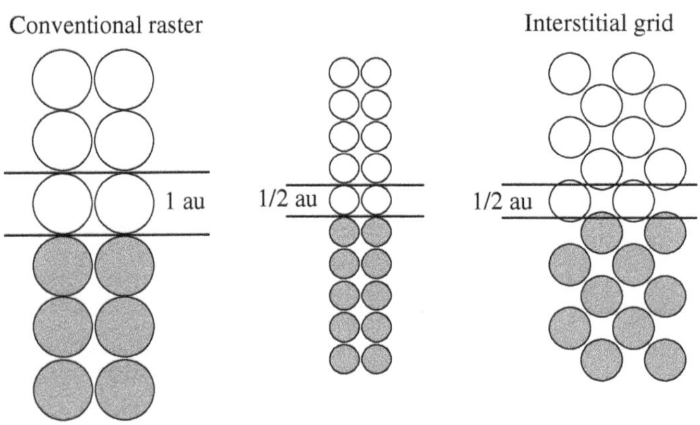

Figure 3.14 Interstital grid improves feature edge resolution with a larger print pixel.

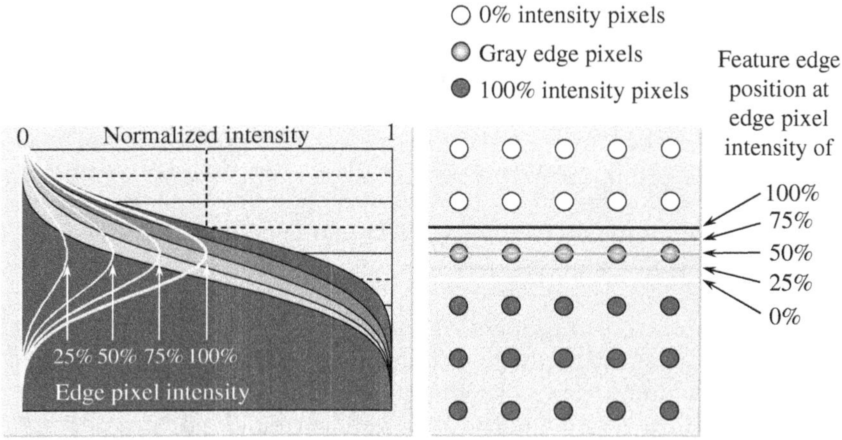

Figure 3.15 Gray level printing. Left graph shows spatial intensity distribution. Right graph shows feature edge position shift as a function of gray level pixel intensity.

achieve the same effect.[158,159] The resulting gray exposure shifts feature edges in proportion to the deposited dose, see Fig. 3.15.

Figure 3.16 shows the progression of edge placement resolution achieved by doubling the pixel density using an interstitial grid, which in combination with 16 nonzero gray levels provide a 32 times improvement in edge placement resolution compared to simple rasterization.

The benefit of combining gray and interstitial printing is that tool address grid snap effects can be greatly reduced. As an example, let the internal address grid unit (I_{AU}) be defined $I_{AU} = P_w/(N_g \times I_g \times V_a)$, e.g., where $P_w =$ 32 nm, $I_g =$ two times reduction in address unit from the four-pass interstitial grid, $N_g = 32$ gray levels per pass, $V_a =$ two times reduction in address because every other pixel along an edge can be turned off. Then $I_{AU} = 0.25$ nm. The 3σ of a beam edge is obtained from a uniform distribution and is $3\sigma_{edge} = 3*(I_{AU})/(\sqrt{12}) = 0.22$ nm . The CD

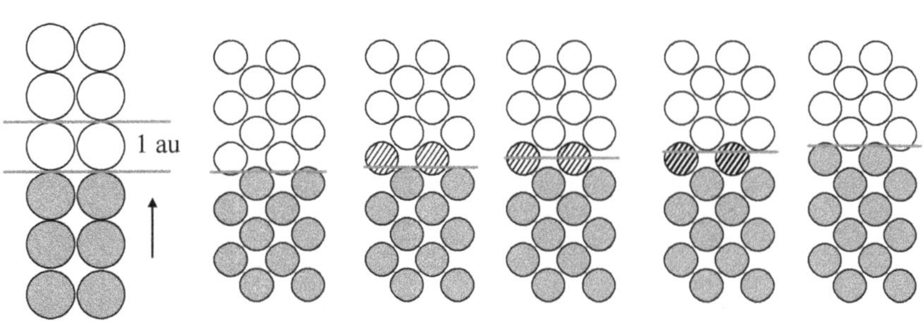

Figure 3.16 Combined interstitial grid and gray printing. Conventional raster printing at one address unit shown on the left.

error is estimated by assuming that the snapping errors for the two edges of a figure are independent $3\sigma_{CD} = 3\sqrt{2}\ \sigma_{edge} = 0.31$ nm. Here four-phase printing on a 32-nm print grid, using 32 gray levels per pass reduces the grid snap to 0.3 nm.

Micronic. Lithography studies in the 1970s at the Royal Institute of Technology in Stockholm led to the formation of a company in 1984 to commercialize photomask technology. Between 1984 and 1988 the company developed several prototypes based on acousto-optic modulation and deflection.

In 1989, Micronic Laser Systems AB was formed by employees of the first endeavor. The first product, introduced in 1990 (LRS-18) was targeted to semiconductor mask production. Micronic first published in 1991 a technical description of the LRS-18 laser mask pattern generator, which used a 442-nm HeCd laser.[160]

Tools were also developed for LCD display masks, and for components such as diffractive optics and SAW filters. By 1996 the company established itself as the leading supplier of laser mask PGs for the flat panel display industry.

The raster system evolution over the last decade is shown in Fig. 3.17.

In 1997, the company announced the LRS 200-25 for semiconductor mask patterning.[161] It was positioned as a mix-and-match tool to cover less critical masking layers. At this time the prospect of rapid data volume resulting from increased use of OPC was anticipated. It was recognized that a scalable raster data path architecture exploited the benefits of high bandwidth graphics architectures with embedded rasterization and figure modification.[162,163] Fast, complex manipulations of the rasterized bitmap can be used to extend lithographic resolution and support the OPC design process.

Omega system. The Omega6000 targeted 180-nm mask technology.[164] The system featured an acousto-optic deflection design and achieved 22-nm (3σ) global CD control.[165]

The Omega6500 product was directed to mask making for 0.5- to 0.15-μm technology.[166] Subsequent to BACUS 2000, the product line was split into four systems as shown in Table 3.10. Variations among

	97	98	99	00	01	02	03	04	05
2 year IC cycle	250		180		130		90		65
MPU gate in resist	200		135		90		53		35
Pilot mask PG intro	180		130		90		65		45
Raster e-beam									
AMAT/Etec	M-4500S	M-5000	M-5500	MEBES-X	eXara		RSB		
Raster laser									
AMAT/Etec/Ateq	ALTA-3500			ALTA-3700	ALTA-3900	ALTA-4000	ALTA-4300		ALTA-4700
Micronic		LRS 200-25	Omega 6000	Omega 6500	Omega-6600E	Sigma7100		Sigma-248	Sigma 7300

Figure 3.17 Raster system evolution over the last decade.

TABLE 3.10 Omega Series Performance Specifications for Mask Quality

Performance specs [nm]	Omega-6600	Omega-6000	Omega-6500	Omega-6050	Omega-6005
Node	150/130	180	250/180	350/250	500/350
Min feature size	400	400	450	550	700
CD uniformity (3σ)	12–18	25	25	30	40
CD linearity (3σ)	25	25	25	25	25
Position accuracy (3σ)	20–25	35	40	50	50
PSM reg (3σ)	60	60	60	60	60

the products were primarily focused on optics, and data path upgrades to the Omega platform.

The specifications for the Omega product line are shown in Table 3.11. For each system the specifications are consistent with the corresponding nodes of the ITRS 1999 roadmap. An exception is that the CD uniformity meets the roadmap for nested features only. The ITRS specification for isolated features is typically tighter by ~30 percent. The minimum feature size is consistent with ITRS mask minimum image size, which is for the primary feature of an isolated gate.

Micronic did not specify OPC features for this system series. However, data exist that show the Omega6500 was able to print OPC features required at the 180-nm node. The difficulty was that a 200-nm OPC feature size required at the 150-nm node is considerably smaller than the 288-nm FWHM beam size, which reduces process latitude.

The Omega6005 was designed for the 350-nm node and cost reduction came from a data path configured for smaller data volumes. The 700-nm

TABLE 3.11 Omega Series Optics Capabilities

Performance specs	Omega6600	Omega6500	Omega6050	Omega6005
Node	150	180	250	350
Optics technologies laser source	CW Kr-ion	CW Kr-ion	CW Kr-ion	CW Kr-ion
Wavelength, λ	413	413	413	413
Number beams	5	5	5	5
Beam formation	Diffractive	Diffractive	Diffractive	Diffractive
Modulation	Acousto optic	Acousto optic	Acousto optic	Acousto optic
NA	0.86	0.86	0.7	0.6
Objective lens reduction ratio	71	71	63	63
Resolution $0.6\lambda/NA$ (FWHM)	288	288	354	413
Focal length [mm]	3	3	3.2	3.2

minimum feature size specification indicated that they also achieved some cost reduction by using a 0.6-numerical-aperture (NA) design from an earlier system. The 413-nm optics resolution $0.6\lambda/NA$ (FWHM) yielded a 1.7 ratio of minimum feature size to beam resolution. Omega series optics capabilities are shown in Table 3.11 and their printing configuration in Table 3.12.

The Omega6050 targeted the 250-nm node. The rated 550-nm minimum feature size indicated that a 0.7-NA design enabled the 413-nm optics resolution ~$0.6\lambda/NA$, a minimum feature size to resolution ratio of 1.6 ratio.

The Omega6500 was specified for the 180-nm node. The rated 450-nm feature size was achieved with a 0.86-NA lens, coupled to 288 nm = $0.6\lambda/NA$ optics resolution, which maintains a 1.6 ratio of feature size to resolution. The system could print 280-nm OPC assist features with limited process latitude to demonstrate the ability to transfer the aerial image into a resist.

The Omega6600 extended the 6500 product to the 150-nm node. The 400-nm minimum feature size was an improvement over the 6500. It used the same 0.86 NA, but improved the lithography by using two-pass exposure to reduce the effective print pixel. A 3-h write time using two-pass exposure had been noted.

The Omega systems print with five-beam raster scanning; see Fig. 3.18. The five-beam brush is generated with a diffractive optics beam splitter. The Omega CW laser used a longer 413 nm compared to the contemporary ALTA 3500, 364-nm source. This choice may have been motivated by the limitations of *acoustoAcousto-optic deflector* (AOD), but enables a simpler optics design due to a wider range of available materials and also reduced contamination.

TABLE 3.12 Print Configuration and Exposure Times for the Omega Series

Performance specs	Omega6600	Omega6500	Omega6050	Omega6005
Node	150	180	250	350
MFS/resolution	1.4	1.6	1.6	1.7
Gray levels/pass	32	16	16	16
Number passes	2	1	1	1
Total gray levels	32	32	16	16
Scan length [mm]	0.12	0.12	0.12	0.12
Print grid [nm]	100	100	125	125
Effective grid [nm]	5	5	10	10
Write time data (135 mm^2) [min]	180	90	75	70
Exposure rate/pass [mm^2/s]	0.8	3.4	4.1	4.3
<Stage velocity> [mm/s]	7	28	34	36

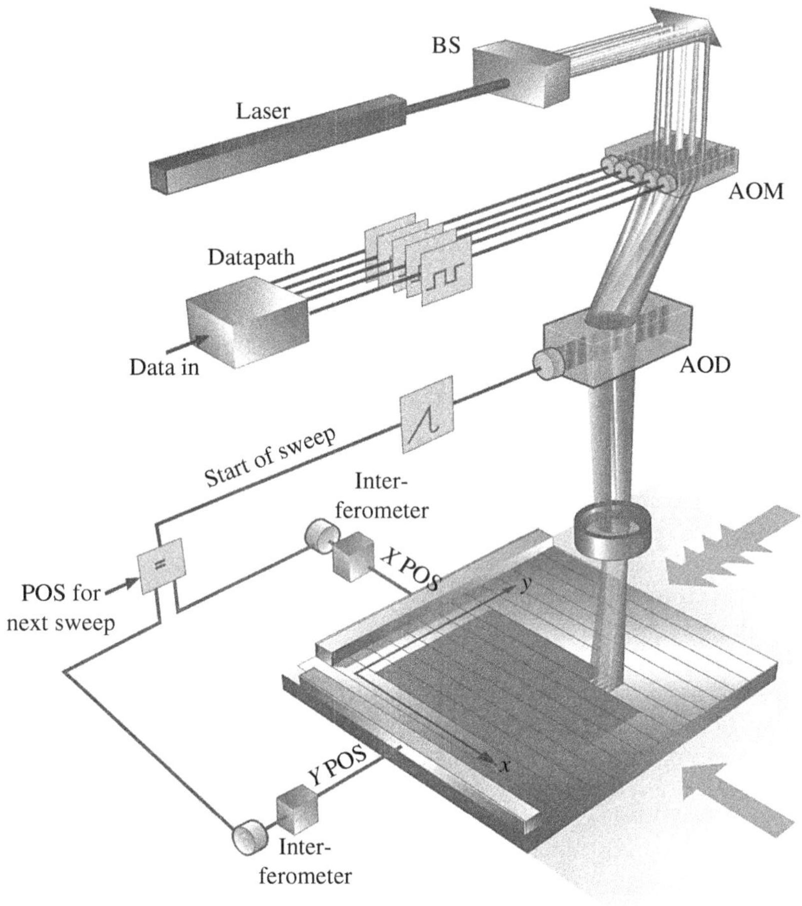

Figure 3.18 Omega system concept.

The multiple laser beams are individually modulated and scanned in unison. Acousto-optic deflector beam scanning is orthogonal to the stage motion and the pattern is built up in stripes. The scanning method is in contrast to the $f \sin\theta$ postscan approach to convert angular scan to linear beam translation.

The beamlet intensity is modulated by a five channel acousto-optic modulator. This piezoelectric transducer couples sound into a crystal, creating periodic variations in the refractive index. This creates a Bragg grating which diffracts the light. The diffracted beam is observed at a given angle by fixing the frequency of the RF signal. The acoustic power is varied to change the diffraction efficiency, providing gray levels.

Although polygonal mirror scanning is common in high throughput laser scanning systems, Micronic chose AOD technology and addressed light loss, nonlinearity, and short scan lengths. A constant acoustic power creates a Bragg grating of fixed efficiency, and the RF frequency is scanned to deflect the beam over a range of angles. Since the acoustic wavelength is about an order of magnitude larger than the optical wavelength, the deflection angles are limited to about 1°.

The nonlinearity of the scan is handled by calibration. The scan is calibrated to a reference grid and the data are used to predistort the frequency ramp. Another calibration ensures that the beam intensity is constant over the length of the scan. This is accomplished with small modulations of the acousto-optic power.

An air bearing *xy*-stage is driven by linear motors. The laser interferometer has a $\lambda/1024$ ~0.6 nm resolution. For a 120-μm stripe height and 90-min print time the average stage velocity is 40 mm/s. *X* motion is at constant velocity with scanning in the *y* axis and there is a retrace at the end of each stripe. Retrace time is an important overhead because short scan lengths result in a large number of stripes and retraces.

The focus system is passive with the objective lens assembly riding on an air cushion to maintain constant focal position. This is in contrast to the ALTA laser PG, which uses a hot-wire anemometer for focus control. Both systems have to deal with gas-flow effects on CD during print. Calibration is performed periodically and the lens height is adjusted within this assembly.

The *x* and *y* beam edges are defined by different mechanisms. Edges parallel to the AOD sweep are positioned by adjusting gray levels of the edge pixels. Orthogonal edges are positioned by adjusting the beam turn on time. This is a specific example of how *xy* bias and CD control issues are manifested in raster scanned systems. *XY* bias can be adjusted in the job deck and global feature biasing can be applied via another parameter.

Printing includes interleaving of successive segmented scans. Segmentation is executed during mask write and is similar but not equivalent to rasterization, which calculates a gray level for each pixel. Instead, the length and intensity of each scan segment is calculated. Data are divided into segments along the scan according to a rule set, see Fig 3.19. At this stage each segment is a simple rectangular geometry. Beam timing and gray levels are determined during writing.

A single-pass strategy was maintained down to the 180-nm node. To compensate for stripe butting a feature termed *swath blending* was used between stripes. Stripes overlap by 10 percent as shown in Fig. 3.20, and the dose is graded using a proprietary algorithm. A method that was described used four intensity steps from minimum to maximum dose. With single-pass writing, errors in stripe overlap raise the issue of feature

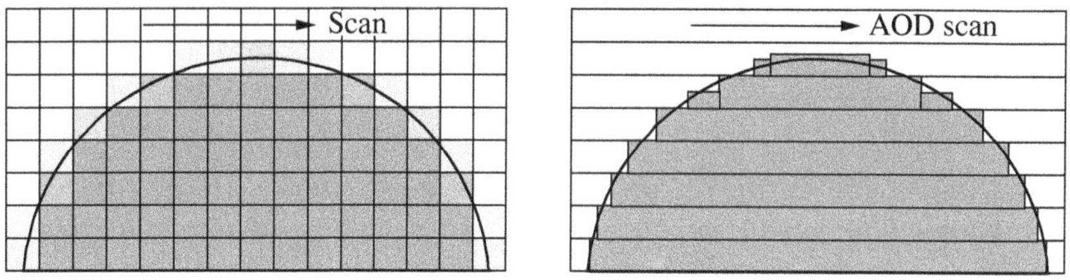

Figure 3.19 Rasterization (left) vs. segmentation (right).

size control at step boundaries. The Omega6600 was the first exception to single-pass printing, which allowed double-pass exposures to improve lithography quality at the expense of throughput.

Sigma systems In the early 1990s Fraunhofer IMS began the development of a system for maskless patterning of semiconductor devices. The system concept uses a novel beam delivery technology comprised of excimer laser illumination of a spatial light modulator (SLM).[167] A prototype using a 308-nm excimer was described in Ref. 168. Around 1998 Sentech Instruments extended the prototype with a larger SLM, a 248-nm excimer, and improved software.[169]

In 2000 Micronic introduced a prototype 248-nm SLM mask pattern generator targeting the 130-nm node.[170,171] The SLM array comprised 512 × 2048 mirrors, using a 1-kHz flash rate, a 248-nm excimer laser, and 0.72 NA/160× reduction optics to produce a 100-nm pixel size. The architecture and alpha tool results were also given for the system.[172] The Sigma7300 had improved CD and registration.[173] One of the design changes increased demagnification to 200 and NA 0.82.[174] The Sigma7300 parameters and capabilities are shown in Tables 3.13 and 3.14.

Writing strategy. The SLM system architecture can be described as a microstepper with a programmable mask. Gray levels are realized with analog modulation of torsional mirror devices and four-pass writing is

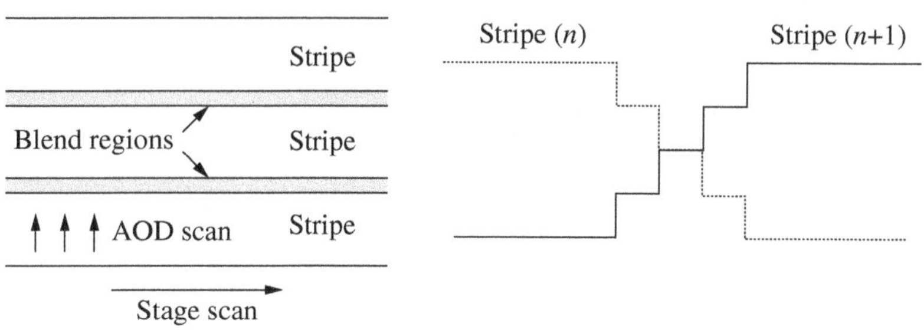

Figure 3.20 Omega stripe blending technique.

TABLE 3.13 System Parameters for Sigma7300

Parameter	Value
Mirror array size	512 × 2048
Pixel grid	80 nm
Address unit	1.25 nm
Number of passes	4
Gray levels	64
Laser flash rate	2 kHz
Pixel delivery rate	2.1 Gpixels/s

used to reduce the impact of defective pixels and other print artifacts. The mask pattern is built up by scanning the stage continuously and triggering the laser flash with the stage interferometer output. The SLM is oriented with the small dimension in the direction of stage motion, the stage moves the width of the image, or 40 μm during the time between flashes. The schematic of Sigma control and beam delivery is shown in Fig. 3.21.

For the 2-kHz laser repetition rate the stage must travel 80 mm/s while maintaining image placement within a fraction of a pixel. This is accomplished in the stage travel direction by precise triggering of the laser flash. At this stage velocity, and with a 20-ns pulse width, which is typical for excimer lasers, the image smears 2 nm during the flash.

The writing strategy is four pass with offsets between passes. The image of a given field is created with four different regions of the optical field. The usual benefits accrue, such as improved stitching, image placement, and CD control. With the conditions described earlier, it takes about 2.5 million flashes to write one pass on a 132-mm square mask area. At 2-kHz repetition rate each pass takes 40 min and the entire plate would be written in under 3 h, ignoring system overheads.

Image formation. A 248-nm KrF excimer laser with 2-kHz repetition rate is used to illuminate an SLM chip with a partial coherence of $\sigma = 0.7$. The SLM chip has a flat mirror like surface composed of a million (512 × 2048) mechanical micromirrors that reflect light onto a mask

TABLE 3.14 Sigma7300 System Lithographic Capability

Parameter	Spec [nm]
Min primary feature size	260
CD uniformity (3σ)	7
CD linearity	10
Positional accuracy (3σ)	15
Second level overlay (mean + 3σ)	30

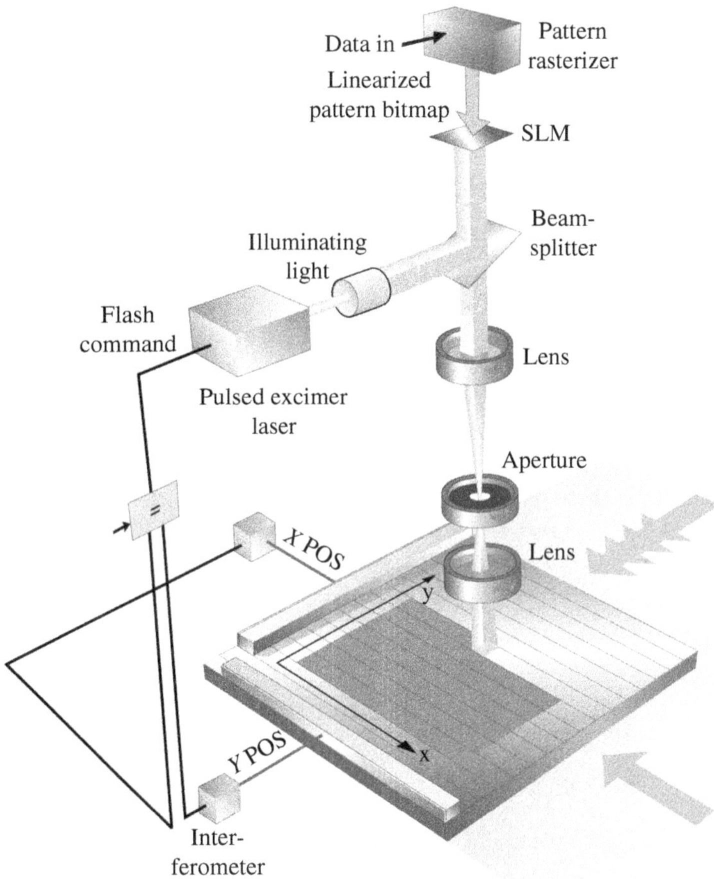

Figure 3.21 Schematic of Sigma control and beam delivery.

blank through a Fourier filter. Underlying *complementary metal-oxide-semiconductor* (CMOS) circuitry on the SLM chip provides an analog addressing electrode for each individual mirror. Individual mirrors are actuated by an electrostatic force through the addressing electrode, yielding a monotonic but nonlinear deflection. The Fourier filter design requires 62-nm deflection to completely blank the light from a mirror.

The 16-μm SLM pixel is demagnified to an 80-nm square at the mask plane. Contrast formation is based on diffraction and so small tilt angles are required. The mirror is tilted ~$\lambda/4$ (62 nm) over its full width to create a dark pixel by applying 21 V to the control electrode. To achieve gray levels the mirror is tilted in increments of the maximum ~$\lambda/4$ by changing the excitation voltage. The switching time is ~10 μs. The optical schematic of SLM system is shown in Fig. 3.22.

For any of the cantilevered SLM devices the SLM array reflects a uniform wave front when the mirror positions are undeflected. When pixels are addressed the phase of the wave front is locally distorted according to $\phi(x) = 4\pi a(x)/\lambda$ where $a(x)$ is the relative vertical deflection of the surface. With reflective optics the phase shift is twice the deflection distance. Hence at a point where the mirror position is deflected by a quarter wavelength, the phase shift is 180°. To create image contrast the phase disturbance must be converted to an intensity variation.

To create an image the laser uniformly illuminates the SLM array following reflection from a beam splitter. The diffracted light from the SLM passes back through the beam splitter and an aperture in the pupil plane blocks the higher orders, while passing the lower orders. The numerical aperture and partial coherence of the optic are chosen so that the periodic mirror structure is effectively blocked.

For a 30 mJ/cm^2 DUV resist the excimer laser's delivered energy per pulse is 0.5 μJ. In comparison the delivered energy per pulse in a high

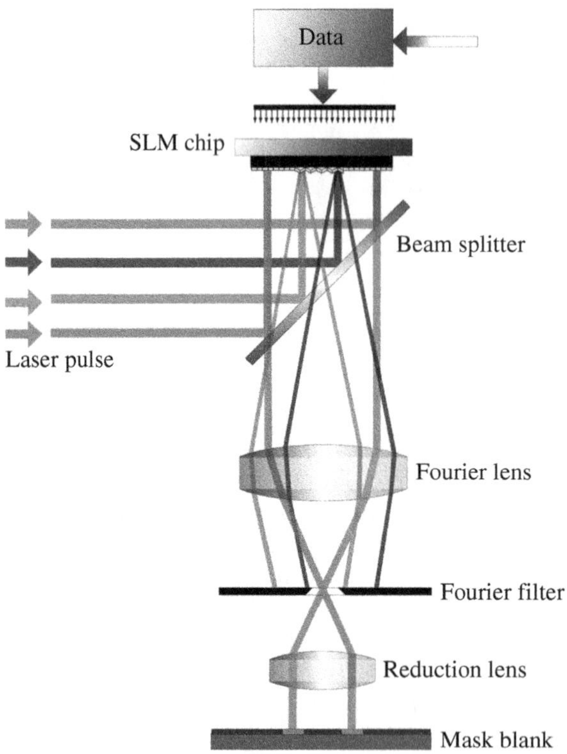

Figure 3.22 Optical schematic of SLM system.

productivity DUV step and scan tool is about 300 μJ. The difference comes from the mask write times, which are about 100 times longer than wafer prints, and a mask writing area which is two times smaller than on 200-mm wafers. On the other hand the power density in the image is much larger for the mask writer, because the exposure area is about 20,000 times smaller than the typical 5 mm × 25 mm slit of a step and scan tool. This would only be an issue if the fluence in the optics were sufficient to result in high contamination or optical materials damage rate.

Variation in laser pulse energy affects mask writing more than in wafer lithography. This is because a wafer litho feature is exposed to 50 or more pulses as a 5-mm slit is scanned, whereas the SLM PG delivers only four pulses per feature. Typical pulse-to-pulse variation for a KrF excimer laser is 0.7 percent (1σ). Four passes reduce this 2×, and in this example would result in a 1 percent (3σ) dose variation. This term is a factor in the CD uniformity error budget.

SLM technology. Various cantilevered micromirrors can be built where the mirrors are deflected by applying a voltage to an underlying electrode, see Fig. 3.23. In the pyramidal device, the points of the four flexible segments are deflected downward, creating a local phase modulation when illuminated. Piston mirrors produce a uniform phase change across the pixel proportional to displacement. Torsional mirrors produce a uniform variation of phase across the pixel. All can be combined with a Fourier filter to modulate intensity. With small $\sim\lambda/4$ deflections, high contrast can be obtained by blocking diffracted light.

The Sigma7300 SLM uses a torsional mirror design with analog modulation, enabling gray scale bitmaps to be transferred having 64 intermediate mirror deflection values. Partial coherent imaging in a diffractive operating mode occurs with complex amplitude modulation of the reflected light.[175] The reflecting elements are deflected to an equilibrium state created by the opposition of an electrostatic attraction and a restoring mechanical force. The mirrors deflect l/4 (62 at 248 nm)

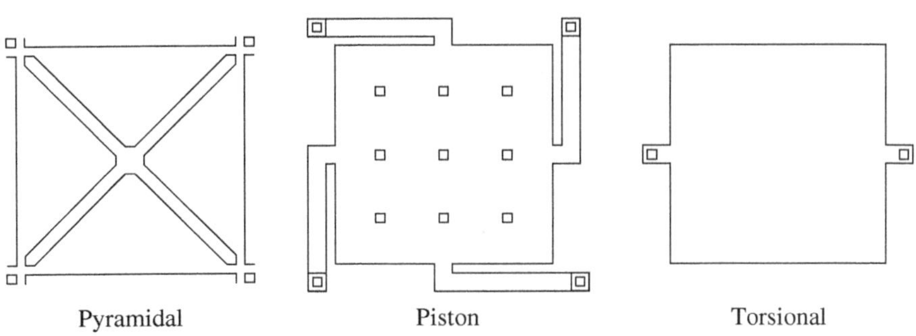

Figure 3.23 Cantilever micromirror types.

to go from the fully-on to fully-off state.[174] The deflection is approximately quadratic with the electrode voltage. At a certain critical voltage the electrostatic force increases more rapidly than the restoring force and the device switches to a full deflected state. For analog modulation the device is operated below this point.

Pixel calibration to 1 to 3 percent of the deflection angle response to torsional voltage characteristic requires additional calibration firmware. Micronic has described a calibration procedure where each bit is individually sampled over its full 64 gray level range with a correction table stored comprising 2.1 million transfer functions, which must be accessed during the exposure flash. The actual calibration procedures and resulting accuracy are proprietary.

Micronic's 6-bit design required 6.3 Mbits to be transferred in 0.5 ms, for an average gray level data rate of 12.5 Gbits/s. The data distribution strategy shown in Fig. 3.24 divides the array into two halves, which are separately addressed. For each half, 64 input lines are demultiplexed on chip to 1024 rows. A 1024-stage shift register is used to distribute the data to each column.

Extendibility of scanned partially coherent diffraction imaging. The three paths to extend resolution are increasing numerical aperture, reducing the effective k_1 factor by modifying the illumination and partial coherence of the optic or via modifications of the raster bitmap, and wavelength reduction.

Wavelength scaling requires the source, optics, and SLM to be substantially reengineered with a change in wavelength. Adoption of low power excimer sources that do not have stability issues is another issue. The extension to a 157-nm source will be an order of magnitude more difficult than a transition to 193-nm printing.

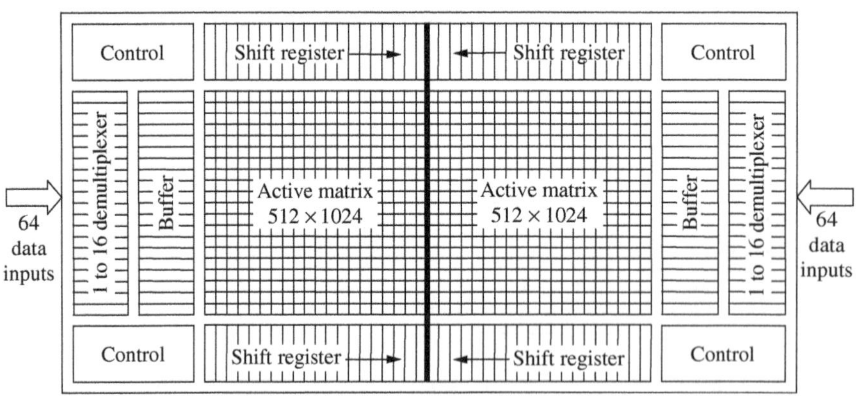

Figure 3.24 SLM data distribution.

A relatively low risk path is to increase the numerical aperture. The Sigma7300 has 0.82 NA, providing a point spread function of ~184 nm. This could be reduced to ~168 nm by increasing to 0.9 NA. Further NA extensions might be conceivable if an immersion optic is developed.

Reducing the k_1 factor exploits the fact that laser bandwidth requirements are less demanding for the small mask PG optical field and presumes mask resolution requirements will never be as challenging as those with 4× stepper demagnification. Driving k_1 to 0.3 with 193-nm illumination of a NA = 0.9 would lead to 64-nm resolution. 193-nm SLM immersion would reduce this to 45 nm. This entails significant new complexity in all aspects of the system design.

As feature sizes shrink, either the mirror density will increase or the image will be demagnified. Since resolution is determined by wavelength and numerical aperture, the resolution to grid ratio effect on lithography will have to be characterized to maintain acceptable lithographic quality.

Ateq/Etec/applied materials. ATEQ Corporation was founded in 1984 by ex-Intel employees and venture capital funding. The primary objective of the company was to develop a high-speed scanned laser pattern generator for direct wafer writing; however, recognizing that penetrating, or more realistically creating, a wafer direct write market would be a slow process, and that the technology could be applied to the existing reticle writing market, the first product was to be a reticle writer, the CORE 2000. The CORE 2000 was announced in October 1985, but acceptance was slow. The first system was shipped in 1987 to Master Images. Although a direct wafer writer was introduced—the WaferWrite 6000—it was unsuccessful in penetrating or creating a market and was soon discontinued. The schematic of the laser CORE 2000 system is shown in Fig. 3.25.[15]

The system used a raster scan writing strategy in which the eight beams were scanned across a 2-mm wide stripe on the reticle by a high speed, multifaceted polygonal mirror while the reticle was translated in the orthogonal direction by a high precision, laser interferometer controlled stage. The beam delivery subsystem used the 363.8-nm line (~100 mW) from a 10-W Argon ion laser, which was split into eight parallel beams, each of which was modulated by separate channels of an eight-channel acousto-optic modulator (AOM). The system used essentially a 0.5-μm fixed writing grid, but a zoom lens in the optics allowed pattern scaling over the range of 80 to 105 percent. The system resolution was 1.2 μm which was capable of writing quality features of 2 μm and larger at very high throughput.

Other features incorporated in the CORE systems were cassette-less mask blank handling and "white box" style system enclosures, which

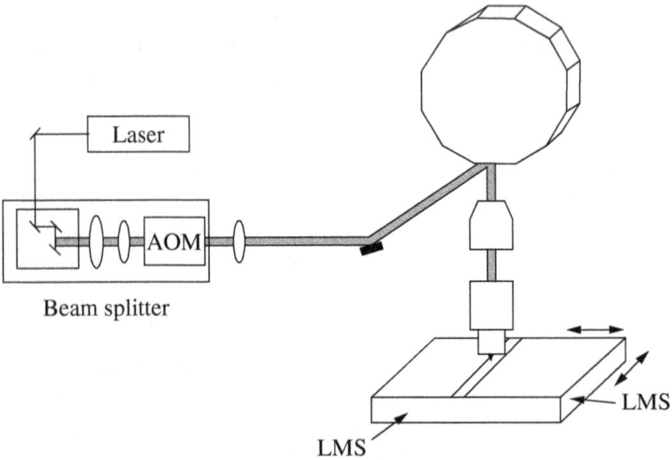

Figure 3.25 Schematic of the laser CORE 2000 system.

provided a clean, controlled environment and limited operator access to only those system components required during normal system use. The scanned gaussian PG resolution, wavelength, and numerical aperture roadmap are shown in Fig. 3.26.

Data path improvements resulted in the CORE 2100, which printed with a 364-nm Ar-ion laser, 0.4 NA and 1.2-μm resolution. Although the 2000/2100 systems were useful mask writers, the fixed writing grid was

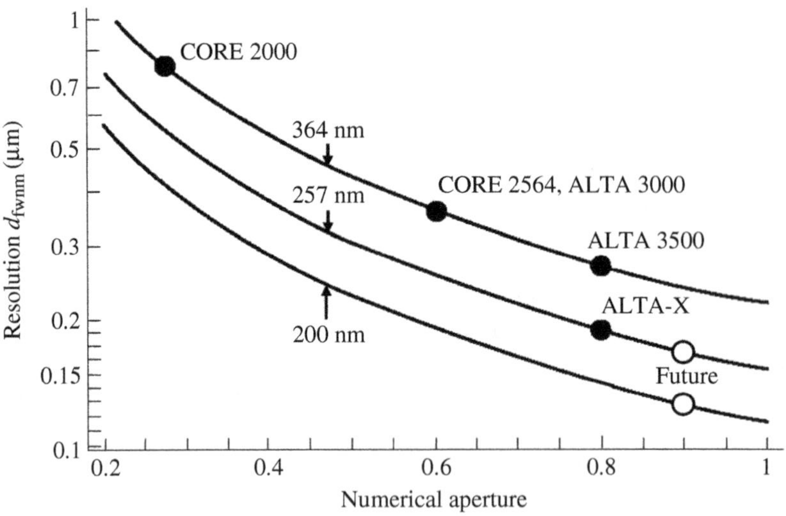

Figure 3.26 Scanned gaussian PG resolution, wavelength, and numerical aperture roadmap.

recognized as a severe limitation. Evolution of CORE/ALTA capability is shown in Table 3.15.

The solution to this limitation—gray-scale writing—was a significant innovation in mask writing. Gray-scale writing was implemented by trading off writing speed for multiple exposures on offset writing grids to subdivide the raw 0.5-μm grid and by fractional modulation of the beam intensity in the AOM. Furthermore, by taking advantage of the multiple exposures to offset writing stripes, many systematic and random errors were reduced. These writing enhancements were introduced in the CORE 2500 and the CORE 2564,[176] which printed 125-nm effective grid with four gray levels and 364-nm beam resolution using a 0.6 NA.

Linear brush, 32 beam, eight-phase printing was introduced with the ALTA architecture. The optics comprises a high speed rotating polygonal mirror, which provides fast deflection of the beam brush through the postscan lens assembly. This is followed by two components, an $f\ \sin\theta$ lens (or $f\theta$ lens for 364-nm systems), which converts the angle of the 32 scanned beams into position in 32 scan lines and a reduction lens that demagnifies the scan lines onto the mask. The ALTA 3000 was designed for the 0.35-μm masks.[177] The 364-nm CW laser and 0.6-NA imaging of the CORE 2564 was retained. The $\lambda/512$ interferometer resolution included velocity of light corrections for travel through the controlled atmosphere. ALTA 3000HT provided a four-phase printing mode to handle 0.5-μm capacity demands. ALTA system architecture is shown in Fig. 3.27.

The ALTA 3500 was designed for 250/180-nm masks by increasing the demagnification from 20× to 33× and the numerical aperture from 0.6 to 0.8. This improved ALTA 3500 resolution (~0.6λ/NA (FWHM)) from 364 nm for the ALTA 3000 to 273 nm.[178] Four-pass-enhanced raster writing strategy with 17 gray-level rasterization and offset multipass

TABLE 3.15 Evolution of CORE/ALTA Optics Technologies

	ALTA 4700	ALTA 4300	ALTA 3900	ALTA 3700	ALTA 3500	ALTA 3000	CORE 2564	CORE 2500
Production node	90/65	130/90	180/130	180/150	250/180	350	600/350	
Wavelength λ	257	257	364	364	364	364	364	364
Number beams	32	32	32	32	32	32	8	8
NA	0.9	0.8	0.8	0.8	0.8	0.6	0.6	0.4
Obj lens reduction ratio	42	33	33	33	33	20	20	20
Resolution ~0.6λ/NA	165	188	273	273	273	364	364	546

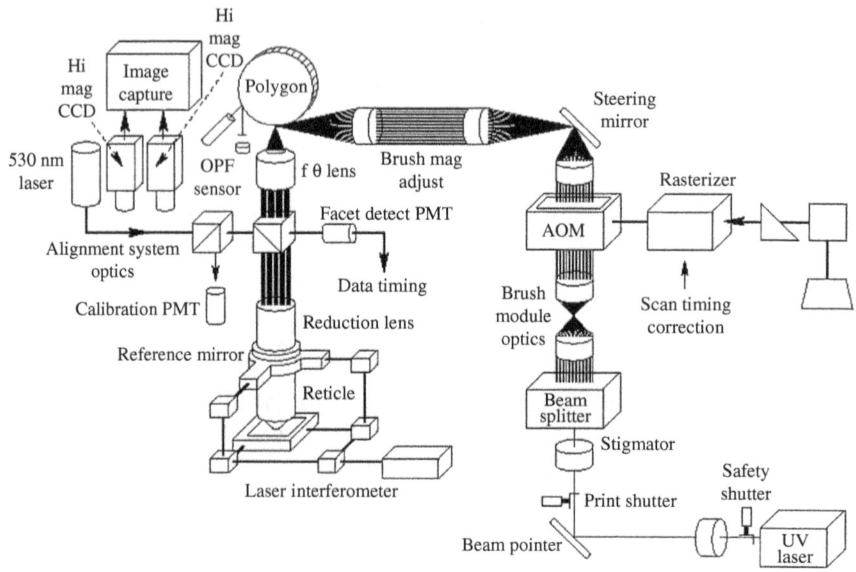

Figure 3.27 ALTA system architecture.

printing enabled 32× reduction of a 160-nm print pixel, which produced a 5-nm system address unit.[179]

The ALTA 3700, delivered 180/150-nm capability. High-precision dose control was achieved using an eight-pass writing strategy and a new acousto-optic modulator. Thermal transients in the AOM were cut in half, which improved scan line CD control and placement accuracy. Improvements to measured optical efficiency and focus system addressed CD control problems including one associated with interactions between the developer and focus system.[180] The evolution of ALTA capability is shown in Table 3.16.

The ALTA 3900 improved the print time 1.2× compared to the ALTA 3700 resulting from a faster set of firmware modules, which affect internal file download and rasterization. Enhancements to system calibration, beam delivery precision, and intensity control improve CD and registration accuracy. The beam optics retains the 33× objective with 0.8 NA to deliver a 273-nm FWHM beamlet size.

The ALTA 4000 series introduced DUV 257-nm CW laser source technology to improve resolution 29 percent compared to the ALTA 3900. Early characterization of the DUV reticle lithography identified areas where the system could be improved to meet 130 nm/90-nm requirements.[181]

The ALTA 4300 system was introduced a year later. It had a new beam brush orientation, rotated perpendicular to beam scanning. Beam scans are antiparallel to the direction of acousto-optic modulator sound propagation—producing a round imaging spot to improve corner

TABLE 3.16 Evolution of ALTA Capability

	ALTA 4700	ALTA 4300	ALTA 3900	ALTA 3500	ALTA 3700	ALTA 3000
Production node	90/65	130/90	180/130	180/150	250/180	350
Min feature size	264	400	500	500	500	600
CD within scan field (range/2)	4	8	12			
CD uniformity (range/2)	6	8	15	20	25	30
CD linearity (range)	10	12	15	25	25	
<*x*-*y*>			12	15	15	20
Corner rounding (radius)	132	205				
LER (3σ)			15	15	15	20
Position accuracy (3σ)	9	12	25	30	35	40
Overlay (3σ)		10	20	20	25	30
PSM reg (± range)	20	35	40	60	60	100

symmetry.[182] *XY* bias due to asymmetries in laser gaussian scanning was further compensated by having a compressed 96-nm *x* axis grid coupled with 160-nm *y* axis printing for each pass.

DUV-resist prototype test results show that this system can print 400-nm patterns, and resolve ~200 nm OPC features.[64] These results and global CD uniformities of 14 and 7 nm (3σ) for logic and memory, respectively, are consistent with the needs of 130-nm mask technology. However, contact feature nonlinearity is controlled only to ~800 nm, whereas 50-kV e-beam printing can achieve ~400 nm under similar conditions.

The laser strategy now reduces the resolution gap compared to 50-kV electron beam by data path precompensation of the aerial image. Aerial image intensity correction enhances the dose on the corners to ensure that the resultant printed corners are not rounded.[183,184] These algorithms can in principle be implemented in any raster data path. Image fidelity is improved by rendering the raster bitmap to improve the aerial image quality. The rendering algorithm has enabled printing of 300-nm contacts with acceptable control.[185] Preprocessed mask data and image quality are shown in Fig. 3.28.[64] Laser PG resolution can be evaluated using through focus CD and wafer yield results to compare laser and e-beam generated mask quality.[186] The latter point has more significance for mask shops resident in vertically integrated process flows where quality assessment can use a broader range of criteria including lithographic performance.

Two new optics components were developed to extend the ALTA 4300 system.[187] Optics resolution is improved by increasing the reduction lens

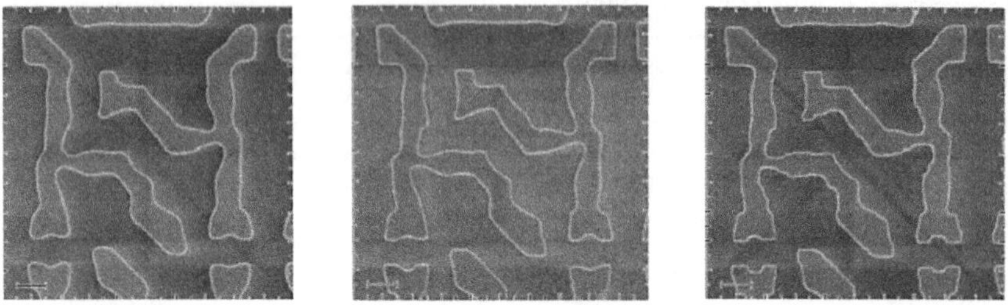

Figure 3.28 Preprocessing data can improve mask image quality.

demagnification to 42× with a 0.9-numerical aperture. Because the increase in reduction ratio is larger than the increase in NA the final spot size is limited by the lens NA and increase in the lens pupil filling. The resulting design spot size is approximately 15 percent smaller or 160-nm FWHM. A high bandwidth AOD detector and servo compensation for beam wander and polygon-facet pointing dose errors were also implemented. This improves CD control in the stripe axis. Higher demagnification also reduces beam placement errors such as beam-to-beam spacing and polygon deflection errors by the same 80 percent ratio.

PSM overlay has been improved through the use of new software algorithms. The new, high-bandwidth, stripe-axis beam position control servo provides for programmed and real time laser intensity correction. As a result, local CD uniformity was 4.3 nm and global CD uniformity reduced to 5.8 nm (range/2). Second layer overlay performance is now 20 nm (maximum error) supporting the 65-nm design node requirements for alternating, tritone, and CPL PSM exposure.[187]

These component technology enhancements have been incorporated in the ALTA 4700, which is designed for 90/65-nm production.[188] Though minimum feature size specifications are not presently available, one can infer potential capability from past data. The ALTA 4300 has a beam resolution ~188 nm. Using aerial image intensity correction enabled minimum printable features to be improved from 360 to 300 nm, which implies a feature size to resolution ratio of 1.6. This ratio has varied over the years from 1.6 (ALTA 3000) to 2.1 (ALTA 4300). Extrapolating aerial image correction performance to the ALTA 4700 with its improved optics indicates that 264-nm feature sizes will be printed with acceptable quality. This minimum feature size estimate is also consistent with early results on corner rounding and pull back data. This extrapolation indicates the machine will be capable of 90-nm production for all mask layers. The 212-nm mask minimum feature size specified for 90-nm critical layers (see Table 3.2) is likely to occur with continuous improvements in process and data path algorithms. Consistent with past history, the system capability is also positioned to access some of the 65-nm first-level layers and be used extensively for second-level PSM. Evolution of printed feature size for Etec laser and electron beam pattern generators is shown in Fig. 3.29.

Etec/perkin elmer/etec/applied materials. MEBES was the first commercially available electron beam lithography system intended for production use.[189] All MEBES mask pattern generators combine raster scanning with continuous stage motion. High patterning accuracy was aided by use of a repeatable beam deflection scan and exposure rate that is independent of pattern density. Automation of system functions and flexibility in having a large print address operating range were longstanding benefits of the system architecture.

MEBES I was introduced in 1977 and replicated the EBES system design.[190] The machine had 125-nm minimum beam size and was designed to assure absolute pattern accuracy over the entire mask. This was achieved via small field printing and automated system control. Innovations included laser interferometer metrology and feedback through the electron beam deflection system, a stable sawtooth generator for raster scan deflection, and periodic beam re-registration using a stage grid. Mirror orthogonality, raster deflection linearity, and stage-deflection errors could be corrected automatically. The system also introduced self-calibration as a coordinate measuring machine to improve calibration and overlay between mask exposures.

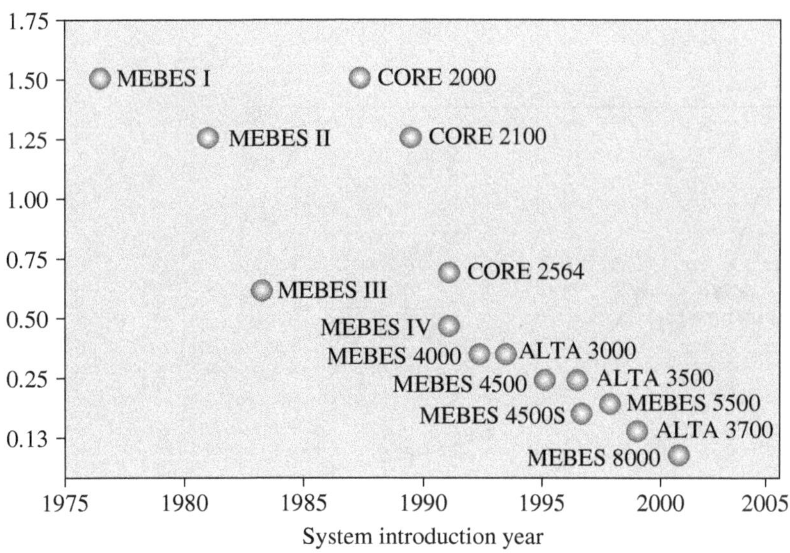

Figure 3.29 Evolution of printed feature size [μm] for Etec laser and electron beam pattern generators.

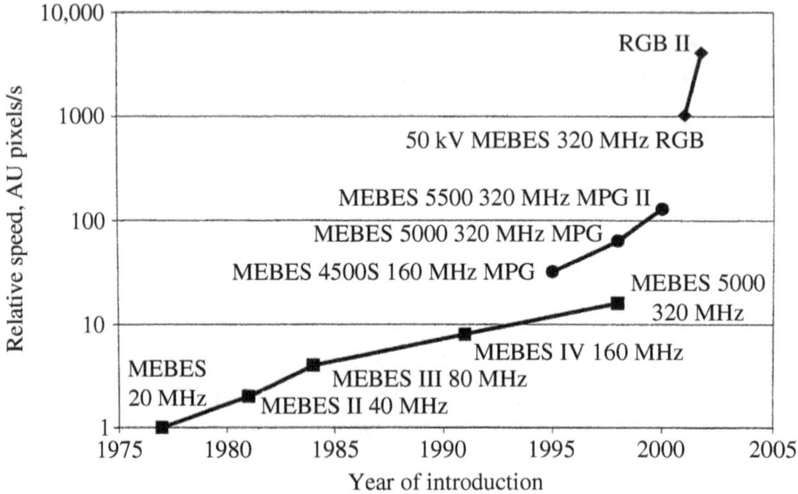

Figure 3.30 Evolution of MEBES pixel delivery rates.

MEBES II doubled the clock frequency to 40 MHz. This began a long trend from 1981 through 1997 where the single-phase print strategy was extended by increasing the blanking rate from 20 up to 320 MHz. Evolutions of MEBES pixel delivery rates and pixel-on time and data rate are shown in Figs. 3.30 and 3.31. Blanker rise and fall time had to steadily decrease to support the ever higher data rates with the general guideline of remaining ~15 percent of the blanker cycle time, see Fig. 3.32. 160-MHz

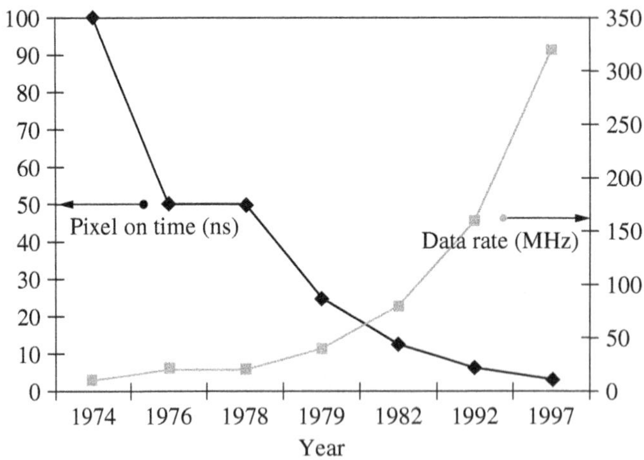

Figure 3.31 Evolution of MEBES pixel-on time and data rate.

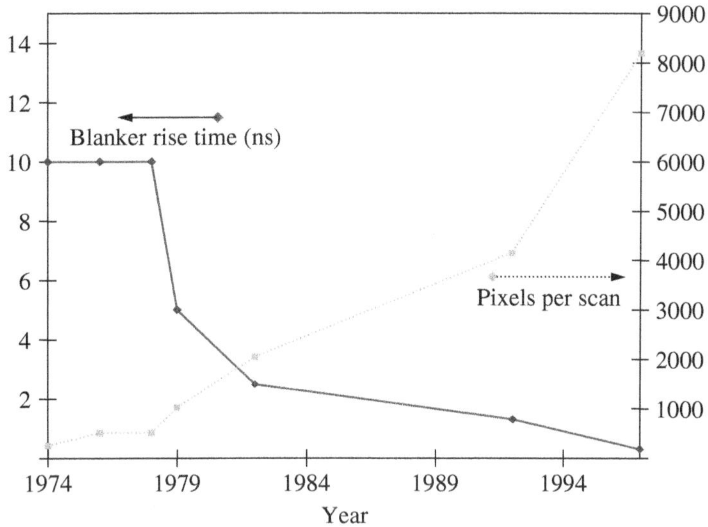

Figure 3.32 Improvements to blanker rise time and scan height to support higher data rates.

blanking required the rise/fall time of the blanker driver to be on the order of 1 ns.

The number of pixels/scan also steadily increased to accommodate finer design grids while maintaining acceptable mask exposure times. The overall scan length, which is a product of pixels/scan and pixel size, shrank over time to ensure high placement accuracy.

MEBES III was introduced in 1983. It printed 0.7-μm features at 80 MHz and extended the system reliability by including automated column setups.[191] The product line had a 20-year run with 100 systems still in operation in 1993 and not until 2003 were the remaining tools discontinued.

MEBES IV featured TFE optics and 160-MHz blanking,[192] which increased resolution, available dose, positional accuracy and CD control.[193,194] The four lens tandem optic provided a continuously variable beam size over a 50 to 300 nm range while maintaining constant current density without adjusting the gun extraction voltage. This latter feature was essential to maintain 1 percent dose control of the Zr/O/W<100> source operated in Schottky emission mode and also provided a wide column operating envelope, which retained the MEBES benefit of flexible and automated system operating range.

Improving CD control at 160 MHz required development of a high bandwidth and high accuracy traveling wave beam blanker.[195–197] Special methods were developed to reduce gun[198] and blanker-induced beam motion, which included use of a second stigmator to ensure conjugate optics with minimal cross talk in the orthogonal direction.[199] Critical dimension control within a range of 40 nm was the limit for 160 MHz, single-phase printing.[194]

Multiphase (2XMPP) printing was also introduced as an option to single-phase printing on MEBES IV. The idea allowed throughput to be maintained while printing four phases at half the system address. The benefits were that four times the dose could be delivered, which allowed the use of less sensitive, high contrast DNQ/novolacs and ZEP7000. The four-phase printing with offset scan voting averaged butting and scan linearity placement errors. No CD improvement resulted as the beam to pixel grid ratio remained one and the pixel print size remained the same as single-phase printing. Four-phase multipass (2XMPP) printing was able to reduce 0.1-μm registration errors on an MEBES IV to 24 nm (3σ).[200] Column schematic of MEBES IV tandem optic (left) and current density and beam size range (right) are shown in Fig. 3.33.

Multipass gray (MPG) printing created four gray levels from offset scanning during four phases. Compared to multipass (2XMPP) printing, the MPG effective grid was cut in half for a given exposure rate, which reduced grid snap errors. As a result CD uniformity went from 50 to 34 nm on the MEBES 4500S.[201]

MEBES 4500S positional accuracy was improved 50 percent via a number of error compensation techniques.[202] A key innovation was automated scan calibration using high-speed blanking to create a stroboscope

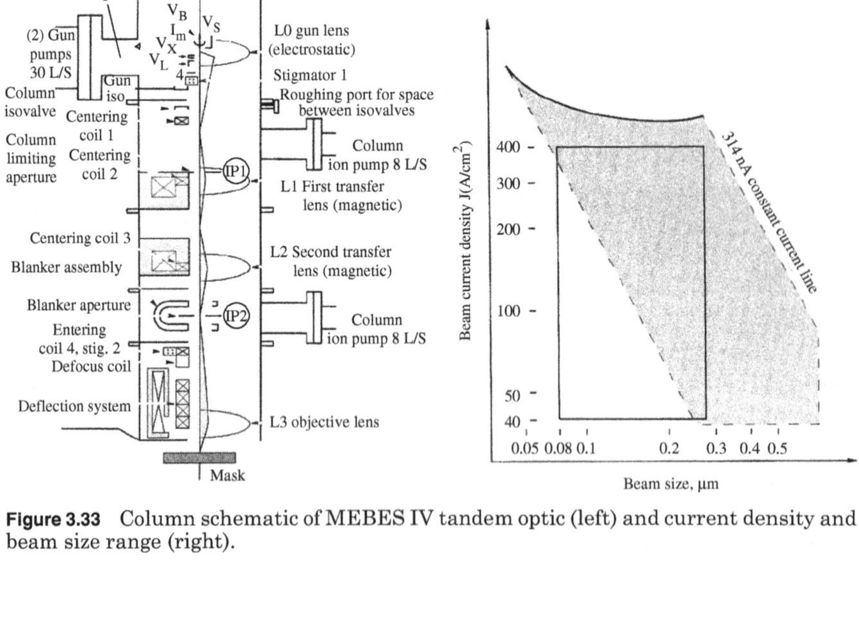

Figure 3.33 Column schematic of MEBES IV tandem optic (left) and current density and beam size range (right).

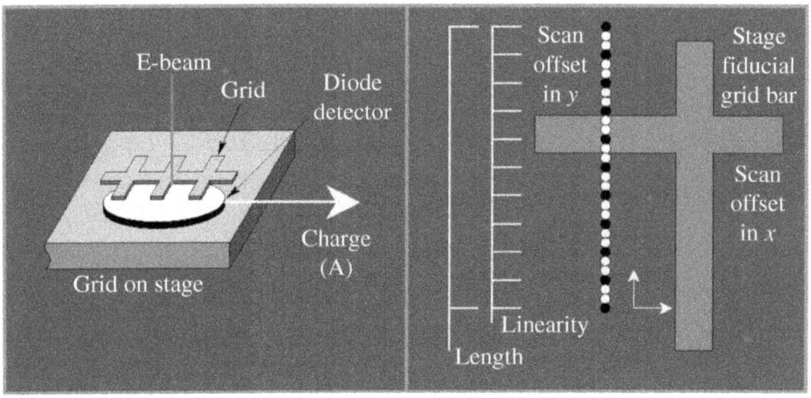

Figure 3.34 Stroboscopic scan calibration.

sample of beam position (see Fig. 3.34).[203] Other improvements included a glancing-angle height sensor, dynamic focus corrections, periodic drift compensation, substrate temperature control, and real time correction of deflection distortions, which reduced single-phase stitching errors from 80[204] to 50 nm. [205]

The relative benefits of VA and MPG were studied. MPG was shown to improve butting accuracy, at the expense of longer print times than VA.[206] The latter technique retained pattern fidelity and linearity at the expense of line-edge roughness for print addresses less than 40 nm.

The MEBES 4700 doubled the blanking rate to 320 MHz, printing on a MEBES 4500 platform. This was twice the MEBES 4500 multipass (2XMPP) writing rate for print pixels in the range 13 to 68 nm. Alternatively it allowed a $\sqrt{2}$ reduction in address unit without reducing exposure time. The blanker driver received 4 bits at 80 MHz, which converts on the fly to a 320-MHz single pixel bit stream. The 3.125-ns pulse width with <300 ps rise and fall time was tap settable in 50 ps increments. The MEBES 4700S provided MPG as an option at 320 MHz. The MEBES 4700 series was repackaged as the MEBES 5000, which combined 320 MHz, MPG, and automated deflection scan correction as standard features.[207,208]

The MEBES 5500 system doubled the beam current density to 800 A/cm^2 to support eight-pass, eight-gray level printing.[209] This was a method to double print rate while maintaining a constant tool-input address. MPG-II doubled gray levels to eight by eight-phase printing.

The first 50-kV raster gaussian MEBES (eXara) was introduced in 2001.[210] The system was built on a new platform with a vacuum air bearing stage. It featured four-phase offset scan printing with four or eight gray levels per phase and automated GHOST proximity effect correction. The feasibility of pulse-width modulation of beam blanking had indicated that gray beam printing could be achieved for each print

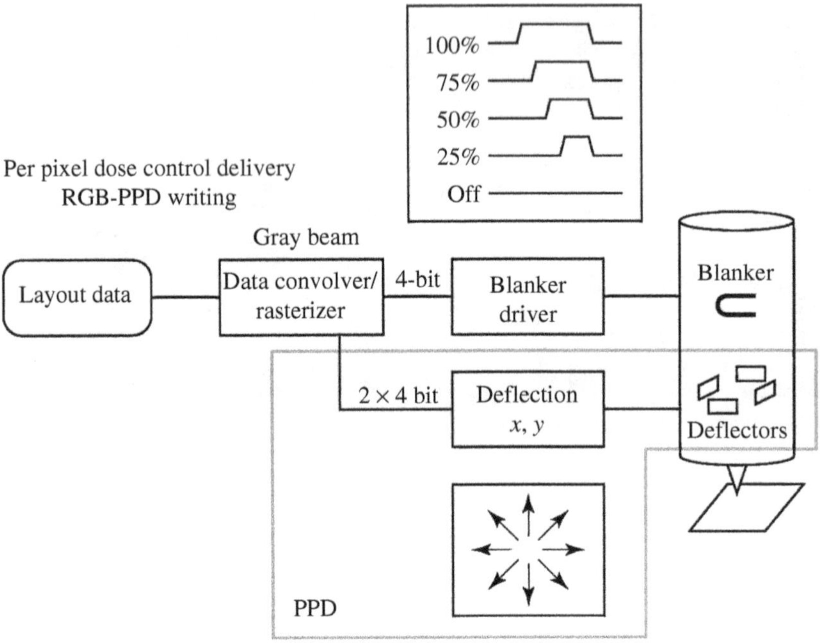

Figure 3.35 Four gray level per phase eXara blanking.

phase.[159] The system printed four gray levels per phase (see Fig. 3.35) with a standard four-phase print strategy. Per pixel deflection (PPD) was never commercialized as it was unable to fully compensate the timing asymmetry of the individual gray levels, which contributed to *xy* bias effects in critical dimensions.

Retrograde scanning improved calibration accuracy and reduced *xy* CD bias. Adding a retrograde scan, shifted 180° in phase to printing, increased the beam dwell at each location, reducing blurring effects. The beam is blanked while the scan moves to the next location. Retrograde scanning is shown in Fig. 3.36.

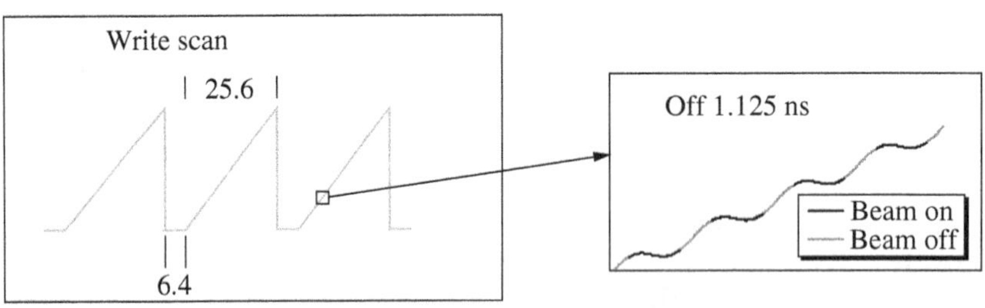

Figure 3.36 Retrograde scanning.

The raster-shaped electron beam concept was introduced as a means to combine the advantages of raster gaussian printing (high data rates and beam current density and accurate and uniform scanning) with those of shaped beam (parallel pixel exposure and a decoupling of beam resolution from the writing grid).[211–213] The maximum variable shaped flash size is set to the minimum feature size to be printed. A limited range of flash sizes allows the use of high-speed, 100-MHz, low-resolution DACs and amplifiers to form shapes. The high flash rate is supported by a high current density beam created by using TFE source technology at high angular intensity. Write time as a function of effective grid size for MEBES eXara and RSB is shown in Fig. 3.37.

RSB exposes patterns with a regular array of 128-nm maximum flashes, using uniform stage motion and a periodic raster scan. Two-phase offset scan printing is used to average scan boundaries. Each variable shaped flash covers a square flash field that is represented in the data path by a group of four gray level pixels. Subpixel edge placement uses an address structure $^1/_{32}$ of the flash size. With two-phase printing the 128-nm flash grid produces a 2-nm effective grid on the mask. A prototype 50-kV raster-shaped beam pattern generation of 70-nm masks was developed on an MEBES platform with real time proximity effect correction.[214]

A unique aspect of the system is use of shadow projection optics,[215] which has a very short column length beneficial for improving resolution, although leading to different depth-of-focus behavior compared to more conventional shaped beam imaging optics, which place the shapes at conjugate points to the mask plane. A schematic of RSB shadow projection optics is shown in Fig. 3.38. The high-speed shaper-blanker

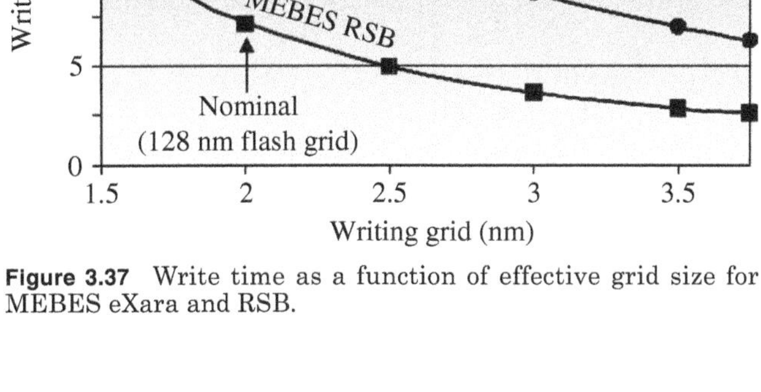

Figure 3.37 Write time as a function of effective grid size for MEBES eXara and RSB.

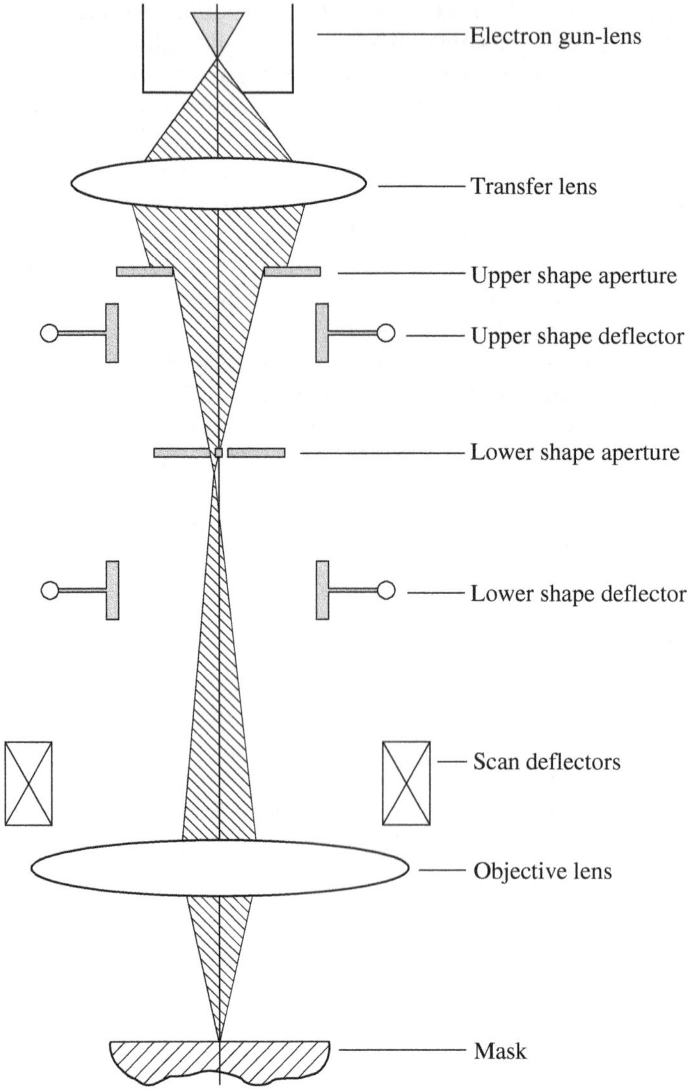

Figure 3.38 Schematic of RSB shadow projection optics.

assembly combines the functions of shaping and blanking with the latter adjusted for a conjugate mode of operation.

The RSB flash could be reduced in size from 128 nm, which enables better lithographic quality. It is conceivable that the data rate could be increased to 200 MHz, which would deliver 800 Mpixels/s burst rates. Other improvements in shaper-blanker technology would also improve CD control.

Extendibility of scanned gaussian multi-beam PG. We consider several variants of a mythical system that delivers 6.4 Gpixels/s data rates combined with high-resolution optics, and accurate scan systems deliver nanometer CD control and pattern placement accuracy to illustrate how scanned gaussian raster multiple beam technologies might extend patterning capabilities to the 32-nm era and beyond.

Higher resolution laser pattern generators are feasible. Past evolution has been through higher numerical aperture optics and with shorter wavelengths. This wavelength reduction trend may continue using pulsed, mode-locked, 198-nm laser sources.[216] Using a 0.9 NA, ALTA 4700, 42× reduction lens would improve resolution (~$0.6\lambda/\mathrm{NA}$) from ~170 to 130 nm. Liquid immersion could further increase $\mathrm{NA} = n_f \sin\theta = 1.3$., where the refraction index $n_f = 1.44$ at $\lambda = 193$ nm.[56] This would drive the beamlet resolution to ~90 nm, which is about the present resolution of 50-kV-shaped electron beam optics.

Under these conditions and using aerial image correction to achieve a feature size to resolution ratio of 1.6, one can project a 198-nm-immersion system to be capable of printing 150-nm features with adequate CD control. This would meet all leading-edge 65-nm mask node requirements and extend well into the 45-nm node. Raster e-beam print strategy evolution is shown in Fig. 3.39.

Printing 6.4 Gpixels/s is feasible, either by extending the number of beams to 64 or doubling the clock rate to 100 MHz. AOM technology might even be extended to support up to 200 MHz operation. Table 3.4 shows an 8-mg acceleration and 20 mm/s stage trajectory profile, which secures 72 percent stage efficiency on a 32-nm print pixel. Improvements in the data path firmware would be engineering intensive, but without a fundamental technical risk.

A natural extension of raster strategy is to combine the resolution capability of e-beam with the speed afforded by multibeam printing.[217] It is feasible to conceive of a scanned gaussian raster multi-e-beam architecture, which results by combining the high resolution benefits of electron beams with the inherent speed of multiple beam printing using

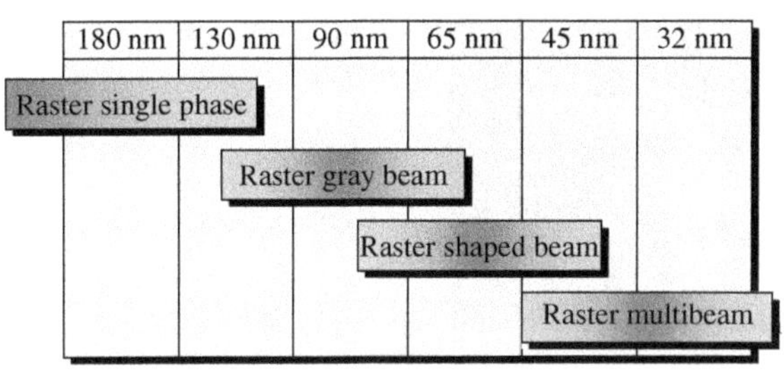

Figure 3.39 Raster e-beam print strategy evolution.

raster bitmapped data path to deliver ~6.4 Gpixels/s data rates and beyond. There are numerous multi-e-beam architectures, a topic which could consume an entire chapter itself. A schematic of the raster multi-e-beam optics is shown in Fig. 3.40.

> A simpler way might be this (though I am not sure it would work): We take light and, through an optical microscope running backward, we focus it onto a very small photoelectric screen. Then electrons come away from the screen where the light is shining. These electrons are focused down in size by the electron microscope lenses to impinge directly upon the surface of the metal. Will such a beam etch away the metal if it is run long enough? I don't know. If it doesn't work for a metal surface, it must be possible to find some surface with which to coat the original pin so that, where the electrons bombard, a change is made which we could recognize later.
>
> There is no intensity problem in these devices—not what you are used to in magnification, where you have to take a few electrons and spread them over a bigger and bigger screen; it is just the opposite. The light which we get from a page is concentrated onto a very small area so it is very intense. The few electrons which come from the photoelectric screen are demagnified down to a very tiny area so that, again, they are very intense. I don't know why this hasn't been done yet![3]

To highlight several of the technical barriers to this approach and list possible solutions, consider an extension of the ALTA print strategy as one means to adopt existing control, data, and calibration architectures. A two-times improvement in resist sensitivity might be expected in the next few years, driven by existing VSB needs and so a 50-kV, 4 $\mu C/cm^2$ resist is assumed with four-phase printing to ensure pattern quality and ~8 h exposure-limited mask print times remains economically viable. The total current required to expose the base dose is then 66 nA, or about 1 nA per beamlet. A beam size to pixel grid ratio of 2:1 requires a

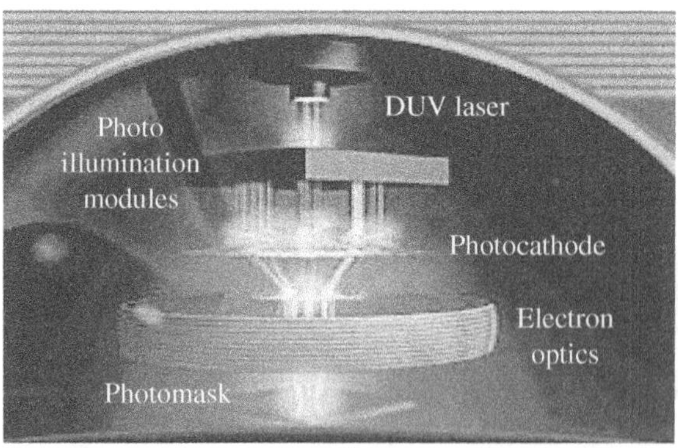

Figure 3.40 Schematic of the raster multi-e-beam optics. (Courtesy DARPA.)

100 A/cm^2 current density at the mask. The optical system should be capable of twice this dose to ensure proximity correction of isolated features. Figure 3.41 shows the characteristic dependence of total current on print time for a range of resist sensitivities and print phases.

The feasibility of a *raster multi-e-beam* (RMB) optic has been demonstrated with a 50-MHz, 50-kV, 32-beam column.[218] The photocathode illumination subsystem modified components from the DUV, 600-mW optic of the ALTA 4300 to produce 32 independently modulated laser beams focused onto a photocathode. A special long working distance objective lens was built to stand off atmospheric pressure to ensure that the laser beamlets could be focused to 300 nm at the photocathode. 50-nm electron beams were measured at the mask and the column optics redesigned for 35-nm electron beams.[219]

A key enabling technology was the development of a stable and robust photocathode.[220] Photoactivation of a wide bandgap CsBr material, which under normal conditions would not photoemit under 257-nm illumination, resulted in photoyields in the ~100 nA/mW range with 100 A/cm^2 current densities and lifetimes per illuminated beam over 50 h. Experimental tests showed it could meet all the electron source requirements of the RMB pattern generator system concept. Print times for various scanned raster gaussian systems are shown in Fig. 3.42.

Further scaling an order of magnitude in RMB optics is conceivable by maintaining existing photoillumination efficiency and without

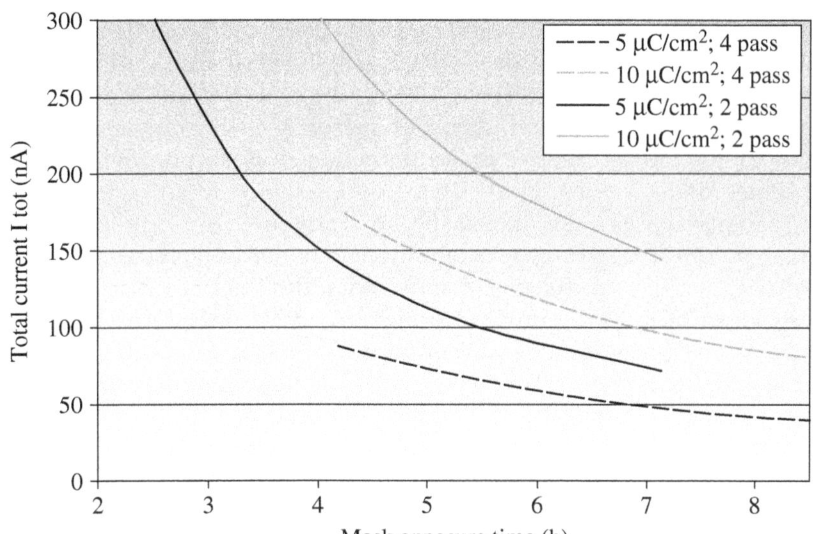

Figure 3.41 Total current required for printing over 88 × 132 mm^2.

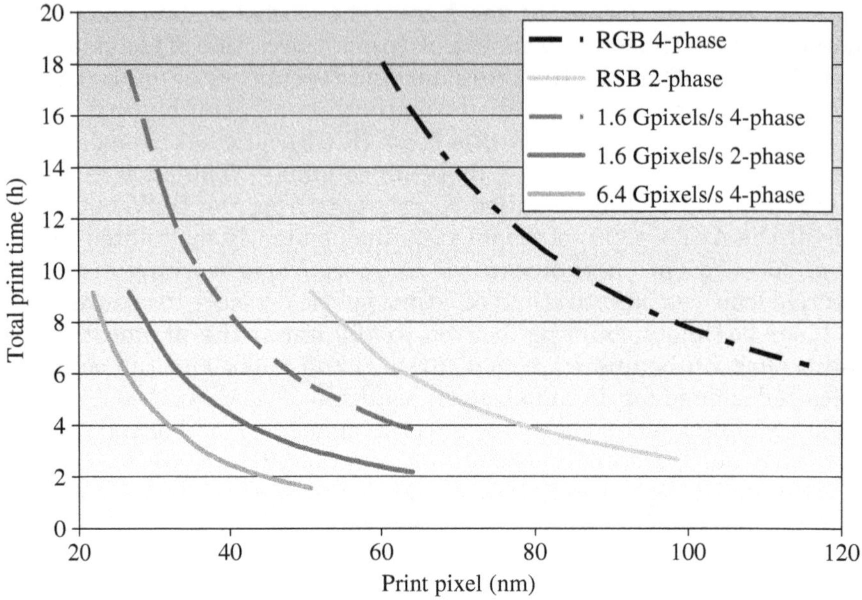

Figure 3.42 Print times for various scanned raster gaussian systems.

increasing laser power above the existing 600 mW.[221] Surface-plasmon-enhanced illumination devices patterned on the photocathode can produce stable electron beams arrays. Based on observed UV light intensity and collection enhancement of these devices, it appears possible to realize a photocathode electron source with a stable small spot size.[222] Two times object reduction by patterning 150-nm spots would quadruple the electron column transmission. Another factor of two times could be expected from steady increase in stable emission with photoyields of 70 to 140 nA/mW.

Dose sensitivity can be increased by reducing landing energy. Retarding potential RMB optics combine the benefits of resolution at high voltages with increased resist sensitivity, though the latter comes at the expense of imaging on thin resist layers.[223] A 4 $\mu C/cm^2$ resist exposed at 50 kV becomes a 32 $\mu C/cm^2$ at 6 kV. A high bandwidth proximity electron correction firmware could conceivably handle the backscatter range going from ~9 back to ~0.4 mm.

The beam delivery subsystem is connected to stage velocity, which is geared to the scan pitch, another aspect of the scan optic and linear brush. This pitch is the cross-scan distance between two scans. For a single beam system this would just be the pixel pitch. Multibeam linear brush optics and beam modulation require that the beams be adequately separated. The gaps between the beams are filled in by having the stage travel less than one brush width for each scan.

In general, this type of interlaced scanning will provide uniform coverage (evenly spaced beams), provided the following relationship holds.

$$m/n = b/q$$

where the scan pitch m is the number of units incremented between scans, n is the number of units between beams, b is the number of beams in the brush, and q is an integer that has no factors common with b. Full exposure with 5:1 scan pitch is shown in Fig. 3.43.

A proof is provided here, though the relation was first stated in Ref. 15. Let N_b = number of beams and P_w = print pixel size [nm]. Other lengths are described as absolute lengths [nm] or in P_w units. The distance measured from one beam to another is the brush pitch B_p. It is commonly expressed as B_p:1. The beam cycle is $Z_p = B_p - 1$. This is the number of pixels between the beams. The term cycle is used as a compact graphical representation and can be developed for analyzing the interlacing scheme using finite groups. The brush length is $L_B = (N_b - 1)$ B_p. And define X_s = scan pitch. This is the distance in x, perpendicular to the scan direction y, which the beam moves from one scan to the next. The array configuration is generally given by $M \times N_b/M$ = array configuration. A linear array has $M = 1$. The boundary conditions on scan pitch are—brush pitch < scan pitch < brush length. Brush length L_B can also be expressed in terms scan pitch.

$$X_s \cdot Z_p + (X_s - B_p) = L_B$$

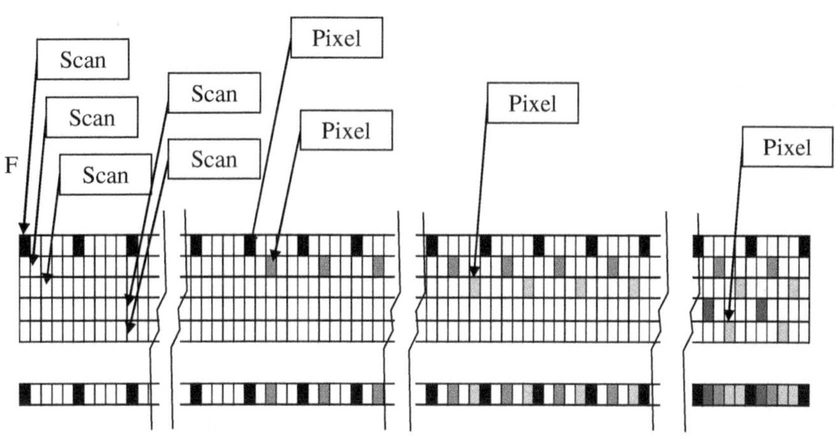

Figure 3.43 Full exposure with 5:1 scan pitch. Brush extent is 160 pixels.

Comparing this equation to the definition for brush length leads to the condition that the scan pitch must be equal to the number of beams:

$$N_b = X_s$$

This condition is easily seen in the case where the brush pitch is 1:1. The parameters for the brush organization must obey the following equation to have uniform dosing at the photocathode:

$$X_s/B_p = N_b/q$$

where q is an integer with no common factors. This completes the proof.

The scan pitch in this example is 64 pixels or 1.024 μm. The stage velocity is then the ratio of the scan pitch to 102-μs scan period, which is 10 mm/s. Other pixel pitches and brush configurations are possible. Finding the right combination depends on the print strategy and error budget along with characteristics of the beam delivery optical components.

A 100-MHz data path printing to 64 beam boards at 6.4 Gpixels/s would generate 44 Tpixels on a 16-nm print-phase address in a 1.9-h print phase over 116 cm^2. To the extent the burst rate can be sustained no overhead is added. A 13.1-μm scan length would be exposed in 81.9 μs when generated from 8K pixels per scan. Allowing 20 μs to scan retrace settling time implies an 80 percent scan efficiency, which drives total exposure time to 9.8 h, and then to 13.5 h with 72 percent stage efficiency.

3.3 Component Technologies and Technical Challenges for Mask Patterning

A brief survey is given of some of the fundamental technical issues that various subsystem and component technologies must address to achieve the pattern quality and throughput needed for future mask making.

Shot noise, line-edge roughness, and metrology. There is a longstanding question as to how shot noise affects lithographic quality.[224] The impact of LER on device performance and the relative contribution of mask and lithographic parameters has been studied by various groups, see Refs. 225 and 226.

High resolution experimental studies of CARs printing 10-nm features using FIB[227] and 50-nm features using EUV[228] have shown the role of Poisson statistics generated from the shot noise process effect process margins.

Stochastic modeling of 50-nm lines is now able to elucidate how different feature types can be related to the statistics of chain scission events from first principles calculations.[229,230]

Experimental studies are beginning to elucidate the relationship between LER, the modulation transfer function, and proximity and process effects.[231] CAR acid diffusion has been shown to have two opposing effects on LER—one is a smoothing effect, which reduced the LER of a 100-nm line using a negative CAR from 6.9 to 3.0 nm; the other is the degradation of deposited energy profile, which occurs when the diffusion length becomes too great due to (e.g.) increased prebake temperatures.[232]

Also, with such small CD variations, the metrology contribution becomes comparable in magnitude to the measured feature. Rigorous calibration and statistical methods are becoming necessary to ensure confidence in measured values as pixel and feature sizes shrink to levels where deposited energy diverges from continuum models. Metrology averaging methods have a significant effect on experimental results.

A brief analysis of shot noise neglecting process effects is given to highlight some aspects of the problem. Shot noise models predict line-edge roughness as a function of resist sensitivity Q_o and pixel size P_w. As a specific example consider $Q_o = 30\ \mu C/cm^2$ at 6.7-kV landing energy, ($4\ \mu C/cm^2$ at 50 kV) and a print pixel $P_w = 16$ nm. The average number of electrons per pixel is $<N_e>/P_{wz}^2 = (1 + 2\eta_e)Q_o$. An expected value for $\eta_e =$ 0.5, so $<N_e>/P_w^2 = 2Q_o$. This is practically equal to the density of the primary dose of an isolated line. Each element of resist is exposed by the same total dose D, which is made up partially from primary electrons and partially from backscattered electrons. The relative fraction of dose from primary versus backscattered electrons at a given location will vary, i.e., an isolated line will receive a dose predominately due to primary exposure, whereas dense features will be dosed by nearly equal amounts of incident and backscattered exposure. Given that both primary and backscattered doses are governed by Poisson processes, then the exposure density corresponding to the primary dose of an isolated feature holds for all pattern densities in these shot noise estimates.

$<N_e>/P_w^2 = 938$ electrons are deposited in the 16-nm pixel. The relative root mean square (rms) fluctuation in Poisson-distributed electrons per pixel is $\sigma/<Ne> = 1/<Ne>^{1/2} = 3.3$ percent.

Linear feature edges are assumed to be shifted from pixel to pixel by the Poisson-distributed dose variations. So the line-edge roughness variation $\sigma_{LER} = P_w/\sqrt{<N_e>}$. Each pixel will be uncorrelated to the others for pixel sizes $P_w \geq$ the primary scattering σ_f range. Some correlation may be introduced with 16-nm pixels, but for the simple example here $3\sigma_{LER} =$ 1.6 nm.

Although tool-dependent noise often correlates feature edges through time-dependent scan effects, the shot noise model assumes feature edges are independent, then the linewidth variation of the feature size is $\sigma_{FS} = \sqrt{2}\sigma_{LER}$. For the example $3\sigma_{FS} = 2.3$ nm.

Now let's consider metrology effects on the measured feature variation. Assume a 0.5-μm metrology window average is taken for the feature, which is a line longer than the window. This means 32 P_w print pixels are averaged for a single metrology datum. The variance of line-edge roughness including metrology $\sigma_{MLER} = \sigma_{LER}/\sqrt{32}$. The variation in CD and linewidth roughness per metrology tool is then $3\sigma_{CD} = \sqrt{2}\ 3\sigma_{LER\text{-metrology}} = 3\sigma_{LER}/4 = 0.4$ nm.

In summary, this model shows that a 4 percent relative variation in 30 μC/cm^2 dose for an isolated line comprised of 16-nm pixels yields an intrinsic feature width variation of 2.3 nm, but this is reduced to a measured value of 0.4 nm due to metrology averaging.

The point being that these statistical effects must also account for metrology effects to be meaningful. Hopefully in the future these factors can be used to guide quantitative determination of how shot noise will limit the variation of features, which require stringent critical dimension control.

3.3.1 Data path

Pattern data file sizes continue to grow at an exponential rate to accommodate Moore's law[233]; see Fig. 3.44. Functional integration of backend IC design through lithography continues with design content moving upstream in the overall process flow. Strategic changes based on open

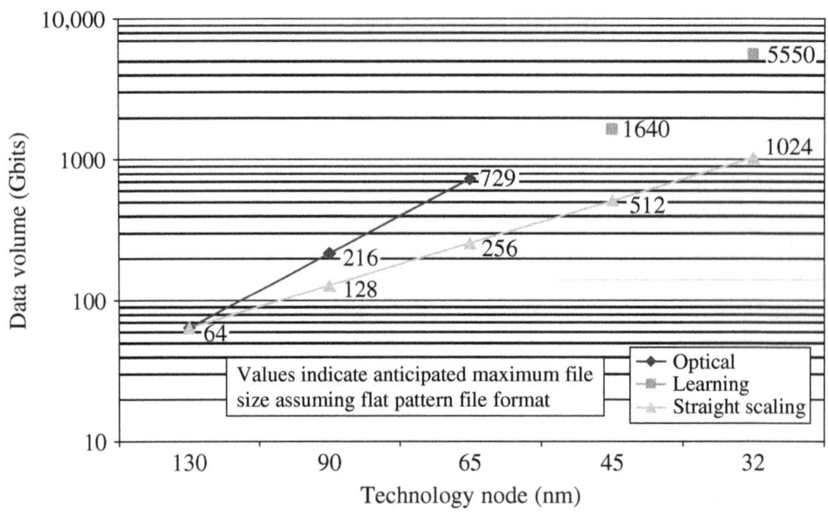

Figure 3.44 Single layer data volume projections. Optical values indicate maximum file size assuming a flat pattern file format. Straight scaling assumes no change in mask complexity other than density over time. Learning represents possible improvement from EDA process integration.

standards and a universal data model are being promoted as a global long-term solution to eliminate interoperational inefficiencies in how IC design and electronic design automation (EDA) tools are presently being used.

A high-level language has been proposed to facilitate interoperation of various design processes.[234] The objective is to reduce the overall cost of design and manufacturing by combining this structure with application protocol interfaces and adopting supporting design languages. Streaming data translation times have become a factor in physical verification of the wafer layout.[235] PSM and OPC have added to this complexity as part of the general EDA trends with these reticle enhancement technologies (RETs) increasingly implemented in physical design. All this drives requirements at the PG data path interface.

The dataflow processes are shown in Fig. 3.45. CAD groups have typically taped out in GDSII hierarchical format. There is a drive to replace GDSII as a means to remove ambiguities in the standard and allow compression by directly representing arrays.[236] Future pattern data input requirements will likely be based on OASIS or a similar variant as the physical design data format for pattern transport. Proprietary formats have also been proposed, which address the redundancy of the GDSII stream files.[237] Widespread adoption of a single format will likely have to be driven by a standard's committee with broad-based industry representation.

OASIS' purpose is to define an interchange and encapsulation format for hierarchical integrated circuit mask layout information.[238] In the fall of 2001, SEMI's Data Path Task Force formed a working group to define a successor to the venerable GDSII Stream format, which had served

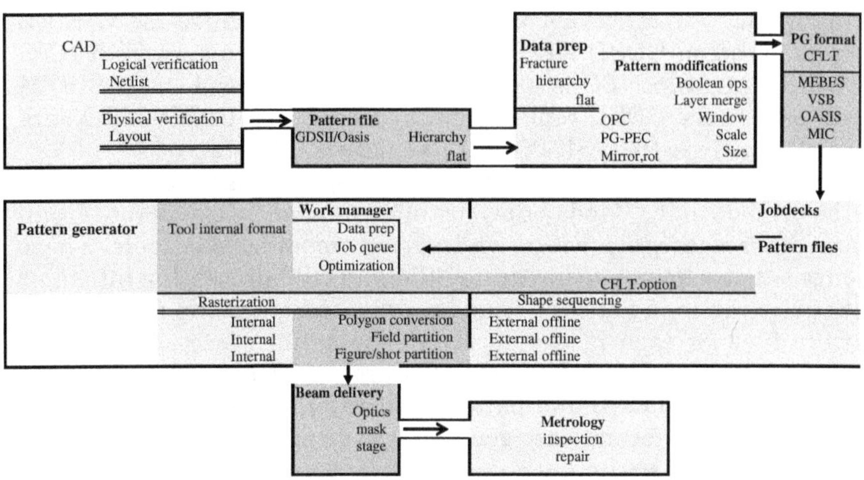

Figure 3.45 Pattern generator process flow chart.

the IC industry as a de facto standard for layout interchange for more than two decades. The old format, limited by 16-bit and 32-bit internal integer fields, by its inefficient representation of cell-native geometric figures and by high structural overhead, was becoming difficult to use for leading-edge designs, and file sizes were becoming unwieldy, in some cases growing to many tens of gigabytes.

Processing 1 Tbyte pattern files will be necessary by the 32-nm era. Data and job preparation modules will have to be converted to a 64-bit operating system, which can address 18 million terabytes of memory. The ability to store an internal representation of a pattern file is limited by the amount of addressable memory, which in turn is limited by the operating system. For a 32-bit operating system, this translates to 4 Gbytes of addressable memory, and this means that GDSII files of size no greater than 400 Mbytes could be processed, assuming an expansion during conversion to tool-internal format occurs. However, based on the type of the pattern files, the corresponding tool-internal or native mode files may be more than 1 Tbyte, as compression provided in some flat file formats is removed in converting to tool-internal files.

The OASIS format was chartered with several overall goals: achieve at least an order-of-magnitude file size improvement compared to GDSII Stream, remove all 16-bit and 32-bit integer width restrictions—make the new format fully 64-bit capable, efficiently represent cells with large payloads of flat native geometric figures, and facilitate interchange of layout-related information between design and manufacturing.

Under the auspices of SELETE and Dai Nippon Printing, JEOL, Hitachi, and NuFlare have developed a unified mask data format named OASIS.NEO1 for variable shaped beam electron beam writers. [239] The data format specification is an enhancement of the e-beam VSB NEO2 data format.[240] Restrictions on OASIS have been defined for VSB EB writers based on GDSII.NEO4, which comprises elements from NEO and GDSII. OASIS.NEO is a pattern data format based on OASISTM 3 released as the GDSII replacement by SEMI. The OASIS.NEO data are fed directly to the VSB PG just like native EB data.

While the trend is for pattern modification operations to be transferred to the CAD domain,[241] today they are often applied by a separate group using other data preparation and pattern modification tools. These groups are also having to integrate PSM and OPC processing into their data preparation modules.[242] The data preparation group receives the pattern files for each IC design layer. The pattern fracturing step is often undertaken with third party software, which acts as an intermediary between the CAD and pattern generator. Pattern modifications and intermediate formats are generated to facilitate the translation.

One example of *streaming data translation* occurs when an MEBES-formatted pattern file is to be printed on, e.g., a Micronic tool. The third party software translates this pattern file to CFLT/CREF, and then to MIC. The latter format being required as input to a Sigma/Omega PG. Other pattern file output formats included MEBES and those dedicated to various vector PGs. There have also been proposals to extend the CFLT format to cover the internal PG tool data, which has some analogy to certain mask inspection configurations. So far this path has not been developed, at least in part due to PG tool vendors being reluctant to give up proprietary control of the internal working of the system, which embodies key intellectual property regarding data transformation.

Complications arise due to hours of processing time and data explosion of two to 10 times, which can occur on conversion from one format to another. The nature of these operations has an effect on pattern size and degree of hierarchy that is retained. Symmetry-breaking OPC serifs, outriggers, and scatter bars can reduce the benefit of a hierarchical format. One publicized early example was a Xilinx pattern for which the original 38-MB GDSII file expanded to 58 Gbytes in MEBES format.[186] The pattern was a *field programmable gate array* (FPGA) with 6 million gates at 150-nm ground rules, which had been decorated with OPC structures by a Mentor Graphics software package. For the large FPGA cells there was essentially no hierarchy retained in the MEBES file. The complexity of the problem is also manifested in some designs being more efficiently represented in MEBES format compared to GDSII.

Pattern modification operations include layer merge, window, boolean, scale, size, mirror, rotate, OPC, and PG PEC. Tone, size, merge, window, mirror, and rotate are transformations of the mask layout. Global, pattern level operations are required to compensate for the prior state of individual patterns. Scale (magnification) and orthogonality are global operations dependent on exposure tool calibrations. Rotate (at multiples of 90°), mirror (around x, y, or both), size (bias), and tone are required for global and pattern operations. Window and merge of pattern files are required pattern operations.

All combinations of the transformations ultimately resolve to a single unambiguous real grid at exposure. Grid snap artifacts, such as reverse tone, can be avoided if the fractured output grid is smaller than the PG-internal tool grid. Some transformations, such as size, must use a healing process to remove gaps or overlaps generated in the transform. The same can apply to merge, which can be logical OR or XOR, because the input patterns to merge may have different native grids. Ideally all these transformations also execute in real time, otherwise this portion of the data process remains a system overhead.

Logical data operations reside in different dataflow locations. These operations include polygon conversion, field partition, and figure or shot partition. Raster architectures can have real time execution of these data operations in the exposure tool. Vector architectures perform these operations off-line and often require pattern refracture for each writing pass. VSB users trade off off-line processing costs for other benefits.

A near-term strategy of extending the hierarchical capability of tool-input formats is being adopted by all PG tool users in one form or another. For example, Micronic users try and maintain hierarchy by converting from GDSII directly to MIC, which can retain hierarchy and cell structure. ALTA also has the capability to process directly from GDSII. GDSII-NEO is another hierarchical extension that is being supported by some Japanese e-beam tool users and vendors.[243]

A long-term vision for future processing would be for the data path interface to be directly with the physical design (GDS/OASIS) database. The user would avoid the fracture process and not need to deal with intermediate forms of the pattern file data. The ultimate goal would be to do this within run time. There would be no distinguishable pattern modification step at all. Data preparation functions would either reside in the design database or be incorporated within the tool itself.

Job deck. The job deck contains instructions on how to print the pattern and defines machine configurations such as beam size, exposure dose, and size of scan field. There may be a single job deck for a given IC device; however, since it is made up of different layers, each layer may be represented by a level in the job deck, requiring a separate mask. On the other hand, levels may represent different chipsets that require different printing configurations but are written on the same mask.

In the future systems might redirect or port modified pattern files to different exposure engines. Run-time fracture of multiterabyte local databases will be needed with support of legacy proprietary PG formats. Installations with different product generations might share an intermediate database. All user interaction could be through the exposure control file (job deck) in a single interaction and through a single interface. Future job control input could be based on an unambiguous mask definition via SEMI P10 or an equivalent nonproprietary standard. Some evolution in this direction is occurring with the MALY VSB job deck specification, which is integrated with the OASIS.NEO data format to provide a unified data preparation flow into the mask pattern generator.[244] Features might include run-time layout composition and pattern modifications. Nonproprietary pattern and job control formats facilitate the goal of a universal data model for design to manufacturing capabilities.

System software architecture. System software handles printing, calibration, self-diagnostics, and recovery functions. The network architecture includes the data path, diagnostics, control, and calibration (Fig. 3.46). Data preparation and job control commands are processed at work stations. A storage area network allows archiving and retrieval of pattern

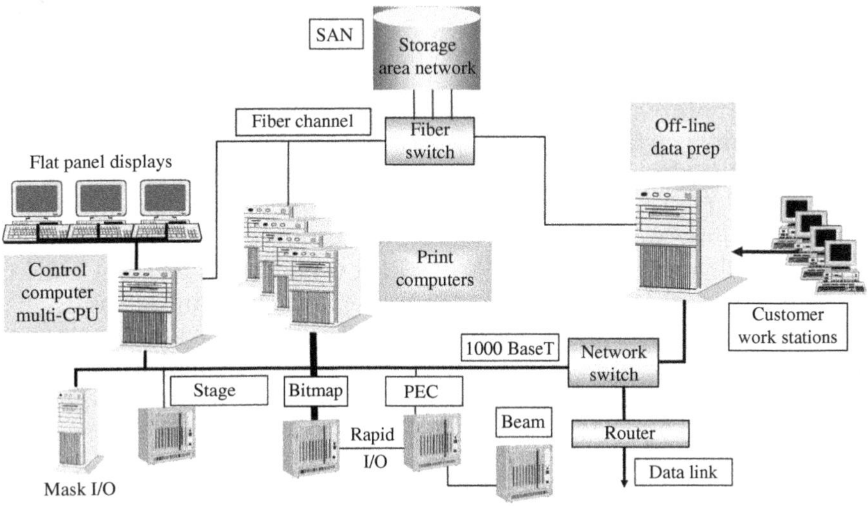

Figure 3.46 A network view of PG software architecture.

files and other mask management data. A work manager within a *front end processor* (FEP) executes the job and data preparation functions. If job preparation time is less than or equal to print time then the operation does not create an operational bottleneck. Ideally, job and data preparation cycles are isolated from, and take no longer than, the print cycle.

Print software reads instructions from a job deck and converts pattern files into tool-internal files. The program loads the data during mask writing, while simultaneously communicating with exposure control, stage, and calibration routines. Pattern files are assembled and processes executed to compute and execute the optimal writing strategy, e.g., an optimized stage trajectory profile is computed to minimize stage motion.

The print computers download PG tool internal format data to rasterizers or corresponding VSB internal sequencing firmware, which renders digital data commands to control signals, which are then transmitted to the beam delivery and stage subsystems for subsequent mask exposure.

Present data path design and cost strategies favor use of commercial off-the-shelf parts. The control systems can be based on commercially available versions Common Object Request Broker Architecture (CORBA). Services to enable GEM/SEC type commands are often available for exception handling, event monitoring, and state change reporting, which can be associated with the GEM/SEC protocols.

Factors such as the compute intensity of microprocessors, disk read/write speeds or I/O operations per second, and bus-Ethernet bandwidth influence implementation of scalable data path designs. These components are upgraded over the course of time to improve ease of use and avoid bottlenecks in data processing when they create unacceptably large overheads in system throughput.

Scalability. Hierarchical data formats such as GDSII/OASIS do not solve the looming problem that pattern files are approaching 1 Tbyte. The transfer of these files, even using 1000_bT network, becomes a time critical step. For example, if 1 Tbyte uncompressed pattern file is transferred over a dedicated Ethernet link (1000_bT), the resultant transfer time will be 1 Tbyte/40 Mbytes/s = 25,000 s ~ 7 h, which would begin to affect mask PG system throughput. Given a 3-h data preparation requirement, then compression of pattern files during transfer, multiple Ethernet links or computers, and using fiber channel connections instead of Ethernet become design issues. The primary limitation in any hard drive storage system is the ability to read and write data to the hard disks. Currently, this is limited to about 60 to 80 input/output per second (IOPs), with the read rate nominally twice as fast the write rate. This is expected to increase with faster disks (30K rpm disk is faster than 15K rpm disk) but the technology curve remains too slow for future 10- to 50-Gbits/s pixel delivery rates where other means will have to be adopted to circumvent the I/O limitation. In most cases, remote disk I/O remains a time critical activity.

Implementing hierarchical data formats is not a panacea. Hierarchy is of little help for ASIC/MPU where there is little repetition of the design cell. Or implementing hierarchical formats will shift the burden to internal operations of the tool's data path, e.g., rasterization of figures into pixels, proximity correction for both raster and vector, or shape sequencing for vector. A data path architecture with an embedded layer is shown in Fig. 3.47.

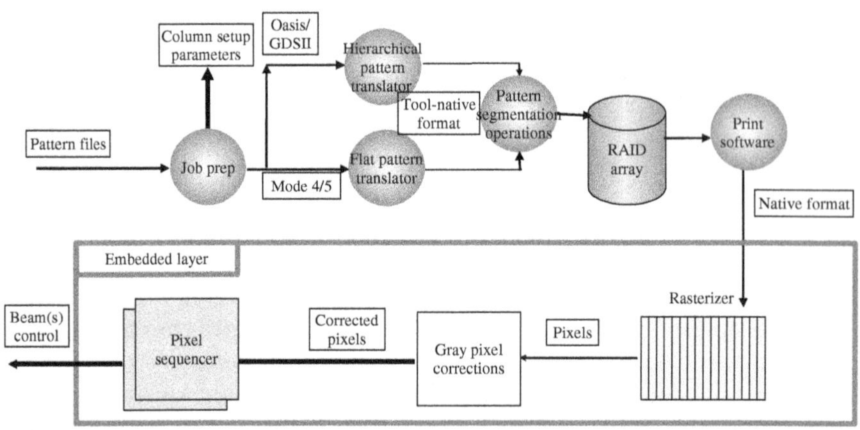

Figure 3.47 Data path architecture. The embedded layer as shown represents a raster system. A vector system does not include rasterization and the pixel sequencer is replaced with a shape sequencer.

In the future the resultant tool-internal files need to be read and written to disk in a compressed form and/or communication bandwidth increased. A key part of most data path development strategies is to have a scalable architecture and implementation plan using commercial off-the-shelf parts. Present data path architectures are not likely to be sufficient to meet the future needs for a 10- to 50-Gpixels/s throughput. However, there are paths to achieve even this performance level by incorporating solid state memory banks to solve the fundamental problem of limited disk I/O. One solution is to rasterize the entire pattern file and store it in memory.

There are three functional requirements. One is to handle the off-line data preparation needs. The second is the pixelized data storage requirement. And the third is the pixel data transmission rate.

Data transfer steps include transfer from pattern file library to pattern generator; transfer ~1 Tbyte pattern file(s) within the pattern generator, such movement of pattern data within tool itself likely internalizes data preparation, job preparation, and translation to native mode; boolean operations (sizing, overlap removal, scaling, mirroring, rotation); and finally download—transfer from the storage media to the rasterizer.

Figure 3.48 shows how such an advanced data path platform might be driven by a parallel processing operating system. Such a platform would handle all data preparation operations in about 1 h for 1-Tbyte file sizes, which is equivalent to ~100 giga-figures, assuming an average of 10 bytes

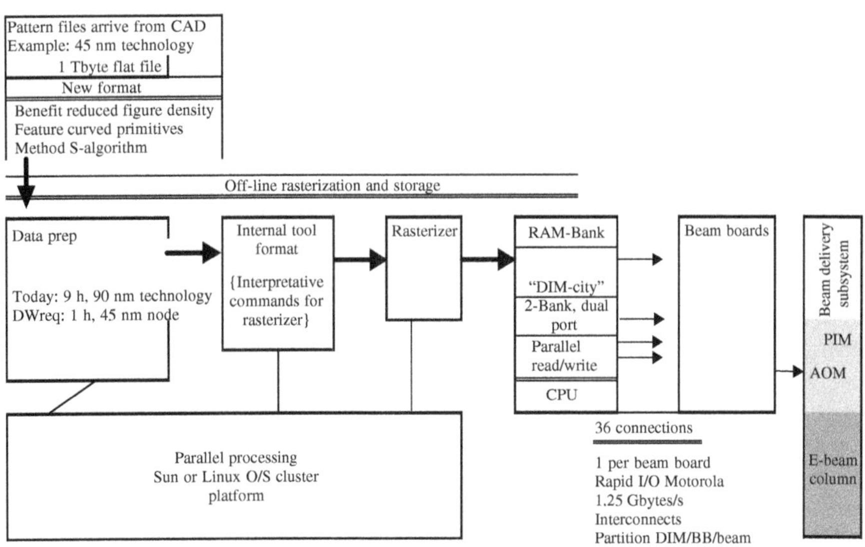

Figure 3.48 Data path architecture schematic for 10 to 50 Gpixels/s era.

per figure. Assuming data preparation is about 50 floating point operations per figure, this is equivalent to 5–tera-operations per pattern.

Today a 1-GHz CPU can ~1 *giga-instruction per second* (GIP). Using four instructions per *floating point operation* (FLOP) implies a 250-mega-FLOP capability. As the processing requirement is 5–tera-operations per hour, this implies 8 CPUs are needed. Hence the reason for a cluster computing platform.

The second issue driving the platform concept is the huge data rate requirement for retrieval and storage. A print strategy with a 12.5-nm interstitial pixel might use two-phase printing and 64 gray levels per phase. Given 8 bits/pixel with 6 bits for gray, 1 for parity, and 1 for zero, then printing a 100 mm × 100 mm mask means 64-T gray pixels per mask must be stored. One storage method could be a RAM bank. Today 1 DIM can handle 1 Gbyte of RAM storage, which implies about 2320 DIMS will reside in the bank for ~2 Tbytes per chip storage capacity. At today's prices of $250 per gigabyte the cost will be on the order of $600,000.

The third issue is handling the I/O requirements. Today a storage area network RAID has a 40 Mbytes/s capacity. Projected improvements in this technology are not sufficient for such a platform. Instead Rapid I/O technology might be used, which today is foreseen as having 1.25-Gbytes/s capability. Assume that actual mask exposure takes 45 min, which allows 15 min for other system overheads at 1 mask per hour throughput. Exposure rate must be fed 17.8 Gbytes/s, which requires the use of 14 channels. These could feed 64 beams using five channels per beam at 100-MHz clock rates.

3.3.2 Precision mask handling

Some history may help frame the topic of ruling engines:[245]

The second engine, installed at the Laboratory has been producing gratings since 1953, was originally built by David W. Mann of Lincoln, Massachusetts. Bausch & Lomb equipped it with an interferometric control system following the technique of Prof. George Harrison of MIT. The Mann engine can rule areas up to 110 × 110 mm, with virtually no ghosts and nearly theoretical resolving power. While the lead screws of the ruling engines are lapped to the highest precision attainable, there are always residual errors in both threads and bearings that must be compensated to produce the highest quality gratings. The Mann engine is equipped with an automatic interferometer servo system that continually adjusts the grating carriage to the correct position as each groove is ruled. In effect, the servo system simulates a perfect screw.

Today's precision mechanical control choreographs substrate handling events between the stage and mask, while providing $\lambda/1024$ interferometry as part of a fast, real time, high precision, and accuracy positioning system. The work chamber-stage assembly interfaces with the beam delivery optic, beam detection assembly, electronics, and data path. This assembly provides the physical environment and mechanical foundation for the x-y precision stage, interferometer, vibration isolation system, mask leveling system, and workstation environmental chamber. Temperature control with vibration and electrical noise reduction ensures precision and repeatability in print quality. The work chamber provides an enclosed structure that supports vacuum (in the case of e-beam), thermal stability, and shielding against EMF/RFI and x-ray emissions.

The substrate positioning system interfaces stage, drive, and interferometer subsystems to receive a mask from the material handling system, locate the mask on a stage, and precisely position the mask in coordination with the position of the beam for printing. Mask handling has been the focus of decades of engineering invested to ensure low distortion of the mask during printing. For many years most systems used mask cassettes with three point clamping.[246] Advanced kinematic designs with no clamping can be used to minimize distortion.[247]

The accurate and precise positioning of the substrate surface relative to the optics mounting centerline and writing plane is affected by the metrology frame; work table with substrate holding hardware and stage assembly, including linear motor drives; stage position sensors; mask and stage height sensors; and leveling mechanisms. The cross section of a stage and metrology frame is shown in Fig. 3.49.

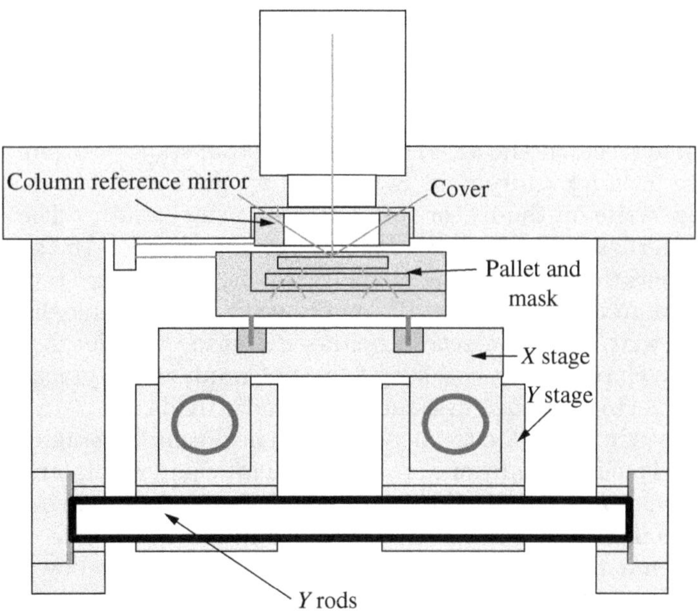

Figure 3.49 Cross section of the stage and metrology frame.

The vibration isolation system supports and levels the write chamber, isolating it from ground vibrations and environmental disturbances. Chamber movement caused by ground noise and surrounding disturbances is detrimental to the mask-writing process as it induces beam deflection, pallet movement, and/or mask slippage. An automated mask handling system must also control AC power distribution, a class 1 microenvironment, thermal and humidity control, chilled water supply and distribution, mask storage and alignment, vacuum control, GUI for material loading, material handling robots, and a SMIF interface.

Environmental control of temperature and ambient atmosphere is key to ensure quality mask printing. Self-contained minienvironments having class 1 particulate and 20-mC thermal control are now common. Reticle SMIF pod interfaces with N2 purging and amine filters to ensure compatibility with chemically amplified resists (CAR) is a necessity. Electron beam systems must accommodate high vacuum. For laser pattern generators air gauge sensing provides unique challenges for control and exposure.[179]

Pattern generators must control mask thermal expansion to ensure accurate overlay. Thermal loading of photomasks has received extensive study. Nonuniform heating of the photomask during multipass e-beam patterning has been studied.[248–250] Finite element heating models have characterized transient mask thermal distributions during simulated patterning.[251] Direct measurement using thermocouples and modeling has shown 70-K temperature rise is possible for 15-kV e-beams using 600-nA beam current.[252]

Stages. The stage performs a scanning motion in the x direction, while the beam prints in the y direction (for raster) or follows a trajectory within a major field (vector). The beam follows the actual stage position. A principal stage design requirement is imposed by the choice of continuously moving "write-on-the-fly" or step-and-scan print strategy. The former reduces overheads but requires faster real time control. In the 1970s pattern generators used rolling element stages, which were typically screw driven and could achieve 10-nm accuracy.[253] The stage did not move during writing. For electron beams the stage was located within the vacuum chamber, while the x-y drive mechanism was placed outside the chamber to minimize associated magnetic fields.[254]

Write-on-the-fly printing led to the need for high performance stages. The vibration produced by traditional rolling element components and friction lead screws led to the development at Bell Labs of air bearing stages driven by ball screws. Air bearings were maintained outside of the vacuum environment through the use of large flexible metal bellows. Screw-driven rolling element stages made of nonmagnetic materials were also used. The continuously moving stage concept was also adopted by Toshiba in the 1980s and Hitachi and Leica in the 1990s. A hydraulic stage with beryllium parts was developed to achieve smooth motion, high bearing stiffness, and low magnetic fields.[255] An *xy* stage using nonmagnetic ultrasonic linear motors reduced position fluctuations to a range of 8 nm for step and repeat electron beam writing with settling time was less than 270 ms for a 650-μm step.[256]

A recent stage innovation has been the vacuum air bearing stage, which was adapted to the Mebes eXara system. A cassetteless mask held on the vacuum air bearing stage is shown in Fig. 3.50. Air bearings in vacuum eliminated bellows allowing the use of linear motors. The need for ball screws or rolling elements of any kind was no longer necessary. A principal benefit was extremely low vibration.[257]

More recent work extends the air bearing in vacuum technology to simpler, more robust topologies reminiscent of technology developments for atmospheric stages used in wafer lithography.[258]

Interferometry and focus control. High-resolution displacement measuring interferometry is used to measure the distance between the desired and actual beam position.[259] A split-frequency laser beam is reflected off a mirror attached to the stage, and the beat frequency from the two lines is measured by high speed electronics, see Fig. 3.51.

Figure 3.50 Cassetteless mask held on a vacuum air bearing stage. (Courtesy Etec.)

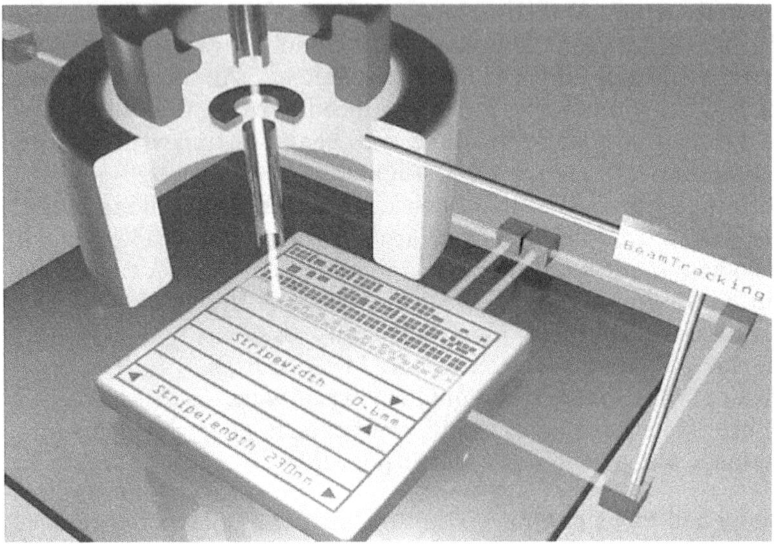

Figure 3.51 Schematic of beam tracking interferometry. (Courtesy Leica.)

When the stage moves, the beat frequency shifts according to the Doppler effect, and the stage position is calculated by integrating the beat counts. The difference between the actual and desired stage position is the following error, which can be fed back to beam scan correction electronics.[190,260] Beam scan correction can compensate for stage errors ~10 μm. In actual practice the open-loop following error can be kept to within ~1 μm beyond which the closed-loop control is via beam scan deflection.

Stage precision is often given in terms of a fraction of the laser's wavelength and the best stages now use $\lambda/1024 = 0.6$ nm. Even though the controller reports the stage location to this precision, the stage metrology accuracy, linearity, and repeatability at the nanometer scale is limited by alignment and path length errors, yaw, mirror bow, and phase mixing due to various birefringent effects. Environmental stability is a major factor in the repeatability and accuracy of measurement. It is difficult to obtain a measurement accuracy of 10^{-7} when working in air. This limitation does not impact vacuum based e-beam systems when properly implemented. These nonlinearities typically limit the absolute placement accuracy to ~2 nm over 100 mm of stage travel. Velocity of light in air must be addressed in laser systems, which are not under vacuum. Multiwavelength interferometry, short- and long-term frequency laser stabilization, optical feedback, and calibration of phase measuring electronics using subfringe interpolation are some of the possible methods that can be used for further improvement.[261]

Analysis of multiple points on the stage mirror allows the measurement of *x*, *y*, and rotation about *z* (yaw). Yaw correction was introduced for step and scan VSB.[262] A planar stage concept was developed to minimize displacement and yaw errors, which had been observed with designs using the conventional two-level principle.[263] The planar stage approach has also been used with Leica's SB350MW, which has a stage guided by ceramic ball linear bearings and mechanical coupling of the drives to the stage. Interferometric position control with a resolution better than 1 nm and a beam tracking system allow the write-on-the-fly exposure with stage speeds up to 30 mm/s.[147]

Stage yaw correction is used to minimize stitching errors, which are proportional to field size.[262,264] Digital servo control with tailored trajectories can be introduced to provide low-impact jerks during continuous stage motion, which ensure high accuracy.[263,265]

Laser PGs have some unique problems associated with focus and atmospheric control. The ALTA system uses a hot wire anemometer focus system, which drives air through the objective lens assembly onto the mask. (See Fig. 3.52.) The interaction of the gas and resist can create unique lithographic signatures, proximity, and process effects that require special focus and atmospheric control.[179]

Mask height offset combined with beam landing errors (nontelecentricity) leads to field stitching errors. To minimize these errors a mask height sensor indicates the height of the substrate and calibration grid relative to the nominal writing plane. Mask position and height data

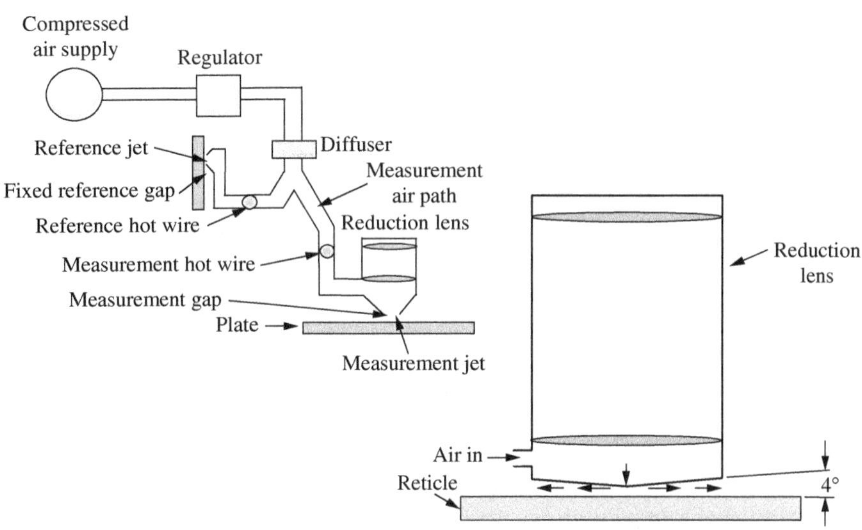

Figure 3.52 ALTA humidity compensated, hot wire anenometer focus system.

support focus calibration, which can be viewed as a two-step procedure. The first calibrates the beam delivery subsystem to the stage grid, which defines the column focal length and enables deflection calibrations. The stage grid also acts as the fiducial reference location for positioning in the (x,y) plane, which may occur during periodic registration.

The second step is to align the substrate to the grid position. The substrate is physically moved in the z axis with the aid of a height sensor to be coplanar with the stage grid. This operation and any subsequent function must not disturb the previously defined working distance for the beam delivery subsystem. Further refinements can account for stage grid tilt.

Calibration research. Scan calibration has become increasingly complex. Deflection alignment, second-order distortion, scan aberrations, and time-dependent thermal effects must be handled. Automation of column setups and dose calibrations with autofocusing, autocentering, magnification and astigmatism correction, as well as periodic reregistration to correct for e-beam drift is now common practice.

Spatial phase locking methods are currently being studied to further improve e-beam positional accuracy. This is a promising technique to achieve sub-10-nm stage control. Closed-loop control of electron beam position is done by monitoring a substrate fiducial grid. 1.3-nm positional accuracy under prototype conditions has been achieved.[266]

Scanning beam interference lithography to pattern grids is another method to achieve nanometer level calibration accuracy via the use of grating-based metrology.[267] Whether and when mask makers are willing to adopt the added complexity of a grid over the writing surface remains an open question.

EUV masks. For mask PG tools the emphasis for EUV mask highlights mask blank quality and handling. Surface flatness, ultralow and repeatable thermal-mechanical distortion, and material handling compatible with inspection and metrology tools sets impose special demands on patterning equipment. The present consensus is that all tools (writing, metrology, exposure) would need to use an ultraflat electrostatic chuck. For the writing tool, this means that backside nonflatness will affect the front-side nonflatness, and the tool must have the means to measure and compensate for these errors so that the final printed image on wafer is exact. The placement accuracy of EUVL exposure tools is extremely sensitive to mask out-of-plane deviations.

The reticle/chuck characteristics and clamping response must be evaluated to meet the 50-nm peak-to-valley flatness requirements. Finite element analysis is providing guidance to how the error budget will be shared between substrate flatness and chuck characteristics.[268]

Height sensing will be critical and present state-of-the-art accuracy of ~0.1 μm will likely be inadequate. It will also likely be a requirement that all reticles have a conductive layer on the backside. E-beam mask writers will need to have a conductive accessible on the top. Although the multilayer stack is conductive, it is not accessible from the top (and some buffer layers under consideration may be nonconductive). A secondary request may be that the system has dual capability—e-chucking for EUVL masks, and conventional fixturing for transmission masks (i.e., –248, 193, 157 nm).

Resist heating of EUV masks has been simulated and experimentally studied at 100 kV. Evidence was inconclusive as to the effect on critical dimensions, but the concern remains due to low thermal conductivity of the substrate and dependence on absorber properties.[269]

3.3.3 Proximity effect correction

Proximity effect correction (PEC) capability is evaluated by lithographic quality and data processing capability. Proximity effects result from unwanted exposure of neighboring pixels. The effect can be approximated as a resist thickness independent, double-gaussian energy distribution.[270] For electrons the cause is scattering of incident (forward), backscattered, and secondary electrons.[271] This scattering depends on the energy of the electrons, the effective atomic number of the substrate materials, and the thickness of the resist and substrate.[272] The distribution of energy due to backscattered electrons is generally approximated as a gaussian—$f(r) = \exp(-r^2/\beta^2)$, where $\beta \approx 9.5$ μm (50 keV on quartz), corresponding to a full width at half maximum (FWHM) of 14 μm. β is proportional to the range of the backscattered electrons.

Proximity effect can be represented in terms of the modulation transfer function.[273] The nominal dose is determined by the grating (equal line/space) pattern, see Fig. 3.53. Dose is increased for isolated lines and reduced for isolated spaces. The required dose range is $(1 + 2\eta \approx 2)$ for a typical value of η, which is the ratio of energy absorbed in the resist layer due to backscattered electrons to that due to incident electrons. Its precise value and function dependence on beam voltage have been debated for years. It is empirically used to refine to PEC lithography quality. It also is a factor in setting the maximum dosage required for exposing isolated features.

Table 3.17 shows some of the key electron beam scattering parameters. The final effective spatial variance σ_f comprises two factors. One is the aerial image of the e-beam optic d_S. The second is the forward scattering of the electrons in the imaging layer α_f. This latter term was experimentally determined to exhibit a power law relationship with resist thickness.[274,275] Solutions of the forward scattering effect are not

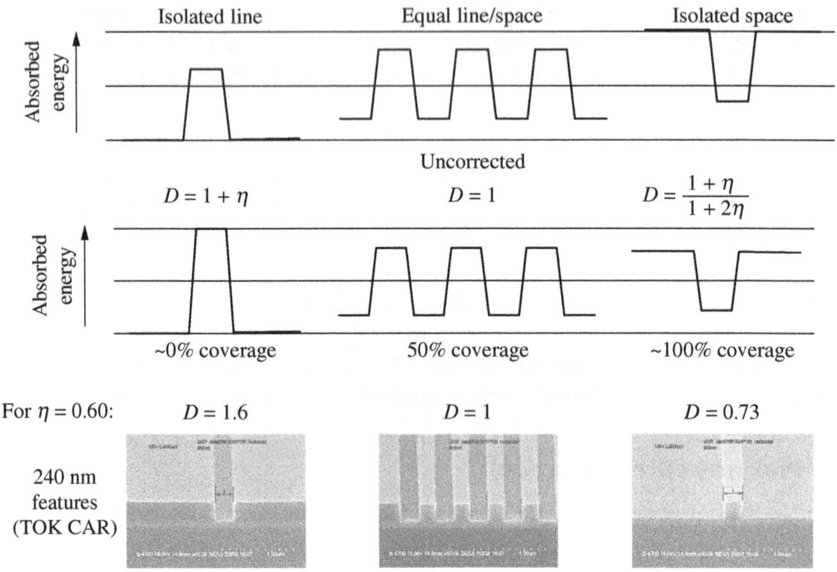

Figure 3.53 Schematic of proximity effect on absorbed energy profiles.

generally solvable as the closed form solutions become unstable as the feature size approaches the range of this parameter.[276] The backscatter values have been confirmed but the forward scattering range is 1.5 times the 50-kV values stated in.[83]

The relationship between beam voltage and scattering parameters has been described in terms of the point exposure distribution.[277] This work was extended to 50 to 100 kV for sub-100-nm features.[278]

The portion of the proximity effect, which has proven correctable, is due to backscattered electrons depositing energy in the resist. A rule of thumb is if the feature size exceeds the forward scatter radius by a factor of three, forward scattering will not be apparent in the lithography.

TABLE 3.17 Energy-Dependent Electron Scattering Parameters

Beam energy [keV]	α_f [nm]	d_s [nm]	σ_f [nm]	σ_b [μm]	η_e
5	24	35	42	0.2	0.55
10	7	35	36	0.6	0.55
20	2	35	35	2.0	0.55
50	0.1	35	35	9.5	0.55

α_f [mm] values are for a 50-nm thick imaging layer.[274]
d_s [mm] = e-beam size (FWHM).
$\sigma f_2 = d_s^2 + \alpha f^2$.

Various PEC approaches have been developed. A classification of most published PEC algorithms has been given.[279] Linear methods lead to exposure even in nonpatterned areas. Nonlinear methods can be further split into those that equalize absorbed energy or some other cost function. Dose correction methods originated with Ref. 280 and pattern bias correction with Ref. 281. The role of secondary electrons can be found in Refs. 229 and 282 to 284.

GHOST is a proximity correction technique which does not suffer from the complexity of dose-dependent geometry compensation. The GHOST correction scheme is based on equalizing the backscattered dose at all points within the pattern. This is achieved by exposing the reverse tone of the pattern with a broad, defocused beam on the order of the backscatter range and a reduced incident correction dose.[284] This method was shown suitable for gaussian and shaped beam systems.[285] The GHOST exposure does affect resist process latitude with thinning of the resist and reduction of sidewall angle.[286]

The relative simplicity of this approach was adopted throughout the 1990s by raster systems. GHOST PEC using 10-kV GHOST for 0.4 μm was shown feasible.[287] The growing complexity of pattern processing by 2000 began to overtake the relative simplicity of the GHOST method, which required in some cases, reverse tone exposures to be combined with primary exposure on multiple chips within the mask pattern. Subsequently the lithographic capability was overcome by reduced dose sensitivity resulting from the reverse tone exposure, and this method was replaced by real time dose modulated PEC.

Present means for correcting these effects require a significant off-line pattern preparation even for those PEC methods touted as running in real time, i.e., without noticeable overhead during exposure. In general these off-line steps are greater for vector scan architectures, since boolean operations cannot be readily transferred to internal digital electronics. Added complexity results from pattern refracture requirements per phase for vector processing.

VSB off-line proximity effect correction has been described for 4-Mbit DRAM LSI, 0.5-μm features on wafer using the EX-7 50-kV system.[288] The use of pattern area density maps for fast proximity correction was introduced in the 64-Mbit DRAM era.[289]

PEC processing was implemented in hardware on Toshiba's EX-11 platform, which fractured input data to flashes using the representative figure method,[290–292] see Fig. 3.54. Data conversion, or fracturing, uses an array compacted, preprocessed interface file. A 4-Gbyte file size with 100:1 array compaction would take about 4 min on a 100-MFLOP workstation. For a four-phase print strategy where each phase must be fractured would be consistent with 16-min processing times for such a mask.

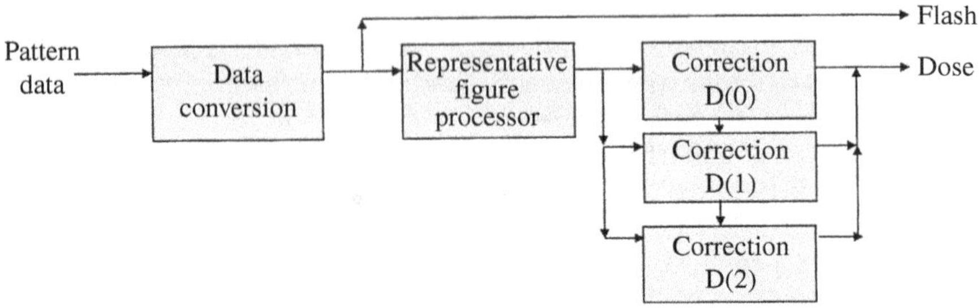

Figure 3.54 EX-11 proximity correction.

The representative figure processor averages the flashes into larger representative figures to calculate backscatter electron effects on reasonable grid sizes. The correction blocks represent the third-order convolution operations of the backscatter kernel with the representative figures. The EX-11 platform used a *finite impulse response* (FIR) method, which implements bandpass filtering. The complexity of the convolution using FIR is proportional to $M \times N$, where N is the number of representative figure pixels and M is the width of the scattering kernel.

Using a 1-μm mesh, the width of the scattering kernel M is about 50 figure pixels, which accounts for slightly more than the backscatter diameter at 50 kV. Commercially available 1999-version FIR chips could process ~30 Mpixels/s or an 8-pixel scattering function and 8 bits/pixel. Based on this information it is reasonable to conclude that the required 50-pixel scattering kernel could be implemented with 6 FIR chips cascaded together. Further simplifications can be made in separating the scattering kernel into two one-dimensional functions, assuming a gaussian scattering distribution. With these assumptions, a correction time of about 30 to 120 min can be estimated for 180-nm-era mask patterns, which was sufficient to be termed "run time" at this node. The data frame approach for RTPEC adapted for raster shaped beam is shown in Fig. 3.55.

Real time PEC for 50-kV backscatter correction is possible with no preprocessing of the data in a raster architecture. The data path clock can run at 100 MHz for both gaussian and shaped beams. Data are first compressed into coarse pixels, each representing 16 × 16 gray pixels. After the calculation, individual flash doses are recovered. A data precision of 256 dose levels can be achieved.[293]

No compression of course pixels is possible when printing in the 5- to 10-kV range as the backscatter range is 0.2 to 0.6 μm. The present use of compression ratios ~100 used at 50 kV would have to be compensated by other data path bandwidth or channel capacity enhancements as

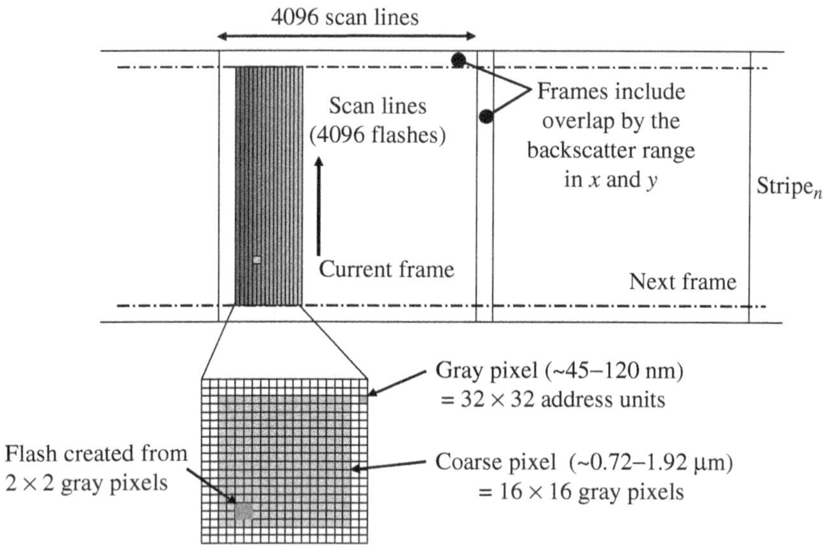

Figure 3.55 Data frame approach for RTPEC adapted for raster shaped beam.

the size of the rasterized print pixel would again be on the order of the correction element.

Process errors are now being bundled with PEC methods. CD linearity improvements have been noted by jointly compensating for proximity and etch loading effects in the scattering kernel. Adjustment of the standard two-gaussian model parameters within commercially available OPC tools was shown feasible as a means of extending the lithographic resolution.[294]

3.3.4 Laser process and proximity effects

Laser proximity effects have been compensated by adjusting focus humidity control[179] and the mask design. [295]Standing waves due to mask substrate reflectivity affect CD control. Substrate reflectivity below 1 percent is desirable for good resist profiles. Various substrates—AR8, NTAR7, and NTAR5—with different reflectivity, 10 to 20 percent, have been tested to reduce standing waves in the resist.[64] However, even exposures of NTAR5, which has the lowest reflectivity in the DUV region, result in significant standing waves.[181]

3.3.5 Long range e-beam fogging effect

Fogging is the result of long ~μm range electron exposure resulting from multiple reflections between the bottom of the electron column and the substrate. Fogging correction is based on a combination of passive, electron-absorbing materials such as a honey-combed grid and active pattern precompensation.[296–298]

Pattern dependent precompensation depends on two factors—a pattern density distribution map and a fogging kernel. The pattern density distribution can be obtained from rasterized data. This approach removes all figure data overlap and handles the issue of positive- versus negative-tone exposures. It yields a map of pattern density distribution down to a cell size of 1 pixel.

The fogging corrections may be additive, subtractive, or multiplicative to the PEC corrections. While the PEC and fogging correction range is orders of magnitude different the same gaussian kernel convolved with pattern density can be used to determine the dose correction. For raster-PEC and fogging corrections, dose is manipulated by changing the gray level values of the affected pixels.

The range difference (cf. 8.5 versus 6000 μm) between backscattering and fogging scattering makes it prohibitively costly to correct fogging simultaneously with backscattering. The pattern density can be computed on a coarse scale and convolved with the fogging scattering kernel ($\sigma \approx 6000$ μm). The combined PEC and *fogging effect correction* (FEC) can be simulated and various implementation methods have been developed. Fogging and loading effects have been corrected by using a second exposure with a defocused beam to reduce global CD errors from 19 to 8 nm (3σ).[299]

3.3.6 Resist heating and charging

The transition to 50-kV e-beam voltage made resist heating a possible error source for critical dimensions. Chrome-on-glass mask resist heating effects were studied at 15, 20, 30,[300] 25, and 50 kV[301] and at 75 kV[302] for VSB e-beam lithography. The resist heating effect is caused by the change in resist dissolution speed by the temperature rise in the resist.[303] VSB optics can deliver 40 A/cm^2 or more, but are typically kept at or below 20 A/cm^2 to avoid this effect.[87] Resist heating could produce as much as a 22-nm error in single-phase printing of a non-CAR. Using four-phase printing this error was reduced to 6 nm. Single-pass error was 8 nm using a CAR.[304] From this information one can infer that 50 kV with four-phase CAR printing can reduce the CD error to 2 nm.

Pattern placement error due to workpiece charging is a significant contribution to total position accuracy. Pattern distortion after a particular exposure sequence has been measured.[305,306] Electron beam resists are insulating polymers. Trapped charges change the surface potential and deflect incoming electrons and lead to beam placement error. The surface potential of the resist can be positive or negative depending on the resist thickness and electron energy.[307] The time dependent charge and discharge mechanisms, which involve electron-beam-induced conductivity, have been modeled and directly measured with a Kelvin probe.[308] Liu's observation of positive surface potential at 10 and 20 keV was experimentally confirmed using a focused beam under a variety of conditions. The mechanism for this phenomenon was subsequently explained in terms of beam-induced conductivity, which resulted from internal secondary electron emission.[309] Thinner resist films and higher beam energies lead to more positive surfaces. The observations indicate that there exists a resist thickness for which the charging is essentially zero.

3.3.7 Resist processes

Resist process development also evolved over the course of time to support the corresponding need for high contrast and stringent CD control. While PMMA might be considered the resist prototype for e-beam lithography,[310,311] PBS was the progenitor of e-beam resists for mask making. The original paper described the synthesis and mechanism.[22] Early R&D work with PMMA was supplanted with PBS as it provided high sensitivity to e-beam exposure.[312]

PBS is in a class of polymers that do not cross-link but undergo chain scission. The fragments of the scission process increase the solubility of the resist. PMMA also does not cross-link, but it does recombine and undergoes other competing reactions that do not create a differential solubility.

PBS has had a remarkable long-lived usefulness within the field as evidenced by a series of work that extends over three decades. There was the early phase.[312–314] This was followed by optimization of various aspects of the resist development process. These optimization studies included the overall manufacturing process,[315] resist profiles,[316] CD control,[317] masks for 256-Mbit DRAM,[318] modified development process for improved CD uniformity,[319] and thin resist.[320]

Other resists were also developed in the 1980s. EBR-9 was a positive resist with improved sensitivity compared to PMMA with good resist sidewall angle and critical dimension control. EBR 9 is an analog of PMMA with chlorine substitutions on the chain which act as electron withdrawers, increasing the probability of the chain scission reaction. ZEP also works this way. Early references include Refs. 321 and 322.

i-line resists. These are novolac resins with a diazonapthoquinone sensitizer. On exposure to i-line radiation, they rearrange to a reactive intermediate, which in turn reacts with water to form a carboxylic acid that is soluble in base.[323] There are many varieties where the type of resin, the concentration of the sensitizer, and other constituents are varied. On e-beam exposure they generally follow the same mechanism, but many other reactions also occur including cross-linking.[324] All novalac-based resists cross-link and become negative after a certain dose. Interrupted development was shown to enhance the e-beam and optical contrast of a diazonapthoquinone, AZ-type resist for 250-nm lithography.[325]

The process of record for mask fabrication on CORE pattern generators was the OCG-895i positive resist.[326] The i-line resist KTI 895i was also used with the CORE and ALTA.[327] It is unique in that it uses a chemical bond to attach the diazonapthoquinone sensitizer to the resist resin. This prevented CD problems due to diffusion of the sensitizer over time of exposure. This was a problem for the laser tools, which expose in air. Unfortunately it cross-linked at about 8.0 μC/cm^2 because of the bond making it is useless for electron beam patterning.[328] I-line resists such as AZ 5206 and SPR 700 were characterized as e-beam resists,[329] but not widely adopted for use as in mask making.

Lithography simulation and line-edge optimization were used to improve process latitude and resolution for i-line optical pattern generators, specifically the ALTA 3500. The use of high resolution resists, high exposure energies, lower developer concentration, long development times, and a postexposure bake improved the process latitude. These process improvements led to better sidewall angles and less CD sensitivity to dose variations. These benefits also led to reduced etch bias, which in turn enabled higher quality printing closer to the isofocal dose.[330]

TOK IP3600 i-line resist was used to evaluate the Micronic Omega6000 tool in 1999. This i-line resist had also been used for ALTA exposures by mask makers and was an indication that a single resist was desired to simplify process complexity and reduce cost. The spectral absorbance curves of IP3600 show that the bleachable and unbleachable terms of the absorbance at the 413-nm Omega wavelength are slightly lower than the values for ALTA at 364 nm. However, the resist functions well at 413 nm and there was some speculation that the lower absorbance may actually provide a benefit in terms of steeper sidewalls and improved resolution. The operating dose range is between 80 and 110 mJ/cm^2. At the isofocal dose, the CD feature size sensitivity is 3 nm per 1 percent dose change.

A positive, dry-etchable i-line photoresist THMR-M100 was used for printing 400-nm features with an ALTA 3500 and achieved a dose latitude of 2.9 nm/mJ/cm^2.[331]

DUV resists. A prototype DUV platform process based on Clariant AZ DX1100P photoresist with AZ Aquatar applied as a stabilizing top coat has been developed in combination with ALTA 4300. The Aquatar top coat was applied to the DX1100 resist layer to improve mean CD control. As was found with e-beam, the topcoat creates defects during the resist development process, which have to date precluded widespread use in manufacturing. ALTA products now use FEP-171 without a topcoat to avoid this problem. PEK-130 resist is also used in a production environment on the ALTA 4300.[64]

Non-i-line resists. 50-kV e-beam resist process development proceeded along two paths—nonchemically amplified and chemically amplified resists. Raster systems are not as sensitive to dose limitations and could exploit non-i-line resists for their contrast and ease of use.

ZEP is another high molecular weight polymer that works by chain scission. It is solvent developed just like PBS. Its sensitivity is ~50 μC/cm^2 at 50 keV or about 8 to 10 μC/cm^2 at 10 keV. It contains aromatic (benzene) rings so as to be dry etchable. It is a copolymer. One part gives it dry etch resistance and the other undergoes chain scission to improve solubility. It worked quite well with good resolution, (200 nm and smaller) and uniformity ~10 nm.[332] However, CARs proved better for dry etch and are far more sensitive.

Chemically amplified resists. Vector scan throughput is critically dependent on resist sensitivity, in part due to heating effects, which drove the need for a fast chemically amplified resist (CAR). Resist requirements are set in the 5 to 10 μC/cm^2 range to support VSB print strategies and take advantage of existing positive- and negative-tone resists.

The invention of CARs[333] has been one of the key enabling technologies for KrF (248-nm) manufacturing.[334] The use of CARs in photomask making has been driven by the need for high sensitivity of VSB systems to improve exposure rates. The development of CARs as a mask resist is another example of where mask makers and equipment suppliers adapt processes and technology from the wafer lithography community, which has greater resources and capability to perform research and development.

CARs work by formation on exposure of a strong acid, which hydrolyzes a protecting group on the resin for positive resists. For negative resists the acid cross-links the resin. In both cases the acid is a catalyst where the acid consumed in the reaction is regenerated after the hydrolysis or cross-link. A *postexposure bake* (PEB) is used to drive the main solubility-changing polymer reaction which is catalytically enhanced. Early versions used a "naked" hydronium ion (H^+). This had problems with neutralization and *T-topping*. It was later replaced by a large molecule Lewis acid producer which diffuses slower improving T-topping and reducing diffusion effects.

The chemical reaction and diffusion transport during PEB shows that the solubility-changing reaction is not localized.[331] The resist blurring effect can be managed by trading off resist sensitivity for resolution by reducing the catalytic amplification factor. The diffusion of the catalytic acid is a factor determining the resolution limit of the resist.[335,336]

These resists have high contrast, show excellent dry etch resistance, and are very sensitive. EP 030, a new experimental resist from TOK, can resolve features at 64 nm. EN 24, a negative CAR from TOK, shows superior performance to NEB 22.

State-of-the-art CAR requirements are driving to 4 μC/cm^2 sensitivity at 50 kV. Controlling the photoactive compound chemistry while maintaining good lithography quality remains a challenge. The positive-tone resist REAP-200 and its congeners have been extensively developed.[337] 3-nm MTT has been shown feasible printing with REAP-200 at 8 to 11 μC/cm^2 at 50 kV.[338] REAP-200 resolution is shown in Fig. 3.56.

KRS was developed for internal use at IBM and characterized as having high performance.[339] The use of EBR900 in e-beam and laser beam lithographies for photomask production was studied.[340] By 1998 EBR900M1, a non-CAR positive-tone resist was compared to CARs. CARs were found to have higher contrast when exposed by 50-kV e-beam.[341,342]

Since 2000 CARs have become adopted with the widespread use of VSB. The acceleration of the critical dimension control and high resolution requirements has led to a range of resists being developed. These include FEP171, NEB-22, the TOK family of REAP 122, 200, and EP030. Positive-tone chemically amplified resists CAP209, EP012M (TOK), KRS-XE (JSR), and FEP171 have also been evaluated.[343] A postexposure effect of 5 nm over 12 h in vacuum was measured along with line-edge roughness for FEP171 for mask making. Currently FEP 171 is being used as part of a 90-nm resist process.[344]

CARs were developed to support VSB exposure rate and total current requirements. Laser PGs now deliver or characterize CARs developed for e-beam to minimize switching costs.

Negative resists. Negative-tone resists are important for vector scan printing where pattern density is a key factor in tool throughput.

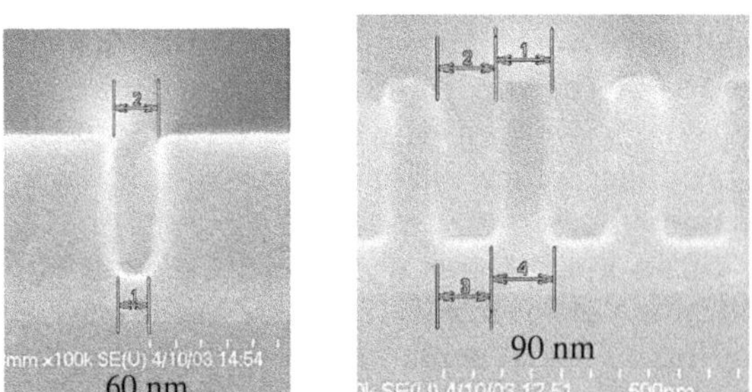

Figure 3.56 REAP-200 resolution in the 60- to 90-nm range with ~10 μC/cm^2 sensitivity at 50 kV.

Negative CARs are of interest as they offer the potential for high resolution,[345] moderate improvements ~20 percent in CD uniformity, and, when employed with VSB pattern generators, can reduce writing times significantly for dense patterns such as DRAM masks.[346]

An IMS negative resist study of FEN270, EN-024M, JSR MES-EN1G, NEB 22 revealed EN-024M as the best negative resist candidate. A negative CAR EN24 (TOK) having 60-nm resolution is shown in Fig. 3.57. FEN270 has resist profile issues and NEB-22 had a high PEB sensitivity of 9 nm/C. Negative CARs: Shipley UVN-2 and Sumitomo NEB-22A5. The latter was found superior for 100 nm l/s array with good PEB and CD control. NEB-22, a negative resist, has been superseded by EN-24. Some maskshops have printed on NEB22 with ALTA to improve CD uniformity compared to results obtained with positive-tone FEP processing. NEB22 has high sensitivity and contrast and CD dependence on postexposure bake is low at least in comparison to SAL 601.[342] Postexposure bake latitude is reduced when using negative CARs.[346,347]

Developer technologies. Various developer technologies are being created to support the tight process control margins. Many of these are adopted from wafer lithography practices. Low impact nozzle dispenser methods and dissolution rate monitoring are effective at maintaining tight CD control.[348]

Puddle development. The advantage of a puddle process is that it slows the development of the resist as the concentration of the dissolved polymer increases in the puddle solvent. This results in improved LER, contrast, resolution, and CD uniformity. The higher concentrations generally make the dissolution more likely to dissolve smaller pieces of the resist. This results in a higher contrast, better sidewalls, and less sloping along the edges. Sidewall angles of 80° using the puddle process compared to 60° or less with spray developers.

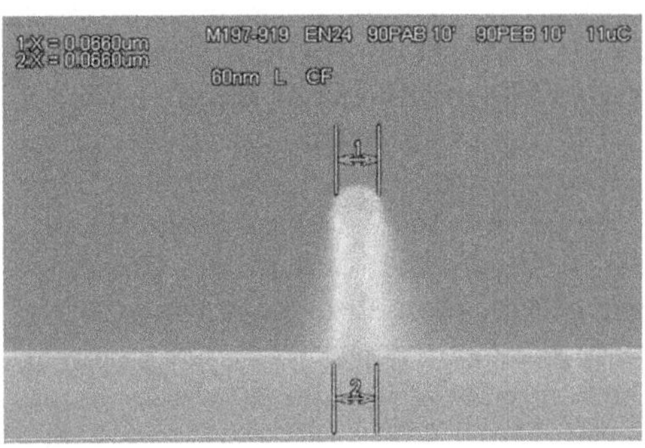

Figure 3.57 A negative CAR EN24 (TOK) showing 60-nm resolution. (Courtesy R. Dean.)

This method has been used with various resists and widely adopted in the industry. The effect of puddle process and PBS has been studied with virtual addressing[349] and integration with MEBES IV. [350] Puddle developers were also used with ZEP 7000.[351]

The 1980s and 1990s saw the puddle develop processes with new solvents to increase resist contrast. Unfortunately this approach was not applicable to dry etch.

In contrast, CARs have been found to work better with a spray develop method, which is correlated with their very high contrast.

Low k_1 lithography and associated mask development processes comprised coat, expose, bake, and development steps require greater monitoring and control to support ever-tighter process control margins. Each of these process steps are subject to random variables, which exhibit statistical effects on the overall objective of tight CD control. For a comprehensive overview of resist process technology the reader is directed to early review articles and the contemporary literature.[352–356]

3.4 PG Trends and Future Prospects

What can we learn from past history with regard to speculating on what future trends may occur with pattern generation? The birth of pattern generation in the 1960s began by combining scanning electron microscope technology with simple pattern generator digital and analog electronics modules. The 1970s saw the initiation of the major system architectures, raster at Bell Labs and initially Toshiba, and vector at IBM, Hitachi, and JEOL. The 1980s established the commercial success of MEBES e-beam raster patterning and also the advent of laser patterning at Ateq and Micronic. The introduction of 4× reduction projection lithography brought about the "mask makers vacation" in the latter half of the 1980s, which relaxed the pattern quality requirements need to print masks. The commoditization of masks augmented the economic difficulties of the equipment makers and laid the groundwork for cost-of-ownership trends for volume manufacturing, which arose in the boom years of the 1990s. The late 1990s brought the need for greater process control and resolution, which drove the adoption of 50-kV e-beam patterning for pilot production. This also began the era of rapid commercialization of phase-shift mask and optical proximity correction as subwavelength lithography was integrated in IC manufacturing. This trend drove even greater needs for enhanced patterning resolution. These technology drivers occurred with a downturn in the business cycle. As a result 50-kV vector patterning tools dominate the market today. 50-kV raster tools were late to market and the laser PGs lack of resolution relegated their diminished share to second-level PSM printing and some aspects of volume production. The pendulum appears to be swinging back to greater use of laser technology in 2005 as mask manufacturing demands for 90 nm grow along with leading-edge development at 65 nm, particularly for metal and via layers.

Given the varied history of patterning and the multitude of technical trends, evolution of requirements, and business imperatives future predictions are of course difficult. Here we venture to describe three scenarios, which provide some outline of possible courses of action, which the market and technology may reveal as we progress in the twenty first century.

The first scenario is one where there is little or no development beyond the existing architectures. Major tool development programs are extremely expensive ~$100 M. The IC industry is overall facing commoditization on one hand and nonlinear difficulties with materials and processes at each successive node below 90 nm. To exacerbate matters there are still numerous PG toolmakers in the market. Even with consolidation, the resources for PG development are under great pressure. Not to mention that the customers, the mask shops, have marginal business models that barely support five to 10 percent R&D themselves.

If this first scenario develops, then what can be done? The focus may turn to serviceability with improved ease of calibration and operation. Improved tool uptime and mask yield could provide value in a service-based business model. Engineering and technical resources might then develop improved feedback and control on machine operation, calibration methods to improve print quality and ability to pinpoint tool-related root causes of mask defect creation.

Possibly the single largest trend, *design for manufacturing* (DFM), could drive this scenario. Front-end data path development, which provides the product data interface to the CAD design and mask shop job deck groups, is a prime area for relatively low cost development with potential leverage of ever-greater process integration occurring at all levels of the IC manufacturing chain. The second area that can support DFM is via greater feedback and control of the system, during both calibration and patterning. Eliminating sources of mask defects that occur through variances in machine operation will continue to be of value, as mask inspection costs continue to be the major CoO contributor along with patterning itself.

Defects can be viewed broadly as a variance outside of specified parameters. Hence tighter calibration with robust operation should reduce the variances in print quality, which is a value add to the overall manufacturing process.

The second scenario is an incremental step in the first with some development to extend PG capability, but also with the same basic tool architectures, which exist today. Incremental advancements, i.e., 10 percent in terms of print quality improvements via resolution enhancements of the optics, data manipulation, and print exposure rates via a combination of electronics advances in various analog and digital scan drivers and modulation components are certainly possible. Their rate of adoption is likely gated by the magnitude of development dollars allocated.

The third scenario is that there will be yet another major development of mask patterning technology. At present it would appear that the source of capital for such a risky venture lies with those organizations which target the elimination of masks altogether. True industrial innovation is revealed when major process steps in manufacturing are completely eliminated. Certainly maskless patterning has long promised this revolution.

Of course this promise has long been in various business plans of both e-beam and laser proponents. The potential reward along with the continued spiraling of mask costs makes this scenario a possibility. A major problem is that the required throughputs of even one wafer per hour are still one to two orders of magnitude more difficult than corresponding mask requirements at the same IC node. One might argue that certain lithographic quality parameters are emphasized, but tight CD control demands for wafer as well as mask, makes this point secondary to the huge advance needed in throughput.

We have explored in summary fashion how each of the major architectures may be extended. Vector e-beam currently dominates the mask patterning market, yet it is very difficult to posit an order of magnitude increase in throughput, both from difficulties in the needed exposure rate as well as ability to handle the enormous data overheads in run time.

One can envision laser raster PGs extending their throughput both in terms of exposure rate (via increase in number of beams) as well as data handling capabilities due to relatively straightforward technical enhancements to rasterization. Further ability to achieve improvements in optical resolution remains a major challenge. In spite of adoption of partial coherent imaging in PGs, the needed resolution may come with a major advance in some type of immersion optics, which at present is only conceived within the university environment.

This leaves one other architecture that combines the speed of multibeam, which exists with laser PGs, rasterization to handle the data volume explosion, and resolution of e-beams. This is the raster multi-e-beam (RMB) concept. A breadboard system was developed at Etec with the support of DARPA. Unfortunately this program has been put on the shelf. A major breakthrough came with the development of a robust photocathode, which combined with a beam delivery optic was shown to achieve the requisite resolution under 32-electron-beam operation. The ability to use low voltage optics with thin layer imaging can overcome total exposure current requirements, which otherwise would limit resolution in the 10 Gpixels/s, 32-nm era.[221]

It may be that sufficient motivation will drive some other group to resume the development of the RMB architecture and address the limits of fast, high current and resolution multielectron beams. Such a path must promise extendibility from mask patterning to direct write on wafer, to interest the venture capitalists or holders of large R&D budgets in vertically integrated IC houses, or possibly as a consortium, which can share the risk and possible rewards of bringing a revolutionary PG architecture to market. If such a scenario is played out, it is highly likely that the first incarnation will be as a mask PG tool. This would allow validation of the system concept and provide early revenue to fund subsequent development, because the probability of success is much greater than the technical and organizational barriers confronting the entry of a new lithography technology within the IC fab.

Acknowledgments

I thank the following contributors for improving the quality of this chapter:

Saul Arnold, Bob Dean (resist), Bill DeVore (Mebes), Tim Groves (Leica), Richard Lozes (PEC), Bob Naber (process integration), Tom Newman (daughter masks, Micronic), Hans Pfeiffer (shaped beam technology), Jim Schoeffel (Ateq), R.L. Smith (market), David Trost (precision mechanical, optics), Mike White (Alta), Seunghune Yang (architecture), Ming Yu (shot noise).

It is also a pleasure to acknowledge the many contributors, colleagues, and mentors who have created and supported the development of the pattern generation technology and business: Frank Abboud, Paul C. Allen, KiHo Baik, Sergey Babin, Chen Baoqin, Charlie Biechler, Mark Berardo, Neil Berglund, George Bertucelli, John Bruning, Peter Buck, Allen Carroll, Jan Chabala, Philip Chang, Parkson Chen, Xiaolan Chen, S.W. Choi, Dave Colby, Damon Cole, Steve Cooper, Steve Coyle, Richard Crandall, Kevin Cummings, Bob Dean, Dan DeBra, Bill DeVore, Steve Dunbrack, Bill Eckes, Carroll Fencil, Jorge Freyer, Scott Gibbs, Brian Grenon, Tim Groves, Hideaki Hamada, Chris Hamaker, Naoya Hayashi, Uli Hofmann, Fritz Hohn, Doug Holmgren, Glen Howard, Sam Howells, Robert Innes, Matt Jolley, Franklin Kalk, Kanemaru-san, Don Kania, Hiroichi Kawahira, Dieter Kern, Asher Klatchko, Kotani-san, Igor Kuzmin, Peter Levin, In Sang Lim, Joe Lin, Richard Lozes, Juan Maldonado, Liu Ming, Chuck Minihan, Eric Munro, Bob Naber, Tom Newman, Tony Novembre, Gerry Owen, Dave Patterson, Henry Pearce-Percy, Fabian Pease, Hans Pfeiffer, John Poreda, Richard Prior, Romin Puri, Jack Radley, Rick Raymond, Steve Rishton, Paul Rissman,

Al Sagle, Norio Saitou, Chuck Sauer, Mike Scheinfein, Jim Schoeffel, Bob Sills, Sass Somekh, R. L. Smith, Jun Min Sohn, Chris Spence, Keith Standiford, John Suzuki, Lyn Swanson, Hitoshi Takemura, Tadahiro Takigawa, Zoilo Tan, Tanaka-san, Al Tasch, Robin Teitzel, Tim Thomas, David Trost, Dave Tuggle, Michael Ungureit, Gil Varnell, Lee Veneklasen, Raman Viswanathan, Al Wagner, Paul Warkentin, Mike White, John Wiesner, Grant Willson, Alan Wilson, Seunghune Yang, Lydia Young, Ming Yu.

> The stone which the builders refused is become the headstone of the corner (*Psalms 118:22*).

To the editors: my appreciation for patience and perseverance.

The following are trademarks or registered trademarks of companies in the United States and other countries:

MEBES, CORE—Registered trademark

ALTA—Trademark

References

1. Stowsky, J., "The weakest link: semiconductor production equipment, linkages, and the limits to international trade," *Berkeley Roundtable on the International Economy (BRIE),* University of California Berkeley, BRIE Working Paper #27, Aug. 1987.
2. Buck, D., and K. Shoulders, "An approach to microminiature printed systems," *Proceedings of the Eastern Joint Computer Conference*, Philadelphia, American Inst. Elect. Engineers, Pa. 3–5 Dec. 1958.
3. Feynman, R., "There's Plenty of Room at the Bottom," *An Invitation to Enter a New Field of Physics*, the transcript of classic talk given on December 29th 1959 at the annual meeting of the American Physical Society at the California Institute of Technology, was first published in the February 1960 issue of Caltech's *Engineering and Science*, which owns the copyright. It has been made available on the web at http://www.zyvex.com/nanotech/feynman.html with their kind permission.
4. Moellenstedt, G., and R. Speidel, "Microminiaturization developments," *1961 International Solid-state Circuits Conference and Phys. Blatter* 16, p. 192 1960.
5. Pease, R. F. W., "High resolution scanning electron microscope, Dissertation, Cambridge University, 1963.
6. Broers, A. N., "Micromachining by sputtering through a mask of contamination laid down by an electron beam," *Proceedings of the first Symposium on Electron and Ion Beam Science and Technology*, Toronto, R. Bakish, ed., Wiley, New York, pp. 3–7, 191, May 1964.
7. Bridges, W. B., "Laser oscillation in singly ionized argon in the visible spectrum," *Appl. Phys. Lett.* 4(7), 128, April 1964.
8. Labuda, E. F., E. I. Gordon, and R. C. Miller, "Continuous-duty argon ion lasers," *IEEE J. Quantum Electron*, QE-1(6), p. 273, 1965.
9. "Great moments in our industry become defining moments in information technology," *Semiconductor Magazine* 1(7), July 2000. Available at http://www.davistown-museum.org/bioMann.html
10. Becker, H., "Early mask making during the 1960's in Dresden," *European Mask Conference on Mask Technology for Integrated Circuits and Micro-Components (EMC);*Sonthofen, Germany; Berlin: VDE-Verlag, GMM-Fachbericht 39, pp. S.103–109 Jan.13–15, 2003.

11. Doherty, J., "Recent advances in electron beam systems for mask making," *Solid State Technol.* 83, May 1979.
12. Speidel, J., and S. Cordes, "Data processing improvements for photomask pattern generators," *Proc. SPIE* 2884, p. 353, 1996.
13. Skinner, J., "Life with EBES," *Internal ATT memorandum,* Bell Labs, Murray Hill, NJ, (unpublished).
14. Howland, F. L. Howland and K. M. Poole, An overview of the new mask-making system, *Bell System. Tech. J.*, 1997, Nov. 1970.
15. Allen, P. C., "Laser scanning for semiconductor mask pattern generation," *Proc. IEEE* 90(10), p. 1653, Oct. 2002.
16. MacDonald, D. B., M. Nagler, C. Van Peski, and T. R. Whitney, "160 Mpx/sec laser pattern generator for mask and reticle production," *Proc. SPIE* 470, p. 212, 1984.
17. Yao, S., and B. Meshman, "Exposing patterns with a scanning laser system," *Proc. SPIE* 633, p. 152, 1986.
18. Varnell, G., D. Spicer, A. Rodger, and R. Holland, *Proc. EBM II E.C.S. Meeting,* San Francisco, May 1974.
19. Varnell, G., D. Spicer, et al., "High-speed, low-overhead electron beam direct slice writing system," *J. Vac. Sci. Technol.* 16(6), p. 1787, Nov/Dec. 1979.
20. Ting. C., R. Anderson, D. Saiki, and A. Kraft, "Chrome mask fabrication with electron-beam lithographic system," *J. Vac. Sci. Technol.* 15(3), p. 948, May/June 1978.
21. Herriott D. R., R. J. Collier, D. S. Alles, and J. W. Stafford, "EBES:A practical electron lithographic system," *IEEE Trans. Electron Devices* ED-22, pp. 385–392, July 1975.
22. Thompson, L. F. and M. J. Bowden, "A new family of positive electron beam resists—-Poly(Olefin Sulfones)," *J. Electrochem. Soc.* 120, p. 12, December 1973.
23. Yew, N., "Electron beam—now a practical LSI production tool," *Solid State Technol.*, p. 86, Aug. 1977.
24. Chang, T. H. P., M. Hatzakis, and A. Broers, "Electron beam lithography draws a finer line." *Electronics* 89, May 1977.
25. Piwczyk, B., "Electron beam and X-ray lithography update," *Electronic Packaging and Production*, p. 36, May 1978.
26. Tatsuno G., M. Fujinami, A. Iwata, and K. Kinamari, *Rev. Phys. Appl.* 13, p. 705, 1978.
27. Fujinami, M., T. Matsuda, K. Takamoto, H. Yoda, T. Ishiga, N. Saitou, and K. Komoda, "Variably shaped electron beam lithography system, EB55: I. System design," *J. Vac. Sci. Technol.* 19, p. 941, 1981.
28. Saitou N., M. Okumura, G. Matsuoka, T. Matsuzaka, T. Komoda, and Y. Sakitani, "A high-speed, high-precision electron beam lithography system (electron optics)," *J. Vac. Sci. Technol.* B 3(1), p. 98, Jan/Feb. 1985.
29. Eidson J., and R. Scudder, "A high speed electron beam lithography system," *J. Vac. Sci. Technol.*, 19(4), p. 932, Nov/Dec.1981.
30. Kelly J., T. Groves, and H. Kuo, "A high-current, high speed electron beam lithography column," *J. Vac. Sci. Technol.* 19, p. 936, 1981.
31. Trotel J., *Proceedings of the nineth International Symposium on Electron and Ion Beam Science and Technology* 137, Electrochemical Society, New York, 1980.
32. de Chambost E., B. Allanos, A. Frechet, and J. Perrocheau, "Experimental results from fast electron beam pattern generator: a variable shaped beam machine," *J. Vac. Sci. Technol.* B 4(1), p. 73, Jan/Feb., 1986.
33. Moriya, S., K. Komatsu, K. Harada, and T. Kitayama, "A large angle electrostatic deflection, variable shaped, electron beam exposure system," *J. Vac. Sci. Technol.* B 1(4), p. 990, Oct.–Dec. 1983.
34. Petric P., N. Taylor, and M. Utlaut, "Design aspects of the optics of the VLS-1000 electron-beam direct-write lithography system," *J. Vac. Sci. Technol.* B 1(4), p. 995, Oct.–Dec., 1983.
35. Riemenschneider P., "EBS-5: A vector scan electron beam lithography system for research and development," *J. Vac. Sci. Tech. B.* 4(1), p. 68, Jan/Feb. 1986.
36. King H., P. Merritt, O. Otto, Ozdemir, J. Pasiecznik, A. Carroll, D. Cavan, W. Eckes, L. Lin, L. Veneklasen, and J. Wiesner, "An electron beam lithography system for submicron VHSIC device fabrication," *J. Vac. Sci. Tech. B* 3(1), p. 106, Jan/Feb. 1985.

37. Veneklasen L., "A high speed EBL column designed to minimize beam interactions," *J. Vac. Sci. Technol. B* 3(1), p. 185, Jan/Feb.1985.
38. Hamaguchi H., J. Kai, and H. Yasuda, "High-precision reticle making by electron-beam lithography," *J. Vac. Sci. Technol. B* 6 (1), p. 204, Jan/Feb. 1988.
39. Reynolds J., "Can electron beam mask making keep pace with advancing technology?" *Solid State Technol.*, p. 119, May 1983.
40. Takigawa T., K. Shimazaki, and N. Kusui, *Proc. SPIE* 632, p. 175, 1986.
41. Pfeiffer, H. C., and K. H. Loeffler, "A high current square spot probe for micro pattern generation," *7th Int. Electron Microscopy Congress*, Grenoble, p. 63, 1970.
42. Goto E., T. Soma, and M. Idesawa, *J. Vac. Sci. Technol.* 15, p. 883, 1978.
43. Pfeiffer H. C., "Variable spot shaping for electron-beam lithography," *J. Vac. Sci. Technol*, 15(3), p. 887, May/June 1978.
44. Pfeiffer H. C., "Recent advances in electron beam lithography for high-volume production of VLSI devices," *IEEE Trans. Electron Devices* ED-26(4), p. 663, 1979.
45. Moore R. D., G. A. Caccoma, H. C. Pfeiffer, E. V. Weber, and O. C. Woodard, "EL3: a high throughput, high resolution e-beam lithography tool," *J. Vac. Sci. Technol.* 19(4), p. 950, 1981.
46. Buckley J., D. Galburt, and C. Karatsas, "Step-and-scan lithography using reduction optics," *J. Vac. Sci. Tech. B* 7(6), 1607, Nov/Dec. 1989.
47. Arnold W., and G. Escher, "Wafer steppers for the 64M and 256Mbit Memory Generations," *64- to 256-Megabit Reticle Generation: Technology Requirements and Approaches*. G. Hearn, ed., Critical Reviews of Optical Science and Technology, SPIE CR51, 1993.
48. Levy K., "The history and future of mask making," *Proc. SPIE*, 2884, pp. 2–5, 1996.
49. Warkentin P. A., and J. A. Schoeffel, "Scanning laser technology applied to high speed reticle writing," *Proc. SPIE* 633, p. 286, 1986.
50. Hattori K., R. Yoshikawa, H. Wada, H. Kusakabe, T. Yamaguchi, S. Magoshi, A. Miyagaki, S. Yamasaki, T. Takigawa, M. Kanoh, S. Nishimura, H. Housai, and S. Hashimoto, "Electron-beam direct writing system EX-8D employing character projection exposure method," *J. Vac. Sci. Technol.* B11(6), p. 2346, 1993.
51. Gesley M., T. Mulera, C. Nurmi, J. Radley, A. Sagle, K. Standiford, Z. Tan, J. Thomas, and L. Veneklasen, "0.25 um lithography using a 50 kV shaped electron-beam vector scan system," *Proc. SPIE* 2437, p. 168, 1995.
52. Cha B., J. Park, Y. Choi, J. Kim, W. Han, H. Yoon, and J. Sohn, "Improved process control of photomask fabrication in e-beam lithography," *Proc. SPIE* 4186, p. 508, 2001.
53. Chen P., "Assessment of mask making in Taiwan and Asia Pacific region," *Proc. SPIE* 3546, p. 2, 1998.
54. See the review of the early days, in Schellenberg F., D. Okimoto, J. Raphael, and N. Shirouzu. Technology Development in the U.S. and Japan: The case of the phase shifting mask. in *64- to 256-Megabit Reticle Generation: Technology Requirements and Approaches*, G. Hearn, ed., Critical Reviews of Optical Science and Technology, *Proc. SPIE CR51*, 1993.
55. Yen A., S. Yu, J. Chen, C. Chen, T. Gau, and B. Lin, "Low-k1 optical lithography for 100 nm logic technology and beyond," *J. Vac. Sci. Technol.* B19(6), p. 2329, Nov/Dec. 2001.
56. Rothschild M., T. Bloomstein, R. Kunz, V. Liberman, M. Switkes, S. Palmacci, J. Sedlacek, and D. Hardy, "Liquid immersion lithography: why, now, and when?" *J. Vac. Sci. Technol.* B 22(6), p. 2877, Nov/Dec. 2004.
57. Suzuki K, et al., "Full-field exposure performance of electron projection lithography tool," *J. Vac. Sci. Technol.* B 22(6), p. 2885, Nov/Dec. 2004.
58. Naber R., A. Muray, and F. Abboud, "Alignment technique, results, and simulation of second level exposure of phase shifting masks using 10keV raster scan electron beam lithography," *Proc. Semicon Korea*, Nov. 1992.
59. Ye J., N. Berglund, F. Pease, G. Owen, and K. Alexander, "How mask errors transfer to the wafer near the optical resolution limit," *BACUS Photomask News* 11(5), May 1995.
60. Randall J., "Optically induced mask critical dimension error magnification in 248 nm lithography," *J. Vac. Sci. Tech. B.* 16(6), p. 3606, Nov/Dec. 1998.

61. Zinn S., S. Lee, S. Choi, and J. Sohn, "Simulation of critical dimension error using Monte Carlo method and its experimental verification," *J. Vac. Sci. Technol.* B20(6), p. 2606, Nov/Dec. 2002.
62. Lu H., R. Hwang, V. Lee, W. Q., J. Fung Chen, T. Laidig, K. Wampler, and R. Caldwell, "Making high performance scattering bar OPC masks with vector scan, variable-shaped e-beam and raster scan laser mask writers," *Proc. SPIE* 3546, p. 368, 1998.
63. Hamaker C., J. Martyniuk, R. Sills, J. Varner, and K. Standiford, "Comparison of reticle placement performance using two-point and multipoint fitting algorithms," *Proc. SPIE* 2884, p. 367, 1996.
64. Jackson C., P. Buck, S. Cohen, V. Garg, C. Howard, R. Kiefer, J. Manfredo, and J. Tsou, "DUV Laser lithography for photomask fabrication," *Proc. SPIE* 5256, 2004.
65. Scott D., Method for fabricating high density DRAM reticles, *US Patent* 5,227,269, 1993.
66. Irie N., K. Muramatsu, Y. Ishii, N. Magome, T. Umatate, S. Kyoh, S. Tanaka, S. Inoue, I. Higashikawa, I. Mori, and K. Okumara, "New photomask pattern generation method based on i-line stepper," *Photomask Japan,* 12 April 2000.
67. Berglund N., "E-beam and optical reticle generation," *64- to 256-Megabit Reticle Generation: Technology Requirements and Approaches*, G. Hearn, ed., Critical Reviews of Optical Science and Technology, *SPIE CR51*, 1993.
68. Rieger M., et al., Image quality enhancements for raster scan lithography, *Optical Microlithography VII* 922, p. 55, 1988.
69. Hamaker H. C., G. Burns, and P. Buck, "Optimizing the use of multipass printing to minimize printing errors in advanced laser reticle-writing systems," *Proc. SPIE,* 2621, p. 319, 1995.
70. Lee S., G. Yeo, J. Lee, H. Cho, and W. Han, "Approach to critical dimension error budget analysis and specification estimation by the Monte Carlo method," *J. Vac. Sci. Tech. B* 21(6), p. 3120, Nov/Dec. 2003.
71. Sauer C., and C. Mack, "Electron-beam lithography simulation for maskmaking, part IV: effect of resist contrast on isofocal dose," *Proc. SPIE* 3748, p. 27, 1999.
72. Yu M., A. Sagle, and B. Buller, "Exploring the fundamental limit of CD control: a model for shot noise in lithography," *Proc. SPIE,* 5751, p. 150, 2005.
73. Munro E., "Electron and ion optical design software for integrated circuit manufacturing equipment," *J. Vac. Sci. Technol. B* 15(6), p. 2692, 1997.
74. Hirumi J., N. Yoshioka, H. Hoshi, H. Ando, S. Tsuchiya, and M. Houga, "The study of high-speed electron beam deflection technology for VSB writers," *Proc. SPIE,* 5446, p. 681, 2004.
75. Hattori K., et al., "Solution for 100 nm-EBM-4000," *PMJ2002* Abstract 11-05 and *Proc. SPIE,* 4754, p. 697, 2002.
76. Nishimura S., M. Shimizu, S. Mitsui, K. Akeno, H. Kusukabe, M. Ogasawara, R. Hayashi, S. Yanaga, N. Kobayashi, and T. Togo, "Development of a mask-scan electron beam mask writer," *J. Vac. Sci. Tech. B.* 20(6), p. 2640, Nov/Dec. 2002.
77. Nakayama Y., et al., "Cell projection system," *J. Vac. Sci. Technol.* B8, p. 1836, 1990.
78. Someda Y., H. Satoh, Y. Sohda, Y. Nakayama, N. Saitou, H. Itoh, and M. Sasaki, "Electron-beam cell projection lithography: Its accuracy and its throughput," *J. Vac. Sci. Technol. B* 12(6), p. 3399, 1994.
79. Hirumi J., K. Kuriyama, N. Yoshioka, R. Yoshikawa, Y. Hojo, T. Matuzaka, K. Tanaka, and M. Hoga, "Shot number analysis at 65nm node mask writing using VSB writer," *Proc. SPIE* 5130, p. 309, 2003.
80. Yoshikawa R., H. Wada, M. Goto, H. Kusakabe, O. Ikenaga, S. Tamamushi, and T. Takigawa, "A high dose and high accuracy variable shaped electron beam exposure system for quartermicron device fabrication," *J. Vac. Sci. Technol. B* 5(1), p. 70, Jan/Feb. 1987.
81. Tamamushi S., H. Wada, Y. Ogawa, I. Sasaki, M. Nakasuji, H. Kusakabe, R. Yoshikawa, and T. Takigawa, "The electron-beam column for a high-dose and high-voltage electron-beam exposure system EX-7," *J. Vac. Sci. Technol. B* 6(1), p. 209, Jan/Feb. 1988.
82. Hattori K., O. Ikenaga, H. Wada, S. Tamamushi, E. Nishimura, N. Ikeda, Y. Katoh, H. Kusakabe, R. Yoshikawa, and T. Takigawa, "..triangle beams, *Jpn. J. Appl. Phys* 28, p. 2065, 1989.

83. Takigawa T., et al., "A high accuracy and high throughput electron beam reticle writing system for 16M dynamic random access memory class and beyond devices," *J. Vac. Sci. Technol. B* 8(6), p. 1877, Nov/Dec. 1990.
84. Takigawa T., H. Wada, Y. Ogawa, R. Yoshikawa, I. Mori, and T. Abe, "Advanced e-beam lithography," *J. Vac. Sci. Technol. B* 9(6), p. 2981, Nov/Dec. 1991.
85. Takigawa T., T. Tojo, Y. Ogawa, K. Koyama, A. Ono, S. Inoue, S. Itho, and M. Goto "Development of an advanced mask and its fabrication system," *Proc. SPIE* 2512, p. 181, 1995.
86. Yoshikawa R., K. Hattori, S. Magoshi, H. Sunaoshi, A. Ando, T. Yamaguchi, S. Mikami, S. Nishimura, H. Housai, S. Hashimoto, and T. Takigawa, "Patterning accuracy improvement of the electron beam direct writing system EX-8D," *J. Vac. Sci. Technol.* 13(6), p. 2493, Nov/Dec. 1995.
87. Tojo T., et al., "Advanced electron beam writing system EX-11 for next generation mask fabrication," *Proc. SPIE* 3748, p. 416, 1999.
88. Takigawa T., "The road to X success: Toshiba Machine equipment support," *X Initiative presentation,* 2001.
89. Kubo S., K. Hiruta, M. Sugiyama, T. Iwamatsu, T. Fujisawa, and H. Morimoto, "Evaluation of an advanced mask writing system," *Proc. SPIE* 3748, p. 426, 1999.
90. Sugiyama M., S. Kubo, K. Hiruta, T. Iwamatsu, T. Fujisawa, and H. Morimoto, "Proximity effect correction for reticle fabrication," *Proc. SPIE* 4066, p. 180, 2000.
91. Yang S., W. Ki. S. Moon, T. Jeong, S. Choi, W. Han, and J. Sohn, "A Study on the Effect of Beam Blur in Mask Fabrication," *Proc. SPIE* 4186, p. 468, 2001.
92. Ohtoshi K., H. Sunaoshi, J. Takamatsu, F. Okabe, K. Ishibashi, S. Yoshitake, H. Yamad, S. Tamamushi, H. Anze, T. Kamikubo, and Y. Ogawa, "Writing accuracy of EBM 3500 electron beam mask writing system," *Proc. SPIE* 4409, p. 238, 2001.
93. Murooka K., K. Hattori, and O. Iizuka, "Initial results of a 50 kV electron beam writer EBM-4000 for a 90 nm node photomask," *J. Vac. Sci. Tech. B* 21(6), p. 2668, Nov/Dec. 2003.
94. Shearer M., H. Takemura, M. Isobe, N. Goto, K. Tanaka, and S. Miyauchi, "Development of nanometric electron-beam lithography system (JBX-5DII)," *J. Vac. Sci. Technol. B* 4(1), p. 64, Jan/Feb. 1986.
95. Nakazawa H., H. Takemura, M. Isobe, Y. Nakagawa, M. Shearer, and W. Thompson, "A thermally assisted field emission electron beam exposure system," *J. Vac. Sci. Technol. B* 6(6), p. 2019, Nov/Dec. 1988.
96. Shearer M., et al., "Improvements in the JBX-6AIII series of shaped beam lithography tool," *Proc. SPIE* 923, p. 290, 1988.
97. Komagata T., H. Takemura, N. Goto, R. Espeseth, and M. Shearer, "Performance and perspective of the newly developed high accuracy mask making EB lithography System, JBX-7000MV, designed for 64M DRAM production," *Proc. SPIE* 1809, p. 14, 1992.
98. Komagata T., H. Takemura, N. Gotoh, and K. Tanaka, "Development of EB lithography system for next generation photomasks," *Proc. SPIE* 2512, p. 190, 1995.
99. Komagata T., et al., "Development of electron beam optics column for mask lithography system," *Proc. SPIE* 3096, p. 125, 1997.
100. Nakagawa Y., T. Komagata, H. Takemura, N. Gotoh, and K. Tanaka, "Development of a next generation e-beam lithography system," *Proc. SPIE* 3546, p. 45, 1998.
101. Komagata T., Y. Nakagawa, H. Takemura, N. Gotoh, and K. Tanaka, "Development of a next generation e-beam lithography system for 1 Gbit DRAM masks," *Proc. SPIE* 3331, p. 313, 1998.
102. Takemura H., T. Komagata, Y. Nakagawa, and K. Tanaka, "Advanced e-beam lithography system JBX-9000MV for 180-nm masks," *Proc. SPIE* 3873, p. 21, 1999.
103. Komagata T., Y. Nakagawa, N. Gotoh, and K. Tanaka, "Performance of improved JBX-9000MV e-beam lithography system," *Photomask Japan* 171, 2001.
104. Komagata T., et al., "Development of a E-Beam lithography system for 100-90nm node reticles," *Proc. SPIE* 5130, p. 328, 2003.
105. Nakagawa Y., T. Komagata, Y. Kawase, and N. Gotoh, "Development of the JBX-3030MV mask making e-beam lithography system," *JEOL News* 38(1), p. 32, 2003.

106. Kimura N., T. Komagata, Y. Nakagawa, N. Gotoh, and K. Tanaka, "High-alignment-accuracy EB writing of phase shift image for 65-nm node mask," *Proc. SPIE* 5446, p. 68, 2004.
107. Komagata T., N. Kimura, K. Funaki, Y. Nakagawa, and N. Gotoh, "Evaluation of overlay accuracy of phase shift image for 65 nm node masks," *Proc. SPIE* 5567, p. 1315, 2004.
108. Nakamura K., Y. Sakitani, T. Konishi, T. Komoda, N. Saitou, and K. Sugawara, "A high speed, high precision electron beam lithography system (system design)," *J. Vac. Sci. Tech. B* 3(1), p. 94, Jan/Feb. 1985.
109. Kawano M., K. Mizuno, H. Yoda, Y. Sakitani, K. Andou, and N. Saitou, "Continuous writing method for high speed electron-beam direct writing system HL-800D," *J. Vac. Sci. Technol. B* 11(6), p. 2323, 1993.
110. Kadowaki Y., K. Kawasaki, K. Mizuno, H. Satoh, M. Hoga, K. Uryu, "High performance and stability reticle writing system, HL-800M," *Proc. SPIE* 3412, p. 58, 1998.
111. Asai S., Y. Kadowaki, K. Kawasaki, K. Mizuno, H. Satoh, M. Horiga, K. Ikeda, E. Iguchi, "The advantage in using the combination of HL-800M and CAR process," *Proc. SPIE* 3748, p. 93, 1999.
112. Nakehara T., K. Mizuno, S. Asai, Y. Kadowaki, K. Kawasaki, and H. Satoh, "Advanced e-beam reticle writing system for next generation reticle fabrication," *Proc. SPIE* 4066, p. 594, 2000.
113. Fujii A., et al., "Advanced e-beam reticle writing system for next generation reticle fabrication," *Photomask Japan* 173, 2001 (*Proc. SPIE* 4400, p. 258, 2001).
114. Satoh H., et al., "New technology to improve throughput of electron beam mask writer," *Photomask Japan,* Abstract 11-04, 2002.
115. Wang Z., H. Satoh, H. Ito, Y. Sohda, H. Ohta, H. Kawano, Y. Kadowaki, K. Mizuno, and T. Matsuzaka, "Recent CD accuracy improvements for HL-7000M," *Proc. SPIE* 5446, p. 657, 2004.
116. Weber E., and R. Moore, "E-beam exposure for semiconductor device lithography," *Solid State Technol.,* p. 61, May 1979.
117. Chisholm T., "Spot-size measurement in an electron-beam pattern generator," *J. Vac. Sci. Technol. B* 6(6), p. 2066, Nov/Dec. 1988.
118. Viswanathan R., A. D. Wilson, J. Lafuente, H. Voelker, and A. Kern, "Electron pattern beam writer for X-ray masks," *Proc. SPIE* 448, p. 130, 1984.
119. Wilson A., T. H. P. Chang, and A. Kern, "Experimental scanning electron-beam automatic registration," *J. Vac. Sci. Technol.* 12(6), p. 1240, Nov/Dec. 1975.
120. Speth A., A. Wilson, A. Kern, and T. H. P. Chang, "Electron-beam lithography using vector-scan techniques," *J. Vac. Sci. Technol.*12, p. 1235, 1975.
121. Chang T. H. P., A. Wilson, A. Speth, and C. Ting, "Vector-scan-I an automated electron-beam system for high resolution lithography," *Proceedings of the seventh International Conference on Electron and Ion Beam Science and Technology,* R. Bakish, ed, 392, 1976.
122. Broers A.. N., "Electron gun using long-life lanthanum hexaboride cathode," *J. Appl. Phys.* 53, p. 1991, 1967.
123. L. Swanson, and N. Martin, "Field electron cathode stability studies: Zirconium/tungsten thermal-field cathode," *J. Appl. Phys.* 46, p. 2029, 1975.
124. Gesley M., F. Hohn, R. Viswanathan, and A. Wilson, "A vector-scan thermal-field emission nanolithography system," *J. Vac. Sci. Technol.* B 6 (6), p. 2014, Nov/Dec. 1988.
125. Munro E., "Design and optimization of magnetic lenses and deflection systems for electron beams," *J. Vac. Sci. Technol.* 12(6), p. 1146, Nov/Dec. 1975.
126. Ohiwa H., "Design of electron-beam scanning systems using the moving objective lens," *J. Vac. Sci. Technol.* 15(3), p. 849, May/June 1978 (The original concept can be found in: H. Ohiwa, E. Goto, and A. Ono, *Electron. Commun. Jpn.* B-54, p. 33, 1971.)
127. Pfeiffer H., and G. Langner, "Advanced deflection concept for large area, high resolution e-beam lithography," *J. Vac. Sci. Technol.* 19, p. 1058, 1981.
128. Yourke H., and E. Weber, "A high-throughput scanning-electron-beam lithography system, EL1, for semiconductor manufacture: General description," *Int. Electron Devices Meeting*, Washington, DC, p. 431, 1976.

129. E. Weber, and R. Moore, "Variable spot-shaped e-beam lithographic tool," *J. Vac. Sci. Technol.* 16(6), p. 1780, Nov/Dec. 1979.
130. Fontijn L., "An electron exposure system for recording and printing," *Ph.D. dissertation*, Delft 1972.
131. Yasuda H., K. Sakamoto, A. Yamada, and K. Kawashima, *Jpn. J. Appl. Phys.* 30, p. 3098, 1991.
132. Sakitani H., H. Yoda, H. Todokoro, Y. Shibata, T. Yamazaki, K. Ohbitu, N. Saitou, S. Moriyama, S. Okazaki, G. Matsuoka, F. Murai, and M. Okumura, "Electron-beam cell-projection lithography system," *J. Vac. Sci. Technol. B* 10, p. 2579, 1992.
133. Itoh H., H. Todokoro, Y. Sohda, Y. Nakayama, and N. Saitou, *J. Vac. Sci. Technol. B* 10, p. 2799, 1992.
134. Sakamoto K., "Electron-beam block exposure system for a 256 m dynamic random access memory," *J. Vac. Sci. Technol. B* 11(6), p. 2357, 1993.
135. Moore R., G. Caccoma, H. Pfeiffer, E. Weber, and O. Woodard, "EL-3: a high throughput, high resolution e-beam lithography tool," *J. Vac. Sci. Technol.* 19, p. 950, 1981; R. Moore, *Electronics* 54, p. 138, 1981.
136. Davis D., S. Gillespie, S. Silverman, W. Stickel, and A. Wilson. "EL-3 application to 0.5 um semiconductor lithography," *J. Vac. Sci. Technol.* B1(4), p. 1003, Oct.–Dec. 1983.
137. Groves T., H. Pfeiffer, T. Newman, and F. Hohn, "EL3 system for quarter-micron electron beam lithography," *J. Vac. Sci. Technol. B* 6(6), p. 2028, Nov/Dec. 1988.
138. Hartley J., T. Groves, and H. Pfeiffer, "Performance of the EL-3+ maskmaker," *J. Vac. Sci. Tech. B* 9(6), p. 3015, Nov/Dec. 1991.
139. Pfeiffer H. C., D. E. Davis, W. A. Enichen, M. S. Gordon, T. R. Groves, J. G. Hartley, R. J. Quickle, J. D. Rockrohr, W. Stickel, and E. V. Weber, "EL-4, a new generation electron-beam lithography system," *J. Vac. Sci. Technol. B*11(6), p. 2332, 1993.
140. Sturans M., et al., "EL5: One tool for advanced x-ray and chrome on glass mask making," *J. Vac. Sci. Tech. B* 16(6), p. 3164, Nov/Dec. 1998.
141. Elsner H., P. Hahmann, G. Dahm, and H. Koops, "Multiple beam-shaping diaphragm for efficient exposure of gratings," *J. Vac. Sci. Technol. B*11(6), p. 2373, 1993.
142. Fortagne O., P. Hahmann, and C. Ehrlich, "We Print 200—The fast e-beam printer with high throughput," *Microelectron. Eng.* 2, p. 151, 1995.
143. Ehrlich C., and J. Gramss, "Large glass reticle writer—exposure strategy and mask handling," *Proc. SPIE* 412, p. 52, 1998.
144. Butschke J, D. Beyer, C. Constantine, P. Dress, P. Hudek, M. Irmscher, C. Koepernik, C. Krauss, J. Plumhoff, and P. Voehringer, "90 nm Mask making processes using the positive tone chemically amplified resist FEP171," *Proc. SPIE* 5256, p. 344, 2004.
145. Y. Laplanche, M. Charpin, L. Pain, J. Todeschini, D. Henry, P. O. Sassoulas, S. Gough, U. Weidenmueller, P. Hahmann, "Mix-and-match capability of e-beam direct write for 65 nm technology," *Proc. SPIE* 5037, p. 572, 2003.
146. L. Pain, et al., "Electron beam direct write lithography flexibility for ASIC manufacturing, an opportunity for cost reduction," *Proc. SPIE-Emerging Lithographic Technologies*, 2005.
147. K. Cummings, L. Schneider-Stoermann, U. Buttgereit, M. Irmscher, D. Mueller, P. Hudek, D. Beyer, B. Brendel, J. M. Whittey, B. G. Eynon, Jr., J. Harsch, C. Constantine, and K. Miller, "Back to square 9, a demonstration of 9-inch reticle capability," *Proc. SPIE* 4889, p. 15, 2002.
148. Beyer D., D. Löffelmacher, G. Goedl, P. Hudek, B. Schnabel, and T. Elster, "Tool and Process Optimization for 100 nm Maskmaking using a 50 kV Variable Shaped E-Beam System," *Proc. SPIE* 4562, p. 10, 2004.
149. Daukscher W., "Fabrication of step and flash imprint lithography templates using a variable shaped beam exposure tool," *Microelectron. Eng.* 75, pp. 345–351, 2004.
150. Alles D., et al., "EBES4: A new electron-beam exposure system," *J. Vac. Sci. Technol. B* 5, p. 47, 1987.
151. Thomson M., R. Liu, R. Collier, H. Carroll, E. Doherty, and R. Murray, "The EBES4 electron-beam column," *J. Vac. Sci. Technol. B* 5(1), p. 53, Jan/Feb. 1987.
152. Kuo H. P., J. Foster, W. Haase, J. Kelly, and B. M. Oliver, "Electron Beam Blanker," *Proceedings of the Electron and Ion Beam Science and Technology, tenth International*

Conference, Montreal, R. Bakish, ed., Electrochemical Society, Pennington, NJ, 83-2, p. 78, 1983.
153. Walker D., D. Fowlis, S. Kugelmass, K. Murray, and C. Rose, "Advanced mask and reticle generation using EBES4," *Proc. SPIE* 2322, p. 56, 1994.
154. Peters D., D. Fowlis, A. von Neida, C. Rose, H. Waggener, and W. Wilson, "EBES4: Performance of a new e-beam reticle generator," *Proc. SPIE* 1924, p. 193, 1993.
155. Biechler C., A. Carroll, R. Graves, and S. Lyons, "Virtual addressing for E-beam lithography," *US Patent* 4,498,010, Feb. 1985.
156. Dameron D., C. Fu, and R. Pease, "A multiple exposure strategy for reducing butting errors in a raster-scanned electron-beam exposure system," *J. Vac. Sci. Technol. B* 6(1), p. 213, Jan–Feb 1988.
157. Warkentin P., M. Rieger, C. Mikalski, and M. Jolley, "Rasterization system utilizing an overlay of bit-mapped low address resolution databases," *US Patent* 4,879,605, 1989.
158. Muray A., F. Abboud, F. Raymond, and C. Berglund, "Feasibility study of new gray-beam writing strategies for raster scan mask generation," *J. Vac. Sci. Technol.* B11, p. 2390, 1993.
159. Muray A., D. Colby, R. Teitzel, and M. Gesley, "Experimental evaluation of an electron-beam pulse modulated blanker (160 MHz) for next-generation electron-beam raster scan systems," *J. Vac. Sci. Technol. B* 13, p. 2488, 1995.
160. Sandstrîm T., et al., "Highly accurate pattern generation using acousto-optical deflection," *Proc. SPIE* 1463, p. 629, 1991.
161. Liden P., "CD performance of a new high-resolution laser pattern generator," *Proc. SPIE* 326, p. 55, 1997.
162. Thuren, "A new approach to data preparation and processing for high density patterns," *Proc. SPIE* 3051, p. 304, 1997.
163. Johansson C., et al., "Scalable pattern generator data path—for the future," *Proc. SPIE* 4066, p. 639, 2000.
164. Vikholm T., L. Kjellberg, P. Askebjer, S. Haddleton, J. Larsson, and M. Bjuggren, "Results from a new laser pattern generator for 180 nm photomasks," *Proc. SPIE* 4066, p. 624, 2000.
165. Liden P., T. Vikholm, L. Kjellberg, M. Bjuggren, K. Edgren, J. Larsson, S. Haddleton, and P. Askebjer, "CD performance of a new high-resolution laser pattern generator," *Proc. SPIE* 3873, p. 28, 1999.
166. Buck P., M. Bjuggren, H. Bunning, V. Garg, J. Larsson, and T. Vikholm, "Integration of the Micronic Omega 6500 into the mask manufacturing environment," *Proc. SPIE* 4186, p. 474, 2001.
167. Kueck H., et al., "New system for fast submicron laser direct writing," *Proc. SPIE* 2440, p. 506, 1995.
168. Seltmann R., "New system for fast submicron optical direct writing," *Microelectron. Eng.* 30, p. 123, 1996.
169. Paufler J., et al., "Continuous Image Writer—a new approach to fast direct writing," *Proc. SPIE Optical Microlithography XIII* 4000, p. 866, 2000.
170. Ljungblad, et al., "New Architecture for laser pattern generators for 130 nm and beyond," *Proc. SPIE* 4186, p. 16, 2000.
171. Ljungblad U., U. Dauderstadt, P. Durr, T. Sandstrom, H. Buhre, and H. Lakner, "New laser pattern generator for DUV using a spatial light modulator," *Microelectron. Eng* 57–58, 23, 2001.
172. Sandstrom T., T. Fillion, U. Ljungblad, and M. Rosling, "Sigma 7100, a new architecture for laser pattern generators for 130 nm and beyond," *Proc. SPIE* 4409, p. 270, 2001.
173. Sjöberg H., J. Chauvet, J. Härkesjö, P. Högfeldt, A. Karawajczyk, J. Karlsson, L. Kjellberg, J. Måhlén, A. Beyerl, J. Vedenpää, R. Goodoree, M. Bjuggren, and J. Aman, "Pattern accuracy and throughput optimization for an SLM-based 248-nm DUV laser mask pattern generator," *Proc. SPIE* 5446, p. 111, 2004.
174. Sandstrom T., and N. Eriksson, "Resolution extensions in the Sigma7000 imaging pattern generator," *Proc. SPIE* 4889, p. 157, 2002.
175. Sandstrom T., and U. Ljungblad, "Phase shifting optical maskless lithography enabling ASIC's at the 65 and 45 nodes," *Proc. SPIE* 5567, p. 529, 2004.

176. Mechtenberg M., and L. J. Watson, "Performance appraisal of the ATEQ CORE 2500 in production," *Proc. SPIE* 1496, p. 124, 1990.
177. Grenon B., D. Defibaugh, D. Sprout, C. Hamaker, and P. Buck. "Technical performance of the ALTA-3000 laser writer," *Proc. SPIE* 2322, p. 50, 1994.
178. Buck P., A. Buxbaum, T. Coleman, and L. Tran, "Performance of the ALTA 3500 scanned-laser mask lithography system," *Proc. SPIE* 3412, p. 67, 1998.
179. Hamaker H. C., G. Valentin, J. Martyniuk, B. Martinez, J. Pochkowski, and L. Hodgson, "Improved critical dimension control in 0.8-NA laser reticle writers," *Proc. SPIE* 3873, p. 49, 1999.
180. Morgante C., and H. C. Hamaker, "The ALTA 3700: extending the application space of the ALTA 3500 laser reticle writer," *Proc. SPIE* 4066, p. 613, 2000.
181. Spence C., C. Tabery, R. Cantrell, L. Dahl, P. Buck, and B. Wilkinson, "A comparison of DUV wafer and reticle lithography—what is the resolution limit?" *Proc. SPIE* 4889, p. 177, 2002.
182. Bohan M., C. Hamaker, and W. Montgomery, "Implementation and characterization of a DUV raster-scanned mask pattern generation system," *Proc. SPIE* 4562, p. 16, 2001.
183. Allen P., A. Buxbaum, S. Howells, B. Kenan, A. Klatchko, P. Pirogovsky, R. Teitzel, and M. White, "DUV ALTA system aerial image enhancement for improved pattern fidelity," *Proc. SPIE* 5256, p. 20, 2003.
184. Klatchko A. and P. Pirogovsky, "Conformal mapping in microlithography," *Proc. SPIE* 5567, p. 33, 2004.
185. Ungereit M., S. Howells, T. Chabreck, J. Hubbard, A. Klatchko, P. Pirogovsky, R. Teitzel, P. Allen, C. Morgante, and M. White, "DUV ALTA aerial image enhancement for improved pattern fidelity: Phase II," *Proc. SPIE* 5446, p. 112, 2004.
186. Buck P., K. Green, K. Ibsen, K. Nakagawa, D. Hong, P. Krishnan, and D. Coburn, "Advanced write tool effects on 100 nm node OPC," *Proc. SPIE* 4889, p. 147, 2002.
187. Allen P., M. Bohan, E. Christenson, C. Hamaker, S. Howells, B. Kenan, P. Pirogovsky, M. Sadiq, R. Teitzel, M. White, M. Ungureit, A. Wickstrom, R. Kiefer, and C. Jackson, "Next generation DUV ALTA mask patterning capabilities," *Proc. SPIE* 5567, p. 32, 2004.
188. Allen P., M. Bohan, E. Christenson, H. Dai, M. Duane, C. Hamaker, S. Howells, B. Kenan, P. Pirogovsky, M. Sadiq, R. Teitzel, and M. White, "ALTA 4700 system mask patterning performance improvements for X-Architecture and wafer electrical performance interchangeability with 50 kV E-beam," *Proc. SPIE*, (in press).
189. Lieberman B., "Quality assurance procedures for MEBES," *J. Vac. Sci. Technol.*, 15(3), p. 913, May/June 1978.
190. Alles D., F. Ashley, A. Johnson, and R. Townsend, "Control system design and alignment methods for electron beam lithography," *J. Vac. Sci. Tech.* 12, p. 1252, 1975.
191. Freyer J., K. Standiford, and R. Sills, "Enhanced pattern accuracy with MEBES III," *Proc. SPIE* 471, p. 8, 1984.
192. Gesley M., "MEBES IV thermal-field emission tandem optics for electron beam lithography," *J. Vac. Sci. Technol. B* 9(6), p. 2949, Nov/Dec. 1991.
193. Abboud F., M. Gesley, D. Colby, K. Comendant, R. Dean, W. Eckes, D. McClure, H. Pearce-Percy, R. Prior, and S. Watson, "Electron beam lithography using MEBES IV," *J. Vac. Sci. Technol. B* 19(6), p. 2734, Nov/Dec. 1992.
194. Gesley M., F. Abboud, D. Colby, F. Raymond, and S. Watson, "Electron beam column developments for submicron and nanolithography," *Jpn. J. Appl. Phys.* 32, p. 5993, 1993.
195. Gesley M., D. Colby, F. Raymond, D. McClure, and F. Abboud, "Electrodynamics of fast beam blankers," *J. Vac. Sci. Techno*l. B 11, p. 2378, 1993.
196. Gesley M., "An electron optical theory of beam blanking," *Rev. Sci. Instrum.* 64(11), p. 3169, 1993.
197. Ogasawara M., H. Sunaoshii, and R. Yoshikawa, "Development of a fast beam-blanking system," *Proc. SPIE* 3412, p. 79, 1998.
198. Gesley M., and D. DeBra, "Mechanically stable field emission gun," *US Patent* 5,304,888, 1994.
199. Gesley M. and W. DeVore, "On the design and effective strength of stigmators for electron beam lithography," *J. Vac. Sci. Technol. B* 9(6), p. 2977, 1991.

200. Ye J., N. Berglund, and F. Pease, "Spatial frequency filtering using multiple-pass printing," *J. Vac. Sci. Technol.* B 12(6), p. 3455, Nov/Dec. 1994.
201. Abboud F., R. Dean, J. Doering, W. Eckes, M. Gesley, U. Hofmann, T. Mulera, R. Naber, M. Pastor, W. Phillips, J. Raphael, F. Raymond, and C. Sauer, "Multipass gray printing for new MEBES 4500S mask lithography system," *Proc. SPIE* 3096, p. 116, 1997.
202. Chandramouli M., J. DeWitt, G. Meyers, J. Millino, and D. Hwang, "Investigation of multiphase printing writing modes on mask performance in MEBES 4500," *Proc. SPIE* 2885, p. 8, 1996.
203. Prior R., "Method and structure for electronically measuring beam parameters," *US Patent* 5,345,085, 1994.
204. Braun C., and F. Peiffer, "On product metrology results from MEBES 4-TFE system," *Proc. SPIE* 2884, p. 37, 1996.
205. Pearce-Percy H., R. Prior, F. Abboud, A. Benveniste, L. Gasiorek, M. Lubin, and F. Raymond, "Dynamic corrections in MEBES 4500," *J. Vac. Sci. Technol. B*12(6) p. 3393, 1994.
206. Kagami I., M. Koyama, and H. Kawahira, "Comparison of multipass gray strategy and conventional writing method," *Proc. SPIE* 3748, p. 83, 1999.
207. Chabala J., F. Abboud, S. Weaver, and D. Cole, "A new MEBES pattern generator," *Proc. SPIE* 3546, p. 74, 1998.
208. Dean R., D. Alexander, J. Chabala, T. Coleman, C. Hartglass, M. Lu, C. Sauer, and S. Weaver, "Investigation of lithographic performance using multipass gray (MPG) with MEBES 5000," *BACUS News Photomask* 15(3), March 1999.
209. Chabala J., et al., "Extension of graybeam writing for the 130 nm technology node," *Proc. SPIE* 3873, p. 36, 1999.
210. Abboud F., K. Baik, V. Chakarian, D. Cole, R. Dean, M. Gesley, H. Gillman, W. Moore, M. Mueller, R. Naber, T. Newman, R. Puri, F. Raymond, and M. Rougieri, "Raster scan patterning solution for 100 nm and 70 nm OPC masks," *Proc. SPIE* 4754, p. 705, 2002.
211. Veneklasen L., "Optimizing electron beam lithography writing strategy subject to electron optical, pattern, and resist constraints," *J. Vac. Sci. Technol. B* 9(6), p. 3063, Nov/Dec. 1991.
212. Rishton S., J. Varner, L. Veneklasen, V. Boegli, A. Sagle, U. Hofmann, H. Kao, and W. Wang, "Raster shaped beam pattern generation," *J. Vac. Sci. Tech. B.* 17(6), p. 2927, Nov/Dec. 1999.
213. Veneklasen L., H. Kao, S. Rishton, S. Winter, V. Boegli, T. Newman, G. Bertucelli, G. Howard, P. Le, Z. Tan, and R. Lozes, "High-throughput electron-beam lithography with a raster-scanned, variably shaped beam," *J. Vac. Sci. Technol. B* 19(6), p. 2455, Nov/Dec. 2001.
214. Newman T., et al., "Raster shaped beam pattern generation for 70 nm production," *Proc. SPIE* 4889, p. 168, 2002.
215. Delong A. and V. Kolarik, "BS500 Electron lithography system—Czechoslavakia academy of sciences Brno," *Czechoslavakia Patent* 221592, 1986.
216. Etec Systems, "Development of a short wavelength pattern generator," *NIST ATP program*, Oct. 2000.
217. Newman T., R. Pease, and W. DeVore, "Dot matrix electron beam lithography," *J. Vac. Sci. Tech. B* 1(4), p. 999, Oct/Dec. 1983.
218. Coyle S., D. Holmgren, X. Chen, T. Thomas, A. Sagle, J. Maldonado, B. Shamoun, P. Allen, and M. Gesley, "Prototype raster multibeam lithography tool," *J. Vac. Sci. Technol. B* 20(6), p. 2657, Nov/Dec. 2002.
219. Maldonado J., S. T. Coyle, B. Shamoun, M. Yu, T. Thomas, D. Holmgren, X. Chen, B. DeVore, M. R. Scheinfein, and M. Gesley, "A raster multibeam lithography tool for sub 100 nm mask fabrication utilizing a novel photocathode," *Proc. SPIE* 5220, p. 46, 2003.
220. Maldonado J., S. Coyle, B. Shamoun, M. Yu, M. Gesley, and P. Pianetta, "Cs halide photocathode for multi-electron-beam pattern generator," *J. Vac. Sci. Technol. B* 22(6), p. 3025, Nov/Dec 2004.
221. Gesley M., "Multi-Electron Beam Pattern Generator for Advanced Prototyping," *Final technical progress report, SPAWAR Contract N66001-03-C-8023*, DARPA Advanced Lithography Research Program, 5 Jan. 2004.

222. Maldonado J., S. Coyle, J. Varner, R. Moore, P. Stark, and D. Larson, "Preliminary evaluation of surface plasmon enhanced light transmission with a scanning 257 nm ultraviolet microscope," *J. Vac. Sci. Technol. B* 22(6), p. 3553, Nov/Dec. 2004.
223. Gesley M., "Retarding electron beams in multiple electron beam pattern generation," *United States Patent Application* 20040090194, 13 May 2004.
224. Smith H., "A statistical analysis of ultraviolet, x-ray, and charged-particle lithographies,"*J. Vac. Sci. Technol. B* 4, p. 148, 1986.
225. Leunissen L., R. Jonckheere, K. Ronse, and G. Derksen, "Influence of gate patterning on line edge roughness," *J. Vac. Sci. Tech. B* 21(6), p. 3140, Nov/Dec. 2003.
226. Kruit P., S. Steenbrink, R. Jager, and M. Wieland, "Optimum dose for shot noise limited CD uniformity in electron-beam lithography," *J. Vac. Sci. Technol. B* 22(6), p. 2948, Nov/Dec. 2004.
227. Rau N., et al., "Shot noise and edge roughness effects in resists patterned at 10 nm exposure," *J. Vac. Sci. Tech. B.* 16(6), p. 3784, Nov/Dec. 1998.
228. Gallatin F., F. Houle, and J. Cobb, "Statistical limitiations of printing 50 and 80 nm contact holes by EUV lithography," *J. Vac. Sci. Tech. B* 21(6), p. 3172, Nov/Dec. 2003.
229. Han G., M. Khan, Y. Fang, and F. Cerrina, "Comprehensive model of electron energy deposition," *J. Vac. Sci. Tech. B* 20(6), p. 2666, Nov/Dec. 2002.
230. Han G., M. Khan, and F. Cerrina, "Stochastic modeling of high energy lithographies," *J. Vac. Sci. Tech. B* 21(6), p. 3166, Nov/Dec. 2003.
231. Ma Y., G. Tsvid, and F. Cerrina, "Line edge roughness of sub-100 nm dense and isolated features: experimental study," *J. Vac. Sci. Tech. B* 21(6), p. 3124, Nov/Dec. 2003.
232. Yoshizawa M., and S. Moriya, "Study of the acid-diffusion effect on line edge roughness using the edge roughness evaluation method," *J. Vac. Sci. Technol. B* 20(4), p. 1342, 2002.
233. International Technology Roadmap for Semiconductors, *ITRS* 2001, http://public.itrs.net/Files/2001ITRS/Home.htm.
234. Grebinski T., "The anatomy of a universal data model," *Proc. SPIE* 4889, p. 127, 2002.
235. Schellenberg F., "Advanced Mask Data Preparation and Design Automation," *Photomask Japan* 4754-13, 2002.
236. Sturgeon R., "The new "Y2K" problem for mask-making (or surviving mask data problems after 2000," *Proc. SPIE* 3748, p. 190, 1999.
237. Ashida I., Y. Sato, and H. Kawahira, "Proposal of new layout data format for LSI patterns," *Proc. SPIE* 3748, p. 205, 1999.
238. SEMI 2004, http://downloads.semi.org/pubs/semipubs.nsf/
239. Suzuki T., J. Hirumi, N. Yoshioka, Y. Hojyo, Y. Kawase, S. Sakamoto, K. Kuriyama, S. Narukawa, and M. Houga, "Enhancement of unified mask data formats for EB writers," *Proc. SPIE* 5567, p. 1186, 2004.
240. Kuriyama K., et al., "Unified mask data formats for EB writers," *Proc. SPIE* 5130, p. 660, 2003.
241. Haruki T., R. Tsujimura, J. Tomida, Y. Machida, S. Asai, and I. Hanyu, "Development of total support CAD system for alternate-type PSMs with optical proximity correction," *Proc. SPIE* 3748, p. 222, 1999.
242. Joshi D., K. Keogan, and J. Falbo, "Integrating CD and lithographic process window analysis with mask data preparation for subwavelength ICs," *Proc. SPIE* 4889, p. 74, 2002.
243. Kuriyama K., et al., "Development of new stream format with GDSII upper compatibility and high-compression rate," *Proc. SPIE* 4889, p. 83, 2002.
244. Suzuki T., et al., "Improvement of unified mask data formats for EB writers," *Proc. SPIE* 5446, p. 414, 2004.
245. Palmer C., *Diffraction Grating Handbook*, 5th ed., Richardson Gratings, Rochester, New York, 2002.
246. Hirano R., K. Matsuki, S. Yoshitae, Y. Takahashi, S. Tamamushi, Y. Ogawa, and T. Tojo, "Pattern positioning error of reticle writing induced by reticle clamping," *J. Vac. Sci. Technol. B* 13(6), p. 2625, Nov/Dec. 1995.
247. Trost D., "Kinematic Stage Assembly," *US patent application publication* 20020154839.

248. Shamoun B., R. Engelstad, and D. Trost, "Assessment of thermal loading-induced distortions in optical photomasks due to e-beam multipass patterning," J. *Vac. Sci. Tech. B* 16(6), p. 3558, Nov/Dec. 1998.
249. Shamoun B., D. Trost, and Frank Chilese, "A predictive and corrective model for bulk heating distortion in photomasks," *Proc. SPIE* 4000, p. 621, 2000.
250. Shamoun B., W. Trybula, R. Engelstad, and E. Lovell, "Effects of material properties on patterning distortions of optical reticles," *Proc. SPIE* 3546, p. 214, 1998.
251. Sohn J., R. Engelstad, and E. Lovell, "Predicting pattern-specific distortions induced during optical mask patterning," *J. Vac. Sci. Tech. B* 19(6), p. 2626, Nov/Dec. 2001.
252. Chu D., D. Bilir, R. Pease, and K. Goodson, "Submicron thermocouple measurements of electron-beam resist heating," *J. Vac. Sci. Technol. B* 20(6), p. 3044 Nov/Dec. 2002.
253. Pasiecznik J., and J. Reeds, "Digitally positioned mechanical stage," *J. Vac. Sci. Technol.* 15 (3), p. 909, May/June 1978.
254. Wilson A. D., "Electron-beam systems for precision micron and submicron lithography," *Proc. IEEE* 71, p. 575, 1983.
255. Nielsen R., J. Bruning, R. Richman, C. Biddick, J. Giachi, G. Kossyk, D. Bush, S. Barna, and D. Alles, "A hydraulic X-Y stage system for application in electron beam exposure systems," *J. Vac. Sci. Technol. B* 5(1), p. 57, Jan/Feb. 1987.
256. Kunioka T., Y. Takeda, T. Matsuda, N. Shimazu, and Y. Nakayama, "XY stage driven by ultrasonic linear motors for the electron-beam x-ray mask writer EB-X3," *J. Vac. Sci. Technol. B* 17(6), p. 2917, Nov/Dec. 1999.
257. Trost D., "Using air bearings in vacuum to control stage vibration," *Semicon. Int.l* 165, pp. 165–168, July 2002.
258. Trost D., "Positioning Device," *US patent* 6,888,619, May 3, 2005.
259. Cahen O. and J. Trotel, *Fourth International Conference on Electron and Ion Beam Science and Technology* R. Bakish, ed., Electrochemical Society, Princeton, N.J., 581, 1970.
260. Wilson A. D. and T. Studwell, "Vibration analysis using the stripe scan method of a laser interferometer controlled X-Y table in an electron beam system," *J. Vac. Sci. Technol.* 16, p. 1819, 1979.
261. Bobroff N., "Recent advances in displacement measuring interferometry," *Meas. Sci. Technol.* 4 p. 907, 1993.
262. Moriyama S., S. Ozasa, and N. Saitou, "Yaw corrected precision X-Y stage for high-throughput electron-beam lithography systems," *J. Vac. Sci. Technol. B* 3(1), p. 102, Jan/Feb. 1985.
263. Kendall R., S. Doran, and E. Weissmann, "A servo guided X-Y-theta stage for electron-beam lithography," *J. Vac. Sci. Technol. B*9, p. 3019, 1991.
264. Innes R., "Yaw compensation for an electron-beam lithography system," *J. Vac. Sci. Technol. B*12, p. 3580, 1994.
265. Reeds J., S. Hansen, O. Otto, A. Carroll, D. McCarthy, and J. Radley, "High speed precision X-Y stage," *J. Vac. Sci. Tech. B* 3(1), p. 112, Jan/Feb. 1985.
266. Hastings J., F. Zhang, and H. Smith, "Nanometer-level stitching in raster-scanning electron-beam lithography using spatial-phase locking," *J. Vac. Sci. Tech. B* 21(6), p. 2650, Nov/Dec. 2003.
267. Konkola P., C. Chen, R. Heilmann, C. Joo, J. Montoya, C. Chang, and M. Schattenburg, "Nanometer-level repeatable metrology using the Nanoruler," *J. Vac. Sci. Tech. B* 21(6), p. 3097, Nov/Dec. 2003.
268. Mikkelson A., R. Engelstad, E. Lovell, K. Blaedel, and A. Claudet, "Extreme ultraviolet lithography mask flatness and electrostatic chucking analysis," *J. Vac. Sci. Tech. B* 21(6), p. 3091, Nov/Dec. 2003.
269. Lu B., J. Wasson, J. Weisbrod, Z. Masnyj, P. Mangat, K. Nordquist, and D. Resnick, "Proximity and heating effects during electron-beam patterning of ultraviolet lithography masks," *J. Vac. Sci. Tech. B* 20(6), p. 3029, Nov/Dec. 2002.
270. Chang T. H. P., "Proximity effect in electron beam lithography," *J. Vac. Sci. Technol.* 12, p. 1271, 1975.
271. Kyser D., "Spatial resolution limits in electron beam nanolithography," *J. Vac. Sci. Tech. B* 1(4), p. 1391, Oct/Dec. 1983.

272. Parikh M., and D. Kyser, "Energy deposition functions in electron resist films on substrates," *J. Appl. Phys.* 50, p. 1104, 1979.
273. Owen G., "Methods for proximity effect correction in electron lithography," *J. Vac. Sci. Technol. B* 8(6), p. 1889, Nov/Dec. 1990.
274. Broers A., "Resolution limits of PMMA resist for exposure with 50 kV electrons," *J. Electrochem. Soc.* 128, p. 166, 1981.
275. Weiss R., and R. Sills, "Application of an electron beam scattering parameter extraction method for proximity correction in direct-write electron beam lithography," *Proc. SPIE*, 1465, p. 192, 1991.
276. Gerber P. D., "Exact solution of the proximity effect equation by a splitting method," *J. Vac. Sci. Technol. B* 6(1), p. 432, Jan/Feb. 1988.
277. Rishton S., and D. Kern, "Point exposure distribution measurements for proximity correction in electron beam lithography on a sub-100 nm scale," *J. Vac. Sci. Tech.* 5, p. 135, 1987.
278. Anderson E., D. Olynick, W. Chao, B. Harteneck, and E. Veklerov, "Influence of sub-100 nm scattering on high-energy electron beam lithography," *J. Vac. Sci Tech. B* 19 (6), p. 2504, Nov/Dec. 2001.
279. Hofmann U., R. Crandall, and L. Johnson, "Fundamental performance of state-of-the-art proximity effect correction methods," *J. Vac. Sci. Tech. B* 17(6), p. 2940, Nov/Dec. 1999.
280. Parikh M., "Self-consistent proximity effect correction technique for resist exposure (SPECTRE)," *J. Vac. Sci. Technol.* 15(3), pp. 931–933, 1978.
281. Nishimura E., T. Takigawa, T. Abe, and Y. Katoh, "Proximity effect reduction in high voltage electron beam lithography by bias exposure method," *J. Vac. Sci. Tech. B* 4(1), p. 164, Jan/Feb. 1986.
282. Murata K., D. Kyser, and C. Ting, "Monte Carlo simulation of fast secondary electron production in electron beam resists," *J. Appl. Phys.* 52, p. 4396, 1981.
283. Wu B., and A. Neureuther, "Energy deposition and transfer in electron-beam lithography," *J. Vac. Sci. Technol. B* 19(6), p. 2508, Nov/Dec. 2001.
284. Owen G., and P. Rissman, "Proximity effect correction for electron beam lithography by equalization of background dose," *J. Appl. Phys.* 54(6), p. 3573, 1983.
285. Owen G., P. Rissman, and M. Long, "Application of the GHOST proximity effect correction scheme to round beam and shaped beam electron lithography systems," *J. Vac. Sci. Technol. B* 3(1), p. 153, Jan/Feb. 1985.
286. Kostelak R., E. Kung, M. Thomson, and S. Vaidya, "GHOST proximity correction technique: its parameters, limitations, and process latitude," *J. Vac. Sci. Tech. B* 6(1), p. 448, Jan/Feb. 1988.
287. Muray A., R. Lozes, K. Milner, and G. Hughes, "Proximity effect correction at 10 keV using GHOST and sizing for 0.4 um mask lithography," *J. Vac. Sci. Tech. B* 8(6), p. 1775, Nov/Dec. 1990.
288. Abe T., N. Ikeda, H. Kusekabe, R. Yoshikawa, and T. Takigawa, "Proximity effect correction for an electron beam direct writing system EX-7," *J. Vac. Sci Tech. B* 7(6), p. 1524, Nov/Dec. 1989.
289. Murai F., H. Yoda, S. Okazaki, N. Saitou, and Y. Sakitani, "Fast proximity effect correction method using a pattern area density map," *J. Vac. Sci. Technol. B* 10(6), p. 3072, Nov/Dec. 1992.
290. Abe T., T. Kamikubo, M. Shimizu, So. Oogi, H. Anze, Y. Hattori, T. Iijima, T. Tojo, and T. Takigawa, "Proximity effect correction for maskmaking," *Proceedings of the International Workshop High Throughput Charged Particle Lithography*, Hawaii, Nov. 1997.
291. Hattori K., "Hardware proximity effect correction for mask writing," *Proceedings of the International Workshop High Throughput Charged Particle Lithography*, Hawaii, Nov. 1998.
292. Abe T., S. Yamasaki, R. Yoshikawa, and T. Takigawa, "The representative figure method for the proximity effect correction [III]," *J. Vac. Sci Tech. B* 9 (6), p. 3059, Nov/Dec. 1991.
293. Boegli V., L. Johnson, H. Kao, L. Veneklasen, U. Hofmann, I. Finkelstein, S. Stovall, and S. Rishton, "Implementation of real-time proximity effect correction in a raster shaped beam tool," *J. Vac .Sci. Tech. B* 18 (6), p. 3138, Nov/Dec. 2000.

294. Kim S., S. Choi, and J. M. Sohn, "Mask process proximity correction for next-generation mask fabrication," *J. Vac. Sci. Tech. B* 21(6), p. 3041, Nov/Dec. 2003.
295. Chang M., A. Yu, J. Chen, J. Lin, J. Huang, F. Hsu, and H. Liu, "Laser proximity correction for advanced mask manufacturing," *Photomask Japan* 113, 2001.
296. Christenson K., R. Viswanathan, and F. Hohn, "X-ray mask fogging by electrons backscattered beneath the membrane," *J. Vac. Sci. Technol. B* 8(6), p. 1618, 1990.
297. Shimomura N., M. Ogasawara, J. Takamatsu, S. Yoshitake, K. Ooki, N. Nakayamada, F. Okabe, and T. Tojo, "Reduction of fogging effect caused by scattered electrons in an electron beam system," *Proc. SPIE* 3748, p. 408, 1999.
298. Ogasawara M., N. Shimomura, J. Takamatsu, S. Yoshitake, K. Ooki, N. Nakayamada, H. Okabe, T. Tojo, and T. Takigawa, "Reduction of long range fogging effect in a high acceleration voltage electron beam mask writing system," *J. Vac. Sci. Technol. B* 17 (6), p. 2936, 1999.
299. Jang S., S. Yang, B. Ahn, W. Ki, J. Choi, S. Choi, and W. Han, "Analysis of dose modulation method for fogging effect correction at 50-KeV e-beam system," *Proc. SPIE* 5446, p. 669, 2004.
300. Murai F., S. Okazaki, N. Saitou, M. Dan, "The effect of acceleration voltage on linewidth control with a variable-shaped electron beam system," *J. Vac. Sci. Technol. B* 5(1), p. 105, Jan/Feb. 1987.
301. Kratschmer E., "Resist heating effects in 25 and 50 kV e-beam lithography on glass masks," *J. Vac. Sci. Tech. B* 8(6), p 1898, Nov/Dec. 1990.
302. Hartley J., and T. Groves, "Chrome on glass mask writing at 75 kV with the IBM EL4+ electron-beam system," *J. Vac. Sci. Tech. B* 17(6), p. 2932, Nov/Dec. 1999.
303. Babin S., and I. Kuzmin, "Experimental verification of the TEMPTATION (temperature simulation) software tool," *J. Vac. Sci. Technol. B* 16, p. 3241, 1998.
304. Kuwahara N., H. Nakagawa, M. Kurihara, N. Hayashi, H. Sano, E. Maruta, T. Takikawa, and S. Noguchi, "Preliminary evaluation of proximity and resist heating effects observed in high acceleration voltage e-beam writing for 180-nm-and-beyond rule fabrication," *Proc. SPIE* 3748, p. 115, 1999.
305. Cummings K., "A study of deposited charge from electron beam lithography," *J. Vac. Sci. Technol. B* 8, p. 1786, 1990.
306. Itoh H., K. Nakamura, and H. Hayakawa, "Charging effects on trilevel resist and metal layer in electron-beam lithography," *J. Vac. Sci. Technol. B* 9, p. 309, 1991.
307. Liu W., J. Ingino, and F. Pease, "Resist charging in electron beam lithography," *J. Vac. Sci. Technol. B* 13, p. 1979, Sept/Oct. 1995.
308. Bai M., F. Pease, C. Tanasa, M. McCord, D. Pickard, and D. Meisburger, "Charging and discharging of electron beam resist films," *J. Vac. Sci. Technol. B* 17(6), p. 2893, Nov/Dec. 1999.
309. Bai M. and F. Pease, "Electron beam induced conductivity in polymethyl methacrylate, polyimide, and SiO_2 thin films," *J. Vac. Sci. Technol. B* 22(6), p. 2907, Nov/Dec. 2004.
310. Hatzakis M., "Electron resists for microcircuit and mask production," *J. Electrochem. Soc.* 116, p. 1033, 1969.
311. Hatzakis M., "Recent developments in electron-resist evaluation techniques,"*J. Vac. Sci. Technol.* 12(6), p. 1276, 1975.
312. Bowden M. J., R. F. W. Pease, L. D. Yau, J. Frackoviak, L. F. Thompson, J. G. Skinner, J. P Ballantyne, "PBS Positive Electron Resist—Capabilities and Limitations," *H. Ahmed, W.C. Nixon, Microcircuit Engineering*, Cambridge University Press, UK, 1980.
313. Bowmer T. N. and J. H. O'Donnell, "Post-irradiation thermal degradation of poly(olefin sulfones)," *Polymer*, 22 January 1981; "Propagation/depropagation equilibrium and structural factors in the radiation degradation of poly(olefin sulfones)s," *J. Polymer Sci.: Polymer Chemistry* 19, pp. 45–50, 1981.
314. Bowden M. J., L. F. Thompson, W. Robinson, and M. Biolsi, "Thermal degradation of poly(1-butene sulfone)," *Macromolecules* 15, p. 1417, 1982.
315. Novembre A. E, R. G. Tarascon, L. F. Thompson, W. T. Tang, C. O. Tange, R. A. Bostic, and D. H. Ahn, "Initial manufacturing performance of an actively controlled PBS resist development process," *Proc. SPIE* 1809, p. 76–84, 1992.
316. Dean R., H. Lem, C. Sauer, and H. Chang, "PBS resist profile studies for sub-micron mask lithography," *Proc. SPIE* 2322, pp. 102–113, 1994.

317. Nakamura K., S. L. Shy, C. C. Tuo, and C. C. Huang, "Critical dimension control of poly-butene-sulfone resist in electron beam lithography," *Jpn. J. Appl. Phys.* 33, p. 6989, 1994.
318. Dean R. and C. Sauer, "Further work in optimizing PBS: is it capable of meeting specifications for 256 Mbit DRAM reticle manufacturing?," *Proc. SPIE* 2621, p. 386–398, 1995.
319. Dean R., and C. Sauer, "Improvement of CD Uniformity with PBS by optimizing the crossover development step," *Proc. SPIE* 2884, p. 473–476, 1996.
320. Dean R., "Characterization and modeling of CD performance with thin PBS," *Proc. SPIE* 3236, 1998.
321. Crawford J. E. and D. N. Gray, "Preparation and properties of Some Poly (alpha-olefin sulfones)," *J. Applied Poly. Sci.* 15, p. 1881, 1971.
322. Stockmayer W. H., A. A. Jones, and T. L. Tredwell, "On Chain Dynamics in Poly(alkene sulfones)s," *Macromolecules* 10(4), p. 762, 1972.
323. Huang J. P., T. K. Kwei, and A. Reiser, "On the molecular mechanism of positive novolac resists," *Proc. SPIE* 1086, 1989.
324. Dean R., and G. Flores, "Process optimization of positive novolac resists for electron-beam lithography, resist characterization using single or multiple development steps with either sodium hydroxide or metal ion free developer," *Proc. SPIE* 1925, 1993.
325. Chiong K., K. Petrillo, F. Hohn, A. Wilson, and W. Moreau, "Contrast and sensitivity enhancement of resists for high resolution lithography," *J. Vac, Sci. Technol. B* 6(6), 1988.
326. Grenon B., D. Sprout, W. Aaskov, and P. Buck, "Investigation of a negative i-line resist with the CORE-2564 laser writer," *Proc. SPIE* 2322, p. 149, 1994.
327. Felmlee T., and V. Raghavan, "A comparative study of MEBES III and CORE 2564 performance in a manufacturing environment," *Proc. SPIE* 1924, p. 234, 1993.
328. Hua-yu Liu, "Characterization of KTI 895i as an electron-beam resist," *BACUS*, 1992.
329. Tan Z. C., P. Lee, and H. Lem, "Evaluation of DNQ/novolac resists for 130 nm device maskmaking," *Microelectron. Eng.*, 41/42, p. 315, 1998.
330. Rathsack B., C. Tabery, S. Scheer, M. Pochkowski, C. Philbin, F. Kalk, C. Henderson, P. Buck, and G. Willson, "Optical lithography simulation and photoresist optimization for photomask fabrication," *BACUS News Photomask* 15 (5), p. 1–4, May 1999.
331. Kobayashi Y., Y. Oppata, Y. Ezoe, F. Shigemitsu, K. Urayama, and K. Doi, "High-contrast i-line positive photoresist for laser reticle writer," *Proc. SPIE* 4066, p. 289, 2000.
332. Kim M. Y., J. H. Lee, Y. J. Yoon, B. Y. Choi, M. Lu, and R. Dean, "A 150 nm mask fabrication using ZEP 7000 resist, GHOST, and dry etch for MEBES 5000 with advanced process," *Photomask Japan* 4066, p 243, 2000.
333. Ito H., "Chemical Amplification Resists: History and development within IBM,"*IBM J. Res. Dev.* 41, p. 69, 1997.
334. Brunner T., "Why optical lithography will live forever," *J. Vac. Sci. Tech. B* 21 (6), p. 2632, Nov/Dec. 2003.
335. Yoshimura T., Y. Nakayama, and S. Okazaki, "Acid-diffusion effect on nanofabrication in chemical amplification resist," *J. Vac. Sci. Technol. B* 10(6) p. 2615, 1992.
336. Kozawa T., A. Saeki, A. Nakano, Y. Yoshida, S. Tagawa, "Relation between spatial resolution and reaction mechanism of chemically amplified resists for electron beam lithography," *J. Vac. Sci. Tech. B* 21(6), p. 3149, Nov/Dec. 2003.
337. Baik K., R. Dean, M. Mueller, M. Lu, H. Lem, S. Osborne, and F. Abboud, "REAP (Raster E-Beam Advanced Process) Using 50 kV Raster E-beam System for Sub-100 nm Node Mask Technology," *Proc. SPIE* 4688, p. 401, 2002.
338. Baik K., H. Lem, R. Dean, M. Mueller, T. Newman, and F. Abboud, "A CAR Process using 50 kV Raster E-beam System," *Photomask Japan*, 4754, p. 215, 2002.
339. Tan Z. C. H., T. Stivers, H. Lem, N. DiGiacomo, and D. Wood, "In search of high-performance resists for 50 kV shaped-beam exposure," *J. Vac. Sci. Technol. B* 13(6), p. 2539, Nov/Dec. 1988.
340. Kurihara M., M. Komada, H. Moro-oka, N. Hayashi, H. Sano, "EBR900 processes in e-beam and laser beam lithographies for photomask production," *Proc. SPIE* 2437, p. 240, 1995.

341. Hayashi N., "Electron beam lithography process for advanced optical masks," *J. Vac. Sci. Tech. B* 16(6), p. 3158, Nov/Dec. 1998.
342. Nordquist K., D. Resnick, and E. Ainley, "Comparison of negative resists for 100 nm electron-beam direct write and mask making applications," *J. Vac. Sci. Tech. B* 16(6), p. 3289, Nov/Dec. 1998.
343. Irmscher M., D. Beyer, J. Butschke, C. Constantine, T. Hoffmann, C. Koepernik, C. Krauss, B. Leibold, F. Letzkus, D. Mueller, R. Springer, and P. Voehringer, "Comparative evaluation of e-beam sensitive chemically amplified resists for mask making," *Photomask Japan*, 4754, 2002.
344. Butschke J., D. Beyerb, C. Constantine, P. Dress, P. Hudek, M Irmscher, C. Koepernik, C. Krauss, J. Plumhoff, and P. Voehringer, "90 nm mask making processes using the positive tone chemically amplified resist FEP171," *Proc. SPIE* 5256, p. 40, 2004.
345. Chiong K., S. Wind, and D. Seeger, "Exposure characteristics of high-resolution negative resists," *J. Vac. Sci. Technol. B* 8, p. 1447, 1990.
346. Tschinkl M., C. Burgel, U. Griesinge, B. Jeansannetas, "A. Vix, Negative chemically amplified resist (nCAR) for DRAM mask fabrications," *Proc. SPIE* 4889, p. 137, 2002.
347. Takahashi N., "Performance of JBX-9000MV with negative tone resist for 130 nm reticle," *Proc. SPIE* 4186, p. 22, 2000.
348. Ito S. and K. Hayasaki, "Monitor and control for development technology," *J. Vac. Sci. Tech. B* 21(6), p. 3177, Nov/Dec. 2003.
349. Dean R., "Virtual address and a diffusion controlled process," *Bacus Monthly Proc. SPIE* 2348, 1993.
350. Dean R., "Advances in PBS Processing with MEBES IV," *Bacus Monthly Proc. SPIE* 2158, 1993.
351. Oaawa K., and N. Abe, "Puddle developers for ZEP 7000," *BACUS*, 1999.
352. Willson C. G., "Organic resist materials—theory and chemistry," *Introduction to Microlithography* L. F. Thompson, C. G. Willson, and M. J. Bowden, eds., *ACS Symposium Series 219*, American Chemical Society, Washington, DC, 1983.
353. Thompson L. F., D. G. Willson, and J. M. J. Frechet, eds., "Materials for Microlithography—Radiation-Sensitive Polymers," *ACS Symposium Series 266*, American Chemical Society, Washington, DC, 1984.
354. Reichmanis E., and L. F. Thompson, "Polymer materials for microlithography," *Annual Review of Materials Science* 17, R. A. Huggins, J. A. Giordmaine, and J. B. Wachtman, Jr., eds., Annual Reviews, Palo Alto, Calif., 1987.
355. Reichmanis E., and A. E. Novembre, "Lithographic resist materials chemistry," *Annual Review of Materials Science* 23, R. A. Laudise, E. Snitzer, R. A. Huggins, J. A. Giordmaine, and J. B. Wachtman, Jr., eds., Annual Reviews, Palo Alto, Calif., 1993.
356. Willson C. G., "Organic Resist Materials," and L. F. Thompson, "Resist Processing," *Introduction to Microlithography*, 2d ed., L. F. Thompson, C. G. Willson, and M. J. Bowden, eds., American Chemical Society, Washington, DC, 1994.

Chapter

4

Photomask Pattern Transfer

Banqiu Wu

4.1 Photomask Pattern Transfer Basics

The technology of photomask pattern transfer can be traced to the very early days of the semiconductor industry. After the transistor was invented in 1947, photoresist was applied as the masking layer for (oxide) etching in 1955. Combined with the traditional wet etch process at that time, the basic technology for photomask fabrication was established. After the integrated circuit was invented in 1959 at Texas Instruments, photomask fabrication including pattern transfer appeared and progressed with integrated circuit (IC) fabrication technology. When plasma etching was introduced into the semiconductor industry in 1968 using a barrel plasma etcher from Branson Co., both wet and dry etching provided photomask pattern transfer alternatives. Original photomask patterns were formed either by contact printing or by light beam exposure until the introduction of electron beam at Cambridge and JEOL in 1969. Now, with a high resolution source, the main processing methods were available for modern photomask fabrication. The progress of the photomask industry was directed by both market and technological challenges. Once automated mask inspection tools were introduced by KLA in 1976, almost all the current photomask fabrication processes, including pattern transfer, were established.

Photomask fabrication and pattern transfer progressed along with semiconductor wafer fabrication until the mid-1980s. From then to the mid-1990s, wafer-printing tools were changed from 1× magnification to 4× or 5× magnification steppers, resulting in *easier* mask fabrication. In those 10 years, the photomask played a lesser role in semiconductor related industries.

In the mid-1990s, IC feature sizes shrank to the submicron level, resulting in technological challenges for wet etch photomask pattern transfer. Then several photomask dry etch systems were introduced to the market, including the PlasmaTherm mask etcher and the Applied Materials Tetra mask etch system.

4.1.1 Pattern transfer processes

Pattern transfer is performed by an etching process in the fabrication of photomask. Pattern can be transferred from the photoresist layer to an absorber layer (Cr layer), the Cr layer to the MoSi layer (phase shifter), or the Cr layer to the quartz (fused silica) substrate. Any changes during etching to original (source) patterns may result in final (target) pattern changes. The pattern transfer method also causes some differences between original and final patterns. One of these examples is the *undercut* in a *wet etch process*, which significantly changes the *critical dimension* (CD).

The basic requirements for pattern transfer have three categories. One is to minimize etching in the lateral direction for critical dimension fidelity, the purpose for which is to maintain the same pattern size and shape after the process. In practical photomask fabrication, it is extremely difficult to achieve perfect pattern fidelity because there usually is some lateral etching. That means that there may be similar pattern shape and pattern signature before and after the etch process, but the dimensions may become larger or smaller. Modern electron-beam pattern writing tools can form a much better pattern on high resolution *chemically amplified resist* (CAR) layers than on the previously used resist, but there are still some systematic errors on resist layer patterns, especially those with high loads because of electron scattering (fogging).

Therefore, another basic requirement for pattern transfer is to correct or compensate for most or all of the systematic pattern errors existing on the masking layer (e.g., resist). One example where this is done is in MoSi etch CD compensation.[1] In positive CAR dark feature patterns, CD values in high-load areas are usually smaller than that in low-load areas on what is called a *high-load photomask* (e.g., the amount of clear area is >35 percent). This phenomenon dictates the use of a MoSi etch process that corrects or compensates in the high or low-load regions, i.e., the MoSi etch CD movement (bias) needs to have a distribution with higher etch CD bias in low-load areas relative to the high-load areas.

The third requirement for pattern transfer is the profile specifications on the pattern layer. The importance of this requirement varies over the mask layers of a semiconductor device. A common profile specification is for absorber sidewall angle. This property not only affects the photomask CD measurement values, but also can affect the wafer printing results. Vertical etch rate uniformity is important for pattern transfer with respect to Cr etching. When the Cr layer is on top of the MoSi layer as for *embedded attenuated phase-shift masks* (EAPSMs), Cr etching influences the trench depth into the MoSi layer during the standard overetch step. It will, with the MoSi vertical etch rate nonuniformity, contribute to the phase angle range error. Because of the low MoSi-to-quartz selectivity, MoSi overetch rate uniformity (affecting quartz etch) is a critical parameter in controlling the overall MoSi layer phase angle errors. The MoSi layer can function as a "stop" layer for Cr etch, and the quartz substrate can serve as a stop layer for the MoSi etch although neither have infinite selectivity. For quartz etch there is no stop layer, so very strict control on the quartz vertical etch rate is required. This is one of the biggest challenges for quartz etching in the fabrication of *alternating aperture phase-shift masks* (AAPSMs) and *chromeless phase lithography* (CPL) masks.

Technologically, the most challenging pattern transfers are in the fabrication of phase-shift masks, including the pattern transfer from the resist to Cr and from Cr to MoSi. Pattern transfer on a binary photomask (resist to Cr layer only) is relatively easier. For the pattern transfers to the Cr layer, the masking (photoresist) layer is an energy sensitive material that can be exposed by energetic beams such as laser or electron radiation. This mask layer is a temporary *soft mask* and will be removed completely after the pattern transfer (etch) by a process called *resist strip*. For the photomask pattern transfer from the Cr layer to the MoSi layer and from the Cr layer to the quartz substrate, the Cr functions as the masking layer, called the *hard mask*. Because the final photomask application requires that the Cr layer is removed (stripped) only in selected regions, usually a "second" pattern transfer is required. It consists of a second resist coating on the patterned Cr layer, resist exposure and development, and a second pattern transfer process. After the second pattern transfer, the final Cr pattern is the result of the two pattern transfer processes. For the pattern transfer to the MoSi layer, there is an alternative way to process where the MoSi layer is etched with the resist remaining on top of the Cr layer. That means the resist strip does not take place until after MoSi etch. There is no significant fidelity difference between these two processes. The main consideration for which to use is based on the level of defects one encounters.

4.1.2 Pattern transfer and photomask fabrication system

Photomask fabrication consists of resist writing, postexposure bake (PEB), resist development, and plasma etching, as well as the backend processes. The main objective for process optimization is to seek the optimal system, or the best final result. Individual processes having minimal systematic errors favor complete fabrication system optimization. However, when the systematic errors in the individual processes cannot be eliminated using the state-of-the-art technology, it is necessary to seek systematic error compensation (correction) between the different processes.

The most common systematic errors include CD radial distribution, CD dependency on local loading, *isolated/dense* (I/D) feature CD bias, phase angle radial distribution, and phase angle side-to-side distribution. In a photomask fabrication process, CD radial distribution is usually due to errors in the writing tool, PEB, resist developing, and etching processes. Because it is difficult to compensate among these processes, each process' systematic error must be controlled to the lowest level. I/D CD bias can affect final photomask pattern fidelity in several process steps, but it can be corrected or compensated using a *proximity effect correction* (PEC) method resident on modern electron-beam write tools, giving more options to the engineers than other process optimization strategies.

CD dependency on local loading is a challenging systematic error because it is very difficult to eliminate. The problem mainly occurs on high-load photomasks (e.g., >35 percent), especially for dark feature CDs. The systematic CD errors in writing and Cr pattern transfer processes have similar distribution signatures. These errors usually make the dark CDs in high-load areas smaller in the resist layer and even worse in the Cr layer. Although this error can be corrected to some extent during MoSi etch,[1] there are still significant residual CD errors in the final MoSi layer. Dose correction during writing can improve this loading-dependent CD distribution, but it is very difficult to reduce this error to a negligible level. Because of this, pattern transfer process optimization of high-load photomasks mainly focuses on reducing adverse effects due to loading.

Up to this point, a positive-tone resist system has been assured. However, the use of negative-tone resists can significantly reduce etch loading effects. The use of a negative resist cannot only save write time and increases production yield, but it can also improve the CD uniformity on high-load photomasks. Historically, positive-tone photoresists were much more widely used in the photomask production because negative resists were difficult to control in a manufacturing process. However, recent progress on negative CAR processing in photomask production shows that photomasks using negative CAR may have better CD uniformity than positive CAR on high-end clearfield photomasks, while maintaining the same stability as seen in positive CAR processes.

CD deviation from nominal [or CD *mean-to-target* (MTT)] is another systematic error. Contributions can come from virtually all processing steps. For low-load photomasks, it is not difficult to control the final average CD to an acceptable level. For example, MTT can be controlled to the ±10-nm range without difficulty by changing process recipes. But for high-load photomasks, it is much more difficult to control to this tight MTT range. It is highly dependent on process variables and process tool designs. Data sizing by adjusting pattern dimensions in the data is a useful method for MTT CD control. In some special situations when data sizing is not allowed, a more creative strategy needs to be undertaken by adjusting process variables in the different process steps. For example, final CD value can be significantly affected by changing the dose value in the writing process, or overetch time in the pattern transfer process.

4.1.3 Etch processing for photomask pattern transfer

Etching is a process that literally eats into any areas laid bare while the rest of the surface is protected by a resistive coating. It is a very old process and has been used in the arts, printing, and material processing for many years. The etching method had been a wet process employing an acid until *dry etching (plasma etching)* was introduced into the semiconductor industry in 1968. Etching in the semiconductor industry is a process by which material is removed from a thin film or substrate surface by (chemical) reaction with a reagent material (etchant).

The science of plasma etching is a special gas-solid (chemical) reaction occurring under vacuum conditions with volatile product(s). The parameters that make it unique lie in the system temperatures, reactive radical properties, and energy supply methods employed. During a plasma etch, there is a significant temperature difference between solid reactants and gases. The solid (etched material) reactant has nearly the same temperature as the ambient, but the temperatures in the gas phase depend on the particles. In a plasma used for etching, electron temperature is much higher than that of ions and neutrals, and ion bombardment supplies the energy with anisotropic properties. The combination of these factors results in an anisotropic etch—the most valuable etch property in a plasma environment.

Since Texas Instruments introduced them in 1959, integrated circuits have been manufactured using wet etching. Around the late 1960s and early 1970s, plasma etching emerged as a replacement for traditional wet etching for integrated circuit fabrications.[2–4] Anisotropy is a highly desired process parameter in IC manufacturing and has been a major force in driving the development of plasma processing technology.

For photomask fabrication purposes, different plasma sources were applied.[5–8] One dominant design is the *inductively coupled plasma* (ICP) design that makes it possible to adjust plasma density and bombardment energy separately when used in combination with bias power.

After the implementation of plasma etching applications in the semiconductor industry, many studies were conducted on plasma fundamentals, process optimization,[9,10] modeling and simulation,[11–14] and kinetics.[15–21] By the mid-1970s, plasma etching and stripping were fully compatible with all stages of *complementary metal-oxide-semiconductor* (CMOS) fabrication, although plasma etching processes were semiempirically developed. The majority were not well understood by that time because of the complexity of the reactions.[22]

Plasma-etched materials in the semiconductor industry include silicon nitride, monocrystalline silicon (111), polycrystalline silicon (poly), silicon dioxide (thermal), silicon dioxide [*chemical vapor deposition* (CVD)], tungsten, molybdenum, titanium, tantalum, aluminum, GaAsP, chromium, aluminum-silicon metallization, gold, Ti-Au alloy, MoSiON, fused silica (glass), and also photoresist (stripping). Some review papers were published on these plasma etch processes.[22–24]

Photomask plasma etching dates back to the very early days of the semiconductor industry in the 1970s.[22,25–27] Although pioneering studies were performed at that time, plasma photomask etches did not have wide application until approximately 20 years later. The main reason may be that before the 1990s, a photomask could be wet etched without any technological challenges for the feature sizes of that era. In the 1990s, when feature sizes dropped to submicron levels, wet-photomask etches could not meet the requirements for small feature size mask fabrication because of the large undercut from the isotropic nature of a wet etch process.

For submicron node photomask fabrication, plasma etch process development for practical applications (not merely for technological potential) were investigated in TRW (1992),[28] AT&T (1993),[29] and Etec and IBM (1993).[30] Since then plasma etching has gradually replaced traditional wet etch processing for essentially all the state-of-the-art (high-end) photomasks. The most important advantage of photomask plasma etching has been its anisotropic etch property leading to higher pattern fidelity.

When traditional Cr photomask resolution came close to its technological limit, *resolution enhancement techniques* (RETs) were proposed and then phase-shift masks appeared (i.e., MoSi and glass etch emerged).

After the mid-1990s, many research papers were published on the etch processing of Cr layers,[31–81] MoSi layers of phase-shift masks,[82–95] fused silica (quartz) of alternating aperture phase-shift masks and chromeless phase lithography photomasks,[96–103] and *next generation lithography* (NGL) masks.[81,104–107] There is a significant difference between wafer and photomask plasma etches concerning the technological specifications and applications. Regarding materials, photomask etches include chrome/chromium etches as well as MoSi and fused silica (quartz) etches. But for wafer fabrication, many etches are silicon- and silica-based material etches as well as other metal etches.

4.1.4 Wet and dry etching of photomasks

Wet etching is a very old technique and existed long before semiconductor industrial fabrication. Although the chemicals (etchants) for different applications vary, the processing principles are very similar.

Currently, about 80 to 85 percent of all photomasks made worldwide are still etched using a wet process (wet etched), but all high-end photomasks today are etched using plasma processes. Because the technological challenges of photomask etching exist in high-end mask processing, almost all publications and industrial emphasis related to photomask etching are focused on plasma etching. Although high-end photomask pattern transfer does not use wet etching, it is widely used for Cr strip after MoSi etch and second pattern transfer in advanced PSM fabrication. Therefore wet etching process still plays some role even for some state-of-the-art photomask fabrication.

Cr wet etching is performed in an aqueous solution containing perchloric acid and ceric ammonium nitrate. The main technological disadvantage of a wet etch is the high degree of undercut (e.g., 100 to 150 nm on each side) and low sidewall profile angle (e.g., 45°). A wet etch also has a faster etch rate for chromium oxide (antireflective, or AR layer) than for chromium in the chrome-AR stack. This makes the sidewall angle profile even worse and creates a faceted edge. The shallow sidewall profile also makes it very difficult to accurately determine the critical dimension size. Other wet etch problems include the wetting problem associated with small geometries during wet etch. Figure 4.1 demonstrates the Cr profile and undercut processes of both dry etch and wet etch.

Some alternating aperture phase shift mask fabrications use *dry plus wet* quartz etch in some high-end photomask fabrication to obtain moderate undercut for the sake of wafer printing benefits. The quartz is wet etched in a hydrofluoric acid (HF) bath with a very strict process control after plasma dry etching. Because HF etch rate is extremely sensitive to temperature and reactant concentration, the temperature and concentration distributions have to be controlled very strictly to ensure high quality.

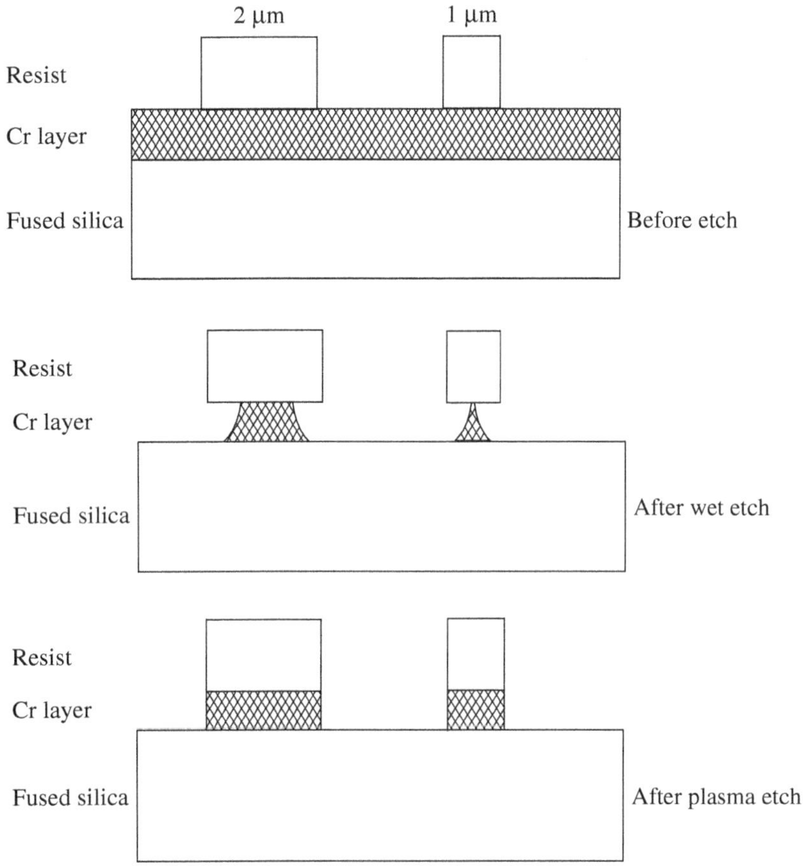

Figure 4.1 Cr profile comparison between wet etch and plasma etch.

4.2 Plasma Etch Fundamentals

4.2.1 Plasma principles

A plasma, the fourth state of matter, is an ionized gas that consists of molecules, atoms, ions, electrons, and photons. Aside from the ions, a plasma is electrically neutral, and this property is called quasineutrality.

In a plasma gas having a mixture of different particles and photons, the particles can be divided into two categories: *heavy particles* (including neutral atoms, neutral molecules, and ions), which travel at low velocities (compared to electrons), and *electrons* with high velocities mainly responsible for inelastic collisions.

A plasma may contain free electric charges by a variety of ionization processes. In a steady state situation, the rate of ionization in the plasma is balanced by the rate of recombination. Many processes can cause ionization, such as high energy radiation, collision processes in electrical discharges, and collision processes in intensely heated gases.

From the definition of particle temperature T_i, we have:

$$\frac{3}{2}kT_i = \frac{1}{2}m_i v_i^2 \tag{4.1}$$

where k is Boltzmann's constant, m_i is the mass of particles i, and v_i is their velocity.

Plasmas can be classified into two types—the *equilibrium plasma* (also called thermal plasma or hot plasma) and the *nonequilibrium plasma* (also called nonthermal plasma or cold plasma). Two of the most important features of thermal plasmas are the equality between electron and heavy particle temperature ($T_e = T_h$) and the existence of chemical equilibrium; however, nonequilibrium plasmas have strong deviations from kinetic equilibrium ($T_e >> T_h$), and the "sensible" temperature T_h is frequently close to room temperature (T_e is electron temperature). Figure 4.2 shows a diagram of electron densities and electron temperatures of various plasmas.[108]

Among the particles, electrons are very light. When the same force (e.g., the electric force in an electric field) acts on an electron and an ion, the resulting velocity is very different. The result is that the velocity of

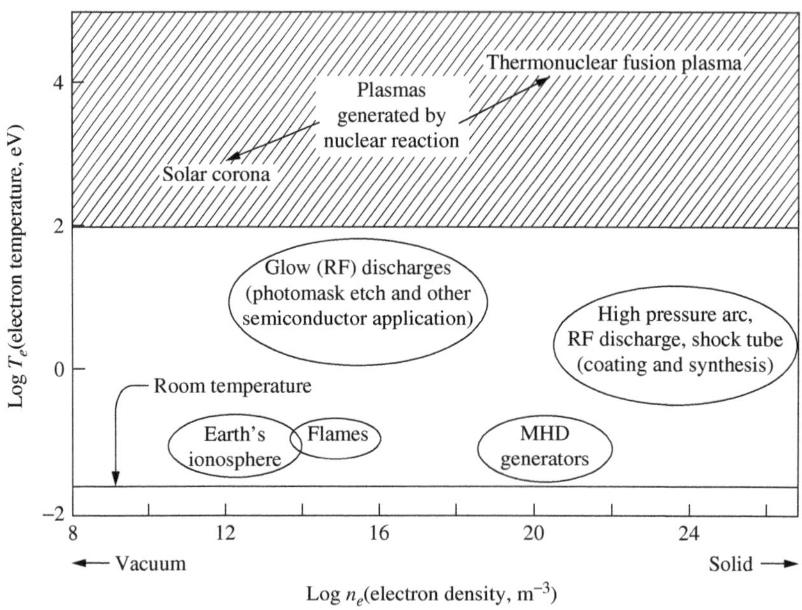

Figure 4.2 Plasma classification.

the electron will be much higher than that of the heavy ion because of the lower mass. If the collisions are not numerous enough at low-density conditions, i.e., low pressure, the mean kinetic energy of the electrons will be higher than that of the much heavier ions (T_h). Under these circumstances, the system is considered to be in a nonequilibrium state, and the plasma is known as a cold plasma. Plasmas for photomask etching are nonequilibrium types as shown in Fig. 4.3.

Plasmas can be obtained by a variety of methods. For certain applications, plasmas can be produced by electrodeless *radio frequency* (RF) discharges, microwaves, shock waves, lasers, or high energy particle beams. The principal use of thermal plasmas began in 1878 when Siemens developed a DC furnace,[108] while industrial application of cold plasma started in 1911 in an *electrostatic precipitator* (ESP) for particle collection using DC corona discharge.[109,110] Plasma applications in the semiconductor industry started around 1968 when the barrel plasma etcher was introduced by Branson.

It is the gas discharge that forms a plasma. The strong attraction between electrons and the remainder of gas atoms or molecules requires an intense energy supply to start and maintain the discharge. This requires that all processes involving high-energy intensity have the capability to create plasma environments, like laser beams, microwaves, radiation, and extremely high temperature. The energy can be coupled through a field, wave heated, and heated directly. On the other hand,

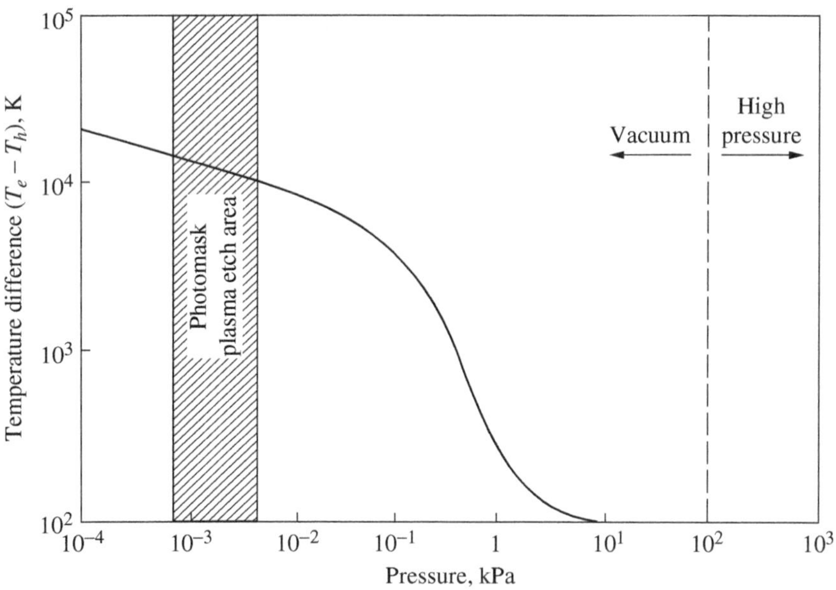

Figure 4.3 Temperature difference between heavy particles and electrons.

the energy can be applied to neutral molecules or atoms through impaction by a charged particle with high enough energy. For impaction effective enough to generate a plasma, neutral particle impaction can be enhanced by providing a long acceleration distance (large mean free path at high vacuum) or by a strong electric field (geometrically asymmetric DC electrodes). Therefore the plasma can have different styles based on the discharge, as shown in Table 4.1.

Inductively discharged plasma is widely used in material processing and can also be referred to as inductively coupled plasmas, *transformer-coupled plasma* (TCP), or *RF inductive* (RFI) plasma. ICP density (10^{11}~10^{12} cm^{-3}) is higher than that of capacitive discharged plasma, but lower than that of wave-heated plasma (>10^{13} cm^{-3}). ICP is a nonresonant plasma, the main advantages of which include structure simplicity, no requirement for DC magnetic fields, and lower RF frequency than microwave. Therefore, ICP combined with capacitive discharged plasmas presently dominate the designs for photomask etch tools.

Photomask and wafer plasma etches use low pressure nonequilibrium plasmas, while higher pressure plasmas are used for other applications such as ESP and resident air cleaners.

Plasma technology applications mainly started in the twentieth century, but plasma behavior can be described using classical physics principles. Plasma fundamentals include electromagnetic field theories, basic conservation equations, and some equilibrium properties. Force laws, laws of motion, particle collisions and dynamics are also very important. Besides, the plasma (DC sheaths) structure, plasma-enhanced chemical reactions, and transport phenomena play very important roles in material processing within plasma environments. Thermodynamics, chemical reaction kinetics (especially nonelementary reaction kinetics), and chemical reaction engineering methods are also very helpful for plasma etch applications in photomask and wafer pattern transfer

TABLE 4.1 Plasma Discharge Varieties

Discharge methods	Pressure	Equilibrium condition
Capacitive discharge	Vacuum	Nonequilibrium
Inductive discharge (ICP)	Vacuum	Nonequilibrium
Wave-heated discharge (ECR and helicon)	Vacuum	Nonequilibrium
DC discharge		
Glow discharge	Vacuum	Nonequilibrium
Corona discharge	Normal	Nonequilibrium
Nontransferred arc	Normal	Equilibrium
Transferred arc	Normal	Equilibrium

processes. Because of the complexity inherent in plasma etching, it is extremely difficult to obtain theoretical equations showing relationships between operational parameters and responses. This makes empirical and semiempirical relationships more significant because they are directly applicable to the process optimization.

The existence of free charged particles in the plasma environment dictates that field theories and related equations can be applied to plasma etch processing. These applications come from Maxwell's equations:

$$\nabla \cdot E = \frac{\rho}{\varepsilon_o} \tag{4.2}$$

$$\nabla \times E = -\frac{\partial B}{\partial t} \tag{4.3}$$

$$\nabla \cdot B = 0 \tag{4.4}$$

$$c^2 \nabla \times B = \frac{j}{\varepsilon_o} + \frac{\partial E}{\partial t} \tag{4.5}$$

The conservation of charge equation:

$$\nabla \cdot j = -\frac{\partial \rho}{\partial t} \tag{4.6}$$

force law:

$$F = q(E + v \times B) \tag{4.7}$$

law of motion:

$$\frac{d}{dt}(p) = F \qquad \text{where } p = \frac{mv}{\sqrt{1 - v^2 / c^2}} \tag{4.8}$$

and gravitation:

$$F = -G\frac{mm'}{r^2} \tag{4.9}$$

where E and B = electric and magnetic fields, respectively
c = speed of light
ρ = charge density
ε_o = permittivity
j = electric current density
t = time
v = velocity
q = charge

F = force
m and m' = masses
r = distance
G = gravitational constant

Particles in the plasma chamber move around and are driven by random motion, electric and magnetic field forces, and inertial forces. Particle motion results in particle collisions among electrons, ions, molecules, atoms, and other neutral particles. In many application conditions, the properties in the bulk plasma can be treated as homogeneous. In boundary regions contacting solid surfaces such as the plasma chamber wall, the difference in particles and their behaviors creates a heterogeneous layer, called a plasma sheath. Compared with the bulk plasma, the volume and dimension of the sheath are very small, but its significance is extremely high. Plasma etching occurs at the sheath border, which makes the sheath much more difficult to understand completely. Heterogeneous properties and gradients of concentration and temperature cause significant transport phenomena in plasmas, especially in the sheath area. Chemical dissociation also makes this already complicated region even more complex.

Electron and heavy particle temperatures in plasmas are different, indicating that the frequency of these particles hitting the solid surface will be different. On the other hand, the charge equity in plasma systems requires the same impacting rates of positive and negative particles. The result is the building up of an electric potential gradient toward the solid surface, which limits the electron bombardment, but enhances positive particle bombardment to balance the charges on the surface.

The bulk plasma consists of charged and neutral particles and can be treated as an electrolyte. The study of electrolytes was originated by Arrhenius in the nineteenth century, and later Ghosh proposed a different view on the nature of electrolyte solutions in 1918 to 1920, stating that the electrolyte is an ordered lattice-like structure of ions. The modern theory of electrolytic solutions was proposed by Debye and Hückel with the intention to improve Ghosh's theory.

In their model, Ghosh's basic idea of the crystal-like distribution of ions in solution was accepted, but the ions in solution were arranged spherically around a central ion due to thermal motion. The sphere surrounding the central ion cannot always contain the same ions so there is a continuous interchange between the ions contained in the sphere and other ions. Such a statistical sphere around the central ion is called the ionic atmosphere, or ionic cloud. The energy associated with the ionic atmosphere is electrical in origin and is a function of the charge density and the potential produced by the ionic atmosphere itself. Therefore Maxwell's equation can be used in the analysis.

Maxwell's first equation shows that the divergence of electric field E is the charge density over permittivity ε_0. For the one-dimensional discussion, the equation can be simplified to:

$$\frac{d^2\phi}{dx^2} = -\frac{\rho}{\varepsilon_o} \tag{4.10}$$

where ϕ is the electric potential and ρ is the charge density. According to statistical mechanics, the density n of particles at the position x is given by:

$$n(x) = n_o e^{-U(x)/kT} \tag{4.11}$$

where $U(x)$ is the potential energy, k is Boltzmann's constant, and T is the absolute temperature. For a single electronic charge, the potential energy of a positive ion can be expressed as:

$$U(x) = q_e \phi(x) \tag{4.12}$$

where q_e is the electron charge. The density of positive ions is then:

$$n_+(x) = n_o e^{-q_e\phi(x)/kT} \tag{4.13}$$

The density of negative ions similarly can be obtained:

$$n_-(x) = n_o e^{+q_e\phi(x)/kT} \tag{4.14}$$

The total charge density is:

$$\rho = q_e n_+(x) - q_e n_-(x) = q_e n_o (e^{-q_e\phi(x)/kT} - e^{+q_e\phi(x)/kT}) \tag{4.15}$$

and after combining Eqs. (4.10) and (4.15), we have:

$$\frac{d^2\phi}{dx^2} = -\frac{q_e n_o}{\varepsilon_o}(e^{-q_e\phi(x)/kT} - e^{+q_e\phi(x)/kT}) \tag{4.16}$$

and by approximation, we have:

$$e^{\pm q_e\phi(x)/kT} = 1 \pm \frac{q_e\phi}{kT} \tag{4.17}$$

Then we have:

$$\frac{d^2\phi}{dx^2} = +\frac{2n_o q_e^2}{\varepsilon_o kT}\phi(x) \tag{4.18}$$

and the solution for the sake of this discussion is:

$$\phi = Ae^{-x/D} \tag{4.19}$$

where A is constant and D is called Debye length expressed as:

$$D = \sqrt{\frac{\varepsilon_o kT}{2n_o q^2}} \tag{4.20}$$

The Debye length was originally formulated for electrochemical analyses in aqueous solutions and now is widely used for the study of plasmas. Its physical meaning is that in the Debye length range of the bulk plasma, there is a significant distribution of properties such as charges and ionic concentration. When the region size is larger than the Debye length in a bulk plasma, the plasma can be considered homogeneous.

In plasmas, the relative still ions may make oscillations of electrons about their equilibrium position. Based on Maxwell's equation, we have a one-dimensional electric field equation:

$$\frac{\partial E_x}{\partial x} = \frac{n_o q_e}{\varepsilon_o} \frac{\partial s}{\partial x} \tag{4.21}$$

where E_x is the electric field component in the x direction, x is the coordinate position, and s is the electron displacement from the equilibrium position. Integrating Eq. (4.21) gives:

$$E_x = \frac{n_o q_e}{\varepsilon_o} s \tag{4.22}$$

The force on an electron in the displaced position is:

$$F_x = -\frac{n_o q_e^2}{\varepsilon_o} s \tag{4.23}$$

This shows a restoring force proportional to the displacement s of the electron, leading to a harmonic oscillation of the electrons. The motion of a displaced electron can be expressed as:

$$m_e \frac{d^2 s}{dt^2} = -\frac{n_o q_e^2}{\varepsilon_o} s \tag{4.24}$$

The s will vary harmonically with the frequency of oscillation called *(electron) plasma frequency*, expressed as:

$$\overline{\omega} = \sqrt{\frac{n_o q_e^2}{\varepsilon_o m_e}} \tag{4.25}$$

Because

$$e^2 = \frac{q_e^2}{4\pi\varepsilon_o} \tag{4.26}$$

the plasma frequency can also be expressed as:

$$\overline{\omega}_p = \sqrt{\frac{4\pi e^2 n_o}{m_e}} \tag{4.27}$$

When its frequency is lower than the plasma frequency, the radiowave cannot penetrate the plasma, but will be reflected back. Usually plasma frequencies for discharges are in the microwave region (1 to 10 GHz).[111]

4.2.2 Plasma sheath

In the boundary area of plasma, the properties and charges are different from the plasma's bulk region. The main reason is due to the mobility difference between negative charges (electrons) and positive charges (ions). The neutral plasma property and fast electron mobility predict accumulation of electrons near the wall area, which results in the existence of a neutral bulk plasma and nonneutral plasma near the wall, i.e., the sheath. The plasma sheath is a nonneutral potential region between the bulk plasma and a solid surface. In order to maintain the continuum of ion flux, a transition layer between the bulk plasma and sheath must exist. This area is called the presheath. There is a potential distribution, but the concentration of ions and electrons is considered to be identical. The plasma sheath structure is shown in Fig. 4.4.

In low-pressure plasma conditions used for photomask etching, the sheath can be treated as a collisionless region. Then there is ion energy conservation such that:

$$\frac{1}{2}m_i u_i^2 = \frac{1}{2}m_i u_{io}^2 + e\phi \tag{4.28}$$

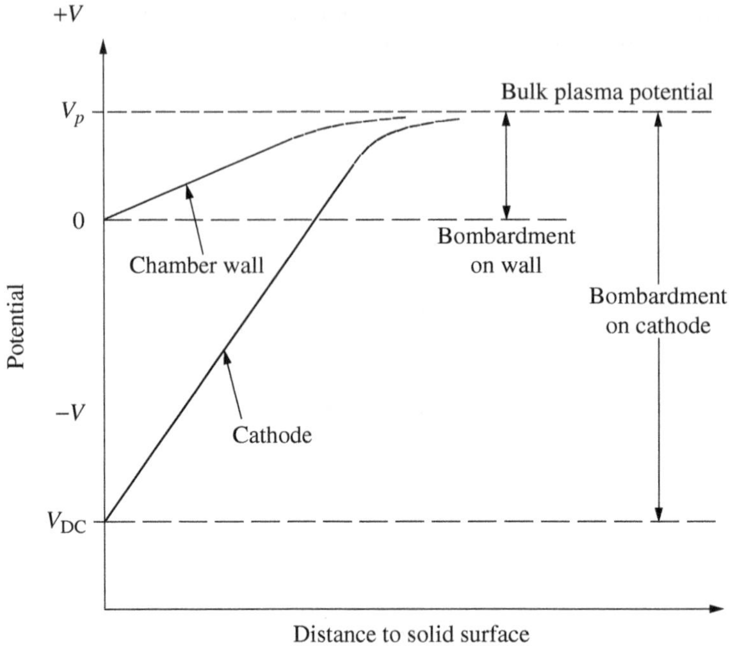

Figure 4.4 Potential distribution in an asymmetric plasma etch chamber.

where m_i is the ionic mass; u_i and u_{io} are ionic velocities in the sheath and in the border position between the sheath and presheath, respectively; and ϕ is the electric potential of the ion. When no ionization in the sheath is assumed, we have the ion flux continuity equation:

$$n_i u_i = n_{io} u_{io} \tag{4.29}$$

where n_i and n_{io} are ion concentrations in the sheath and on the sheath edge, respectively. For the sheath potential at a floating wall, we have: [111]

$$\phi_w = -\frac{kT_e}{2}\ln\left(\frac{m_i}{2\pi m_e}\right) \tag{4.30}$$

where m_e is the electron mass and T_e is the electron temperature. This formula shows that the floating wall potential is proportional to the logarithm of the mass ratio.

For a capacitively coupled plasma with asymmetric areas on both electrodes, the sheath potential drop is significantly different. It is usually expressed as:[6]

$$\frac{\phi_a}{\phi_b} = \left(\frac{A_b}{A_a}\right)^n \tag{4.31}$$

where the n value is usually greater than or equal to 2, mostly depending on the geometric design of the plasma reactor and the plasma conditions; the A_a and A_b are electrode areas on electrode a and b, respectively; and ϕ_a and ϕ_b are potential drops on sheath a and b, respectively. In a plasma etch chamber for a photomask process, the cathode on which photomask rests is usually much smaller than the other electrode's surface area, including the chamber wall itself. This means that the potential drop on the cathode is much larger than that on the chamber wall, reducing the ionic bombardment on the chamber wall and inherent contamination due to process on the chamber wall.

The high velocity of electrons due to their small mass and quasineutral properties in a plasma is responsible for the low electron density in the plasma sheath and the potential drop toward the surface. The potential drop across the sheath is a function of relative masses of electrons and ions, electron temperatures, and reactor chamber design. The results are that the ions have an accelerating voltage across the plasma sheath, enabling an anisotropic plasma etch. When a DC voltage or a capacitively driven radio frequency voltage is applied on a surface (cathode), the potential drop across the plasma sheath is enhanced. Plasma etching under these conditions is called *reactive ion etching* (RIE). Actually the etch is not limited to ionic process. In many situations, the etching occurs with neutral radicals. Therefore, RIE is not necessarily an accurate term, but due to historical reasons, the term RIE is used regardless.

Because direct current (DC) power for bias voltages can be only used for conductive materials, RF voltages with 13.56-MHz frequencies are widely used for plasma etching in photomask fabrication. The potential drops across the sheath on solid surfaces are similar to those without using bias power. After RF bias power on the processed surface (cathode) is applied, the potential drop increases asymmetrically. The potential drop increases due to bias power on the cathode results in different potential-drop-increase values on the cathode and chamber walls, because the potential drop is nearly inversely proportional to the surface area squared. In commercial plasma etch equipment, the chamber wall is usually grounded. To simplify the facilitation of potential drops, DC bias is widely used instead of the potential drop across the plasma sheath. DC bias voltage is defined as the potential difference between the cathode and the plasma chamber wall (grounded).

The potential drop on the sheath (or DC bias) is one of the most important characteristics of plasma etching. It supplies anisotropic bombardment energy, which significantly reduces the undercut compared with wet etching. In practical photomask etching processes, the anisotropic etch is not perfect, especially in the Cr etch case. The lateral etch rate is significantly high relative to vertical etch rate, and the control of anisotropy is critical to mean-to-target CD result success.

Among the most common etches for photomask fabrication, DC bias voltages are usually in the range of 100 to 250 V for Cr etch, 50 to 150 V for MoSi etch, and 200 to 500 V or even higher for quartz etch.

An important plasma property is its anisotropic energy supply to the etch surface since the reactive radicals may or may not be the bombardment ions during photomask etching. When the reactant is a neutral radical, ion bombardment supplies only energy.

The etch reaction rates in different directions depend on etch reaction conditions. In wet etch processing, the etch reaction is isotropic. The reason is that the lateral and vertical surfaces have the same kinetic parameter values, such as temperature and reaction concentration—the former overcoming activation energy and the latter supplying reactants.

In a plasma etch environment, ionic bombardment overcomes the activation energy. Anisotropic bombardment determines the directional distribution on energy transport. In a bulk plasma chamber, the ion angular distribution is isotropic, and there is no significant property distribution when the domain dimension is larger than Debye length. In the presheath, ion angular distribution is not isotropic and after ions pass the sheath, this anisotropic property is enhanced by the electric field acceleration on the ions. Generally, the average angle of incidence θ can be expressed as:[5]

$$\theta = \tan^{-1}\sqrt{\frac{T_i}{V_{\mathrm{DC}}}} \tag{4.32}$$

where T_i is the ion temperature and V_{DC} is the DC bias voltage.

4.2.3 Transport phenomena in plasmas

Transport phenomena play an important role in photomask plasma etch processes. Among the three main transport phenomena—mass transport, energy transport, and momentum transport—mass transport is the most critical to photomask plasma etching. Photomask etch reactions are usually not perfectly activated controlled rate processes, so etch performance is strongly affected by mass transport phenomena.

One of the examples is the etch rate dependence on the etch surface area (loading).

Mass transport in plasma environments involves neutral molecules, atoms, radicals, ions, and electrons. In the plasma sheath, the potential gradient is responsible for the movement of charged particles, while the potential distribution is relatively weak in bulk plasma. The potential gradient plays an important role in the movements of charged particles and their mass transport. According to statistical physics, all of these phenomena are dependent on the collisions between particles such as those between neutral and charged particles.

For one particle, the collision frequency N over time period T can be expressed as:

$$N = \frac{T}{\tau} \tag{4.33}$$

where the constant τ is the average time between collisions. In a system with N_o particles, let $N(t)$ be the number that have not collided up to the time t, then we have:

$$N(t) = N_o e^{-t/\tau} \tag{4.34}$$

Then for average, we have:

$$\text{Average time until the next collision} = \frac{1}{N_o}\int_0^\infty t\,\frac{N(t)dt}{\tau} \tag{4.35}$$

For molecules having a mean velocity υ, we can express the average distance between collisions, i.e., the mean free path λ as the product of τ and υ:

$$\lambda = \tau\,\upsilon \tag{4.36}$$

When forces are applied on some particles in the system, like the ions in neutral particles in plasma environments having a potential gradient, there is net progress in the direction of force F. This means that a drift is superimposed over the random motion, and the speed of this drift is called drift velocity. It can be expressed as:

$$\upsilon_{\text{drift}} = \frac{F\tau}{m} \tag{4.37}$$

where m is the mass of the particles. The drift velocity is proportional to the force, and the ratio of the drift velocity to the force is the mobility:

$$\mu = \frac{\tau}{m} \tag{4.38}$$

The mobility is proportional to the mean time between collisions and inversely proportional to the mass. When there is a gas in the plasma etch chamber, the electric field will produce a force on the ions, so that they drift based on the field direction, and the ionic gas will behave like a resistor.

When the background species are approximately still and at a steady state condition, the force equation is:

$$qn\vec{E} - \nabla p - \frac{m}{\tau} n\vec{v}_{\text{drift}} = 0 \tag{4.39}$$

where p is pressure; n is concentration; and vector E and v_{drift} are the electric field and drift velocity, respectively. For isothermal plasmas, we have:

$$\nabla p = kT\nabla n \tag{4.40}$$

and

$$\vec{v}_{\text{drift}} = \mu q\vec{E} - \frac{\mu kT\nabla n}{n} \tag{4.41}$$

Then the particle flux can be expressed as:

$$\Gamma = n\vec{v}_{\text{drift}} = \mu nq\vec{E} - D\nabla n \tag{4.42}$$

where the diffusivity D can be expressed as:

$$D = \mu kT \tag{4.43}$$

For free diffusion, the particle flux equation [Eq. (4.42)] becomes Fick's law

$$\Gamma = -D\nabla n \tag{4.44}$$

With D independent of position, we can obtain from continuity equation:

$$\frac{\partial n}{\partial t} - D\nabla^2 n = 0 \tag{4.45}$$

For ambipolar diffusion in plasmas consisting of electrons and ions, we have:

$$\frac{\partial n}{\partial t} - D_a \nabla^2 n = 0 \tag{4.46}$$

$$D_a = \frac{\mu_i D_i + \mu_e D_e}{\mu_i + \mu_e} \tag{4.47}$$

where μ_i and μ_e are the mobilities of ions and electrons, respectively; and D_i and D_e are the diffusivities of ions and electrons, respectively.

The gross mass transport in a plasma etch chamber is a product of diffusion due to concentration differences and by convection currents due to density differences. The latter can be much stronger than the former; therefore the gas flow pattern is very critical to the gross mass transport. Etch reactor design also plays an important role in overall mass transport effects.

Mass transport occurs in the bulk plasma and in the plasma sheath. It is usually assumed that the properties throughout the bulk plasma are identical, but there is a significant difference in the sheath. The plasma sheath has a significant effect on charged particles, including electrons and ions. In photomask plasma etch processing, electronegative gases such as fluorine, chlorine, and oxygen are usually used. These gases may form negative charge carriers. The behavior of negative ions plays a very important role in mass transport within the sheath region. Neutral particle mass transport is also affected by charged particles in the plasma environment.

4.2.4 Chemical reaction and equilibrium in plasma

The basic requirement for plasma etches is a nonzero anisotropic etch rate and the presence of volatile etch products. The combination of the plasma sheath structure and the DC bias voltage produces these etch properties with a higher etch rate in the vertical direction than in the lateral direction. It is the anisotropic etch performance that enables the reduction of CD etch bias. The nonequilibrium plasma used in the semiconductor industry requires a high vacuum, which favors the production of volatile products. Halogen group etchants are usually preferred for the photomask plasma etching because of their superior volatile properties.

The etchants in halogen groups are mainly fluorine- or chlorine-containing chemicals. Chlorine or chlorine with oxygen usually has acceptable etch rate selectivity to the materials beneath. It can etch metallic materials with moderate etch rates, but it etches nonmetallic materials such as ceramic or glass materials more slowly. This characteristic enables the MoSi layer to function as a stop layer for the Cr etch in phase-shift photomask fabrication. On the other hand, the fluorine-containing etchants have a strong etch rate on ceramic or glass materials, but a low rate on metallic materials. These fundamental chemical properties make for the basis of etchant selection and optimization.

One mechanism that occurs during plasma etching is physical sputtering. Sputtering does not display good etch selectivity, and the products do not have to be volatile. Because it is difficult to control, pure sputtering etch is not used for pattern transfer in photomask fabrication. The other "competing" plasma etch mechanism is pure chemical etching where reactive radicals created in the plasma environment (such as in the downstream etcher) are used. Actually this etch reaction may not occur in plasma environment. Due to the isotropic property of chemical etching, it is not widely used in photomask pattern transfer.

However, almost all photomask etches belong to the *chemical etching* family with energy supplied by ionic bombardment. Chemical dissociation, equilibrium, electron and ion concentrations, and plasma conditions play important roles in photomask etch performance. Actually, in nonequilibrium plasmas used for photomask etching, there is no "real" thermodynamic equilibrium in the plasma, especially in the regions close to the chamber boundary, such as in the sheath region. In the bulk plasma region (excluding the sheath), *chemical equilibrium* is often assumed. It can be thought of as a *quasiequilibrium*. Therefore the thermodynamic principles can be used for plasma etch analysis in many applications. The most powerful method is still the physical chemical method as seen in thermodynamics, electrochemistry, and chemical kinetics. Electron and ion concentrations in plasma conditions are low. For high-density plasmas, ion and electron concentrations are only at 10^{-4} level. This means that most neutral molecules and atoms are not ionized under the plasma conditions, making it reasonable to use the chemical equilibrium approximation for theoretical analysis in a plasma environment. Therefore, the Gibbs energy minimization method is used to calculate the equilibrium composition due to the fact that the system has minimal Gibbs energy under equilibrium condition.[112–115]

Gibbs energy and its thermodynamic analysis for chemical reactions have wide applications for reaction tendency indication at constant temperature and pressure conditions. An important application is the Ellingham diagram, showing metal oxide reduction capability by carbon or carbon monoxide under industrial operation conditions over a wide temperature range.[116] For complicated industrial systems with multiple products, it is usually not convenient to indicate reaction direction tendency using a single chemical reaction Gibbs energy change. Therefore, the Gibbs energy minimization method found use in several applications for complicated industrial chemical reaction systems. Since the Thermochemistry System software was introduced by *Commonwealth Scientific and Industrial Research Organization* (CSIRO) during the 1970s to the 1980s, a few commercial software packages were developed, such as SOLGASMIX-PV and Outokumpu HSC Chemistry. Original applications were in the field of industrial metallurgical processes. Presently they are applied in plasma syntheses,[112,113] chemical vapor depositions,[114] and astronomy.[115]

Gibbs energy is a state function, and its change within a system from one state to another is dependent only on the two state conditions. Gibbs energy is defined as:

$$G = U + PV - TS = H - TS \tag{4.48}$$

where U = internal energy
P = pressure
V = volume
H = enthalpy
T = temperature
S = entropy

Gibbs energy indicates the direction that the reaction trend proceeds spontaneously. When a reaction Gibbs energy change at constant temperature and pressure is less than zero, the reaction can occur spontaneously, while it is nonspontaneous if it is greater than zero. If the reaction Gibbs energy change is equal to zero, the reaction is at equilibrium. A negative Gibbs energy change for a process does not necessarily mean that the process will take place. It is merely an indication that the process can occur provided that the conditions are right.

For a multiproduct system having predetermined reactant input and thermodynamic conditions of temperature and pressure, the system energy can be expressed as:

$$G = \sum_{i=1}^{m} G_i n_i \tag{4.49}$$

where G is the system Gibbs energy; G_i and n_i are molar Gibbs energy and molar number of species i, respectively; and m is the total species number in the system. Gibbs energy is a function of temperature, pressure, and composition, so when a system attains equilibrium, the system has the minimum Gibbs energy according to which Gibbs energy minimization method is used to calculate the system equilibrium composition.

Plasma density, or the ion-to-neutral ratio varies over a very large range in different plasmas. For the plasma conditions of interest, the ion-to-neutral ratio is very low (even for high-density plasmas). In other words, the majority of particles in the plasma etch chamber are neutral atoms, molecules, and radicals. The radio frequency plasma for photomask etching is not strong enough to break most chemical bonds. Therefore, the chemical behavior follows thermodynamic principles, and the plasma only changes the kinetic process, i.e., the etch rate. Gibbs energy changes for chemical reactions in plasma environments can still be used as a direction indicator for whether a reaction will proceed spontaneously.

Figures 4.5 and 4.6 show examples of calculated equilibrium compositions in predetermined systems consisting of Ta and Cl_2 (Fig. 4.5) and Ta, Cl_2, and O_2 (Fig. 4.6). When Ta is etched using Cl_2 gas, the equilibrium composition in Fig. 4.5 indicates that the reaction produces the volatile etch product $TaCl_{5(g)}$. In this system, there is no nonvolatile product, so the etch process can proceed to completion.

Figure 4.6 shows the composition of a Ta, Cl_2, and O_2 system. The stable condensation product Ta_2O_5 can stop the etch process, so that the etch process using gases containing O_2 cannot proceed to completion.

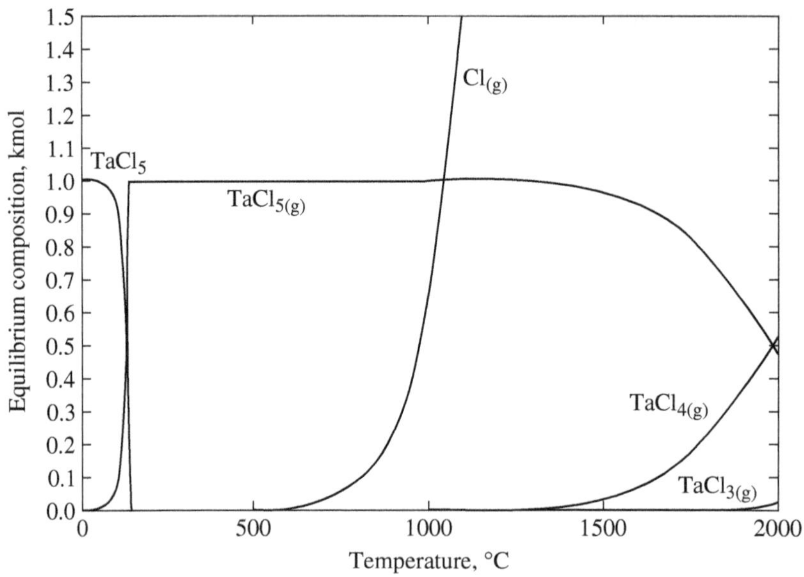

Figure 4.5 Gibbs energy minimization of Ta + Cl_2.

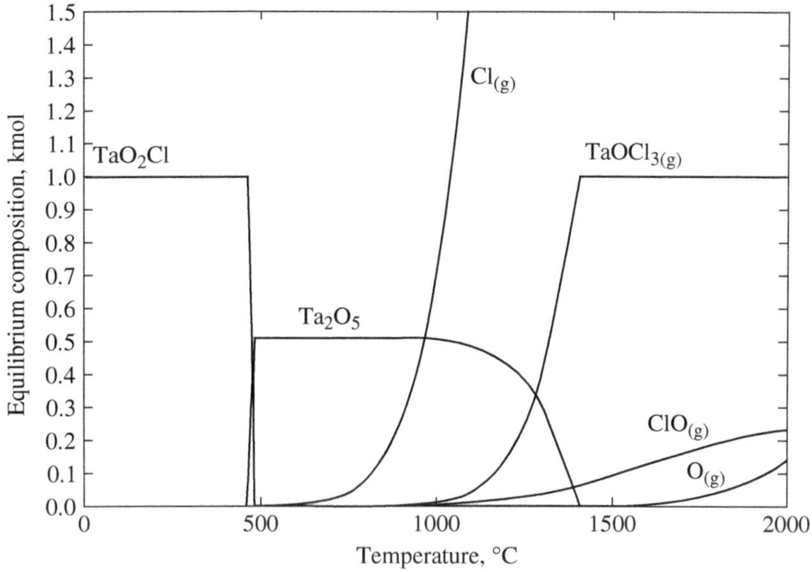

Figure 4.6 Gibbs energy minimization of Ta + Cl_2 + O_2.

4.2.5 Plasma diagnosis and the Langmuir probe

Among all plasma diagnostic methods, the Langmuir probe is the most common. Structurally, the probe is only a metal surface having variable shapes such as planar, cylindrical, or spherical. Theoretically, the Langmuir probe is an electrode, or also called a "working" electrode, and so requires another electrode, i.e., a reference electrode used together. The reference electrode can be the cathode, anode, or another probe, which forms a "double probe." The application of the Langmuir probe is to measure the plasma electron temperature and density by placing the probe directly into the plasma. The ion or electron concentrations near the probe surface will ultimately be recorded. Because of the complexity, a general theoretical solution to cover all applications was not available using this method. Theoretical solutions are only applicable to the specific situations for which they were derived, such as plasmas with or without magnetic fields. The application space is also determined by the relative dimensions between the probe's equivalent radius and the Debye length. In many situations, the probe radius is so large that the original Langmuir's approximation for a simple plane probe can be used. The main advantage of using this electric probe is its simple applicability. The actual measured concentration is of the bulk plasma rather than in plasma sheath. Therefore, the concentration of electrons and ions are identical. The Langmuir probe reports data by plotting a current-voltage (I-V) curve on the working-reference electrodes connected to a variable-voltage power supply in series, as shown in Fig. 4.7. The typical I-V curve is shown in Fig. 4.8.

Figure 4.7 shows that the probe potential is produced by the variable-voltage power supply. When the probe potential is biased positively (V_p increases) with respect to the local plasma potential, the flux of negative particles to the probe is collected until current saturation occurs at a probe potential Φ_p. Because the negative charges are usually electrons, this potential region is called electron saturation. The saturation of electrons can result in a substantial current that can perturb the plasma. When the probe potential is adjusted in the negative direction, negative charge collection on the probe will decrease to a point where there is no negative charge collection present. For any probe potential more negative than this value, only positive charges will be collected. This potential region is called the (positive) ion saturation region.

Since Langmuir's original study on this topic in the 1920s, a large number of investigations have been published. It is not difficult to formulate the ion collection, but the solution to the resultant equation is very difficult. Therefore, different assumptions and approximations have been proposed, including assumptions of cold ions, monoenergetic ions, and monenergetic ions.

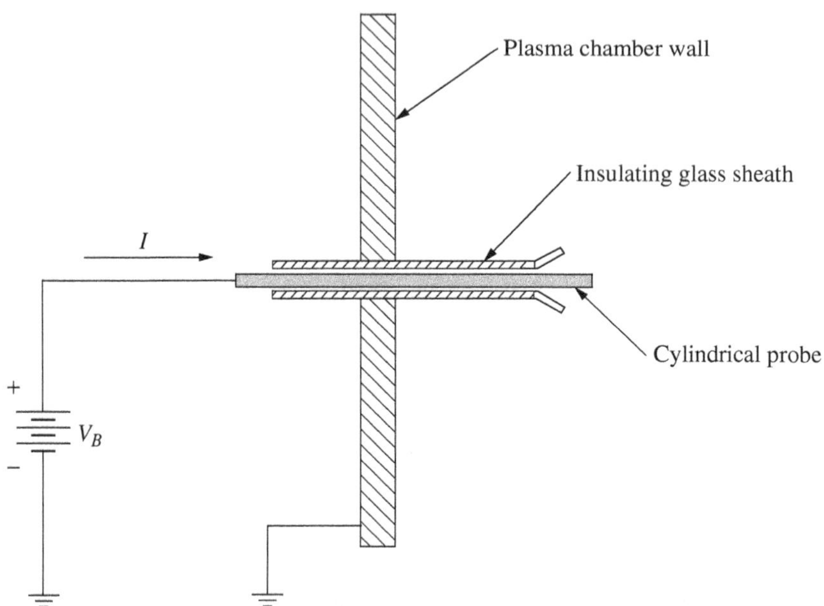

Figure 4.7 Langmuir probe circuit.

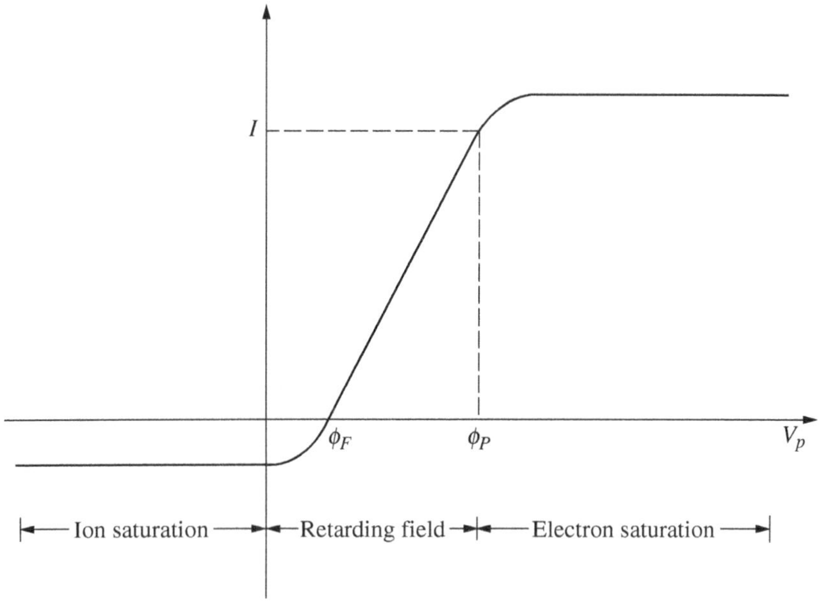

Figure 4.8 Typical I-V curve of Langmuir probe.

Theoretical work provides the solutions to electron temperature and charge density (concentration). The following is a simple example showing how plasma density is measured and calculated using the Langmuir probe method. The relationship among probe current I_p, electron temperature T_e, and probe potential V_p can be expressed as:[117,118]

$$\ln(I_p - I_s) = \text{const} + \frac{eV_p}{kT_e} \tag{4.50}$$

This shows that electron temperature can be obtained from the slope of the logarithmic probe current as a function of probe potential because ion saturation current I_s and probe potential can be experimentally measured. Then the plasma density n_e can be calculated using the following equation:

$$\frac{I_s}{A} = en_e\left(\frac{kT_e}{m_e}\right)^{1/2} \exp\left(\frac{1}{2}\right) \tag{4.51}$$

where A is the probe area.

Single plane Langmuir probes are very simple, but their application is sometimes limited by their performance. Results and accuracy may be affected by low ionization plasma conditions and by the disturbance they cause to the main plasma. Because of this, double Langmuir probes are sometimes used to improve measurement performance.

4.2.6 Plasma etch equipment

Before photomask plasma etching became a mainstream technology, it was performed mainly on modified wafer etchers.[71,119–122] In these applications, the plasma was created by using a 13.56-MHz (barrel-type) RF discharge or a diode-type BMP mask etcher,[71,119] or other reactive ion etch tools such as the Nextral NE110 RIE system,[120] or the Tegal and LAM RIE system.[71,119–122] In the early days of photomask plasma etching, RIE-type and RF (barrel-type) etchers were widely used.[28,29,31,119,120] In the mid 1990s, the ICP-bias power photomask etcher was introduced to photomask etching world.[32] Superior etch performance made this type of etcher dominant in the photomask industry. *Magnetically enhanced RIE* (MERIE) configuration etchers also have found applications since the late 1990s.[36,37,47] The comparison between ICP-bias power and MERIE configurations were performed,[40] showing the former etch performance is better than the latter. The etch performance comparison among ICP-bias power, RIE, and *plasma etch* (PE) configurations was reported, indicating that the ICP-bias power configuration has the best results.[123] This study may be one of the main reasons that ICP-bias power applications are more widely used currently than the rest of the configuration choices. The *neutral loop discharge* (NLD) plasma etcher was introduced to photomask fabrication in the late 1990s,[48] and then several investigations were performed.[53,55,80] Systematic comparisons between ICP-bias power and NLD-configuration plasma etchers have not been published to date.

Photomask etch systems can function using RIE only, but better etch performance can be obtained by combining RIE (bias power) with other high-density plasma scenarios using a variety of coupling methods such as *electron cyclotron resonance* (ECR), inductively coupled plasma, MERIE, and magnetic neutral loop discharge plasma.

Currently, major photomask etch systems have the combination configuration of bias power with inductively coupled plasma. Photomask etch system performance depends not only on the etch tool hardware design, but also on the degree to which the etch process (recipes) has been developed and optimized. Good understanding of the photomask etch process by etch tool manufacturers, and cooperation between tool makers and mask makers becomes critical for the success of the state-of-the-art photomask etching.

Figure 4.9 shows a typical ICP plasma etcher diagram. The RF bias power input into the bias electrode (cathode) goes through the *automatic matching network* (AMN). The RF bias power travels through

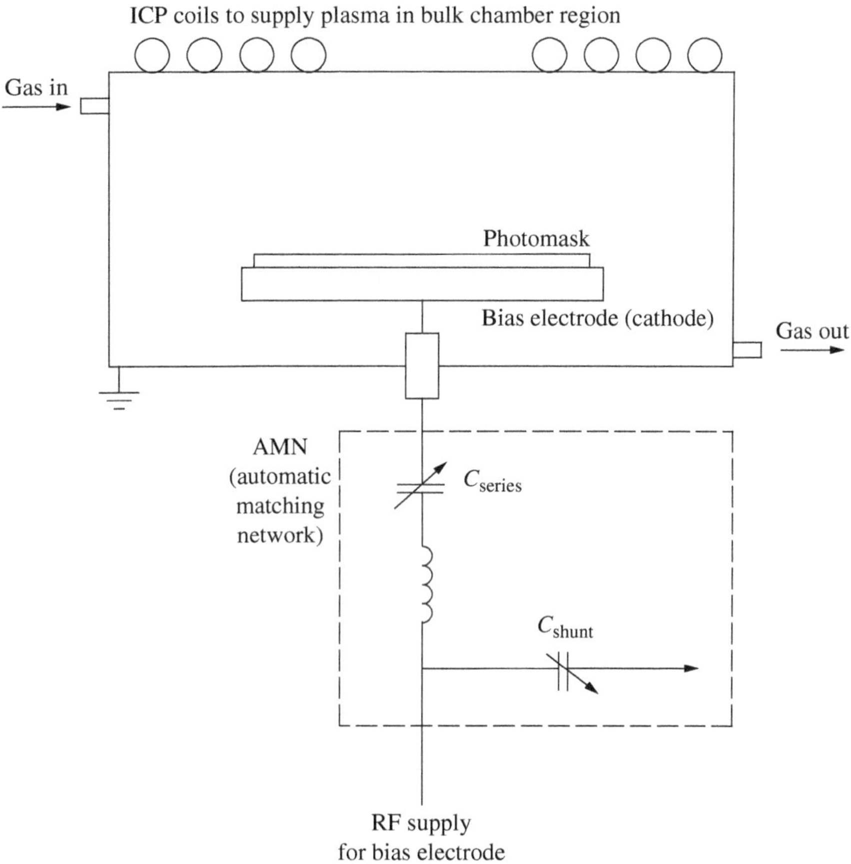

Figure 4.9 ICP plasma etch system diagram.

the chamber and returns to the grounded chamber wall to complete the circuit. The ICP power is inductively coupled to the chamber bulk gas to create a plasma through the ICP coils on top of or outside of the cylindrical chamber, which is like the secondary coil of a transformer. Common RF bias power supplies have a standard 13.56-MHz frequency, while the RF ICP power source frequency usually is either about 2.0 or 13.56 MHz. High-density plasma power supplies are not limited to an ICP strategy since other energy coupling methods, such as microwaves, can also be used. The main reason to use ICP is for operational flexibility and etch performance. Both bias and ICP powers are used together for photomask etching so that high etch performance can be ensured. Only ICP power for Cr etch can function with an adequate etch rate, but bias power can significantly improve other etch properties such as CD etch bias and I/D bias. The most popular photomask etchers are made by

Unaxis and Applied Materials. Other mask etchers include MAPS and NLD by ULVAC, and CleanPlasma by COMTECS.

4.2.7 Photomask test patterns

Photomask quality is dependent on the fabrication processes and their optimization. Etch processes are optimized by using a group of specially designed photomasks, commonly called "test vehicles." The pattern designs on test vehicles are very critical and must be customized for the photomask fabrication process being optimized.

Basic optimization requirements on test vehicles dictate that the etch performance can be accurately evaluated. Etch results on test vehicles must provide information that clearly indicates the key problems so that the optimal etch processes can be developed and applied to the production line. Etch loading effect source affecting CD uniformity is a critical issue for high-load mask etching. Therefore, the test vehicle for this purpose should have a high load, large pattern area, with corresponding performance indicators. On the other hand, responses not resulting from the loading effect (noise signal) should be controlled to as low a level as possible. Test patterns should have the capability to separate main response parameters from unimportant (noise) parameters. Because of the almost infinite pattern variety among production photomasks, it is important to determine the number and type of test patterns that can cover all possible production photomask types. For this reason, two major test pattern categories are usually used to cover all production photomask cases. One is for high-load (e.g., 20–50 percent) pattern situation having isolated dark CDs as the main response parameter (e.g., gate polylayer photomasks). The other one is for low-load (e.g., 10 percent) patterns with space CDs as the main response parameter (e.g., contact layers or other similar patterns such as *dynamic random access memory*, or DRAM).

Regarding loading effects on CDs and other response parameters, Cr etch rate and local CD etch bias are functions of etch loading, implying that large pattern area combined with high etch load situations are the most challenging pattern types. Because the maximum area for 6 in. by 6 in. photomask is about 132 mm by 104 mm (scanner field dimensions), and a high etch load in the pattern area contributes about 80 percent, the maximum total etch load across the entire positive-tone photomask is about 50 percent. For the more typical photomasks with lower load and smaller area than this test pattern, the etch performance is relatively easier to control. The key response parameter is the etch changes evidenced by CD value differences between the higher load regions (in the pattern) and lower load regions (on and outside of the pattern region). Other response parameters include CD etch bias and I/D etch bias.

If the test pattern area is significantly larger than the maximum size that any pattern area can become, or the pattern area is located on one side of masks instead of the center, it is possible to introduce some systematic errors (noise signal). For regions very close to photomask edges (e.g., <7 mm from the edge), etch performance in this region may be totally different from the central area. The main reason for this phenomenon is that resist properties may be different due to resist thickness variation, e-beam writing, and PEB variation at the mask edges. Etch contributions for this region are also different because of the difference of reactant mass transfer and plasma properties. Therefore, the test pattern area should not be significantly larger than the maximum production photomask pattern size, and the pattern location should be centered. Separate loading effect tests can be run with nonstandard patterns as needed. The position of test pattern area is critical for the etch optimization, because when it is too close to one corner or one edge, the test pattern itself can introduce process noise that affects the etch process optimization.

There is still no universal standard etch test pattern in the photomask industry. Most photomask and tool manufacturers use their own test patterns, and even the response parameters they choose vary. This means that reported etch performance from different sources are not totally compatible. There is a need to establish an international standard for test photomask patterns to make the etch results from different sources compatible.

4.2.8 Photomask etch optimization

For etch process optimization, the selection of proper etch response parameters (the objective function) is critical. Misunderstanding exists of the relationship between etch rate uniformity and etch CD uniformity. Taking Cr etch as example, etch rate uniformity usually means vertical etch rate uniformity if not specified, but etch CD uniformity is the uniformity of the *point-to-point* (PTP) CD difference between the resist before etch and Cr CDs after etch due to the plasma etch. Both of them can be evaluated by simply looking at the standard deviation (σ) or 3σ, but good etch rate uniformity does not mean good etch CD uniformity. For most Cr etching of photomasks with high etch load (e.g., >30 percent), the vertical etch rate at center is slower than the outer portions of the photomask, but the CD etch bias (i.e., the CD difference between the resist and Cr) at the center is higher (larger absolute value) than the outer portions. Actually, vertical etch rate uniformity for Cr etch is not as important as the etch CD uniformity. Etch CD uniformity should always be the main etch response parameter considered before etch rate uniformity.

Since plasma etching was introduced to the semiconductor industry originally for wafer fabrication, photomask etching was affected by wafer etching in terms of fundamentals and process optimization. What was found though, was that there are significant differences between wafer and photomask etches. Wafer etching pays attention to three-dimensional feature control and etch responses such as etch rate uniformity, *aspect ratio dependent etch* (ARDE), notching effect, RIE lag, resist charging effects, loading effects, shading effects, and selectivity. In photomask etching, the most challenging aspects are focused on two-dimensional CD properties such as uniformity, I/D bias, and deviation to nominal value. This is because most photomask etches have a *stop layer*. The phase-shifter layer etch needs tight vertical etch rate control on quartz etch, but the production volume is much smaller than Cr and MoSi etching.

Photomask plasma etching criteria include CD etch uniformity, CD etch bias, isolated/dense etch contributions, cross-sectional profiles, and phase angle uniformity and transmission for phase shift masks. Final photomask CD values come as a result of all the processing steps including writing, post exposure bake, resist development, Cr etch, MoSi etch, fused silica (quartz) etch, and cleaning. Since photomask fabrication became an "enabling" technology for integrated circuit fabrication in the late 1990s, tight specifications and an accelerated roadmap push photomask fabrication capabilities close to their technological limits. This situation requires individual processing step optimization as well as overall processing system optimization, seeking possible system error compensation among all fabrication steps.

State-of-the-art photomasks utilize resolution enhancement techniques in the fabrication of EAPSMs and AAPSMs. Final MoSi CD uniformity is determined by MoSi etch, Cr etch, and resist CD uniformity, which is dependent on essentially all of the processing steps. Of these two etches, the Cr etch plays a more significant role in determining final MoSi CD values.

Typical plasma etch optimization variables and etch properties are listed in Table 4.2. The trend to use thinner Cr layers can also result in different optimal etch conditions for both the Cr layer (e.g., 73 nm thickness) of binary masks and the Cr layer (e.g., 59 nm thickness) for EAPSMs. An example CD result of a test vehicle for the 90-nm technology node is shown in Table 4.3.

CD uniformity obviously depends on photomask pattern density and distribution. The etch process for high etch load (high etch area fraction) photomasks is usually more difficult to control than that for low load. Among the different features in high-load photomasks, such as isolated, dense, dark feature, and clear feature, controlling isolated dark feature CD uniformity is the most difficult. CD nonuniformity results from local

TABLE 4.2 Example of Plasma Etch Optimization Variables and Etch Properties

Parameters in etch optimization and tool setting	Etch properties
ICP power	Isolated dark feature CD 3σ
Bias power	Dense dark feature CD 3σ
Pressure	Isolated space feature CD 3σ
He flow rate	Dense space feature CD 3σ
SF_6 flow rate	Isolated dark feature CD etch bias
O_2 flow rate	Dense dark feature CD etch bias
Cl_2 flow rate	Isolated space feature CD etch bias
Dome temperature (if applicable)	Dense space feature CD etch bias
Chamber wall temperature	Dark feature I/D CD etch bias
Cathode temperature	Space feature I/D CD etch bias
Overetch parameters (if different from main etch step)	Isolated dark feature etch linearity
	Dense dark feature etch linearity
	Isolated space feature etch linearity
	Dense space feature etch linearity
	Serif etch properties
	SRAF (subresolution assistant feature) etch properties

load differences, and the majority of the processing steps have systematic errors in similar distributions, e.g., writing, PEB, and Cr etch processes. These systematic errors can be compensated to some extent by the MoSi etch process and e-beam *global dose correction* (GDC).

The Cr layer, consisting of chromium and a chrome oxynitride AR layer, is the most popular absorber for most binary photomasks. It is also the masking layer for MoSi and quartz etching. The pattern transfer fidelity from the resist to the Cr layer depends on etch tool design, etch recipe, pattern type, resist profile, and Cr layer thickness. The traditional Cr layer thickness is about 105 nm, but recently thinner Cr layers are finding more applications because of their higher pattern transfer fidelity. Examples are 59-nm Cr layers for PSM and 73 nm for binary masks.

State-of-the-art photomasks are usually phase-shift masks whose final CDs are on the MoSi layer or directly in the fused silica. Final CD

TABLE 4.3 Some CD Examples of 90-nm Node Photomasks (Feature Size: 360 nm)[124]

Tone	Load %	Tone	Iso-CD 3σ, nm	Dense CD 3σ, nm
Positive CAR	9	BIM Cr space CD	3.1	3.5
		193 MoSi space CD	6.1	6.9
Negative CAR	75	BIM Cr dark CD	3.5	3.7
		193 MoSi dark CD	4.7	4.0

values depend on the interaction of all process steps. For etch processing, pattern transfer errors come from both Cr and MoSi (or fused silica) etches, but the largest impact is from the Cr etch.

Cr etch gases usually include Cl_2, O_2, and He. The overall etch reaction for the chromium layer (AR layer is different) is:[124]

$$Cr_{(s)} + Cl_{2(g)} + O_{2(g)} = CrO_2Cl_{2(g)} \tag{4.52}$$

Although the overall reaction looks very simple, it is not an elementary reaction.

Etch CD uniformity is the most important etch property and is determined by local etch rates in both the vertical and lateral directions. Accurate identification of etch rate effect variables is critical to effective etch process optimization. The local lateral etch rate, rather than the average vertical etch rate is required to understand most etch properties. Its distribution depends on etcher design, etch operational parameters (recipes), and the photomask pattern (e.g., etch load). It is difficult to investigate local etch rates as well as their distribution, but the average rate effect variables versus their responses can supply useful kinetic information.

The loading effect is a common phenomenon in plasma etching, and the vertical etch rate in high-load areas is lower than that of low-load areas. The mechanism for this is the depletion of the rate controlling radical. This vertical etch rate distribution can be measured using a profilometer on a partially etched photomask. Actually, the loading effect on pattern CD is more important for photomask plasma etching because it contributes most of the etch nonuniformity for high-load photomasks. The mechanism of this phenomenon is complicated although it is known that etch CD change with local load moves opposite to vertical load. The local resist vertical etch rate distribution also plays an important role in determining lateral etch rate distribution on resist and Cr layers.[74,75]

Isolated/dense CD etch differences lead to I/D etch bias. One reason for this is due to the reactive ion etch lag from the resist charging and the shadowing of neutral and ionic radicals. It makes the etch rate in dense areas slower than that in isolated areas, resulting in higher CD etch bias for isolated features than for dense features (i.e., the I/D etch bias). Therefore this mechanism is called the aspect ratio dependent etch.

MoSi, consisting of molybdenum, silicon, oxygen, nitrogen, and carbon is the most common phase shifter for resolution enhancement applications. Because of the more complex composition (relative to Cr) and the uncertain elemental ratio, its etch product identification is difficult. The final etch products can include MoF_6, SiF_4, $MoOF_4$, and $SiOF_6$.[94] Common MoSi etch gases include SF_6, CF_4, O_2, and He.

Final MoSi CD values come as a result of all processing steps including writing, postexposure bake, development, Cr etch, MoSi etch, resist strip and cleaning processes. Among them, the writing, PEB, development, and etches have a stronger influence on final CD values. When feature sizes drop down to about 240 to 260 nm (65-nm node technology) on the mask, final CD uniformity and mean-to-target specifications become very strict, e.g., MoSi CD uniformity at about 6- to 10-nm range and mean-to-target at 5 to 8 nm. In reality, resist CD uniformity of high-load photomasks is already at this level or very close to the final MoSi specification CD values based on current writing-PEB-development technologies. The challenge for all etch processes is to keep the MoSi CD uniformity at the same level as the resist CD or to make it even smaller than the resist.

In process optimization, Cr and MoSi etches are considered together as an interdependent system. Once the Cr etch process is fixed, MoSi etches should be designed to produce the best final MoSi CD rather than the best etch contribution (small point-to-point 3σ) by itself. Low CD etch biases in small local load areas can be obtained within certain process windows. Because of the known loading effects during Cr etch, the point-to-point MoSi etch contribution (3σ) will be high in order to achieve better MoSi CDs using CD compensation. It is reported that final MoSi CD uniformity is better than Cr (even better than the resist) CD uniformity when using the CD compensation method.[73,92]

For alternating aperture phase shift masks and chromeless phase lithography masks, fused silica (quartz) substrate etching is needed. Fused silica etch is similar to MoSi etch in many respects. Usually fluorine-containing gases are used while the Cr or resist layer is still on the mask. In both MoSi and fused silica etches, polymers that deposit on the mask are usually formed during etch and need to be removed. A significant process control problem for the fused silica etch is that there is no stop layer or suitable endpoint detection system available. Therefore, phase angle (or etch rate) uniformity control is very difficult. It is also very challenging to repair a glass defect generated during the etch process. Therefore, the minimization of these defects should be emphasized in fused silica etch optimization.

Fused silica CDs are mainly dependent on Cr CDs that mask them, but vertical etch rate uniformity has been the major challenge. Because there is no stop layer for the etch, phase angle uniformity is mainly determined by vertical etch rate uniformity. Due to the fact that there is a distribution of etch variables such as plasma density, neutral radicals, and mass transfer rates, it is very difficult to control the AAPSM phase angle uniformity in the range typically seen in EAPSMs. Besides the etch rate uniformity, etch surface roughness and sidewall profile control must also be considered.

Design of experiments (DOEs) methodology is a common tool for experimental investigation and process optimization. It can significantly reduce the number of trials necessary to make a decision thereby reducing experimental costs. Even so, photomask etch optimization is very costly. Take Cr etch as an example (from Table 4.2). When the variables are ICP power, bias power, pressure, O_2 flow rate, Cl_2 flow rate, and He flow rate, the DOE has six factors. A very simple fractional factorial DOE for 2 levels with three main factors needs eight trials and the full factorial needs 64 trials. After a 2-level fractional factorial DOE, more tests are necessary for refining the process. Therefore, theoretical kinetics and empirical relationships are necessary to highlight the optimal process window area, reduce costs, and improve etch optimization efficiency.

Plasma etch processing now has found wide application in photomask fabrication, and many process optimization investigations have been published. However, accelerated roadmaps and strict requirements from end users (customers) push the photomask etch processes close to their technological limits, and further process improvement of photomask etches will depend on even better understanding.

Plasma etch performance is mainly determined by etch system design (tool) and etch operational parameter arrangement (recipes). It is not expected that there will be break-through progress in the near future regarding plasma etch tool design and theory, so etch optimization progress will depend mostly on the development engineers' or scientists' knowledge, experience, and optimization strategies. This means that more flexible etch optimization methods and groups of recipes (instead of one recipe for all mask patterns) may be necessary to accommodate the different requirements for major pattern categories such as DRAM, logic circuits, isolated feature only, dense feature only, high-load varieties, and negative-tone patterns.

4.2.9 Endpoint detection techniques

Etch endpoint detection is an important technique used to control etch performance. For photomask etch processes, Cr and MoSi etching each have stop layers. This makes endpoint detection relatively easier than quartz etch that does not have a reliable endpoint detection method thus far. For Cr and MoSi etching, there are two options for endpoint detections, i.e., the reflectivity endpoint detection and *optical emission spectrometer* (OES) endpoint detection. Both of them rely on etch property differences between the etched layer and the stop layer beneath. The former detects the difference in reflectivity and the latter detects optical emission intensity. When the difference between the etched layer and the stop layer beneath is not significant, the reliability of the endpoint detection will be poor. For example, the reflectivity difference between the MoSi layer and the quartz layer beneath is not significant, so detecting MoSi etch endpoint by measuring reflectivity is not very reliable. There is some dependence on etcher design and end user's experience and technique, however. Several investigations on plasma etch endpoint detection have been reported.[58,62,65,67,125,126]

The advantage of the reflectivity endpoint detection method is its nondependency on photomask load, while its disadvantage is that a specific etched area (e.g., large barcode area) is needed for accurate endpoint detection. It is an excellent endpoint technique for Cr etch when the endpoint area design and automatic operation are both optimized.

The OES endpoint detection has an intrinsic disadvantage in low-load photomask etch situations because this method depends on emission signal intensity. Low-load photomask etches have a weak signal, especially when the load is less than 2 percent. Regardless of whether utilizing a multiwavelength source, or even when utilizing the full wavelength spectrum, the reliability of low-load photomask etches may still prove inaccurate. The main advantage of OES is its automation and user friendliness. For OES endpoint detection, chromium and molybdenum atomic emission spectra are used for Cr and MoSi etch endpoint detections, respectively. Common wavelengths for photomask Cr and MoSi are listed in Table 4.4. A typical OES endpoint curve is also shown in Fig. 4.10.

State-of-the-art photomask etchers are also equipped with automatic matching networks and automatic pump rate adjustment systems, as well as other automatic adjustment systems. When the photomask etch reaches endpoint, the etch condition and plasma environments change. The automatic system adjusts its parameters to ensure prior setting values (e.g., pressure) remain constant. These parameter changes can also be used as a form of endpoint detection. Examples are the AMN tune operation (Fig. 4.11) and load changes, DC bias change, reflective power changes, and throttle valve opening changes. These endpoint signal changes can be displayed or extracted from log files during Cr and MoSi etching. To obtain the endpoint signal from the opening of the throttle valve, pressure is maintained at a constant level while the throttle valve

TABLE 4.4 Example Line Spectra of Elemental Chromium and Molybdenum[127]

Cr atomic spectra wavelength, Å	Relative intensity	Mo atomic spectra wavelength, Å	Relative intensity
3578.69	19,000	3132.59	14,000
3593.49	17,000	3798.25	29,000
3605.33	13,000	3864.11	29,000
4254.35	20,000	3902.96	19,000
4274.80	16,000		
4289.72	10,000		
5208.44	11,000		

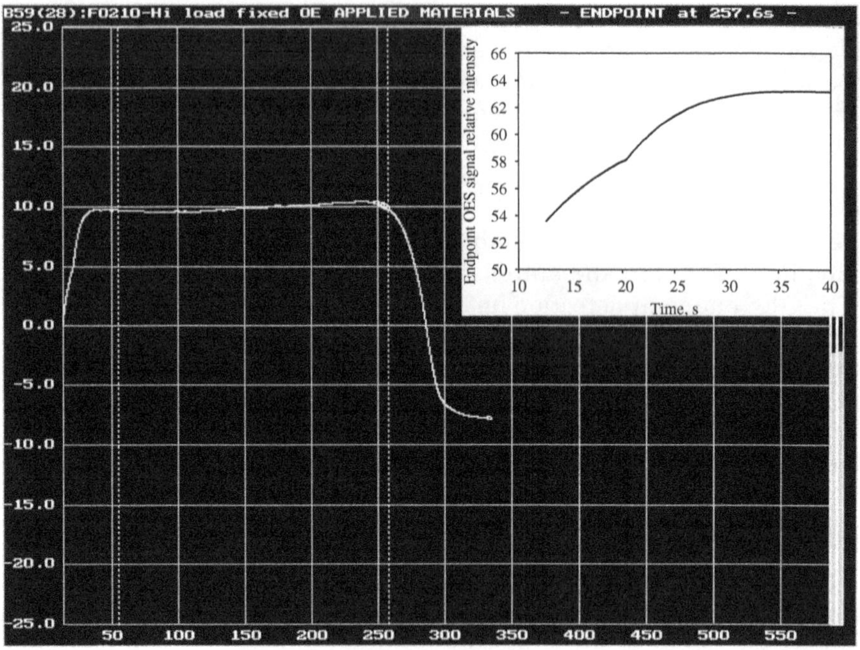

Figure 4.10 Endpoint signal and its changes at AR layer.

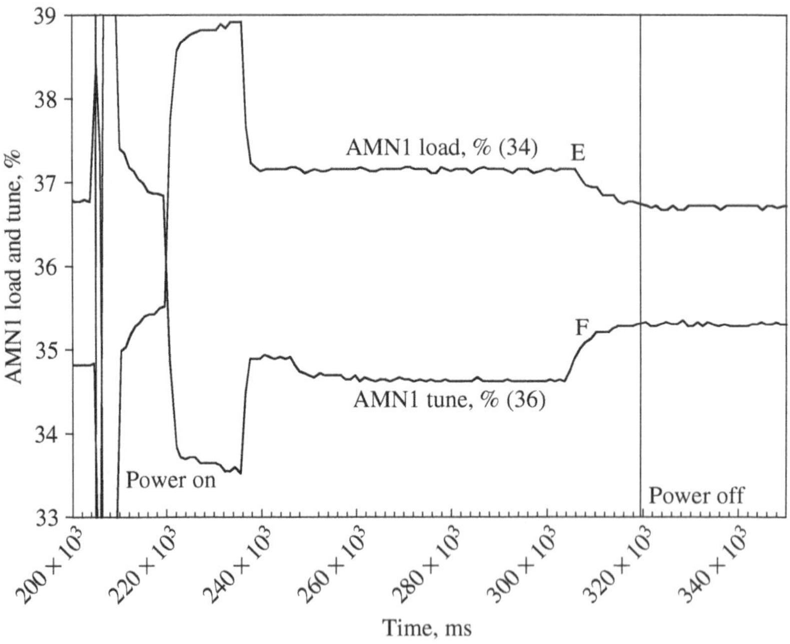

Figure 4.11 Tune and load endpoint signal changes.

opening is continuously adjusted to obtain this constant pressure. When the endpoint is reached, the gas flow rate out of the chamber is different from before and after endpoint. Therefore, for a constant inflow rate and pressure, the pump-out rate needs to be changed, which can be adjusted by opening the throttle valve. For Cr etching, the pump-out rate becomes higher after the main etch passes endpoint, then the throttle valve opening will need to be increased. It can be seen clearly during high-load Cr etch processing, so the opening changes can be used as a reliable etch endpoint detection method.

In production photomask Cr etching, very low-load photomask etches usually need a longer overetch percentage relative to the main etch than high-load mask etches. It is not surprising to find the dark-field photomask overetch percentage to be in the range of 50 to 100 percent, or even up to 150 percent. The overetch percentage of low-load photomask etches are not as critical as high-load etches. This means that timed Cr etching (including main etch and overetch) is usually acceptable for very low-load photomask etches. The etch rate at a predetermined etch load can be predicted fairly easily and used in manufacturing setting.[74]

For MoSi etch endpoint detection in very low-load photomasks, endpoint usually can be detected using the OES method when pattern loading is greater than 2 percent. For photomasks with loads in the range of 1 to 2 percent, endpoints can be observed and detected, but not very reliablely for production control. Actually the MoSi etch rate dependence on load is not as sensitive as that for Cr etch. This means that timed MoSi etches for low-load photomasks generally do not have technological challenges.

4.2.10 Plasma etch kinetics and modeling

Over the last 30 years, the semiconductor industry has reached a large production scale, but plasma etching as a critical fabrication step has been semiempirically developed. In other words, dry etching as currently practiced is still one part art and one part science.[2] This analogy implies that both empirical relationships and scientific theories are important for plasma etch applications. For photomask plasma etching, studies are not as mature and intensive as have been performed in the wafer etch area.

Plasma etch kinetics is a science of etch rate processes and mechanisms. It is dependent primarily on chemical principles that were developed by both chemists and physicists.[128] The former relies on thermodynamics and transport phenomena, but the latter mainly employs statistical mechanics. To date, chemical kinetics has been applied successfully in chemical engineering, chemical reaction engineering, and material processing, but statistical mechanics and statistical thermodynamics are mainly used to explain chemical phenomena.

Thermodynamic properties are state functions, determined by state parameters like temperature, pressure, and chemical composition. Pressures for photomask plasma etches are usually in the range of 2 to 30 mtorr, with a Knudsen number $1 < \text{Kn} < 110$. This indicates that the gas flow is somewhere (transition) between viscous ($\text{Kn} > 110$) and molecular ($\text{Kn} < 1$), implying that some transport theories derived based on the newtonian viscous theory may not apply to this pressure regime if not modified, e.g., motion equations and diffusion laws.

In nonequilibrium plasmas used in photomask etching, the electron and heavy particle temperatures are not in equilibrium. It is not very clear whether the chemical reaction and composition will follow thermodynamic equilibrium rules in the system having low charge concentration ($\ldots 10^{-4}$), or whether the nonequilibrium plasma conditions will limit the application of the nonelementary reaction kinetic method in photomask etches.

Thermodynamic studies on photomask plasma etching have been reported.[112] Thermodynamically calculated results agreed with experimental results, showing the applicability of thermodynamics to photomask etching. It has advantages for new etch gas system selections and for the determination of etch conditions with the help of commercial thermodynamic software such as CSIRO Thermochemistry System, SOLGASMIX-PV, and Outokumpu HSC Chemistry.

An equation for wafer plasma etch rate versus wafer number was proposed in 1977, showing the relationship between operational parameters (wafer number) and etch properties (etch rate).[15] This demonstrated the successful application of chemical reaction kinetics to plasma etches. For photomask etching, a theoretical kinetic equation was proposed,[75] showing the relationship between operational parameters (ICP power, oxygen flow rate, and oxygen partial pressure) and etch properties (etch rate):

$$r_{\text{cr}} = \frac{\left(Q_{\text{O}_2} p_{\text{O}_2}^{1/2} W_{\text{icp}}\right)}{a + b\left(Q_{\text{O}_2} p_{\text{O}_2}^{1/2} W_{\text{icp}}\right)} \tag{4.53}$$

$$a = \frac{Vk_v + A_w k_w + AL_d k_{\text{cr}}}{\left(10^{10}\frac{k_{\text{cr}}k_1}{\beta_{\text{cr}}\rho_{\text{cr}}}\right)} + \frac{(1 - L_d)A\rho_{\text{pr}}\beta_{\text{pr}}}{k_1 S_e} \tag{4.54}$$

$$b = \left(\frac{\beta_{\text{cr}}\rho_{\text{cr}}}{10^{10}k_{\text{cr}}}\right)\left(\frac{k_2}{k_1}\right) \tag{4.55}$$

where r_{cr} = Cr etch rate (Å/s)

Q_{O_2} = oxygen flow rate (sccm)

p_{O_2} = oxygen partial pressure (mtorr)
W_{icp} = ICP (source) power (W)
k_v (s^{-1}), k_w (m/s), and k_{cr} (m/s) = oxygen radical consumption rate constants in the bulk chamber, chamber wall, and Cr surface, respectively
V = etch chamber volume (m^3)
L_d = Cr load ($0 < L_d < 1$)
A_w and A = chamber wall and photomask areas (m^2), respectively
ρ_{pr} = resist molar density (mol/m^3)
β_{pr} = resist etch stoichiometric constant
S_e = Cr-to-resist etch selectivity
β_{cr} = Cr etch reaction stoichiometric constant
ρ_{cr} = chrome molar density (mol/m^3)
k_1 and k_2 = constants

At most experimental conditions, a and b are assumed to be constants. The main advantage of using this equation is that it can be tested experimentally and can be used for etch process optimization. It also highlights the main-effect parameters for further theoretical studies, such as etch rate.

Etch rate is a function of ICP power, pressure, and gas compositions and flow rates. In most studies, oxygen concentration was used as a main variable. However, recent experimental results show that etch rate is mostly determined by oxygen flow rate rather than oxygen concentration.[75] Oxygen partial pressure does have an effect on Cr etch rate, but its effect is not as strong as oxygen flow rate. The reason may be that the etch rate control mechanism is the depletion of oxygen in the bulk chamber (in most Cr etch processes) or that there is equilibrium composition limiting the etch rate.

The etch rate control step seems to be the result of oxygen depletion due to the strong effect of oxygen flow rate on overall etch rate. However, the experimentally measured oxygen conversion is only a few percent,[75] much lower than the effect from depletion. This result implies that equilibrium conversion of oxygen may also control the etch rate, as shown in thermodynamic studies on photomask plasma etches. At predetermined conditions, equilibrium conversion agrees well with experimentally measured oxygen conversion.[112]

MoSi etch rate at constant pressure can be expressed as:[94]

$$r_A = AW_{icp} \exp\left\{-\frac{1}{V_{DCB}}\left[\frac{E_a}{2eN_A}\cdot \ln\left(\frac{m_i}{2.3m_e}\right)\right]\right\} C_{SF_6}^{1/5} C_{O_2}^{1/2}$$

$$= 0.6784W_{icp} \exp\left\{-\frac{79.55}{V_{DCB}}\right\} C_{SF_6}^{1/5} C_{O_2}^{1/2} \qquad (4.56)$$

where A = constant
W_{icp} = ICP power
V_{DCB} = DC bias voltage
E_a = activation energy
e = elementary charge
N_A = Avogadro's number
m_i and m_e = ion and electron masses, respectively
C_{SF_6} and C_{O_2} = SF_6 and O_2 concentrations, respectively

The MoSi etch activation energy was obtained using linear regression and found to be 1350 kJ/mol. This activation energy, by definition, is the enthalpy of vaporization and is very close to that of $MoSi_2$ (1674 kJ/mol).

Statistical mechanical models have been developed over the last 30 years. Fundamentals on plasma physics were studied and etch kinetic theories were proposed. Because of the lack of analytical techniques for reactive radical measurement and corresponding kinetic data, most kinetic theories are *descriptive* and cannot be experimentally proven. This means that many plasma etch theories are actually hypotheses. Therefore these kinetic models cannot supply relationships between operational parameters and etch property responses for industrial practice, although these "niche" theories are helpful toward understanding the plasma etch process.

A plasma model can be thought of as a computer program that numerically solves a series of equations describing the physics and chemistry occurring in the plasma. Many research[129–131] and review[132–134] papers covering plasma models have been published, most of which were developed using the Monte Carlo method.[132,134]

The Monte Carlo method is a technique for obtaining an approximate solution to certain mathematical and physical problems, characteristically involving the replacement of a probability distribution by sample values. Kinetic models such as *particle-in-cell* (PIC) simulations[135,136] and "convective schemes"[137] fully resolve the *electron energy distribution* (EED) and *ion energy distribution* (IED) as a function of space and time. These results are obtained by integrating Boltzmann's equation, either statistically using Monte Carlo techniques for collisions, or directly as in the convective scheme. The main disadvantage of kinetic models is that they are computationally intensive. Therefore most present kinetic models are "hybrid models" that can speed up computations while retaining important nonequilibrium aspects of the EED.

Besides the Monte Carlo method, the analytic method was also examined for photomask plasma etching.[138] These studies have just begun, so a full evaluation of this method cannot be completed until more data are available.

4.2.11 Etch property number method

Experimental design and result analysis using statistical software for photomask plasma etching have been applied since the early 1990s.[28] Experimental response parameters were analyzed using regression and contour plots. The single variable regression usually cannot give satisfactory regression results because a photomask plasma etch response is a function of multiple variables, and contour plots are limited to only two variables. Understanding the relationship between one etch property (response) and the multiple variables is necessary for photomask etch process optimization, however.

Photomask etch optimization to date has been limited by the lack of theoretical and empirical relationships. This forces most photomask etch optimization efforts to rely on design of experiments analyses and related computer software. DOE is mainly based on mathematical statistics, which does not involve the etch process and mechanisms. However, analytic relationships between operational parameters and etch properties are useful and do play an important role in photomask etch optimization.

Because of the high costs of photomask raw materials (e.g., PSM blanks) and write tool time, the number of experimental samples is strictly controlled for etch optimization DOE. In plasma environments, more samples are really needed because experimental parameters not only affect etch properties, but also may affect each other. Even though photomask plasma etching is complicated, only two-level DOEs are usually authorized due to cost control purposes. Fractional factorial DOEs limited to three or four factors are commonly used instead of full factorial designs in order to reduce etch optimization costs. Because of these restrictions, relationships between variables and responses usually have some degree of uncertainty.

In order to obtain a useful etch recipe using a two-level DOE, the variables' ranges should not be set too large. Kinetic theories and empirical relationships are helpful for setting variable domains; however, there is virtually no mature theory showing the relationship between operational parameters (variables) and photomask etch properties (responses) for photomask etch optimization. The empirical relationship between operational parameters and etch properties then become extremely important.

Analytical (theoretical or empirical) relationships between the operational parameters and response results are very helpful for industrial process optimization. In plasma environments, there are many parameters to consider, such as ICP and bias powers, pressures, and gas flow rates of oxygen, chlorine, and helium. Even with the help of DOE software, it is extremely difficult to express the relationship between these variable and response functions in a simple analytic equation. To solve this problem, *etch property number* method was proposed by defining a measurable etch property number having an analytical relationship to one or more etch responses.[1,74,94] An etch property number consists of measurable operational etch parameters.

CD etch movement (also called CD etch bias), combined with the pattern formation processes, determines final CD values and their deviation from nominal (also called *mean to target*). The etch property number "CD movement, W_3" was then defined for Cr etch.[74]

$$W_3 = \frac{W_{\text{bias}}^{0.18} W_{\text{icp}}^{1.38} Q_{\text{He}}^{0.16}}{p^{0.88} Q_{\text{Cl}_2}^{0.96}} \tag{4.57}$$

where Q_{He} is the helium flow rate and p is the pressure. The relationship between CD Cr etch bias and W_3 is shown in Fig. 4.12.

W_3 is related to CD etch bias as well as other etch properties such as I/D etch bias and *pattern edge CD bias* (PECDB). PECDB is defined as the CD value difference between the pattern edge (including outside pattern) CDs and CDs in the pattern area.[74] It is an indicator of loading dependent CD error. PECDB etch contribution is a critical etch property for maximizing high-load photomask CD uniformity control, especially for the dark feature CD. W_3 influences these etch properties as shown in Fig. 4.13. This figure indicates that when W_3 increases, CD etch bias and I/D etch bias also increase, but PECDB decreases, meaning that

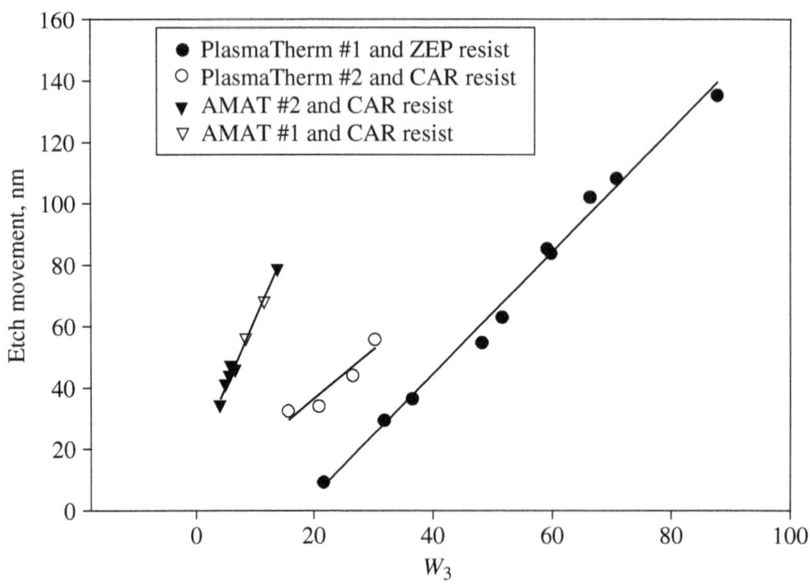

Figure 4.12 Relationship between etch CD movement and W_3.

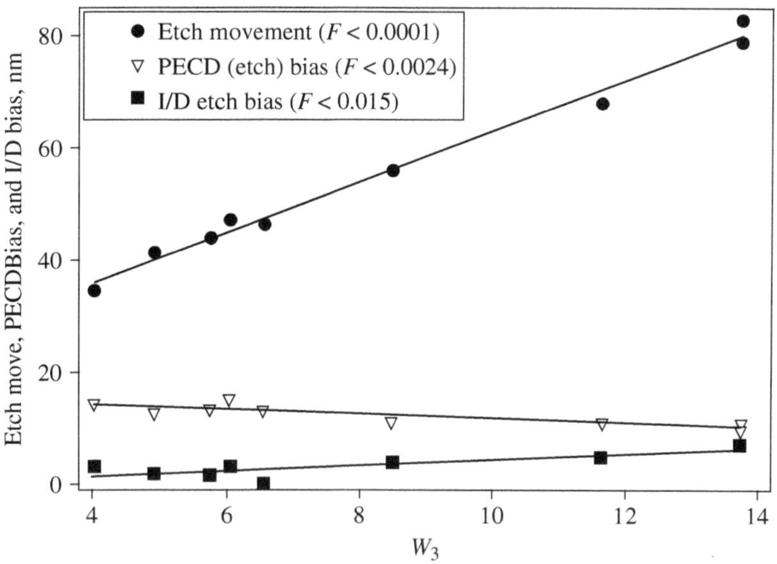

Figure 4.13 W_3 influences on several etch properties.

good etch uniformity for CD values (low PECDB) is always accompanied by a high CD etch bias. Therefore, striking a compromise is necessary when emphasizing this specific etch property. Usually, isolated dark feature CDs have the highest contribution toward high-load photomask Cr etch optimization, so the recipe employed usually displays a low-loading effect, but a relatively high I/D bias and CD etch bias.

The I/D etch bias number W_4 is the etch property number concerned with aspect ratio dependent etch phenomenon. Wafer etch ARDE has been the subject of intense study,[24] but the study of photomask I/D etch bias is relatively immature. One empirical relationship between I/D Cr etch bias and etch parameters at constant pressure was proposed using the etch property number method:[74]

$$\text{I/D etch bias} = aW_4^b \tag{4.58}$$

$$W_4 = \frac{W_{\text{icp}}^{0.8}}{W_{\text{bias}}^{0.3} Q_{\text{Cl}_2}^{0.5}} \tag{4.59}$$

where W_4 = I/D etch bias number
W_{icp} and W_{bias} = ICP and bias powers, respectively
a and b = constants
Q_{Cl_2} = chlorine gas flow rate

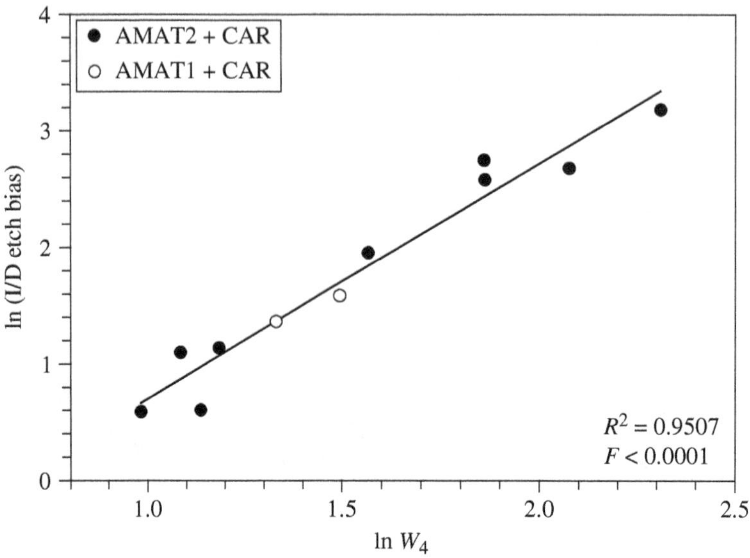

Figure 4.14 Relationship between I/D etch bias and W_4.

The experimental results and their relationships with Eqs. (4.58) and (4.59) are shown in Fig. 4.14. The strong correlation indicates that ICP power enhances isotropic behavior and so favors I/D etch bias. Conversely, the bias power and chlorine flow rate enhance small I/D etch bias.

Definitions of etch property numbers and their relationships with etch property responses are shown in Table 4.5. For empirical relationships applicable only to certain etch tools, the relationship may need further

TABLE 4.5 Etch Property Numbers and Their Relationships

Etch property number	Relationship with etch properties	References
$W_1 = \dfrac{W_{\text{icp}}}{W_{\text{bias}}} \cdot X_{\text{SF}_6}$	MoSi I/D CD etch bias $= aW_1 + b$	94
$W_3 = \dfrac{W_{\text{bias}}^{0.18} W_{\text{icp}}^{1.38} Q_{\text{He}}^{0.16}}{p^{0.88} Q_{\text{Cl}_2}^{0.96}}$	Cr etch CD bias $= aW_3 + b$	74
$W_4 = \dfrac{W_{\text{icp}}^{0.8}}{W_{\text{bias}}^{0.3} Q_{\text{Cl}_2}^{0.5}}$	Cr I/D CD etch bias $= aW_4^b$	74
$W_q = \dfrac{W_{\text{icp}}^{1.01} V_{\text{DC}}^{1.51} p_{\text{SF}_6}^{0.10} Q_{\text{He}}^{0.37} \left(1 - X_{\text{O}_2}\right)^{0.7}}{p^{0.52}}$	Qz etch rate $= aW_q^b$	138

investigation to extend applicability to other etch tool configurations. For applicability to etchers with different configurations, the relationship may provide some fundamental theories so that customized studies can be pursued as required.

4.3 Chromium-Based Material Plasma Etch

4.3.1 Cr plasma etch

Chromium-based material (Cr layer) plasma etch has a fairly long history. After Cr masks began to be etched using the gas plasma method starting around 1976,[71,121] several investigations on Cr photomask plasma etching followed, but did not have commercial application for photomask fabrication at that time.[119,120] Until 1992, a dry etch process was used at TRW for fabricating practical photomasks for Ultratech 1500 1× optical steppers in wafer manufacturing lines.[28]

From feasibility studies to practical applications in industry, many problems with Cr plasma etching still remain to be solved, such as defectivity, process stability, CD control, and resist etch selectivity. Besides using Cr as the absorber layer, other absorber materials (e.g., molybdenum silicide) were also investigated.[140] In the early days of Cr plasma etching, the etch resistance of photoresist materials, i.e., the Cr-to-photoresist etch rate selectivity, was of primary concern.[120] Before 1990, almost all high-end e-beam mask fabrication technologies were using *polybutene sulfone* (PBS) resists combined with wet etching processes. This PBS resist has very poor dry etch resistance. In the early 1990s, plasma etching had evolved to such a level that the line and space resolution capability for 100 to 200 nm was possible[120] and industrial production using plasma etching began.[28] Since then, the main development objectives for photomask etching have been to improve continuously etch tool designs, process performance controls (recipes), and defectivity.

After about 10 years of slow progress by the photomask industry due to the transition from 1× to 5× steppers (also called the "mask maker's vacation"), an ICP-bias power photomask etch system came to the market in the mid-1990s.[32] Its superior etch performance (especially with respect to etch uniformity) made it possible to now employ plasma etching in standard photomask fabrication for half micron and below node integrated circuits. Since the mid-1990s, many studies on Cr etch process improvements have been published.[33–75,77–81,119–120,123,125,126,141–148] At about the same time, that ICP-bias power configuration etcher emerged, the magnetically-enhanced reactive ion etch system was studied,[36,37] but in the end, ICP-bias power configuration tools found more application for photomask etching. The advances in photoresist and

Cr layer technologies also drove the application of dry etch in photomask fabrication processes. The combination of the ICP-bias power etcher with OCG-895i resists,[30,31] ZEP7000 resists,[37] and later, with CAR resists has been widely used now in the photomask industry. Early plasma etch capability studies also drove the adoption of thin Cr layer strategies for their superior resolution.[36,149]

The effect of oxygen on etch uniformity was reported in 1992—low O_2 partial pressure provides better etch uniformity.[28] Pressure effects on etch uniformity were also investigated in the early days.[28,31,32] It was found that the best etch uniformity was achieved at the lowest pressure (10 mtorr) studied.[28] This pressure falls in the range of current optimal Cr etch conditions. Low RF power in early parallel plate etchers[28] and low ICP power in ICP-bias power etchers display better uniformity.[32] Defectivity in plasma etches and etch performance dependence on mask loading were also observed.[32] Since the OCG-895i resist was used as the mask for photomask plasma etching,[30,31] Cr-to-resist etch rate selectivity was not a significant issue. By 1996, many major etch variables and responses were better understood, leaving further plasma etch optimization refinement to the future.

Lower ICP power tends to provide better etch CD uniformity, but the reduction of ICP power is limited to where the plasma loses stability due to the strong electronegative property of chlorine gas.[39] A large amount of helium can significantly lower the ICP power stability level, e.g., from 400 to 250 W, and also improve etch uniformity.[39] Resist etch performance has been investigated to understand its relationship to Cr etch performance, and it was observed that the Cr etch rate was dependent on photomask load (using the same etch recipe), but the resist etch rate was not.[39,62] Later, differences in resist etch rate signature compared to Cr etch was observed, showing that resist etches center faster, but Cr etches edge faster.[42] In order to reduce resist property fluctuation effects, a method employing point-to-point subtraction of Cr CD to resist CD was used. For this, it was found that the CD change signature due to the plasma etch directly correlated to the resist erosion signature.[42,70] The mechanisms affecting etch rate uniformity distribution were investigated,[43,49] and it was found that neutral radical diffusion may control the Cr etch rate, and neutral radical uniformity, rather than plasma density uniformity, affects etch rate uniformity. Neutral radical uniformity cannot improve resist etch uniformity, but plasma uniformity can. For etch performance improvement, alternative chemicals were also tested. It was reported that the replacement of oxygen with carbon dioxide in $Cl_2 + O_2 +$ He gases increased etch selectivity by 50 percent, reduced CD etch bias, and lowered etch rate dependence on pattern load.[46]

Etch performance on different pattern and feature types are very important for photomask etch process optimization. Isolated lines CD etch uniformity,[47] especially in very large pattern load conditions (e.g., overall mask load 50 percent),[74] is very difficult to control. The main reason is due to the difference in local CD etch bias over regions of varying local loading on one photomask. CD etch uniformity degradation due to pattern loading is a widespread problem.[31,39,41,50–52,59–61,73–74,78,141,143] Resist erosion actually determines the signature of postetch final CDs,[61] where the local CD etch bias is proportional to resist thickness removal rate.[1,74]

For a while, it was believed that etch rate was a function of the Cl_2-to-O_2 ratio, not the gas flow rates. Experimental results indicate that the Cr etch rate changes significantly with flow rate at a constant gas molar ratio.[69] More accurately, the Cr etch rate is only a function of O_2 (not Cl_2) flow rate and concentration in most regimes of photomask etching.[75] This indicates that the Cr etch rate is controlled by depletion of oxygen-containing radicals, and the average concentration of the radical in the etch chamber is affected by oxygen flow rate and photomask etch load. This also explains why etch rate is a function of load, agreeing with the etch rate-load relationship.[74]

Based on the overall etch reaction stoichiometric chemistry, the volume ratio of chlorine and oxygen gases is unity for chromium etch. In practical Cr etch process windows, this ratio is much higher (e.g., 7–15 times), resulting in the starvation of oxygen. Therefore, Cr etch rate is dependent on its mass flow rate due to O_2 depletion.

Regarding thin Cr layers, advantages include low CD etch bias, low I/D CD bias, and high pattern transfer fidelity. Besides the Cr etch itself, thinner Cr layers allows for thinner resist layers, corresponding to better resist CD uniformity and sidewall profiles. Therefore the use of thin Cr layers produces better etch performance, better resist CD and sidewall profile, and better final pattern fidelity in both the Cr layer in BIMs and in the MoSi layer of PSMs.

The composition difference in Cr sublayers does cause some degree of etch performance difference layer to layer. Figure 4.10 shows a Cr layer etch OES endpoint signal change with etch time. The etch gases consist of oxygen, chlorine, and helium, and the OES endpoint signal is the optical emission relative intensity of the chromium atomic spectrum at 360-nm wavelength. At the top-right corner, the small figure shows a signal changes at about 21 s, corresponding to the transition from AR to chromium sublayers. Based on the data in this figure, the chromium etch rate is 0.351 nm/s and the AR sublayer etch rate is about 0.881 nm/s. Because average Cr layer etch rate is strongly dependent on oxygen flow rate and partial pressure,[75] the etch rate difference in the chromium and AR sublayers may result from the oxygen content difference in the two Cr sublayers. This endpoint signal change was observed only at low oxygen flow rate using gaseous etchants. The etch rate difference was also reported in other plasma etch environments using oxygen and argon,[71] showing that chromium etch rate is higher than its oxide when oxygen concentration is higher than a certain threshold.

Cr etch rate and related mechanisms were studied by Nakata in 1980 using $CCl_4 + O_2 + Ar\ (N_2)$ gases.[71] Cr actually exists as chromium, CrO_x, and CrO_y, where CrO_y contains impurity tungsten. Etch rate changes with varying oxygen gas concentrations are shown in Fig. 4.15, showing that metallic chromium etch rate is higher than that for both chromium oxides with and without impurity (tungsten) when oxygen concentration is higher than a certain threshold value (about 15 percent). When oxygen concentration is below 15 percent, chromium etch rates

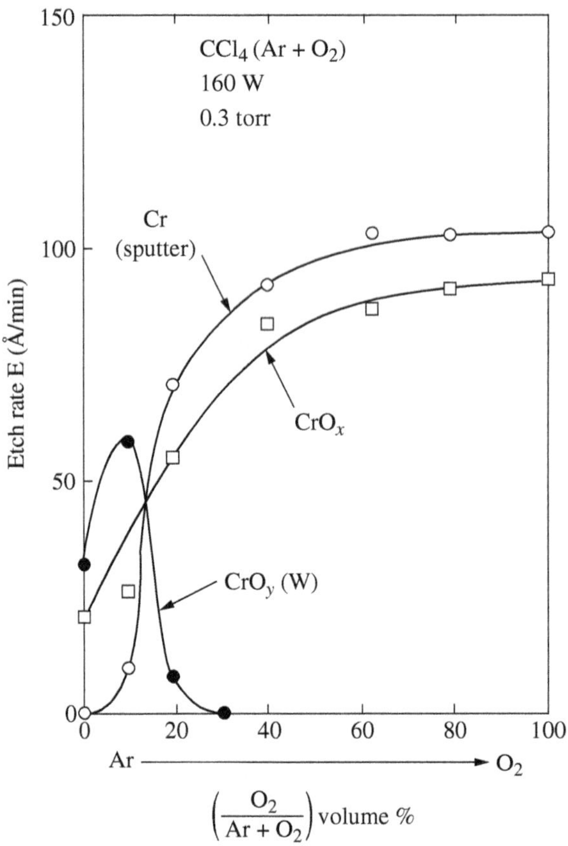

Figure 4.15 Etch rate changes with gas composition for Cr and CrO_x. (Reproduced with permission of American Institute of Physics and figure courtesy of H. Nakata, K. Nishioka, and H. Abe.)

are lower than that for chromium oxide. For the etch conditions seen in Fig. 4.10, the relative oxygen concentration is 9 percent so the chromium oxide etch rate is higher than that of chromium, which agrees with Fig. 4.15. For these two studies, inert gas, plasma etcher configurations, and chlorine concentrations are different, but the effect of oxygen is very similar. It indicates that the relationship between etch rate and oxygen concentration does not have significant dependence on inert gas properties or etcher configurations. A recent study on the photomask Cr layer shows that the etch rate is strongly determined by oxygen mass flow rate and oxygen partial pressure, but not chlorine flow rate or concentration.[75] It also shows that the oxygen mass flow rate has a stronger effect on etch rate than oxygen partial pressure in these experimental conditions.

The volatile Cr etch product is CrO_2Cl_2 and overall etch reaction can be expressed as:

$$\mathrm{Cr} + \mathrm{O}_{2(g)} + \mathrm{Cl}_{2(g)} = \mathrm{CrO_2Cl}_{2(g)} \tag{4.60}$$

Similarly, the chromium oxide reaction is expressed as:

$$2\mathrm{CrO}_x + (2-x)\mathrm{O}_2 + 2\mathrm{Cl}_{2(g)} = 2\mathrm{CrO_2Cl_2} \tag{4.61}$$

Chlorine and oxygen as well as inert gases (e.g., helium, argon, and nitrogen) can be thought as a standard gas system for Cr etching, while other etchant gases are alternatives. In order to improve the etch performance, alternative gas systems differing from the gas system (described in the section) were investigated, as shown in Table 4.6.

TABLE 4.6 Cr Etch Systems

Number	Gas system	References
1	$Cl_2 + O_2 +$ He (or Ar)	71
2	$CCl_4 +$ air	71
3	$CCl_4 + O_2$ (or adding inert gas Ar or N_2)	28, 71, 150
4	$Cl_2 + NF_3$	151
5	$Cl_2 + O_2 + H_2$	41
6	$Cl_2 + O_2 + HCl$	41, 152, 153
7	$Cl_2 + O_2 + NH_3$	41
8	$Cl_2 + O_2 + CO_2$	46
9	$Cl_2 + HCl$	154
10	$Cl_2 + HCl + H_2$	154
11	CO + chlorine containing gas	155
12	Cl_2 (halogen) + O_2 + reducing gas	156
13	$HCl + O_2 +$ He	147

Cr etch chemical reactions using alternative gases can be thermodynamically calculated. One example is the chromium etch reaction using $CCl_{4(g)}$, as shown in the equation:

$$Cr + CCl_{4(g)} = CrCl_{4(g)} + C \tag{4.62}$$

There are two main considerations for Cr etch using alternative gases. One is CD etch performance, and the other is the defectivity control. CD etch performance considerations include CD etch uniformity, CD etch bias, loading effects, I/D etch bias, and Cr-to-resist etch selectivity. An intrinsic disadvantage of ICP-bias power configuration is low selectivity, so one objective for using alternative gases is to achieve an increase in selectivity. ICP-bias power configurations have better loading effect impact than other configurations such as RIE, but strict requirements of state-of-the-art photomask etching demand better results on this etch property. One simple alternative gas strategy is to use standard $Cl_2 + O_2$ gases plus reducing agents. Because of the oxidation mechanism that takes place during resist erosion, an alternative gas approach can usually improve selectivity. Defectivity, however, is dependent on whether all of the etch products are volatile.

Most experiments for alternative gases were based on trial-and-error methods. Recent thermodynamic studies on photomask plasma etching make it possible to select an appropriate gas system and control gas compositions by using thermodynamic calculations,[112] which can significantly save time and resources expended during photomask plasma etch optimization.

Average etch rates are usually calculated by dividing the standard Cr layer thickness (including AR) by the main etch time. The main reason for this simplistic method is due to photomask raw material availability and study result applicability to etch process optimization.

Etch rates and main etch times vary with Cr thickness, chemical compositions, loading, and recipe parameters. For photomasks with Cr thickness from 59 to 110 nm, the main etch time can vary from 30 to 300 s, corresponding to etch rate variation from about 0.37 to 2.2 nm/s.

4.3.2 Etch pattern dependency

In both dark and clear field recipe groups, a basic etch recipe is usually developed and some other recipes with minor changes from this baseline are subsequently developed for other specific applications. Actually, the decision to use a high or low-load etch recipe is also dependent on photomask specifications. It is common knowledge that high-load photomasks specify dark feature CDs and low-load photomasks specify clear feature CDs. This information is helpful to determine which recipe to use for photomasks with pattern loading around 20 percent.

For photomask Cr etch, the low-load photomask etch process is relatively easier to optimize than that for high-load photomasks due to the CD etch bias dependency on local load. Besides the loading effects on CD etch uniformity, CD mean-to-target value control for high-load photomasks is also more difficult than for low-load photomasks. Therefore it is wise to focus most of the Cr etch optimization effort on clear field photomask error reduction.

In I/D etch bias studies,[50,74] pressure was seen to be a very sensitive parameter. Lower pressure usually exhibits a lower CD loading effect and better etch uniformity for a given feature type (isolated or dense), but a higher I/D etch bias. In most metrology specifications, isolated and dense CDs are not required on the same photomask. For example, LOGIC product photomasks usually require isolated CDs and DRAM photomasks require dense. These properties force the etch recipe regime to be located in a relatively low pressure process space in order to achieve lower CD loading effects and better CD uniformity.

CD etch load effects, i.e., the CD etch bias being faster at highly loaded mask area, has been observed and found that this systematic error is similar to the resist CD pattern signature.[51] Resist CD systematic errors may be created by e-beam fogging during the writing process.

Another etch pattern dependency occurs among photomasks with different loading. This particular CD etch bias dependence on pattern (load) affects final CD mean-to-target values. Using the same etch recipe, high-load masks have a longer etch time with higher CD etch bias than low-load masks. Final CD values depend on all process steps combined, including writing, PEB, development, and etching. Although it is not common to change etch conditions to adjust final CD values, the capability to control final CDs by adjusting etch conditions is available if necessary for further process control. Final CDs can be controlled by etch by adjusting the pressure, oxygen concentration, and overetch time.[57] A ready-to-use formulation was proposed using a special relationship between CD etch bias and a number W_3 [see Eqs. (4.63) and (4.64)]:[74]

$$W_3 = \frac{W_{\text{bias}}^{0.18} W_{\text{icp}}^{1.38} Q_{\text{He}}^{0.16}}{p^{0.88} Q_{\text{Cl}}} \tag{4.63}$$

$$\text{CDBias} = a + bW_3 \tag{4.64}$$

where W_{bias} = bias power
W_{icp} = ICP power
Q_{He} and Q_{Cl} = helium and chlorine flow rates, respectively
p = pressure
CDBias = CD etch bias
a and b = constants, derived from experimental data

The equation shows the three main parameters affecting CD etch bias are ICP power, chlorine flow rate, and pressure. Among them, ICP power may also affect etch CD uniformity and selectivity. Chlorine flow rate is usually already high, and further increase will make it difficult to keep pressure low and constant, so pressure is the best parameter to adjust for CD mean-to-target control. Therefore we have:

$$\frac{\partial \text{CDBias}}{\partial p} = \frac{\partial \text{CDBias}}{\partial W_3}\frac{\partial W_3}{\partial p} = b\frac{\partial W_3}{\partial p} = bW_3\frac{\partial \ln W_3}{\partial p} = -\frac{0.88bW_3}{p} \tag{4.65}$$

and after rearrangement, we have:

$$\Delta \text{CDBias} = -\frac{0.88bW_3}{p}\Delta p \tag{4.66}$$

This equation can be used to adjust final CD values in plasma etch processes. It should be noted that the adjustment for low-load photomask recipes can usually produce satisfactory results, but it may degrade CD uniformity when high-load photomask recipes are used.

4.3.3 Low-load photomask Cr etch

For dark field Cr etch optimization, the main etch responses include CD etch uniformity, I/D etch bias, CD etch bias, CD mean-to-target value, and CD linearity. The main systematic CD errors in the resist layer before etching consist of CD radial distribution, CD side-to-side distribution, and random CD errors.

One of the main dark field Cr etch challenges is in resolving very small features such as contact holes, where the resolution limit may already have been reached in the resist layer before the etch step is started. The most common method for solving this problem is to use a process called *descum*, which is actually a type of etch process using only bias power and oxygen gas to remove residue resist at the base of the features before Cr etch begins. This process dramatically improves minimum feature resolution capability, and is also helpful to remove some particles before etch. Because there is no chlorine in this etch step, the Cr etch rate in the descum step is very low.

The endpoint detection of dark field Cr etch can be problematic. Common photomask etch endpoint detection methods include the *laser reflectivity endpoint* (LRE) technique and the use of optical emission spectra. The LRE method measures laser reflectivity on an etched surface, which usually requires a specially designed area outside the main pattern such as the barcode region. One disadvantage of using this method is that the tool operator needs to locate the laser spot manually, which is not user-friendly and is prone to error. The advantage of using LRE is that it is not pattern load dependent. Conversely, the OES endpoint method does not require manual operation and is very user-friendly. However, when the photomask pattern load is below 2 percent, OES endpoint detection is not very reliable and OES method use below 1 percent pattern load generally has detection problems caused by an inadequate signal-to-noise ratio. Fine-tuning the OES algorithm, or using a multiwavelength OES can improve endpoint detection performance, but there can still be difficulty in photomask endpoint detection if loading is less than 1 percent.

Another approach to the technical challenges associated with low-load endpoint detection is to calculate the etch rate for the low-load photomask based on theoretical analysis. For a given recipe, the vertical etch rate of a photomask depends on its global load,[1,39,74] while the resist etch rate is not dependent on the photomask load.[39,62] Based on experimental results, a theoretical etch rate equation was proposed as:[74]

$$r_{\mathrm{cr}} = \frac{a_{\mathrm{ld}}}{1 + b_{\mathrm{ld}} L_d} + c_{\mathrm{ld}} \tag{4.67}$$

where a_{ld}, b_{ld}, and c_{ld} are constants for a given recipe. The experimental data in Fig. 4.16 agree well with predictions made by the theoretical equation, and the etch rate versus load curves are very similar compared to other experimental curves.[39,62] The theoretical etch rate equation can be used to calculate main etch times and over etch time for extremely

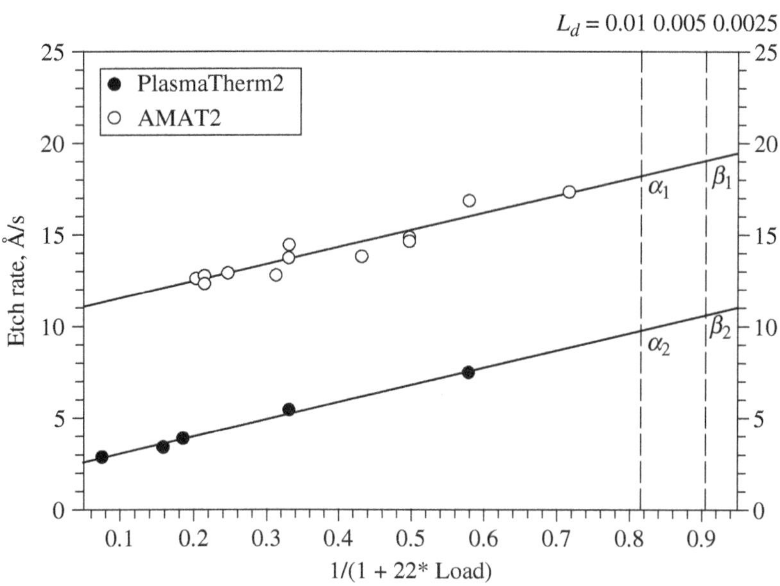

Figure 4.16 Etch rate changes with load.

low-load photomasks without the use of any other endpoint detection scheme. In commercial production, this method has proven to be very effective. The overetch percentage relative to main etch time can vary over a large range for low-load masks (e.g., 50–150 percent) without affecting overall etch performance significantly.

4.3.4 High-load photomask Cr etch

Unlike low-load photomasks, etch process optimization for high-load photomasks is much more difficult. The main technological challenge is that local etch performance varies greatly due to the dependence on pattern variation, i.e., the local loading. High loading and large pattern areas tend to be more difficult for the etch engineer to maintain good quality control. Isolated Cr (not space) feature CDs are harder to control than dense feature CDs. Therefore, the following discussion will focus on isolated Cr features primarily.

As has been discussed previously, the resist layer already contains some degree of systematic error as a result of the pattern generation processes. For state-of-the-art high-load photomasks, patterns are usually printed in chemically amplified resist by a 50-KeV electron beam (e-beam) writing system. The fogging effect inherent in e-beam lithography produces significant CD differences among the various loading areas. Therefore, dose correction on the writing tool is currently employed in order to reduce this systematic CD error. Due to the fact that the fogging problem can affect several centimeters of mask surface area, photomasks with large pattern areas and high loads are affected most. Another reason is that high-load photomasks usually require strict CD uniformity on dark feature patterns (especially isolated dark features), making it more difficult to control from pattern center to edge. Hence, to design high-load test vehicles for the worst case, the largest pattern area (about 104 mm × 132 mm) and highest pattern load (e.g., 80 percent) should be used. In this scenario, the overall photomask load percentage is $(104 \times 132)/(150 \times 150) \times 0.8 = 49$ percent. If an etch recipe has adequate etch performance on this test vehicle across the entire pattern area, including the pattern edges, it will be suitable for almost all commercial photomasks.

CD differences between those on pattern edges and those in the pattern result from the specific pattern loads in these two areas. Negative-tone photoresists are more suitable for CD uniformity improvement of high-load photomasks. Table 4.7 shows a comparison between positive and negative resists for the same pattern, demonstrating superior high-load Cr CD uniformity when the negative resist is used. However, one negative CAR photomask etch issue it has now is Cr-to-resist etch selectivity. Because of the high etch load on negative resist photomasks (e.g., 70–85 percent) in order for the portion outside the pattern area to be completely etched out, the etch time must increase relative to positive-tone photomasks with same pattern.

TABLE 4.7 Comparison of CDs of a Photomask with 80 percent Load in the Pattern Area

	Positive CAR		Negative CAR	
360-nm dark feature (NTAR7)	Resist CD	Cr CD	Resist CD	Cr CD
Uniformity 3σ, nm				
Isolated line	13.1	12.7	5	3.5
Dense line	10.3	11.7	9.4	3.7

This means that etch process selectivity also (including the overetch contribution) will be very low if the same recipe is used as for positive-tone photomasks. Therefore, special attention should be paid to Cr-to-resist etch selectivity during negative-tone etch process optimization.

Negative-tone photomask processing is also more difficult to control because negative-tone resist chemistry is more sensitive to environmental conditions and process parameters than positive-tone resist chemistry. Recent progress on making negative resists more robust has helped to overcome this disadvantage. The benefits of lower writing time and better etch uniformity predict that negative-tone resist strategies will find more application for high-load photomask processing in the future.

Etch CD loading dependency mechanisms are complicated since the nonuniformities may come from a variety of sources.[74] One major contributor is from vertical resist removal during etch. In practice, resist profiles are not perfect right angles (90°). When the resist erodes in the vertical direction, the lateral dimension changes as well. Since the vertical resist etch rate depends on the local load, local CD etch bias is also determined by local load. For example, even when the resist profile angle is in the range of 85° to 87°, there is significant CD nonuniformity contribution from vertical resist etch rate nonuniformity. Averaging over many different photomasks, the resist etch rate is nearly independent of the photomask pattern load,[62] but within one photomask, small etch rate differences in regions having different local loads should not be neglected. Local resist response in different area of loading may vary due to write tool fogging, resulting in the resist etch rate differences.

Etch CD nonuniformity also results from lateral resist erosion and Cr etch rate nonuniformity. But so far, there are no publications claiming that lateral etch rate nonuniformity has the same loading distribution as that for vertical resist etch rate.

Figure 4.17 shows the change in resist erosion in the vertical direction with changes in local loading, and the corresponding CD etch bias change for the same location. The resist thickness nonuniformity range on the two lines of the figure is not significantly different, but there is a significant difference on the slopes. This suggests that the range of the vertical resist conditions as related to the CD values is different.

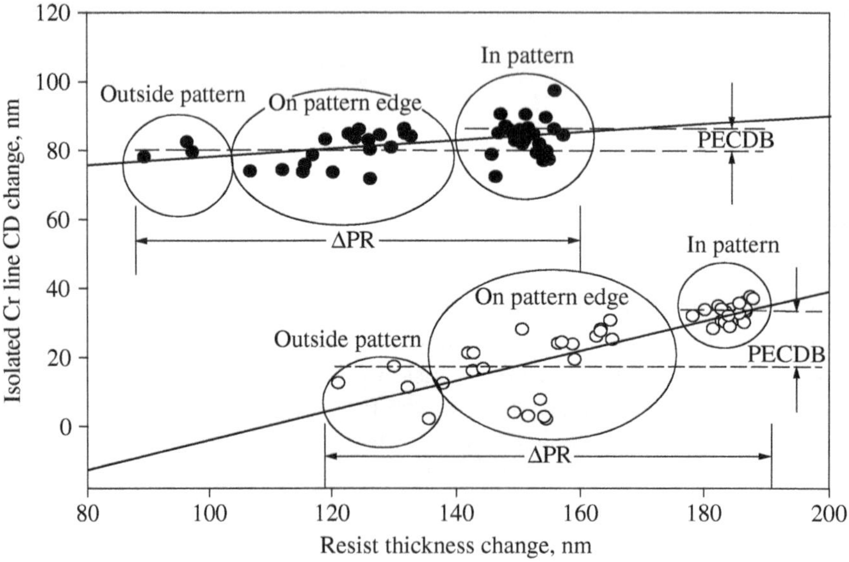

Figure 4.17 Cr etch CD movement versus resist thickness change on one photomask.

The recipes for these two lines are significantly different in pressure, ICP and bias powers, gas composition, and gas flow rate.

Etch chamber pressure not only affects the CD etch loading dependency (uniformity), but also I/D etch bias and CD etch bias. Figure 4.18 shows the effects of pressure on I/D etch bias and PECDB, defined as

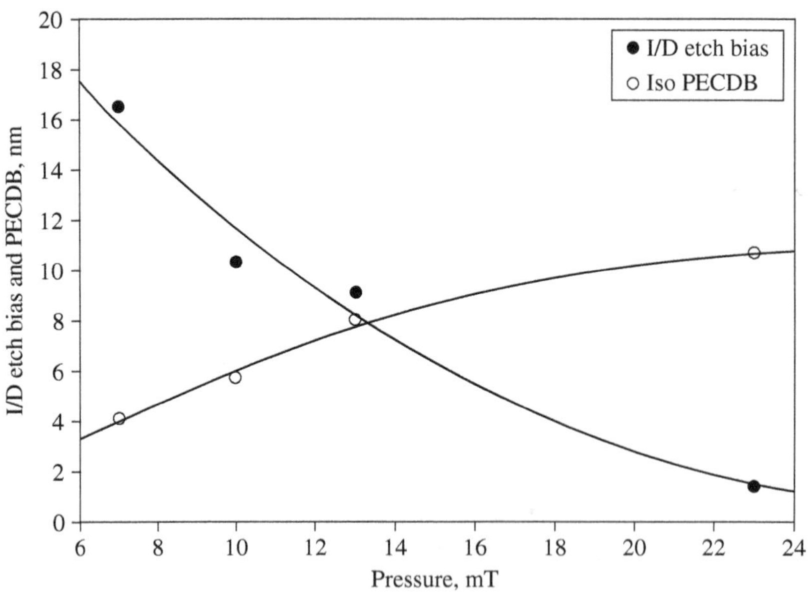

Figure 4.18 Pressure versus etch load effects and I/D etch bias.

TABLE 4.8 Comparisons of Different Cr Layer Thicknesses

Cr thickness, nm	I/D CD etch bias, nm
73	9.0
59	7.1

the etch contribution to the CD difference between the main pattern and pattern edges. When pressure decreases, the relative lateral-to-vertical Cr etch rate ratio increases, resulting in an increase in CD etch bias. Due to the fact that Cr etch is controlled by mass transport rather than bombardment energy, the relationship between pressure and CD etch bias shows that at low pressure, mass transport is more isotropic. Of course, isotropy also means that there is a higher lateral Cr etch component in the total CD etch bias, and etch CD nonuniformity by vertical resist etch rate nonuniformity is diluted. Therefore, in Fig. 4.17, the top condition has better etch CD uniformity although the resist vertical range is nearly the same.

The isotropic mass transport at low pressure also makes the I/D CD bias more of a factor, especially for smaller features and for thicker Cr layers. I/D CD bias mainly results from the phenomenon called the *neutral radical shadow*. Table 4.8 shows this effect as Cr layer thickness varies, and Table 4.9 shows I/D etch bias differences as feature size varies. Table 4.10 shows the overall effect of resist and Cr thickness on I/D etch bias.

Tables 4.8 to 4.10 clearly show that small feature size and thick Cr and resist layers tend to have high I/D CD etch bias.

4.3.5 Other issues in high-load Cr photomask etching

One important variable for high-load photomask Cr etching is the proper choice of overetch time. High-load photomask Cr etching has little to no endpoint detection challenges, but etch performance strongly depends on the time of overetch and related parameters. Because Cr-to-resist process selectivity of high-load photomask etching is significantly lower than that for low loads, excessive overetch times may induce an

TABLE 4.9 I/D Etch Bias Difference on Different Feature Sizes

Feature size, nm	I/D CD etch bias, nm
320	16.2
400	11.4

TABLE 4.10 Resist and Cr Thickness Combined Effects on I/D Etch Bias

Resist and Cr thickness, nm	I/D CD etch bias, nm
400 nm resist and 105 nm Cr	9.8
300 nm resist and 59 nm Cr	2.8

unwanted radial CD error as well as excessive CD etch bias. High-load photomask overetch times therefore, need to be established and monitored carefully. Overetch time of about 25 to 50 percent of the main etch times is common, depending on the individual situation. And to assist the etch engineer, those parameters unique to the overetch step (compared to the main etch) provide another degree of freedom for process optimization.

As gas flow parameters change from the main etch to the overetch step, it should be remembered that it will take some time for gas to flow from the *mass flow controller* (MFC) to the chamber. For example, if it takes 10 to 20 s to flow from the MFC to the etch chamber, gas composition and gas-flow stability will experience a transition period (instability), which will also affect plasma stability. If ICP and bias powers also change from main etch to overetch steps, the transition time is much shorter to reestablish plasma stability. Because vacuum conditions are required for photomask Cr etching, the resident time of gaseous reactants in the etch chamber is very short—usually less than 1 s.

While Cr etch optimization is obviously critical for binary photomask fabrication, it is also fundamental to successful phase-shift mask fabrication. For EAPSM, AAPSM, cPSM, and CPL photomasks, final CD qualities (MoSi layer or quartz layer) such as uniformity, I/D bias, and mean-to-target values are mainly determined during Cr etch. For EAPSM, Cr overetch usually "trenches" into the underlying MoSi layer because of limited selectivity (e.g., Cr-to-MoSi etch selectivity is about 4–6), which contributes to MoSi thickness nonuniformity. If the error is significant and systematic, it needs to be controlled during Cr etch in order to achieve acceptable phase angle range control in the MoSi layer. The removal of some quartz during the Cr overetch step is usually minimal because of the high Cr-to-quartz selectivity, so is usually not a source of significant phase error.

4.4 MoSi-Based Material Plasma Etch

4.4.1 Phase shifter layer plasma etch

MoSi-based materials are the most common phase shifters used in EAPSM strategies. The main constituents are molybdenum and silicon, but other elements include oxygen, nitrogen, and carbon. Plasma etching of this material was developed earlier for a similar material, molybdenum silicide ($MoSi_2$).[157,158] Etching of this material was studied extensively for potential applications in gate and interconnecting levels in *very large scale integration* (VLSI) circuits, as well as in the photomask industry as an absorber (an alternative to the chrome absorber),[92,140,159,160] when there were still many technical difficulties with the plasma etching of Cr in the early days of photomask plasma etching. NF_3 and Ar (or He) gases were tested in the original $MoSi_2$ etch studies of using planar plasma reactors. It was found that the etch rates were strongly affected by RF current, pressure, and NF_3 concentration.[157] The use of $SF_6 + O_2$ and $CF_4 + O_2$ for molybdenum silicide plasma etching was already gaining ground by the early 1980s as it was observed that oxygen accelerates etch rates and improved resulting profile angles.[158] $MoSi_2$ etching using Cl_2 was also tested, but has not been used extensively in the photomask process.[158] Gas systems for MoSi-containing materials are shown in Table 4.11.

For molybdenum silicide plasma etching using both $SF_6 + O_2$ and $CF_4 + O_2$, the etch rate increases with increases in oxygen concentration up to a point (e.g., 20 percent), then the etch rate decreases with any further oxygen increases.[158,159] It was thought that the oxygen gas enhanced the formation of fluorine radicals in the plasma environment when oxygen concentration was low, but excessive oxygen concentration dilutes the fluorine concentration and forms less volatile molybdenum oxifluoride, resulting in an etch rate decrease as oxygen concentration increases.[92,159] Recent thermodynamic calculation results support this hypothesis.[112]

TABLE 4.11 MoSi Composition and Gas Systems

Number	MoSi composition and gas system	References
1	$MoSi_2 + NF_3 + Ar$ (or He)	157
2	$MoSi_2 + SF_6 + O_2$	158
3	$MoSi_2 + SiF_4 + Cl_2$	158
4	$MoSi_w + CF_4 + O_2$ where $w = 1.9, 2.4$, and 3.0	159
5	$MoSi + CF_4 + O_2$	140
6	$MoSiN_x$ (or MoSiON) $+ CF_4 + CHF_3 + O_2$	93
7	MoSiONC (or MoSiOC) $+ CF_4 + O_2$	161
8	MoSi materials + Cl containing gases (e.g., Cl_2 or HCl) + F containing polymerizing materials (e.g., CHF_3 or CH_3F)	162
9	$MoSiON + Cl_2 + CF_4 + O_2 + He$	84
10	MoSi-containing materials $+ Cl_2 + SF_6 + O_2 + He$	86
11	MoSi-containing materials $+ Cl_2 + CHF_3 + O_2 + He$	86
12	MoSi-containing materials $+ HBr + SF_6 + O_2 + He$	86
13	MoSi-containing materials $+ HBr + CF_4 + O_2 + He$	86
14	$MoSiON + Cl_2$	90

Actually, the composition of commercial MoSi absorber materials varies. The Si-to-Mo ratio changes from 1.8[92] to 1.9, 2.4, or even 3.0.[159] Some MoSi materials contain oxygen[92] and sometimes an antireflective layer was deposited onto the MoSi absorber to reduce unwanted reflection during wafer printing.[160]

When phase-shift masks were called on by the semiconductor industry in the mid-1990s, many different embedded phase-shifter materials were investigated, including MoSi-based,[93,163] Cr-based,[164,165] W/Si thin films,[166] SiN_x layers,[167] $(LaNiO_3)_x(Ta_2O_5)_{1-x}$ oxide thin films,[168] and Ta-Si-O absorptive films.[169] Due to the superior properties of MoSi-based phase shifters, plasma etching of MoSi shifters having moderate nitrogen content was studied extensively.[1,52,82–91, 94–95,170,171]

MoSi etch investigations begins with the determination of the necessary reactive gases. The most studied gas systems are SF_6+ He + O_2 and CF_4+ He + O_2. EAPSM MoSi etch conditions leverage ICP-bias power configuration plasma etches and most often leverage SF_6+ He gases.[91] The addiction of He gas dilutes the reactant concentration and reduces the etch rate, making the etch process easier to control. The MoSi profile resulting from these etching conditions is usually greater than 80°, which is better than the RIE-only etchers that usually only achieve sidewall angles less than 70°. In this study, a zero-CD-etch-bias condition was obtained. Originally, EAPSMs were used solely for contact layer photomasks, which have low etch loading. Later, EAPSMs found application for critical line/space layers (e.g., gate layer), which have larger etch loading variability. The MoSi etching of high-load-MoSi-based photomasks is more difficult than for low load.[91] Gas systems based on CF_4 and SF_6 chemistries were compared experimentally while it was found that the CF_4 gas system outperformed SF_6 regarding etch CD uniformity.[82] This particular comparison was only based on predetermined experimental conditions so it is very difficult to conclude authoritatively that one gas system has better etch performance over the other. However, the CF_4 gas system introduces more polymer than the SF_6 system because of its carbon component. Carbon-containing defects can then be deposited in the etch system, which are very difficult to remove. With feature dimensions shrinking and defect control requirements becoming more demanding for state-of-the-art photomasks, SF_6 systems seem to have more overall advantages at this point. Other gases such CHF_3 and Cl_2 have also been tested,[83,84,90] but MoSi etching using SF_6 and CF_4 are by far the most common processes used in commercial EAPSM fabrication.

The function of oxygen has been discussed in several studies.[52,85,86,89] Adding a small amount of oxygen during etching improves MoSi/quartz etch selectivity,[85,86] may also improve sidewall profile angles[158] and polymer formation,[91] but it reduces MoSi/Cr selectivity.[89] Oxygen effects on sidewall profile and anisotropic properties are still not well understood as not enough validation has been performed to support these.

One significant difference between Cr and MoSi etching is the difference seen in the effect of bias power changes. In MoSi etch, the bias power affects not only the etch rate, but also the CD etch bias. Different etch parameters in the main etch and overetch steps can also be used to improve the etch performance. The lower bias power used during overetch than main etch allows a longer overetch time without losing phase angle control, but it can also increase the CD etch bias. Because of the much higher costs of MoSi photomask raw materials relative to binary raw materials, MoSi etch optimization has not been studied as thoroughly as it has for Cr.

MoSi material is a nonstoichiometric amorphous compound synthesized using *plasma-enhanced chemical vapor deposition* (PECVD) method. Its stoichiometric composition varies across mask blank manufacturers, and will vary according to which wafer exposure tool wavelength will be employed. The MoSi layer thickness for 248-nm wavelength ranges from 90 to 130 nm, but resulting phase angle and transmission percentage are 180° and 6.0 percent, respectively. For 193-nm MoSi films, the thickness is around 72 nm, thinner than that for 248-nm wavelength, but having the same phase and transmission characteristics (180° and 6.0 percent, respectively) at exposure wavelength.

When Cl_2 (without oxygen) is used for the MoSi etching, the bias power is kept relatively high to ensure adequate etch rate. There are several volatile etch products formed when fluorine-based etchants are used—mainly MoF_6, SiF_4, $MoOF_4$, and Si_2OF_6. The main by-product is an unwanted polymer that needs to be removed usually by cleaning the mask after the MoSi etch step.

At present, a near-zero MoSi CD etch bias has been achieved and is common in the production of both high and low-load photomasks.

4.4.2 Low-load photomask MoSi etch

Low-load photomask MoSi etching has an endpoint detection problem almost as bad as was seen in Cr etching, especially when the photomask pattern load is below 2 percent. As discussed before, when loading is less than 1.5 percent, the OES endpoint detection method is generally not reliable. The laser reflectivity endpoint detection method is also not reliable because there is no sharp change of laser reflectivity when the etch progresses from the MoSi layer to the quartz material underneath. Therefore, endpoint detection of low-load (<1.5 percent load) photomasks must seek other alternative methods.

$$r_{\text{mosi}} = \frac{a_{\text{mo}}}{1 + b_{\text{mo}} L_d} \tag{4.68}$$

where

$$a_{mo} = \frac{10^{10}}{\beta_{mosi}\rho_{mosi}}\left(\frac{k_{mosi}R_{cm}}{Vk_{vm} + A_w k_{wm} + Ak_{crm}}\right) \tag{4.69}$$

$$b_{mo} = \frac{A(k_{mosi} - k_{crm})}{Vk_{vm} + A_w k_{wm} + Ak_{crm}} \tag{4.70}$$

R_{cm} is the MoSi etch reactant consumption rate in the chamber (mol/s), k_{vm} (s^{-1}), and k_{wm} (m/s), k_{mosi} (m/s), and k_{crm} (mol/m^2s) are reactant consumption rate constants in the bulk chamber, chamber wall, on the MoSi surface, and on the Cr mask, respectively. Because of the high selectivity of MoSi to Cr and the relatively low Cr etch rate, b_{mo} can be simplified as:

$$b_{mo} = \frac{Ak_{mosi}}{Vk_{vm} + A_w k_{wm}} \tag{4.71}$$

Now b_{mo} expresses a reactant consumption rate ratio on the MoSi surface comprehending the chamber volume and wall effects. Compared to Cr etch load effects, the b value in MoSi etching is much smaller than Cr (e.g., 2.12 vs. 22.00), indicating that fluorine chemical etching exhibits a lower mask-to-chamber selectivity. The low b value in the MoSi etch means that MoSi etch rate dependency on load is not as strong as Cr etch, and it is possible to use Eq. (4.68) to estimate the etch rate. In reality, the use of a timed MoSi etch provides an acceptable phase angle mean-to-target value for very low-load photomasks if a suitable recipe with high enough MoSi-to-quartz etch rate selectivity is chosen. Under high selectivity conditions, a timed etch can make the decision to extend the overetch time slightly feasible without incurring significant phase angle error. If inadequate bias power is used for this purpose, a long overetch will affect CD values significantly, but not phase angle. All of these factors must be considered when attempting to optimize low-load MoSi photomask etching, especially using a timed etch.

Because Cr layer CD values for dark-field photomasks can be tightly controlled, the main objective for MoSi etch optimization is usually to obtain high fidelity pattern transfer by strictly controlling CD uniformity and mean-to-target CD responses. In MoSi etch optimization, however, one must satisfy many other responses, including phase angle ranges, phase angle mean-to-target, cross-sectional profile angles, and final transmission. Due to the limited MoSi-to-quartz etch rate selectivity in a MoSi etch, special care of etch selectivity should be taken in order to obtain a predicted and controllable phase angle mean-to-target values.

Regarding profile angles, right angles are usually preferred. In the commercial mask condition, profile angles greater than 80° usually meet requirements. Ranges of 70° to 80° can even satisfy less-critical layer specifications. It should be noted that the profile angle is determined not only by the etch recipe, but also by etch tool design and hardware settings, as well as the MoSi layer properties themselves. For example, even using the same etch tool and etch recipe, 193 and 248 nm MoSi may have different profile angles, due to stoichiometric differences in the materials.

4.4.3 High-load photomask MoSi etch

The main reason that high-load MoSi etching is more difficult than low load is due to signature CD uniformity requirements. The Cr CD signature for high-load photomasks contains systematic errors, especially for the CD dependency on local loading. With a suitable etch recipe, MoSi etch can compensate (correct for) these CD errors. The results of using a CD compensation strategy is that the MoSi CD uniformity 3σ value can be smaller than what was seen in the Cr CD signature, or even smaller than that of the resist CD signature.

Typical systematic CD errors before MoSi etch manifest themselves as smaller high-load dark CDs compared to CDs in load areas. In order to compensate for this error, the MoSi CD etch bias should have the same magnitude but opposite direction from the resident CD distribution. That means the MoSi CD etch bias should be smaller in high-load areas than in low-load areas. The solution is to use gaseous etchants that "starve" gaseous reactive radicals in high-load areas, resulting in low vertical etch rates in high-load areas and high vertical etch rates in low-load areas. For MoSi etch, vertical etch rate is proportional to lateral etch rate as shown in Fig. 4.19. Therefore, using this technique will

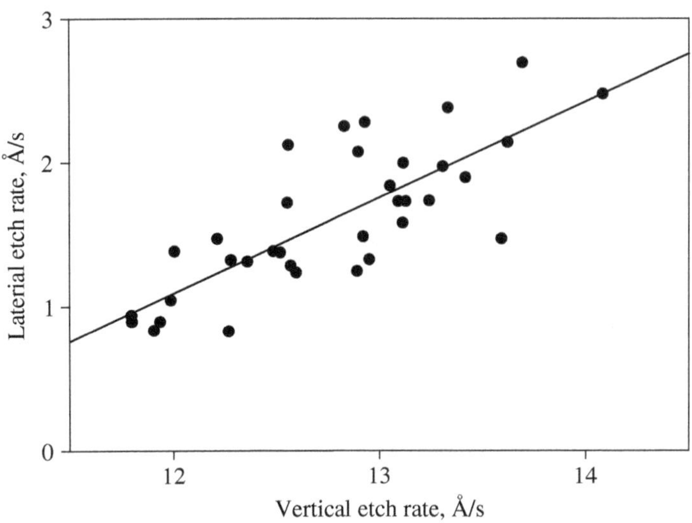

Figure 4.19 Local vertical MoSi etch rate versus lateral etch rate.

TABLE 4.12 MoSi Etch CD Compensation of Dark Feature CDs for 50 percent Load Photomasks

Pattern layer	Isolated CD 3σ, nm	Dense CD 3σ, nm
Resist (positive CAR 300 nm thick)	13.8	10.3
Cr (59 nm thick)	15.1	12.3
MoSi (193 nm wavelength)	10.5	9.0

make the MoSi CD etch bias higher in low-load areas than high-load areas (i.e., MoSi etch CD compensation). Table 4.12 demonstrates this compensation, and Fig. 4.20 shows how EAPSM CD changes with changes in local load.

MoSi etch CD compensation may cause some problems if not used correctly. One of the drawbacks that can result is the presence of a CD etch bias side-to-side distribution error. Because the gaseous etchants are being starved, the gas flow pattern may induce a CD etch bias nonuniform distribution. This error is related to the etch tool design and the etch recipe selected. For the same reasons, phase angle may also have a side-to-side distribution signature. Usually the phase angle range can be controlled within 3° without major technological challenges.

A critical parameter to observe and thereby control MoSi CD etch compensation is to determine the relationship between surface etch rate and etchant depletion. MoSi etch rate kinetics can provide useful information for this activity. MoSi etch rates have been studied theoretically

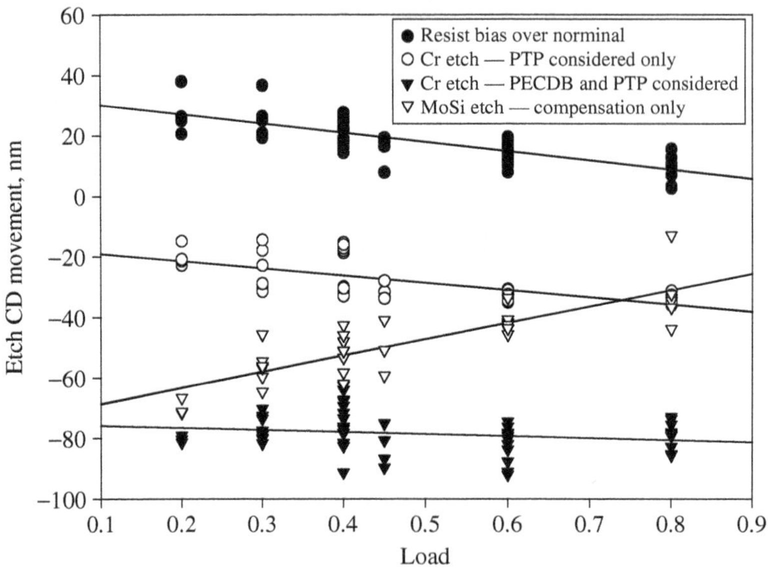

Figure 4.20 Cr and MoSi etch CD movement versus load.

and the etch rate equation was experimentally validated, as shown in Eq. (4.72):[94]

$$r_A = AW_{\text{icp}}\, e^{-\frac{1}{V_{\text{DCB}}}\left(\frac{E_a}{2eN_A}\cdot\ln\left(\frac{m_i}{2.3m_e}\right)\right)} C_{\text{SF}_6}^{1/5} C_{\text{O}_2}^{1/2} \tag{4.72}$$

where r_A = etch rate (Å/s)
C_{SF_6} and C_{O_2} = SF_6 and O_2 concentrations, respectively
A = apparent frequency factor
E_a = activation energy (J/mol)
R = gas constant
e = elementary charge value
N_A = Avogadro's number
m_i and m_e = ion and electron masses, respectively
W_{icp} = ICP power
V_{DCB} = DC bias voltage (V)

The activation energy is at the same level as the surface binding energy for many refractory materials (i.e., roughly the enthalpy of vaporization).[2]

For high-load photomasks using negative-tone resists, Cr CD dependence on loading is minimal. Therefore, CD compensation during MoSi etch is usually not necessary to achieve the highest pattern transfer fidelity.

4.4.4 Special considerations for MoSi etch

Special attention should be paid to several issues. One of them is the etch selectivity of MoSi to quartz and MoSi to Cr. The selectivity of MoSi to quartz directly affects the resulting phase-angle range because this range is strongly influenced by quartz trench nonuniformity during MoSi overetch. Higher selectivity allows for longer MoSi overetch that provides for larger process margins in a commercial setting.

The selectivity of MoSi to Cr is also important because inadequate selectivity can cause significant removal of the anti-reflective layer. Considering the entire photomask fabrication process, this missing AR can degrade the adhesion of the second resist coating, resulting in more AR loss during the Cr strip process.

Another issue to consider is the cleaning process. The MoSi layer erodes in standard photomask cleaning processes because basic solutions aggressively attach the MoSi layer, especially those solutions containing ammonium. The erosion of the MoSi layer due to standard cleaning should be considered when establishing the MoSi etch process in order to obtain acceptable average phase angle values. It is common to see phase angle reduction of 0.5° to 1.0° in the cleaning process, but these values can vary due to MoSi chemical resistance differences as the MoSi material properties fluctuate. Usually, 193-nm MoSi has better chemical resistance than 248-nm MoSi given the same cleaning process.

4.5 Fused Silica (Quartz) Plasma Etch

4.5.1 Fused silica etch fundamentals

Fused silica, also called quartz, comprises silicon oxide. The plasma etching of silicon oxides and silicon has been investigated in the interest of wafer fabrication long before phase-shift photomask fabrication. Many fundamentals and applications have been reported,[172] but specific photomask etch requirements differ from wafer applications. Fused silica etching for PSM applications was reported after the ICP-bias power plasma etcher became available.[97] Since then, many quartz etch optimization process studies have been published.[98,102,103,139,173–175] The etch chemistry used for quartz etch is a fluorine-containing gas of some type.

Fused silica (quartz) is the substrate material for present photomasks. For AAPSM and CPL photomasks, the quartz is etched without a stop layer. The main challenge for controlling the quartz etch process is to maintain tight etch depth uniformity and vertical quartz etch rate uniformity. Another challenge is to produce a recipe that results in accurate trench depth control. Obviously, optical emission spectrometry is not suitable for endpoint detection as in Cr and MoSi etches. In situ endpoint detection of quartz etch has been attempted,[174] but so far no mature endpoint detection method for quartz etch is available.

Because of this, using a timed etch seems to be the only option. Etch rate dependency on photomask pattern load causes the etch rate to change on virtually every commercial AAPSM. For engineering investigation, the full etch depth can be obtained in two steps where the *finishing* etch time can be calculated after the etch rate is measured based on the *partial* etch. But this method is difficult for manufacturing because it not only adds process complexity, but also may introduce defects. The other option is to find the correlation between etch rate and photomask loading, and then calculate a customized etch time according to the given photomask load. This also makes production control difficult. The best way to deal with this problem is to develop a quartz etch recipe whose etch rate is almost independent of photomask load.

Because of its excellent stability, fused silica is etched at a moderate etch rate using a fluorine-containing gaseous system. The most common reactive gases used for photomask etching are sulfur hexafluoride (SF_6) and hexafluoroethane (C_2F_6). Other reactive etchants include carbon tetrafluoride (CF_4), trifluoromethane (CHF_3), and perfluoroisobutene (C_4F_8), or combinations of the these gases. Carrier gases such as helium (He) and argon (Ar) are usually used as well in photomask quartz etching. Oxygen (O_2) has also been found to be helpful for etch performance control purposes.

Fluorine-containing chemicals will dissociate as fluorine radicals are released in high-vacuum plasma environments. They react with quartz under ion bombardment conditions where the main overall etch reactions using SF_6 and CF_4 gases can be expressed as:

$$SiO_2 + SF_{6(g)} = SiF_{4(g)} + SO_2F_{2(g)} \tag{4.73}$$

$$SiO_2 + CF_{4(g)} = SiF_{4(g)} + CO_{2(g)} \tag{4.74}$$

The main by-product of quartz etch is a polymer that deposits onto the mask surface during etching. Therefore, a cleaning process must follow the quartz etch step to remove the polymer. For carbon-containing reactants, the polymer formation rate is influenced by the carbon-to-fluorine ratio. Because of this, SF_6 forms less polymer than carbon-containing chemicals.

In order to improve etch depth uniformity, a stop layer below the phase-shift layer was proposed.[176] Technological and economic feasibility needs to be verified for this strategy. Since, this photomask raw material is not widely used or even available for commercial applications.

Nonuniform etch rates are the result of different mechanisms, ICP power and bias power being the most significant. ICP power favors the center-fast etch pattern and bias power induces the opposite effect,[97,139] but the influence of ICP power is much stronger than bias power. Because of this, small ICP power increments need larger bias power increments in order to maintain similar etch rate uniformity distributions. To avoid needing excessive bias power, a small amount of ICP power is recommended for uniform etch rate, but small amounts of ICP power in photomask quartz etch processes tend to result in unstable plasma conditions. Therefore, to balance the nonuniform etch rate caused by the use of ICP power, bias power is usually fairly high (e.g., 100–350 W). Low pressure is also commonly used in order to achieve uniform etch rates. High bias power and low pressure produce such high quartz etch rates that these rates can be much higher than the reactant mass transfer rate relative to the etch surface. As a result, etch rate nonuniformity may be present from the mass transport phenomenon. For example, etchers having a gas exhaust outlet on only one side of the chamber may exhibit side-to-side etch rate differences. To reduce side-to-side etch rate errors, photomasks can be rotated 180° at the etch halfway point, but this technique is not user-friendly for commercial production, and really serves only as a temporary solution.

A more reasonable solution to the etch rate nonuniformity problem is to use low ICP power by overcoming plasma instability such that a relatively high mass transfer rate can be maintained. This reduces both the side-to-side etch rate error and the etch rate dependence on photomask pattern load. It was reported that by using this strategy the etch rate difference among photomasks with loads of 9 to 50 percent can be controlled to 1 percent error, and phase angle ranges for photomasks in this wide range of pattern load variation is about 3°.[139] When low bias power is used, it should be noted that the ion incidence angle will increase, possibly causing significant ion reflection off the side walls, resulting in microtrenching.

The effects of etch rate were experimentally investigated before AAPSM and CPL masks were used in the photomask industry.[97] These studies show that the quartz etch rate increases with the increase in ICP power, bias power, and helium flow rate, but with a decrease in pressure. Like most Cr and MoSi etch processes, quartz etch results were also investigated using the DOE methods. An empirical analytical equation has been proposed to describe the quartz etch rate as:[139]

$$r_{\mathrm{qz}} = 1.650 \times 10^{-6} \frac{W_{\mathrm{icp}}^{1.13} V_{\mathrm{DC}}^{1.69} p_{\mathrm{SF_6}}^{0.11} Q_{\mathrm{He}}^{0.42} \left(1 - X_{\mathrm{O_2}}\right)^{0.79}}{p^{0.58}} \tag{4.75}$$

where W_{icp} = ICP power
V_{DC} = DC bias voltage
$p_{\mathrm{SF_6}}$ = SF_6 partial pressure
$X_{\mathrm{O_2}}$ = oxygen molar fraction
p = total pressure

Experimental results were plotted in Fig. 4.21, showing good agreement between experimental data and the empirical equation, in which W_q is defined as:

$$W_q = \frac{W_{\mathrm{icp}}^{1.01} V_{\mathrm{DC}}^{1.51} p_{\mathrm{SF_6}}^{0.10} Q_{\mathrm{He}}^{0.37} \left(1 - X_{\mathrm{O_2}}\right)^{0.70}}{p^{0.52}} \tag{4.76}$$

Equation (4.75) implies that the quartz etch rate is affected mostly by ICP power, DC bias voltage (bias power), and pressure. It also shows that plasma density nonuniformity (caused by ICP coil variations) can result in etch rate nonuniformity. A nonzero ion bombardment distribution by electric field nonuniformity also contributes to etch rate nonuniformity. Variability distributions of pressure and oxygen concentrations, as well as the helium flow pattern are also responsible for the etch rate nonuniformity.

Although many silicon dioxide plasma etch studies have been carried out,[96,101,172,177,178] the aims for successful photomask etching are quite

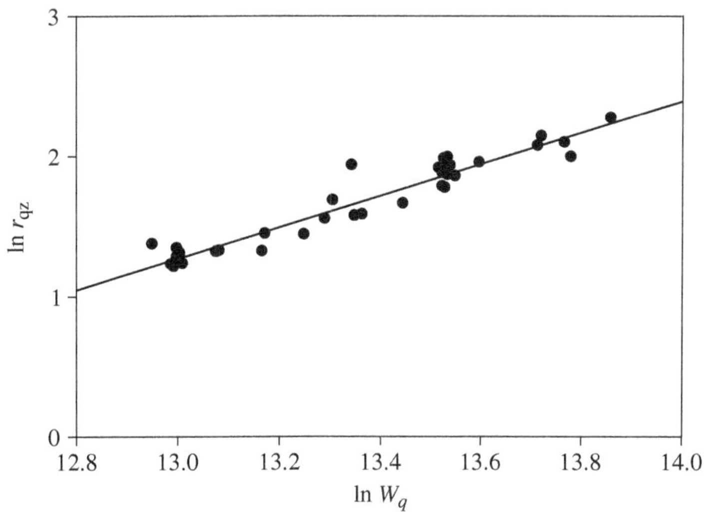

Figure 4.21 Quartz etch rate versus W_q.

different from other applications. Until recently, quartz etching of photomasks has not been studied intensively. There is no direct evidence showing that one gas system is better than the rest either. A variety of fluorine-containing gases has been used over many investigations (e.g., CHF_3,[97] CF_4,[173] SF_6,[103,106] and CHF_3/CF_4[175]). Usually, carbon/fluorine-containing gases form more polymer during etching, so for the purpose of defectivity control, SF_6 is preferred. Quartz etch performance dependence on the masking layer (e.g., resist or Cr layer) is intuitive and has been observed in practice.[139]

4.5.2 Fused silica etching of various mask types

AAPSM quartz etching usually leverages either a resist or Cr (or both) as the etch mask. Due to second writing alignment inaccuracy, small areas of Cr may be left exposed during quartz etch. Although a resist is the primary masking layer for the quartz etch, there are still small areas where Cr itself functions as the masking layer. AAPSM pattern loading is usually low (e.g., less than 15 percent). For 248-nm wavelength wafer exposure, the quartz etch depth is about 240 to 250 nm, while for 193-nm wavelength, the depth is about 170 to 180 nm.

The use of the resist versus Cr as the mask for AAPSM quartz etching is more beneficial to etch rate uniformity control. Experimental results indicate that for the same test vehicle and etch recipe, there is a significant etch rate uniformity difference between using a Cr versus resist as the masking layer. For example, the phase angle range of an

AAPSM etch using a resist masking layer is 3°, but it climbs to over 10° if a Cr hard mask is used. As with Cr and MoSi, etch performance also depends on etch tool design and recipe construction.

Defectivity control in quartz etch has been of major concern because of the inability to repair quartz defects perfectly. Although commercial repair tools are now available, quartz repair is still immature in commercial mask making, and only a quartz removal process is available.

Quartz-to-resist etch rate selectivity is also difficult, especially for 248-nm AAPSM due to the deeper trenches required. Therefore, attention must be given to both etch rate uniformity and quartz-to-resist etch rate selectivity for AAPSM etch process optimization. Regarding etch rate selectivity and defect control, a multistep quartz etch is sometimes used.

Chromeless phase lithography is a promising technology, but the quartz etch step in CPL mask fabrication is challenging. One reason is that CPL photomask quartz etching usually uses Cr as the masking layer, making etch rate (etch depth) uniformity control difficult. Another reason is that the load of CPL photomasks varies over a wide range, from a few percent up to 40 or 50 percent. Etch rate independence to load is required over this wide load range. In addition, quartz-to-Cr etch rate selectivity needs to be high for CPL mask quartz etch since low selectivity may cause significant missing AR (antireflective layer damage), limiting the effectiveness of the photomask itself.

4.5.3 Special considerations for quartz etching

Special considerations for photomask quartz etching include etch surface roughness and flatness, cross-sectional profile angle, and the microtrenching phenomenon. Surface roughness and flatness are dependent on chemistry and process conditions (recipe), but if the process space area is chosen properly, these factors should not cause a problem. Cross-sectional profile angles over 80° are common and acceptable in most applications. In the future, photomask specifications may tighten regarding this etch property, forcing the requirement of near 90° angle sidewalls. Quartz microtrenching may result due to an improper quartz etch recipe setting, especially where there is low ion bombardment. If microtrenching is not very severe, there is no significant influence on wafer printing.

4.6 The Future of Photomask Plasma Etching

4.6.1 Plasma etching of 157-nm photomasks

The wavelength of 157 nm was planned for 45-nm technology node lithography, but recent progress on 193-nm immersion lithography technology successfully postponed the introduction of 157-nm scanners and the photomasks associated with them. While still difficult to speculate, the use of 157-nm lithography may never reach commercial production. Even now, 157-nm photomask raw materials are not finalized although several options have been studied. If required someday, the plasma etching of 157-nm photomask will be dependent on final material stoichiometries and new photomask design challenges.

4.6.2 Plasma etch for next generation photomasks

Next generation lithography includes x-ray lithography, *step-and-flash imprint lithography* (SFIL), *e-beam projection lithography* (EPL), *ion projection lithography* (IPL), and *extreme ultraviolet lithography* (EUVL). Although x-ray lithography was studied intensively (mainly in Japan) for several years,[179] EUVL has been the most dominant process for NGL with less attention now on EPL or IPL. SFIL is gaining more support, but still relatively.

EUVL reflective technology is one of the leading candidates for the 32-nm-node-and-below technology. An EUVL photomask consists of several layer categories. The top layer is an absorbing layer, usually chromium (Cr) or tantalum (Ta) or tantalum nitride (TaN or TaN_x).[105,106] Its main function is to absorb EUV radiation during exposure. The layer beneath is the buffer layer such as silicon dioxide (SiO_2) or silicon oxinitride (SiON). The layers below comprise the reflector, and are made up of alternating the MoSi multilayers (e.g., 40 pairs).[107] Their function is to reflect EUV radiation. There is sometimes a silicon layer between buffer and multilayers[104] as a protective capping layer.[180]

Chlorine-containing gas is used to etch the absorber. When chromium is the absorber material, chlorine, oxygen, and helium are used. Oxygen is not needed for Ta-containing absorber etching. Instead, a chlorine + $BrCl_3$ mixture is used for the Ta absorber.[104,105] These etchant gases are not new, but the etch specification may be much stricter because EUVL will be used for the 32-nm-node-and-below technology. The main concern of chromium absorber etching is the excessively high CD etch bias for EVUL masks,[180,181] although a *near* zero etch bias process has been reported.[181]

NGL photomask blank and absorber materials are not finalized, but there are several options. Because of the technological challenges of using chromium etch for masks having very strict requirements, Ta-based material studies may find more popularity as a possible replacement for the chromium absorber. Ta-containing materials mainly consist of tantalum (Ta) and tantalum nitride (TaN or TaN_x), and etch performance differences between Ta and TaN_x have been reported,[180] showing better profile capability and less undercut when using TaN_x materials.

Ta etch using chlorine gas produces a volatile product $TaCl_5$. The overall etch reaction can be expressed as:[112]

$$2\mathrm{Ta} + 5\mathrm{Cl}_{2(g)} = 2\mathrm{TaCl}_{2(g)} \tag{4.77}$$

For Ta etching, oxygen should be avoided in the gas system, because the nonvolatile product Ta_2O_5 serves as an etch reaction inhibitor, as shown in the following reaction:

$$6\mathrm{Ta} + 3\mathrm{Cl}_{2(g)} + 6\mathrm{O}_{2(g)} = 2\mathrm{Ta}_2\mathrm{O}_5 + 2\mathrm{TaOCl}_{3(g)} \tag{4.78}$$

TaN can also be etched in a plasma environment using chlorine gas. Volatile products are also produced and overall etch reaction is:

$$2\mathrm{TaN} + 5\mathrm{Cl}_{2(g)} = 2\mathrm{TaCl}_{5(g)} + \mathrm{N}_{2(g)} \tag{4.79}$$

The etching of the buffer layer being a glassy material, is similar to that of quartz etch for AAPSM and CPL masks. The primary difference lies in the stricter etch requirements for EUVL masks. For example, the buffer layer, SiO_2 can be etched using C_4F_8 and O_2.[104] Etch rate selectivity between the buffer and multilayers needs to be very high for the buffer layer etch so as not to damage the reflector below.

References

1. Wu, B., "Photomask Cr-MoSi etching," *J. Vac. Sci. Technol.* B22(3), pp. 1150–1159, 2004.
2. Manos, D.M., and D.L. Flamm, eds., *Plasma etching—an introduction*, Academic Press, San Diego, Calif., 1989.
3. Waits, R., "Evolution of integrated-circuit vacuum processes: 1959-1975," *J. Vac. Sci. Technol.* B18(4), pp. 1736–1740, 2000.
4. Irving, S. M., K. E. Lemons, and G.E. Bobos, "Gas plasma vapor etching process," *US Patent* 3615956, 1971.
5. Shul, R. J., and S. J. Pearton, ed., *Handbook of Advanced Plasma Processing Techniques*, Springer, Berlin, Germany, pp. 57–65, 285, 2000.
6. Wolf, S., and R. N. Tauber, *Silicon Processing for the VLSI Era, Vol. 1—Process Technology*, Lattice Press, Sunset Beach, Calif., pp. 455, 655–719, 2000.
7. Kota, G. P., J. W. Coburn, and D. B. Graves, "The recombination of chlorine atoms at surfaces," *J. Vac. Sci. Technol.* A16(1), pp. 270–277, 1998.
8. Balooch, M., M. Moalem, W. E. Wang, and A. V. Hamza, "Low-energy Ar ion-induced and chlorine ion etching of silicon," *J. Vac. Sci. Technol.* A14(1), pp. 229–233, 1996.
9. Yu, H. N., R. H. Dennard, T. H. P. Chang, C. M. Osburn, R. Dilonardo, and H. E. Luhn"Fabrication of a miniature 8K-bit memory chip using electron-beam exposure," *J. Vac. Sci. Technol.* 12(6), pp. 1297–1300, 1975.
10. Coburn, J. W., and H. F. Winters, "Utilizing saturated and unsaturated halocarbon gases in plasma etching to increase etch of SiO_2 relative to Si," *US Patent* 4162185, 1979.
11. Ashida, S., C. Lee, and M. A. Lieberman, "Spatially averaged (global) model of time modulated high density argon plasmas," *J. Vac. Sci. Technol.* A13(5), pp. 2498–2507, 1995.

12. Meeks, E., and J. W. Shon, "Effects of atomic chlorine wall recombination: Comparison of a plasma chemistry model with experiment," *J. Vac. Sci. Technol.* A13(6), pp. 2884–2889, 1995.
13. Meeks, E., P. Ho, A. Ting, and R. J. Buss, "Simulations of $BCl_3/Cl_2/Ar$ plasmas with comparisons to diagnostic data," *J. Vac. Sci. Technol.* A16(4), pp. 2227–2239, 1998.
14. Meeks, E., R. S. Larson, P. Ho, C. Apblett, S. M. Han, E. Edelberg, and E. S. Aydil, "Modeling of SiO_2 deposition in high density plasma reactors and comparisons of model predictions with experimental measurements," *J. Vac. Sci. Technol.* A16(2), pp. 544–563, 1998.
15. Mogab, C. J., *J. Electrochem. Soc.* 124, pp. 1262–1272, 1977.
16. Flamm, D. L., D. N. K. Wang, and D. Maydan, "Multiple-etchant loading effect and silicon etching in ClF_3 and related mixtures," *J. Electrochem. Soc.* 129, pp. 2755–2760, 1982.
17. Chang, J. P., and H. H. Sawin, "Kinetic study of low energy ion-enhanced polysislon etching using Cl, Cl_2, and Cl^+ beam scattering," *J. Vac. Sci. Technol.* A15(3), pp. 610–615, 1997.
18. Jurgensen, C. W., R. S. Hutton, and G. N. Taylor, "Resist etching kinetics and pattern transfer in a helicon plasma," *J. Vac. Sci. Technol.* B10(6), pp. 2542–2547, 1992.
19. Ahlrichs, P., U. Riedel, and J. Warnatz, "Spatially-averaged model for plasma etch processes: Comparison of different approaches to electron kinetics," *J. Vac. Sci. Technol.* A16(3), pp. 1560–1565, 1998.
20. Chang, J. P., A. P. Mahorowala, and H. H. Sawin, "Plasma-surface kinetics and feature profile evolution in chlorine etching of polysilicon," *J. Vac. Sci. Technol.* A16(3), pp. 217–224, 1998.
21. Mahi, B., Y. Arnal, and C. Pomot, "The etching of silicon in diluted SF_6 plasmas: Correlation between the flux of incident species and the etching kinetics," *J. Vac. Sci. Technol.* B5(3), pp. 657–666, 1987.
22. Poulsen, R. G., "Plasma etching in integrated circuit manufacture—A review," *J. Vac. Sci. Technol.* 14(1), pp. 266–274, 1977.
23. Armacost, M., P. D. Hog, R. Wise, W. Yan, J. J. Brown, J. H. Keller, and G. A. Kaplita et al., "Plasma-etching processes for ULSI semiconductor circuits," *IBM J. Res. Dev.*, 43(1/2), pp. 1–30, 1999.
24. Gottscho, R. A., and C. W. Jurgensen, "Microscopic uniformity in plasma etching," *J. Vac. Sci. Technol.* B10(5), pp. 2133–2147, 1992.
25. Abe, H., *Suppl. Jpn. J. Appl. Phys.* 44, p. 287, 1975.
26. Bersin, R. L., *Solid State Technol.* 19, p. 31, 1976.
27. Brusic, V. A., and P. M. McCaffrey, *IBM Tech. Disc. Bull.* 17, p. 306, 1974.
28. Flack, W. W., K. E. Tokunaga, and K. D. Edwards, "Chrome dry-etching for photomask fabrication," *Proc. SPIE* 1809, pp. 85–96, 1992.
29. Novembre, A. E., D. A. Mixon, C. Pierrat, C. S. Knurek, and M. W. Stohl, "Dry etch patterning of chrome on glass optical masks using P(SI-CMS) resist," *Proc. SPIE* 2087, pp. 50–56, 1993.
30. Buck, P. D., and B. J. Grenon, "A Comparison of wet and dry chrome etching with the CORE-2564," *Proc. SPIE* 2087, pp. 42–49, 1993.
31. Coleman, T. P., and P. D. Buck, "Plasma etching of chromium films in the fabrication of photomasks," *Proc. SPIE* 2621, pp. 62–72, 1995.
32. Coleman, T. P., and P. D. Buck, "Plasma etching of Cr films in the fabrication of photomasks: II . A comparison of etch technologies and a first look at defects," *Proc. SPIE* 2884, pp. 92–102, 1996.
33. Constantine, C., D. Johnson, R. Westerman, T. Coleman, and T. Faure, "Plasma etching of Cr photomasks: parametric comparisons of plasma sources and process conditions," *Proc. SPIE* 3096, pp. 11–18, 1997.
34. Constantine, C., D. U. Johnson, R. U. Westerman, T. P. Coleman, T. Faure, and L. F. Dubuque, "Plasma etching of Cr photomasks: optimization of process conditions and CD control," *Proc. SPIE* 3236, pp. 94–105, 1997.
35. Coleman, T. P., D. Alexander, and M. Lu, "Comparison of EBR-900 M1 and ZEP 7000 with plasma-etch processing for MEBES 4500S," *Proc. SPIE* 3236, pp. 397–404, 1997.

36. Inoue, T., Y. Matsuda, and Y. Tanaka, "Application of dry etching to 1-Gb DRAM mask fabrication," *Proc. SPIE* 3412, pp. 138–148, 1998.
37. Tsai, W., F. Chen, M. Kamma, S. Chegwidden, S. Labovitz, J. Farnsworth, and G. Dao, "Plasma etch of AZ5206/Cr and ZEP7000/Cr for 0.18- to 0.25-μm-generation advanced mask fabrication," *Proc. SPIE* 3412, pp. 149–162, 1998.
38. Park, K., and K. Lee, "Application of dry etching process on high-end Cr photomasks," *Proc. SPIE* 3412, pp. 246–251, 1998.
39. Lu, M., T. P. Coleman, and C. A. Souer, "A 180 nm mask fabrication process using ZEP7000, multipass gray, GHOST < and dry etch for MEBES 5000," Proc. SPIE 3546, pp. 98–111, 1998.
40. Chen, F., W. Tsai, S. Chegwidden, S. Yu, M. Kamna, J. N. Farnsworth, and T. P. Coleman, "Impact of the loading effect on CD control in plasma etching of Cr photomasks using ZEP7000 resist," *Proc. SPIE* 3546, pp. 429–439, 1998.
41. Aoyama, S., S. Sakamoto, T. Koike, N. Yoshioka, N. Harashima, A. Hayashi, and T. Sasaki, "Advanced Cr dry etching process," *Proc. SPIE* 3748, pp. 137–146, 1999.
42. Constantine, C., R. U. Westerman, and J. Plumhoff, "Plasma Etching of Cr: A Multiparameter Uniformity Study Urilizing Patterns of Various Cr Loads," *Proc. SPIE* 3748, pp. 153–157, 1999.
43. Constantine, C., R. J. Westerman, and J. Plumhoff, "Plasma Etch of Binary Cr Masks: CD Uniformity Study of Photomasks Utilizing Varying Cr Loads," *Proc. SPIE* 3873, pp. 93–97, 1999.
44. Handa, H., S. Yamauchi, K. Hosono, and Y. Miyahara, "Dry etching technology of Cr films to produce fine pattern reticles under 720 nm with ZEP-7000," *Proc. SPIE* 3873, pp. 98–106, 1999.
45. Hochmuth, J., G. Ruhl, and T. P. Coleman, "Control Methodology of off-target for varying pattern densities with chrome dry etch," *Proc. SPIE* 3873, pp. 297–306, 1999.
46. Albrethsen-Keck, B., M. Lu, and C. A. Sauer, "Improving CDs on a MEBES system by improving the ZEP 7000 development and dry etch process," *Proc. SPIE* 3873, pp. 592–599, 1999.
47. Handa, H., S. Yamauchi, and H. Shirai, "Improvement of Cr dry etching characteristics with MERIE system," *Proc. SPIE* 4066, pp. 226–234, 2000.
48. Iwamatsu, T., T. Fujisawa, K. Hiruta, H. Morimoto, N. Harashima, T. Sasaki, M. Hara, K. Yamashiro, Y. Ohkubo, and Y. Takehana, "Evaluation of loading effect of NLD dry etching," *Proc. SPIE* 4066, pp. 235–242, 2000.
49. Constantine, C., R. U. Westerman, and J. Plumhoff, "Plasma etch of binary Cr masks: CD uniformity study of photomasks utilizing varying Cr load—part III," *Proc. SPIE* 4186, pp. 85–96, 2001.
50. Ruhl, G. G., R. Dietrich, R. Ludwig, N. Falk, T. B. Morrison, B. C. Stoehr, "Chrome dry etch process characterization using surface nano profiling," *Proc. SPIE* 4186, pp. 97–107, 2000.
51. Handa, H., S. Yamauchi, K. Hosono, and H. Shirai, "Investigations of CD variation in CR dry etching process," *Proc. SPIE* 4186, pp. 513–521, 2000.
52. Kwon, H. J., K. S. Oh, B. S. Chang, B. Y. Choi, K. H. Park, and S. H. Jeong, "Dry etching technology of Cr and MoSi layers using high-density plasma source," *Proc. SPIE* 4186, pp. 532–539, 2000.
53. Fujisawa, T., T. Iwamatsu, K. Hiruta, H. Morimoto, N. Harashima, T. Noriyuki, M. Hara, K. Yamashiro, Y. Okubo, and Y. Takehana, "Evaluation of loading effect on NLD dry etching (2)," *Proc. SPIE* 4186, pp. 549–552, 2000.
54. Kwon, H. J., D. S. Min, P. J. Jang, B. S. Chang, B. Y. Choi, K. H. Park, S. H. Jeong, "Dry Etching of Cr layer and its loading effect," *Proc. SPIE* 4409, pp. 382–389, 2001.
55. Fujisawa, T., N. Yoshioka, T. Sasaki, and K. Yamashiro, "Improvement of NLD mask dry etching system," *Proc. SPIE* 4409, pp. 390–395, 2001.
56. Lee, C. J., H. S. Bang, J. W. Choi, H. S. Jung, C. Shin, and H. S. Kim, "Effect of clear field ratio on critical dimension on dry etching process," *Proc. SPIE* 4409, pp. 396–400, 2001.
57. Erber, F., G. Ruhl, C. Ebi, R. Dietrich, J. Mathuni, and P. Nesladek, "Development and characterization of a new plasma etching process for mask manufacturing," *Proc. SPIE* 4409, pp. 401–408, 2001.

58. Faure, T. B., C. Huynh, M. J. Lercel, A. Smith, and T. Wagner, "Utilization of optical emission endpoint in photomask dry etch processing," *Proc. SPIE* 4562, pp. 68–78, 2001.
59. Kwon, H. J., K. S. Min, P. J. Jang, B. S. Chang, B. Y. Choi, and S. H. Jeong, "Loading effect parameters at dry etcher system and their analysis at mask-to-mask loading and within-mask loading," *Proc. SPIE* 4562, pp. 79–87, 2001.
60. Buxbaum, A. H., M. J. Buie, B. C. Stoehr, W. Montgomery, and S. E. Fuller, "Characterization of an integrated multibeam laser mask-pattern generation and dry-etch processing total solution," *Proc. SPIE* 4562, pp. 338–352, 2001.
61. Lee, J. Y., S. Y. Cho, C. H. Kim, S. W. Lee, S. W. Choi, W. S. Han, and J. M. Sohn, "Analysis of dry etch loading effect in mask fabrication," *Proc. SPIE* 4562, pp. 609–615, 2001.
62. Buie, M. J., B. C. Stoehr, A. H. Buxbaum, and G. G. Ruhl, "An Endpoint Solution for Photomask Chrome Loads Down to 0.25%," *Proc. SPIE* 4562, pp. 616–623, 2001.
63. Huang, Y. C., M. J. Buie, B. C. Stoehr, A. H. Buxbaum, and G. G. Ruhl, "Extended chamber matching and repeatability study for chrome etch," *Proc. SPIE* 4562, pp. 624–632, 2001.
64. Buie, M. J., B. C. Stoehr, and Y. C. Huang, "Chrome etch for <0.13 μm advanced reticle production," *Proc. SPIE* 4562, pp. 633–640, 2001.
65. Anderson, R., N. Sandlin, M. Buie, C. Su, A. Agarwal, C. Brooks, Y. Huang, and B. Stoehr, "In-situ optical emission spectroscopic examination of chrome etch for photomasks," *Proc. SPIE* 4754, pp. 312–322, 2002.
66. Chou, W., F. Tsai, C. Tuo, C. Yoo, T. Tsai, and L. Shue, "CAR dry etching technology to produce 0.13 μm reticle," *Proc. SPIE* 4754, pp. 323–331, 2002.
67. Min, D., P. Jang, H. Kwon, B. Choi, and S. Jeong, "Applications of multiple-wavelength absorption endpoint system in photomask dry etcher," *Proc. SPIE* 4754, pp. 341–349, 2002.
68. Mueler, M., S. Komarov, and K. Baik, "Dry etching of chrome for photomasks for 100-nm technology using chemically amplified resist," *Proc. SPIE* 4754, pp. 350–360, 2002.
69. Anderson, R. B., G. Ruhl, N. L. Sandlin, and M. J. Buie, "Study of the role of Cl_2, O_2, and He in the chrome etch process with optical emission spectroscopy," *Proc. SPIE* 4889, pp. 641–652, 2002.
70. Wu, B., and D. Y. Chan, "Cr photomask etch performance and its modeling," *Proc. SPIE* 4889, pp. 667–678, 2002.
71. Nakata, H., K. Nishioka, and H. Abe, "Plasma etching characteristics of chromium film and its novel etching mode," *J. Vac. Sci. Technol.* 17(6), pp. 1351–1357, 1980.
72. Ruhl, G., R. Dietrich, R. Ludwig, N. Falk, T. Morrison, and B. Stoehr, "Optimizing the chromium dry etch process," *Semicond. Int.*, Vol. 24, pp. 239–246, 2001.
73. Wu, B., and D. Chan, "Cr and MoSi photomask plasma etch," *Proc. SPIE* 5038, pp. 1053–1064, 2003.
74. Wu, B., and D. Chan, "Cr photomask etch," *J. Microlith. Microfab. Microsyst.* 2(3), pp. 200–209, 2003.
75. Wu, B., "An investigation of Cr etch kinetics," *Proc. SPIE* 5256, pp. 701–712, 2003.
76. Granik, Y., "Dry etch proximity modeling in mask fabrication," *Proc. SPIE* 5130, pp. 86–91, 2003.
77. Clevenger, J. O., M. J. Buie, and N. Sandlin, "Effect of Chamber Seasoning on the Chrome Dry Etch Process," *Proc. SPIE* 5130, pp. 92–100, 2003.
78. Chang, S. M., C. C. Chin, W. C. Wang, C. L. Lu, and S. C. Chin, "Study of Loading Effect on Dry Etching Process," *Proc. SPIE* 5130, pp. 228–233, 2003.
79. Anderson, R., G. Ruhl, P. Nesladek, G. Prechtl, W. Sabisch, A. Kersch, and M. Buie, "Improvement of Chrome CDU by Optimizing Focus Ring Design," *Proc. SPIE* 5130, pp. 264–274, 2003.
80. Harashima, N., T. Sasaki, K. Kuwahara, T. Hayashi, Y. Tanaka, N. Yoshioka, M. Hara, and Y. Ohkubo, "Advanced NLD mask dry etching system for 90 nm node technology," *Proc. SPIE* 5130, pp. 275–280, 2003.
81. Smith, K. H., J. R. Wasson, P. J. S. Mangat, W. J. Dauksher, and D. J. Resnick, "Cr absorber etch process for extreme ultraviolet lithography," *J. Vac. Sci. Technol.* B19(6), pp. 2906–2910, 2001.

82. Constantine, C., D. J. Johnson, R. U. Westerman, and A. C. Hourd, "Inductively coupled plasma etch of DUV MoSi photomasks: a designed study of etch chemistries and process results," *Proc. SPIE* 3546, pp. 88–97, 1998.
83. Lee, K. Y., L. U. Kim, K. H. Nam, K. T. Park, Y. M. Ku, S. S. Ku, and I. B. Hur, "Study on ICP (Inductively Coupled Plasma) dry etch of DUV MoSi HTPSM," *Proc. SPIE* 3748, pp. 158–165, 1999.
84. Choi, S. J., H. S. Cha, S. Y. Yoon, S. M. Jung, S. S. Choi, and S. H. Jeong, "Dry etching characteristics of attenuated phase-shifting masks using $Cl_2/CF_4/O_2/He$ plasmas," *Proc. SPIE* 4562, pp. 561–570, 2001.
85. Plumhoff, J., C. Constantine, C. Strawn, and J. Shin, "Investigation of MoSi etch processes for embedded attenuating phase shift mask applications utilizing a next generation ICP source," *Proc. SPIE* 4562, pp. 694–703, 2001.
86. Plumhoff, J., C. Constantine, J. Shin, and E. Rausa, "Improvements in MoSi CD bias and iso-dense linearity plasma etch results utilizing design of experiments process optimization of Gen III ICP plasma source," *Proc. SPIE* 4754, pp. 291–302, 2002.
87. Choi, S., H. Cha, S. Yoon, Y. Kim, D. Lee, J. Kim, and J. Kim et al., "Highly anisotropic etching of phase-shift masks using ICP of CF_4-SF_6-CHF_3 gas mixtures," *Proc. SPIE* 4754, pp. 303–311, 2002.
88. Yoon, S., S. Choi, Y. Kim, D. Lee, H. Cha, J. Kim, S. Choi, and S. Jeong, "Etching selectivity and surface profile of attenuated phase-shifting mask using $CF_4/O_2/He$ inductively coupled plasma (ICP)," *Proc. SPIE* 4754, pp. 332–340, 2002.
89. Min, D. S., B. S. Chang, H. U. Kwon, and B. Y. Choi, "Higher Anisotropy and Improved Surface Conditions for 90 nm MoSiON ICP Dry Etch," *Proc. SPIE* 4889, pp. 694–701, 2002.
90. Chang, B. S., D. S. Min, H. J. Kwon, and B. Y. Choi, "Characteristics of Selective MoSiON Etching in a Chlorine Plasma," *Proc. SPIE* 4889, pp. 702–712, 2002.
91. Peng, S., and W. Adair, "High-density plasma dry etch for DUV attenuated phase-shifting masks," *Proc. SPIE* 3412, pp. 228–232, 1998.
92. Pierrat, C., R. G. Tarascon, M. L. Peabody, L. R. Harriott, and S. Vaidya, "Dry etched molybdenum silicide photomasks for submicron integrated circuit fabrication," *J. Vac. Sci. Technol.* B9(6), pp. 3132–3137, 1991.
93. Ushida, M., M. Mitsui, K. Okada, Y. Ohkubo, H. Suda, H. Kobayashi, and K. Asakawa, "Development of deep UV MoSi-based embedded phase-shifting mask (EPSM) blanks," *Proc. SPIE* 2884, pp. 403–410, 1996.
94. Wu, B., and D. Chan, "MoSi etch of phase-shift masks," *J. Microlith. Microfab. Microsyst.* 2(1), pp. 54–60, 2003.
95. Mueller, M., S. Komarov, and K. H. Baik, "High-resolution etching of MoSi using electron beam patterned chemically amplified resist," *Proc. SPIE* 5130, pp. 234–245, 2003.
96. Sankaran, A., and M. J. Kushner, "Fluorocarbon plasma etching and profile evolution of porous low-dielectric-constant silica," *Appl. Phys. Lett.* 82(12), pp. 1824–1826, 2003.
97. Constantine, C., and L. Hecherd, "ICP quartz etch uniformity improvement for phase-shift mask fabrication," *Proc. SPIE* 3412, pp. 220–227, 1998.
98. Westerman, R. U., C. Constantine, J. Plumhoff, and C. Strawn, "Plasma etching of quartz for the fabrication of alternating aperture phase shift photomasks: etch rate uniformity study utilizing a next generation ICP source, *Proc. SPIE* 4186, pp. 316–324, 2000.
99. Wu, H. M., L. He, J. Farnsworth, and G. Liu, "CF_4/O_2 plasma simulation and comparison with quartz etch experiment," *Proc. SPIE* 4409, pp. 409–417, 2001.
100. Rhyins, P. D., M. Fritze, D. Y. Chan, C. Carney, B. A. Blackowicz, M. Vieira, and C. A. Mack, "Characterization of Quartz Etched PSM Masks for KrF Lithography at the 100 nm node," *Proc. SPIE* 4562, pp. 486–495, 2001.
101. Bretoiu, S., K. Di Mola, E. Fioravanti, and S. Visona, "Inductively coupled plasma etching for arrayed waveguide gratings fabrication in silica on silicon technology," *J. Vac. Sci. Technol.* B20(5), pp. 2085–2090, 2002.
102. Jang, I. Y., J. Y. Lee, Y. H. Kim, S. W. Choi, and J. M. Sohn, "Development of etch rate uniformity adjustment technology for photomask quartz etch in manufacturing the 100% attenuated PSM," *Proc. SPIE* 5130, pp. 246–252, 2003.

103. Plumhoff, J., C. Constantine, J. Shin, B. Reelfs, and E. Rause, "Evaluation of 193 nm alternating aperture phase shift mask dry etch processes," *Proc. SPIE* 5130, pp. 253–263, 2003.
104. Hoshino, E., T. Ogawa, M. Takahashi, H. Hoko, H. Yamanashi, N. Hirano, and S. Okazaki, "Damage control during dry etching of EUV mask-1-control of surface roughness," *Proc. SPIE* 3873, pp. 786–791, 1999.
105. Hoshino, E., T. Ogawa, N. Hirano, H. Hoko, M. Takahashi, H. Yamanashi, A. Chiba, M. Ito, and S. Okazaki, "Dry etching of Ta absorber for EUVL masks," *Proc. SPIE* 4186, pp. 749–755, 2000.
106. Racette, K. C., C. T. Williams, and M. J. Lercel, "Characterization and etching of sputter deposited absorber films for extreme ultraviolet lithography," *Proc. SPIE* 4562, pp. 883–892, 2001.
107. Zhang, G. and P. Y. Yan, "Cr and TaN absorber mask etch CD performance study for extreme ultraviolet lithography," *Proc. SPIE* 4889, pp. 1092–1098, 2002.
108. Wu, B., "Synthesis of zirconium carbide from zircon using high temperature plasma," Ph.D. Dissertation, The University of Idaho, Moscow, Idaho, 1996.
109. Stern, A. C., *Air Pollution Control*, Academic Press, New York, Vol. 4, 1984.
110. Hesketh, H. E., "Fine particles in gaseous media," Lewis Publishers, Boca Raton, Fla., p. 113, 1986.
111. Lieberman, M. A., and A. J. Lichtenberg, *Principles of Plasma Discharges and materials Processing*, Wiley-Interscience, New York, pp. 98, 154–191, 1994.
112. Wu, B., "Thermodynamic study of photomask plasma etching," *Proc. SPIE* 5567, pp. 1195–1206, 2004.
113. Wu, B., P. R. Taylor, and E. E. Vidal, "Zirconium Carbide Synthesis in a High Temperature Plasma Reactor," *PED Congress by TMS*, pp. 245–254, 1997.
114. Rode, E. J., and W. R. Harshbarger, "Thermodynamic analysis of the W-F-Si-H and W-F-Si-H-Cl systems, *J. Vac. Sci. Technol.* B8(1), pp. 91–94, 1989.
115. Burrows, A., and C. M. Sharp, "Chemical equilibrium abundances in brown dwarf and extrasolar giant planet atmospheres," *Astrophys. J.* 512, pp. 843–863, 1999.
116. Ohring, M., *The Materials Science of Thin Films*, Academic Press, New York, pp. 23–33, 1992.
117. Tokiguichi, K., N. Sakudo, K. Suzuki, and I. Kanomata, "Plasma uniformity of microwave ion sources," *J. Vac. Sci. Technol.* 17(5), pp. 1247–1251, 1980.
118. Lochte, W., *Plasma Diagnostics*, North-Holland, Amsterdam, The Netherlands, p. 681, 1968.
119. Bell, G., and H. Spierer, "Plasma etching of chrome masks using PBS resist," *Proc. SPIE* 1264, pp. 446–451, 1990.
120. Tedesco, S., et al., "Dry etching for high resolution mask making," *Proc. SPIE* 1264, pp. 144–157, 1990.
121. Abe, H., K. Nishioka, S. Tamura, and Nishimoto, *Suppl. Jpn. J. Appl. Phys.* 15, p. 25, 1976.
122. Conley, W., et al., "Chemically amplified negative tone photoresist for sub-half micron device and mask," *Proc. SPIE* 1466, pp. 53–66, 1991.
123. Abe, T., T. Yokoyama, S. Kyoko, H. Miyashita, and N. Hayashi, "Comparison of etching methods for sub-quarter micron rule mask fabrication," *Proc. SPIE* 3412, pp. 163–173, 1998.
124. Wu, B., "Plasma etch in photomask fabrication," *Technology Review-Photronics Internal* 2(2), pp. 1–21, 2003.
125. Collard, C., S. A. Anderson, R. B. Anderson, J. O. Clevenger, M. Halim, C. B. Brooks, M. J. Buie, and T. Sahin, "Examination of various endpoint methods for chrome mask etch," *Proc. SPIE* 5256, pp. 744–748, 2003.
126. Johnson, D., J. Plumhoff, J. Shin, and E. Rausa, "A comparison of endpoint methods in advanced photomask etch applications," *Proc. SPIE* 4889, pp. 40–49, 2002.
127. Lide, D. R., *Handbook of Chemistry and Physics*, 75th ed., CRC Press, Boca Raton, Fla., pp. 10-1–10-128, 1995.
128. Feyman, R. P., R. B. Leighton, and M. L. Sands, *The Feyman Lectures on Physics*, Addison-Wesley, Reading, Mass., Vol. 1, pp. 3-1–3-2, 40-1–40-10, 1989.
129. Sommerer, T. J. and M. J. Kushner, "Numerical investigation of the kinetics and chemistry of rf glow discharge plasma sustained in He, N_2, He/N_2/O_2, He/CF_4/O_2, and

SiH_4/NH_3 using a Monte Carlo-fluid hybrid model," *J. Appl. Phys.* 71(4), pp. 1654–1673, 1992.

130. Sommerer, T. J. and M. J. Kushner, "Monte Carlo-fluid model of chlorine atom production in Cl_2, HCl, and CCl_4 radio-frequency discharges for plasma etching," *J. Vac. Sci. Technol.* B10(5), pp. 2179–2187, 1992.
131. Hartig, M. J. and M. J. Kushner, "Monte Carlo hydrodynamic simulation of neutral radical transport in low pressure remote plasma activated chemical vapor deposition," *Appl. Phys. Lett.* 62(14), pp. 1594–1596, 1993.
132. Kline, L. E. and M. J. Kushner, "Computer simulations of materials processing plasma discharges," *CRC Critical Reviews in Solid State and Materials Science,* CRC Press, Boca Raton, Fla., Vol. 16, pp. 1–35, 1989.
133. Graves, D. B., *Am. Inst. Chem. Eng.* 35, p. 1, 1989.
134. Lister, G. G., "Low-pressure gas discharge modeling," *J. Phys.* D25(12), pp. 1649–1680, 1992.
135. Boswell, R. W., and I. J. Morey, *Appl. Phys. Lett.* 52, p. 21, 1988.
136. Surendra, M., and D. B. Graves, *IEEE Trans. Plasma Sci.* PS-16, p. 144, 1991.
137. Sommerer, T. J., W. N. G. Hitchon, R. E. P. Harvey, and J. E. Lawler, *Phys. Rev. A*43, p. 4452, 1991.
138. Babin, S., and S. Okulovsky, "Software to simulate dry etch in photomask fabrication," *Proc. SPIE* 5567, pp. 183–189, 2004.
139. Wu, B., "Quartz etch optimization," *Proc. SPIE* 5446, pp. 94–105, 2004.
140. Shigetomi, A., S. Matsuda, K. Moriizumi, H. Kusunose, H. Onoda, T. Imai, and Y. Watakabe, "Fabrication technologies for advanced 5X reticles for 16M-bit dynamic random access memory," *J. Vac. Sci. Technol.* B8(2), pp. 117–121, 1982.
141. Plumhoff, J., C. Constantine, J. Shin, B. Reelfs, E. Rausa, "Improvements in binary chrome CD performance utilizing an optimized 4th-generation reactor platform," *Proc. SPIE* 5256, pp. 736–743, 2003.
142. Brooks, C. B., R. B. Anderson, J. O. Clevenger, C. Collar, M. Halim, T. Sahin, and A. Mak, "Optimization of chrome dry etch in Tetra II using asymmetrically loaded patterns," *Proc. SPIE* 5256, pp. 749–757, 2003.
143. Kotani, J., T. Yanagihara, E. Umeda, T. Senou, Y. Kikuchi, T. Tanaka, and Y. Okuda, "Mask CD uniformity improvement by dry etching loading effect correction," *Proc. SPIE* 5256, pp. 758–765, 2003.
144. Coleman, T., et al., "Plasma etching of chrome films in the fabrication of photomasks for electron beam and laser beam exposure systems," *GMM Proc.* 21, p. 43, 1997.
145. Yang, K., "Two-Step Etching Progress for Small Size Pattern," *Proc. SPIE* 3412, pp. 233–236, 1998.
146. Sato, Y., H. Hand, Y. Kushida, S. Asai, H. Maruyama, Y. Miyahara, and M. Naito et al., "Process bias control with thin Cr film blanks for 90 nm-node reticle fabrication," *Proc. SPIE* 4889, pp. 50–58, 2002.
147. Jang, I. Y., J. Lee, S. Moon, S. Choi, and J. Sohn, "New method for approaching to the loading free process for photomask Cr etching," *Proc. SPIE* 5256, pp. 59–65, 2003.
148. Handa, H., S. Yamauchi, H. Maruyama, S. Ishimoto, M. Kosugi, Y. Miyahara, T. Uryu, T. Yokoyama, and A. Naito, "Process monitoring of chrome dry-etching with RF sensor for reticle production beyond 90-nm node," *Proc. SPIE* 5256, pp. 85–92, 2003.
149. Cha, B. C, S. W. Choi, J. M. Kim, H. J. Cho, and J. M. Sohn, "Resolution enhancements with thin Cr for chrome mask making," *Proc. SPIE* 2793, pp. 53–61, 1997.
150. Yamazaki, T., Y. Suzuki, and H. Nakata, "Gas plasma etching of ion implanted chromium films," *J. Vac. Sci. Technol.* B17(6), pp. 1348–1350, 1980.
151. Rothschild, M., J. H. C. Sedlacek, J. G. Black, and D. J. Ehrlich, "Visible-laser photochemical etching of Cr, Mo, and W," *J. Vac. Sci. Technol.* B5(1), pp. 414–418, 1987.
152. Buie, M. J. and B. C. Soehr, "Methods and apparatus for etching metal layers on substrates," *US Patent Application Publication,* Mar. 13, 2003, US 2003/0049934 A1, 2003.
153. Johnson, D. J., S. Onishi, and C. Constantine, "Method and apparatus for the etching of photomask substrates using pulsed plasma," *US Patent Application Publication,* 37756, US 2003/0089680 A1, 2003.

154. Naeem, M. D., S. M. Burns, R. Christie, V. Grewal, W. W. Kocon, M. Narita, B. Spuler, and C. H. Yang, "Low pressure and low power Cl_2/HCl process for sub-micron metal etching," *US Patent*, 5976986, 1999.
155. Buie, M., B. Stoehr, and G. Ruhl, "Etch process for photo lithographic reticle manufacturing with improved etch bias," *US Patent Application Publication*, January 2, 2003, US 2003/0003374 A1, 2003.
156. Sasaki, T., N. Harashima, S. Aoyama, and S. Sakamoto, "Dry-etching method and apparatus, photomasks and method for the preparation thereof, and semiconductor circuits and methods for the fabrication thereof," *US Patent*, 6391791, 2002.
157. Chow, T. P., and A. J. Steckl, "Plasma etching of sputtered Mo and $MoSi_2$ thin films in NF_3 gas mixtures," *J. Appl. Phys.* 53(8), pp. 5531–5540, 1982.
158. Zhang, M., J. Z. Li, I. Adesida, and E. D. Wolf, "Reactive ion etching for submicron structures of refractory metal silicides," *J. Vac. Sci. Technol.* B1(4), pp. 1037–1042, 1983.
159. Watakabe, Y., S. Matsuda, A. Shigetomi, M. Hirosue, T. Kato, and H. Nakata, "High performance very large scale integrated photomask with a silicide film," *J. Vac. Sci. Technol.* B4(4), pp. 841–844, 1986.
160. Chiba, A., S. Matsud, and Y. Watakabe, "Antireflective MoSi photomasks," *J. Vac. Sci. Technol.* B10(6), pp. 2480–2485, 1992.
161. Inazuki, Y., T. Maruyama, M. Kojima, H. Kaneko, M. Watanabe, and S. Okazaki, "Phase shift blank, phase shift mask, and methods of manufacture," *US Patent* 2003, 6641958, 2002.
162. Coleman, T. P., Y. C. Huang, M. J. Buie, L. C. Sheu, B. C. Stoehr, and P. L. Jones, "Methods for etching photolithographic reticles," *US Patent Application Publication*, Apr. 15, 2004, US 2004/0072081 A1, 2004.
163. Tokoro, Y., S. Kawada, T. Yamamoto, Y. Saito, A. Hayashi, and A. Isao, "Attenuated phase-shifting mask blanks for the deep ultraviolet," *Proc. SPIE* 2322, pp. 387–391, 1994.
164. Mohri, H., M. Takahashi, K. Mikami, H, Miyashita, N. Hayashi, and H. Sano, "Chromium-based attenuated phase shifter for DUV exposure," *Proc. SPIE* 2322, pp. 288–298, 1994.
165. Kalk, F. D., R. H. French, H. U. Alpay, and G. P. Hughes, "Chromium-based attenuated shifter preproduction," *Proc. SPIE* 2322, pp. 299–304, 1994.
166. Mitsui, H., H. Sakai, and Y. Yamaguchi, "Development of a W/Si thin film for the single-layered attenuated phase-shifting mask for 248-nm lithography," *Proc. SPIE* 2512, pp. 343–347, 1995.
167. Kawano, K., M. Asano, S. Tanaka, T. Iwamatsu, H. Sata, and S. Ito, "Lithographic performance of SiN_x single-layer halftone mask," *Proc. SPIE* 2512, pp. 348–355, 1995.
168. Cheng, C. C., T. B. Wu, J. Y. Gan, L. C. Tue, and J. J. Wang, "$(LaNiO_3)_x(Ta_2O_5)_{1-x}$ oxide thin films for attenuated phased-shifting mask blank," *Proc. SPIE* 2793, pp. 146–154, 1996.
169. Yang, Y. S., C. C. Cheng, C. L. Lin, J. Y. Gan, T. B. Wu, L. C. Tuo, and J. J. Wang, "Ta-Si-O absorptive shifter for the attenuated phase-shifting mask," *Proc. SPIE* 2793, pp. 155–164, 1996.
170. Nemelka, J. O., "Use of lower-end technology etch platforms for high etch loads," *Proc. SPIE* 5256, pp. 724–735, 2003.
171. Sahin, T., C. Collard, S. A. Anderson, A. W. Mak, C. B. Brooks, and M. J. Buie, "Integrated phase shift measurements for advanced mask etch process control," *Proc. SPIE* 5256, pp. 76–84, 2003.
172. Yeom, G. Y., and M. J. Kushner, "Si/SiO_2 etch properties using CF_4 and CHF_3 in radio frequency cylindrical magnetron discharges," *Appl. Phys. Lett.* 56(9), pp. 857–859, 1990.
173. Strawn, C., C. Constantine, J. Plumhoff, and R. Westerman, "Investigation of quartz etch rate uniformity for alternating phase shift mask applications utilizing a next generation ICP source," *Proc. SPIE* 4347, pp. 37–45, 2001.
174. Brooks, C. B., M. Buie, N. Waheed, P. Martin, P. Walsh, and G. Evans, "Process monitoring of etched fused silica phase shift reticles," *Proc. SPIE* 4889, pp. 25–31, 2002.

175. Anderson, S. A., R. B. Anderson, M. J. Buie, M. Chandrachood, J. O. Clevenger, Y. Lee, N. Sandlin, and J. Ding, "Optimization of a 65 alternating phase shift quartz etch process," *Proc. SPIE* 5256, pp. 66–75, 2003.
176. Lassiter, M., M. Cangemi, and D. Taylor, "Improved phase uniformity control using a new AAPSM etch stop layer technique," *Proc. SPIE* 5256, pp. 412–422, 2003.
177. Morikawa, Y., N. Mizutani, M. Ozawa, T. Hayashi, and W. Chen, "Etching characteristics of porous silica (k = 1.9) in neutral loop discharge plasma," *J. Vac. Sci. Technol.* B21(4), pp. 1344–1349, 2003.
178. de Beeck, M. O, K. Ronse, K. Kamon, M. Fujinaga, H. Kusunose, and H. Morimoto, "Influence of shifter errors on the printability of L/S structures using the alternated phased-shifting design: Simulations and experiments," *J. Vac. Sci. Technol.* B10(6), pp. 2468–2479, 1992.
179. Lercel, M., E. Fisch, K. Racette, M. Lawliss, C. Williams, L. Kindt, and C. Huang, "Mask availability for next generation lithography," *18th European Mask Conference on Mask Technology for Integrated Circuits and Micro-components* 36, pp. 15–19, 2002.
180. Rau, J., H. Wendt, J. Mathuni, C. Stepper, A. Ehrmann, and F. M. Kamm, "EUV mask development: material and process," *18th European Mask Conference on Mask Technology for Integrated Circuits and Micro-components* 36, pp. 23–28, 2002.
181. Smith, H. I., "Japan could dominate industry with X-ray lithography, *Semicond. Int.* 24(2), pp. 67–72, 2001.

Chapter

5

Photomask Metrology

Richard M. Silver and András E. Vladár

5.1 Introduction

Metrology is an integral element in the fabrication and use of photomasks. Historically, stand-alone metrology tools were used to calibrate feature dimensions and feature-to-feature positions to ensure accurate fabrication of features on photomasks. Metrology in the early days was largely accomplished using optical metrology tools, and the focus was on ensuring accurate dimensions for chrome on quartz binary photomasks. This was acceptable as the 4× or 5× reduction by the photolithographic steppers meant that features were several times larger than the measurement or exposure wavelength, and only binary pattern generation techniques were used. Nowadays, however, with sub-90-nm features common in mass production and 65-nm-sized critical dimensions well along in development, even with 4× stepper reduction, features on the reticle are routinely 250 nm in size. With the widespread use of reticle enhancement techniques, phase-shifting features, and optical proximity corrections the smallest features or physical attributes of interest are now less than 100 nm in lateral dimension, and the depth and phase have become increasingly important aspects.

Furthermore, there has been a recent move in the design and manufacture of semiconductors to more tightly close the loop in the design, fabrication, measurement, and evaluation of final device performance sequence of semiconductor manufacturing. This entails that accurate design information be used as the input for the *computer-aided design* (CAD) data used in the reticle fabrication. It then requires that the features be accurately written on photomask blanks and that the resulting patterns on the photomask be measured to ensure a properly written reticle. The reticle is then used in the lithography process and the resulting features on the wafer are again measured with the metrology output used for feedback to the design process.

Of course there are many smaller control loops in the wafer and photomask fabrication process, which may operate in a feedback or feedforward mode with varying degrees of time lag. These loops are commonly used to control individual processes more tightly and are often tuned for optimum sensitivity in a specific process control application that may have quite different requirements than those metrology outputs used by larger cells of the fabrication facility (fab). There is also a distinction between integrated metrology and a more tightly linked fab wide use of metrology outputs. The more manufacturing cell boundaries one crosses (e.g., litho output to design) with the metrology output, the more important the accuracy and traceability of that measurement become.

The effort to more broadly link metrology for use as inputs in related process steps has become increasingly more relevant and necessary with the widespread use of reticle enhancement techniques and phase-shift applications. Optimization of optical proximity correction features requires tight feedback between the design, features actually fabricated on the reticle, and those patterns which result from the lithographic process. As an example, detailed feedback is now more critical to ensure that accurate proximity features or enhancement features are written for optical proximity corrections and resolution improvement or feature integrity. This is also now closely linked to the optical electromagnetic scattering models used to predict the shapes of enhancement features.

In this chapter we will focus on the different metrology techniques used to measure features on photomasks. In view of the preceding discussion, we will focus on the importance of accurately measuring features and developing traceability.[1] The metrology techniques broadly include measuring feature positions, feature dimensions, defect printability, optical phase shifts, and the additional elements that contribute to the final image transferred into the resist. Also, evaluation of the dimensions and flatness of photomask blanks will be discussed.

5.1.1 The key measurement requirements

Perhaps the most critical measurement is that of feature size. In a photomask production environment, this is accomplished primarily using an optical microscope or scanning electron microscope. The complete determination of the dimension of a feature, however, involves measuring the feature width as well as an estimate of the line-edge roughness, edge runout, and surface flatness characteristics. The line-edge roughness is characterized by individual edge "wandering" in which the two edges of the line may wander in phase (correlated) or out of phase and consequently affect the measured linewidth value and ultimately device performance differently. The vertical edge runout can be thought of as a variation in the linewidth when looking at different slices of a line in cross section. The optical effects from edge runout, however, must be coupled with the fact that conventional photomask chrome lines have an intrinsic variation in their optical properties due to the depth dependence of the antireflective chrome oxide layer, which is formed during the chrome deposition process. There is also a flatness measurement of the underlying quartz substrate and the resulting top surface of the chrome. Together these elements define the geometrical shape fully, but not necessarily its optical properties since these may depend on nearby features, the stepper wavelength, numerical aperture and the like.

Simply measuring the geometrical shape of a feature is one of the primary metrology requirements and will be discussed extensively in this chapter. In order to accurately estimate or measure feature dimensions, models are applied to extract the physical feature dimension from the measurement tool output, and at least with optical tools, this involves knowing several characteristic materials properties. The importance of these models in accurately measuring a feature's geometrical shape cannot be underestimated and will be discussed at length. The treatment presented here is meant to introduce the importance, complexity, and necessity of modeling but not be a comprehensive treatment as optical modeling and electron beam scattering are in-depth subjects and complete books have been written on these techniques individually.[2,3]

Although knowing the optical properties is essential to accurate linewidth metrology when using optical tools, these properties are also fundamentally important in the application and use of the photomasks in the lithography process. These properties are generally parameterized by n and k, the real and imaginary parts of the index of refraction or, alternatively, the complex dielectric constant. These parameters are dependent on the wavelength of light used in the exposure or lithography process and are more fully characterized or defined by a dispersion relationship.[4] The complexity introduced by the frequency dependence of the indices of refraction makes this a somewhat difficult measurement to make accurately and a significant amount of research and development over the years has gone into developing accurate measurement techniques for these quantities. These optical constants have a direct impact on the lithography process, phase-shifting characteristics, optical linewidth metrology, and methods for measuring film thickness.

Phase-shifting and optical proximity effects from *optical proximity correction* (OPC) features are of increasing importance in lithography. As a result there is now an increased emphasis placed on measuring the optical phase and phase-shifting attributes. Phase is measured optically and is an important input parameter for the optical models used in metrology and lithography. Determining the phase may include a geometrical measurement, for example, of an etched quartz phase shifter involving lateral and vertical measurements and an estimation of the resulting phase or a direct measurement of the feature of interest optically at the wavelength of interest.

Accurately placing features on a photomask is critical for several reasons including accurate feature overlay on the wafer, proper optical proximity corrections, and ultimately final device performance. Although the final printed or etched wafer feature-to-feature position values may be influenced by several things such as stepper lens distortion, step and scan performance, or process related shifts, their measurement on the reticle is quite important. This measurement captures several errors such as low frequency, photomask wide components, stripe-to-stripe exposure electron beam writing errors and offsets, local distortions as well as global errors relative to alignment marks and fiducial marks. This measurement is one of the standard photomask fabrication metrology requirements and is typically specified by a number of measurement sites and their acceptable errors and tolerances. There is also usually a requirement to define the method of fitting or data reporting, which specifies the global fitting of the measured data set to the design values.

Another typical attribute measured is defects. These may be particulate defects with tolerances specified in the photomask fabrication process for the allowed defect density and their sizes. Alternatively, these may be pattern defects. Due to the large area, which must be scanned for defects, these measurements are usually made optically and are routinely made in transmission using an optical microscope in the aerial imaging mode.[5] This allows defects to be classified as to whether or not they will print at a specified wavelength. These types of measurements and the metrology tools used to make them, outlined next, will be discussed at length in this chapter. When a comprehensive treatment is not warranted or feasible, the appropriate references to a more thorough examination will be provided.

5.1.2 The primary metrology tools

There is a range of metrology techniques applied to the measurement of photomasks. One important set of measurements is that of feature dimensions. This is the measurement of the physical size of a feature on the photomask. It more specifically involves the measurement of the feature width, height, and perhaps the variation of linewidth along the feature, known as line-edge roughness. This generally involves the imaging of a feature over a small field of view, and the evaluation of a profile acquired by a tool. The image profile, which represents that feature, is in general a complex combination of the physics and interaction of the measurement probe and the target and other effects known as the measurement tool transfer function. The most important point here is that the accurate extraction of the width of a feature from the measured profile necessarily involves the deconvolution of the instrument effects and a detailed understanding and model of the scattering or interaction process. This is in general true for optical measurements, scanning electron systems, and more recently atomic force microscopes.

Although each method will be covered in depth in the following sections, a brief overview of each measurement method and the primary application of those methods are given in this section.

Transmission mode optics. Since most photomasks are used in transmission to expose wafers at visible or ultraviolet (UV) wavelengths, transmission mode microscopy at or near the exposure wavelength is a particularly useful metrology method since it is the method that measures the photomasks closest to the way they will be used. Commercially available aerial image measurement system tools and transmission mode UV photomask inspection and metrology tools are often the tools of choice for the reasons just described.

Reflection mode optics. Reflection mode optical mask metrology is often viewed as more difficult due to a greater sensitivity to variations and uncertainties in such parameters as chrome thickness and Cr and quartz (Qz) surface reflectivity. As a consequence, transmission mode measurements are generally more robust and are usually recommended since they more closely emulate the wafer exposure conditions. An advantage of reflection mode microscopes, however, is their ability to image wafer features as well as mask features.

Currently, transmission and reflection mode light microscopes are used extensively by the photomask manufacturing industry and *scanning electron microscopes* (SEMs) are used less frequently, in about 10 percent of the measurements. The *atomic force microscope* (AFM) is used occasionally for height and film thickness measurements only. A combination of the techniques is also being implemented more frequently in the quest for better dimensional metrology.

Scanning electron microscopy. In the past the use of optical tools for mask metrology was adequate for meeting the measurement requirements. This has changed with the introduction of advanced masks which have optical assist features that are much smaller than the primary features. As with wafers, SEMs are a natural choice for measurements of these small structures. SEMs have very high spatial resolution and the magnification can be changed easily. Therefore mask samples from large structures to small (several nanometer size) features can be relatively easily observed and measured.

The mask metrology SEMs which are currently available are modified CD SEMs that were originally developed to fit the dimensional metrology requirements on wafers. These are primarily low electron landing energy tools, optimized for the distinct requirements of optical mask metrology.[6,7] Photomasks are fabricated on a thick quartz substrate, which is an excellent insulator. As a result, one of the biggest problems with SEM photomask metrology is sample charging. Similar to the situation with SEM wafer metrology, there are several possible ways to reduce charging to an acceptable level. With high-speed imaging and proper electron dose and energy control SEMs can deliver acceptable results on most mask samples.

Atomic force microscopy. Recently, the atomic force microscope has been applied to photomask metrology. The AFM has high resolution and is, in fact, well suited for dimensional measurements on photomask structures. For some time the AFM method (or its predecessor, profilometry) has been successfully applied to depth and height measurements, and acceptable accuracy can be easily attained for these measurements. For linewidth measurements, on the other hand, accuracy is usually more challenging due to the convolution of the tip shape in the measurement result.

The universal challenge, however, with AFM is the low throughput and potential reliability problems, particularly with tip stability and longevity. Nevertheless, the method has gained some favor in the growing application of reference metrology systems as opposed to standalone metrology tools.

5.1.3 Introduction to calibration strategies

Although this chapter goes into specific details for the comprehensive calibration of optical and scanning electron metrology tools, in this section we develop the general calibration procedures and guidelines which can be directly applied to the calibration of numerous metrology tools and measurements encountered in photomask manufacturing. In addition, since multiple strategies for calibrating metrology tools can be implemented, it is essential to identify the difference between developing a tool to achieve good repeatability performance and a calibration procedure, which yields an accurate, traceable answer. To achieve an accurate answer necessarily implies an unbroken traceability chain, which means that each measurement is linked by a complete uncertainty statement. It should be noted, however, that the traceability chain may be to a set of company-specific standards or to the generally accepted SI units of length through traceable calibration procedures and calibration artifacts.[8] It should also be recognized that an accurate measurement can be achieved without good repeatability by using an extensive sampling strategy and, alternatively, good measurement repeatability in no way implies that a measurement output is accurate and traceable. Although acceptable measurement performance can often be achieved with an extensive sampling strategy, in general, this may not be acceptable in a manufacturing environment since the desire for high throughput implies single or limited measurements.

Accuracy and traceability. To clarify this discussion it is important to specify and define the terms accuracy, traceability, repeatability, and measurement uncertainty. Although accuracy is defined in the Semiconductor Equipment and Materials International (SEMI) and International Sematech (ISMT) nomenclature, it is only recognized as a qualitative term by the U.S. Guide to the Expression of Uncertainty in Measurement (GUM) and the ISO equivalent.[9,10] The GUM is widely followed and recognized by the international measurement and metrology institutions and is, in fact, ordinarily the required methodology to evaluate measurements and their uncertainties within the ISO framework. Although the specific elements required in the preparation of a complete uncertainty statement will be presented in more detail later in this chapter, a traceable measurement necessarily has a complete combined uncertainty. This measurement uncertainty includes all aspects of the measurement required to achieve an unbroken measurement path from the quantity to be measured to the recognized international SI unit (of length in this case). Although measurement repeatability is a part of this sequence, it is only a portion of the uncertainty and does not include systematic bias and a traceable scale. That is, *one can make measurements, which very nearly always give the same answer, but that answer may have a significant bias and not be the true value.*

Calibration strategies. There are different strategies for calibrating tools using a calibration artifact. An artifact can be used as a calibration artifact or a verification/test artifact. When using an artifact as a calibration artifact, a series of measurements is performed using the calibrated artifact to determine the tool's statistical performance (repeatability). The mean of those measurements is adjusted to match the "certified" mean of the calibrated artifact. When using this method, the uncertainties of the certified value of the calibration artifact and the repeatability of the measurements from the tool being calibrated are combined (as root sum of squares) to form a combined uncertainty. This method results in a measurement traceable to that artifact which in turn may represent the traceability path to a company or to international standards.

An alternative approach is to independently, comprehensively calibrate the metrology tool. In this implementation the tool is completely evaluated by measuring or scientifically estimating all of its components of uncertainty. They are combined to form the combined uncertainty.[9] In a typical sequence, subsequent to calibrating the tool, a calibration artifact is then measured several times by the calibrated tool. The distribution of those measurements is then compared to the calibration value and uncertainty of the calibration artifact. A test is then applied to verify if the measured values from the calibrated tool are in compliance.

Calibrating with an artifact. This method of calibration can be implemented on an individual tool or a tool set. To calibrate an individual tool, a series of measurements is performed to properly evaluate the mean and variance for a series of measurements performed on a calibrated artifact. The mean value of the measurements from the tool and the calibration value from the calibration wafer are compared and an offset (calibration) is applied to make the results match. This is shown schematically in Fig. 5.1. The metrology tool can then be considered calibrated and traceable to the SI unit of length with an appropriate combined uncertainty.

Although this results in a calibrated tool with an expanded uncertainty, there are two central concerns to address. The first concern is that the calibration is only valid for samples (the measurand) with characteristics very nearly the same as those of the calibration artifact and, if there are significant deviations between future measured samples and the calibration artifact, they must not alter the calibration in any measurable way. That is, the new sample must not exercise the tool in such a way

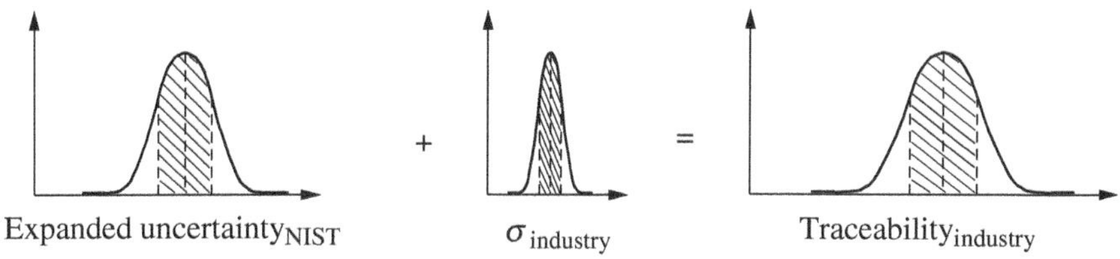

Figure 5.1 The calibrated artifact has an expanded uncertainty. The individual industry tool has a repeatability distribution. The two uncertainties are combined resulting in a traceable industry tool. The mean of the industry tool is shifted (calibrated) to match the traceable mean value.

as to introduce a bias or systematic error not accounted for in the uncertainty statement and calibration previously determined.

The second concern is that, although the tool precision or repeatability has not been changed, the combined uncertainty is larger. The initial, uncalibrated tool variance is increased by the addition of the uncertainty component from the combined uncertainty of the calibration artifact. This increased measurement uncertainty is a result of including all of the systematic error components in a combined uncertainty that are not captured in the original repeatability measurements of the metrology tool. The new combined uncertainty includes the repeatability of both the metrology tool under calibration and the tool used to calibrate the artifact. This circumstance is not encountered in the comprehensive calibration, however, this latter method involves considerably more work as described in the next section. The situation of a tool set is also more complex and will be addressed next.

If, for manufacturing purposes, the tool or tool-set precision is most critical, then the calibration offset may be applied while still using the tool repeatability as the most essential manufacturing parameter. Still as will be seen, removing the measurement bias will improve the tool-set precision.

A comprehensive tool calibration. The alternative approach to a calibration based on a specific calibration artifact is to perform an exhaustive tool evaluation which includes the scale calibration, an evaluation of the complete set of tool-related error components, and uncertainty contributions from the physics-based model used in interpreting the measurement tool output as well as errors introduced into the model from inputs such as uncertainties in optical parameters. Each component is carefully evaluated by either a Type A or Type B determination of uncertainty.[9] The Type A components are defined as those whose components can be evaluated through statistical means. This includes many components which may have a systematic bias on the measurement, but which can be evaluated through a series of measurements characterized by a mean and variance. The Type B uncertainty components are those whose variances are estimated through scientific judgment based on a body of the best available knowledge and information. This is the current internationally accepted method for determining uncertainty. It differs from previous methodology which divided the uncertainty into systematic and random components. This change was implemented to avoid confusion over evaluation of those previous systematic components, which could, or should, be evaluated through statistical means.

This comprehensive uncertainty analysis requires the scientific estimation of the uncertainty and its confidence level for many of the uncertainty components. Although some of the "systematic" error components may be determined through a sequence of measurements, characterized by a standard deviation and mean, several errors and uncertainty components may have to be estimated with expert analysis or scientific judgment. The errors inherent to a specific technique will be discussed in the sections which deal exclusively with the evaluation and discussion of those tools which includes errors specific to hardware and those errors attributed to the control software and analysis algorithms. Knowledge about the sample and its asymmetry is also essential in this analysis as it has a direct effect on the resulting measurements of a feature. The effects of sample asymmetry are discussed in the respective sections as well.

Following the evaluation of the uncertainty of the different components, their variances are combined as the *root sum of squares* (RSS). Different components may have different distributions as illustrated in Fig. 5.2. The variances are then combined as seen in Fig. 5.2. Following the complete tool error evaluation and RSS combination of the components, a combined uncertainty is obtained. A coverage factor can then be applied resulting in an expanded uncertainty. The coverage factor or

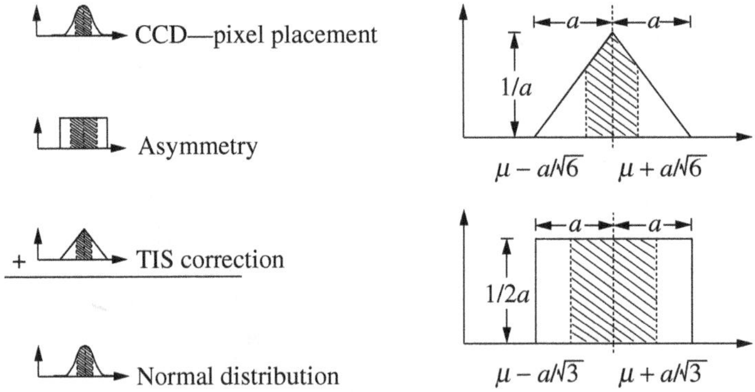

Figure 5.2 These are representative distributions for illustration. Each individual error component has an associated distribution and confidence level which varies depending on the type. As the different distributions are combined by RSS, they quickly approach a normal distribution.

confidence level determines the statistical probability as to whether any given measurement will fall within the expanded uncertainty.

After completing a comprehensive calibration or at some time later, the tool calibration can be validated. This is accomplished by taking the mean value and measurement distributions from the calibrated artifact and the calibrated industry tool and comparing the results using a *t* test. A schematic of the *t* test is shown in Fig. 5.3 with its corresponding mathematical expression.[11] The *t* test determines if the industry tool is accurately calibrated by comparing the mean values within a confidence level (coverage factor). This test queries whether the measurements agree to within the statistical probability and ascertains whether the measurement differences are statistically significant or not. A level of confidence can be chosen to quantify the desired degree of agreement. If the measurements are within the statistical limits, then the tool is currently calibrated correctly and continues to yield a traceable measurement. The advantage of the comprehensive calibration method is that there is no contribution in the final expanded uncertainty of the industry tool from the repeatability of the calibration tool used in preparing the calibrated artifact.

Tool set calibration. This same statistical methodology can be applied to a tool set. This framework can be used to establish the tool set as a module which can then be treated as a single statistical object. As an artifact goes in, a measurement is made, and a measurement result is returned without regard to which tool it was actually measured on. The motivation for this is utilization and cost-of-ownership benefits. To do this effectively, the tools should be matched (calibrated as a set) as closely as possible. Figure 5.4 shows a graphical representation of how to combine the independent tool repeatabilities. The algorithm can then be applied to determine the complete tool set precision (repeatability) and subsequent combined uncertainty (accuracy).

Figure 5.4 shows the measurement distributions from each tool taken with the same calibrated wafer. Also shown are the mean values from

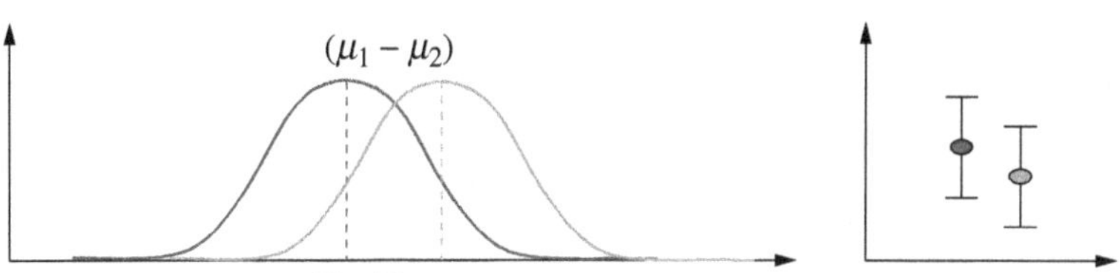

$$T = \frac{\bar{X} - \bar{Y} - (\mu_1 - \mu_2)}{\sqrt{(S_X^2/n_1) + (S_Y^2/n_2)}}$$

Figure 5.3 The mean and distributions from the calibrated artifact and the calibrated industry tool are compared using the *t* test. This verifies whether the industry tool is properly, accurately calibrated.

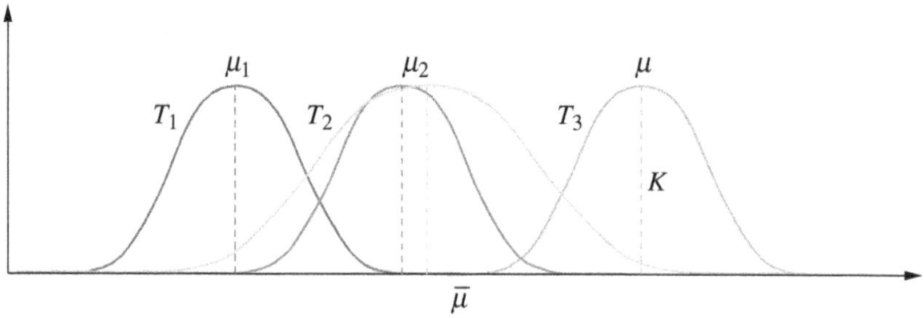

$$\bar{\mu} = \frac{1}{N}\sum_{1}^{N} \mu_K \qquad \sigma^2(\bar{\mu}) = \frac{1}{N}\sum_{1}^{N} \sigma_K{}^2 + \frac{1}{N}\sum_{1}^{N} (\mu_K - \bar{\mu})^2$$

Figure 5.4 This illustration shows graphically how to combine the tool set repeatabilities. The algorithm can be used to investigate the tool set precision (repeatability) and accuracy.

each tool. If the tools have been calibrated with the comprehensive calibration sequence or using the calibrated artifact method, then the mean values should be identical or very close. In any circumstance, the equations show that the mean from the tool set and the variance is a combination of the individual tool distributions and a sum of terms representing the distribution of means about the tool set mean. The individual distributions can be a simple repeatability or a complete combined uncertainty. The end result is that the algorithm allows the tool set to be treated as a single unit with a single precision and accuracy. We will now go through a broad evaluation of the most significant error components.

The reference measurement system approach. Recently, with the advent of low-throughput, high-resolution tools, such as the atomic force microscope, gaining wider use in the metrology community, a new approach combining several measurement tools is being adopted. Since one particular tool may have a particular aspect of the overall measurement, which it performs best, such as edge roughness, height, or side wall runout, multiple measurement techniques can be combined to achieve the ultimate measurement result. This methodology has become known as the reference measurement system approach.[12] It affords the metrology engineer the possibility of linking several different metrology tools together, some of which may have very low throughput but good accuracy.

An example of this methodology was used in recent photomask calibrations at the National Institute of Standards and Technology (NIST). In this work a cross-section AFM was used to accurately measure the widths of the smallest features on a control photomask. A large series of measurements were made optically on the set of photomasks to be calibrated that included multiple measurements on the control mask, which were measured using the AFM. The control photomask was also extensively measured with a SEM along with cross-section SEM and top down AFM measurements to evaluate and verify edge runout and chrome height. The end result was numerous calibrated artifacts based on measurements using high-throughput optical methods that were referenced or linked to other higher resolution methods more suitable to evaluating specific aspects of the features under calibration.

The role of standards. Since universally accepted standards which comprehensively cover the range of process levels and geometries are not routinely available, and will likely never be due to process and sample specific issues, current day-to-day practice dictates calibration strategies which achieve tool-to-tool matching and repeatability or local traceability requirements without regard to international traceability and accuracy. In fact, a local traceability is often achieved in many manufacturing protocols through the use of in-house "golden standard artifacts" or recent moves to in-house reference measurement systems. This predicament is driven by process variation and the need for specific calibrations for a given process level or geometry and the significant sensitivity of the calibrations to the process level.

However, even in this current manufacturing situation, standards can play a very useful role. Quantitative metrology tool comparisons and tool evaluations in the procurement process can be achieved in part by a comparison of results acquired from an artifact with known values. Although tool-to-tool matching can be achieved with the application of a relative offset (i.e., a corrective offset), understanding the cause of the offset accurately helps to improve the range of calibration validity. Verifying tool accuracy and scale calibration are of increased importance with any attempt to move toward tighter closed loop control. Likewise traceability gives consistency among the photomask house, wafer fabs, and/or foundries and can decrease ramp up times for new processes, transferred processes, or new tool installations.

Another strong motivation to have accurate samples available is the increasing reliance on simulations and model-based metrology.[13,14] Often, the inputs required for reliable simulation results or when performing experiment to model comparisons must be as close as possible to the correct values to yield meaningful results. The move to critical dimensions below 90 nm is seeing an increasing dependence on model-based metrology.

Certain aspects of the measurement process can also be independently evaluated and improved with calibrated standards. Verifying tool performance on difficult test targets having known values can assist in improving tool performance or in monitoring its long-term performance. Well-designed standards are also very useful in testing the effectiveness of the measurement tool algorithm sets. Tall features, features in proximity, or asymmetric features can often cause measurement errors due to incorrect analysis by the measurement algorithms.

Currently there are two primary photomask standards available from the National Institute of Standards and Technology, the Standard Reference Materials (SRM) 2059, SRM 5001, and there is also the related SRM 2800.[15] The SRM 2059 is the latest generation in a series of photomask linewidth standards, which includes pitch standard targets as well. This standard includes a series of calibrated linewidth features and space width features with different widths and proximity. The current standard covers the range of linewidths from 0.25 to 10 μm, the latter of which is used to model and measure chrome-on-quartz features with largely independent noninteracting edges. The SRM 2059 layout is shown in Fig. 5.5. There is also a fully calibrated two-dimensional (2D) grid standard known as SRM 5001. This SRM has a series of targets throughout the 6-in. field of view in which lateral positions for each target have been calibrated. This SRM is discussed in more depth in the section on registration metrology.

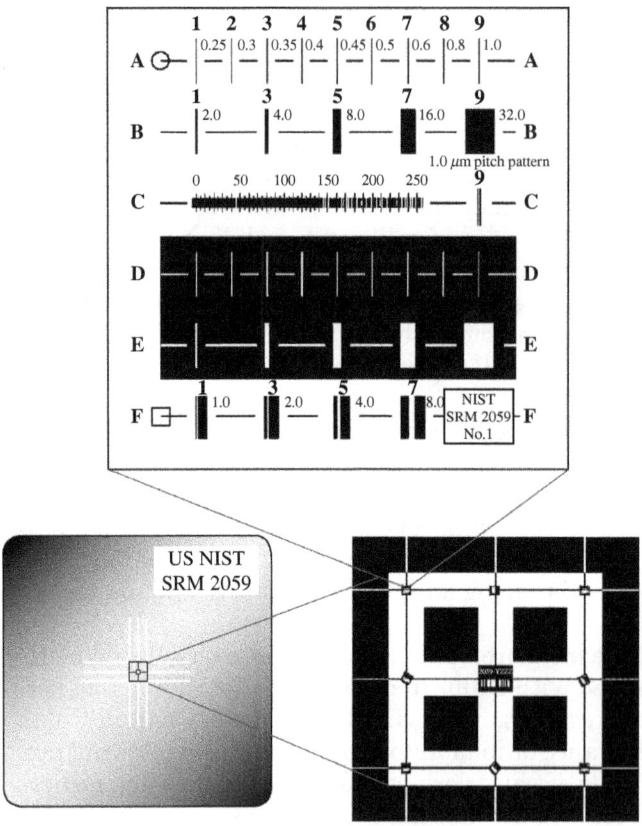

Figure 5.5 NIST Standard Reference Material 2059 Photomask Linewidth Standard.

5.1.4 Introduction to modeling

Modeling will be discussed at some length in several sections of this chapter. As modeling is referred to here, its key use is to extract a physical feature dimension or attribute from a metrology tool output. In general the metrology tool output is a combination or convolution of the measurement probe, the sample, and the imaging device. This is the case regardless of the imaging tool used. If it is an AFM, SEM, or optical instrument, they all at some level of desired accuracy require modeling to properly interpret an image.

In the case of an optical tool, a well defined electromagnetic illumination field is scattered off the sample and then collected through imaging optics and captured via a *charge-coupled device* (CCD) camera, as an example. In this case, the illumination field is usually modeled by plane wave electromagnetic fields at appropriate incoming angles of illumination. The sample scatters the electromagnetic fields forming the reflected, refracted, and scattered fields. The scattered and reflected fields which make it through the imaging system result in a two-dimensional image captured by the CCD. This image is then extensively processed and the resulting cross-section profile can be analyzed to determine the linewidth or pitch. To model this physical problem, a complicated set of field equations must be solved in order to analyze the scattered electromagnetic fields. This is discussed in Sec. 5.2.4 in more length.

The experimental profiles can be analyzed through a process that compares theoretical-based scattering results and a simulated optical instrument with the actual experimental results. If the simulation accurately represents the experimental configuration including the optical properties of the sample, then an estimation of the physical feature, which created the resulting profile, can be made. This somewhat complex process has become an integral part of critical dimension measurements at the nanometer scale for many metrology tool applications in photomask metrology.

5.2 Optical Methods for CD Metrology

Due to the high throughput and lower cost of ownership, optical methods have been the primary technique for photomask metrology. Optical methods generally are a nondestructive technique, which can be operated in a flexible ambient environment. Until recently the resolution of optical methods was generally considered adequate and with the advantages, as described in the chapter, over other methods such as SEM or AFM, optical methods were in widespread use. Even though optics is still widely used, other higher resolution methods are gaining acceptance. Although optical methods work well for isolated lines, accurate measurements become potentially more difficult to obtain when features are placed in close proximity. Also, feature dimensions are now generally small enough that modeling a measurement result to determine the physical feature size is a requirement. That is, a simple edge threshold algorithm is inadequate for many current optical metrology applications. It has now evolved to the point where a similar trend is occurring in electron-beam-based measurements.

5.2.1 Introduction to optical microscopy

When performing optical measurements which require the determination of a linewidth as opposed to a pitch value, reliable image modeling is a requirement to accurate optical metrology of photomask feature dimensions. There are several optical scattering models now available commercially, as well as a few from academic institutions. The optical models used in metrology applications are the same basic models used for analyzing and predicting intensity patterns in a resist during the lithographic exposure process. The models generally solve Maxwell's equations with boundary conditions that are appropriate for the sample and optical configuration and then give the resulting scattered electromagnetic intensity pattern or image in a range near to the best focal plane of the microscope. As features become significantly smaller than the measurement wavelength used, the relative accuracy of the optical models becomes less reliable. A significant amount of research is currently underway to evaluate and improve the models and their performance in this regime. In this chapter we present some comparisons which give a good estimate of the state of the art in model performance for metrology applications.

In addition to comparing the optical scattering simulation packages to one another, it is essential to obtain good agreement between the models and experimentally acquired images. To achieve satisfactory agreement it is essential to have models which correctly simulate the optical microscope configuration. This is routinely accomplished by using Fourier optical modeling methods to simulate the collection optics and multiple angles of plane wave illumination to simulate Köhler illumination. Likewise, it is essential to properly represent the sample and its geometry since these required model inputs can have significant effects on the model outputs as the model may be very sensitive to particular input parameters such as the optical properties or materials constants. The sensitivity to the optical parameters must be evaluated to determine the uncertainty in the modeled results.

The resulting microscope image depends on several sample parameters such as chrome thickness, topography, chrome optical constants n and k, substrate n and k, and on several tool parameters such as illumination wavelength, illumination homogeneity, and condenser and objective *numerical apertures* (NAs). These parameters are the input parameters to the microscope imaging model. Some of these can be quite difficult to measure (e.g., chrome n and k vary continuously from substrate to top of chrome on binary masks), and errors in the model input parameters propagate into model output errors.

There are also a number of measurement and modeling issues associated with the three-dimensional (3D) character of the features being measured. Although we often use two-dimensional approximations, the actual situation is in fact a three-dimensional sample. Although the two-dimensional models do work well in many applications, as resolution becomes more critical, the detail of the three-dimensional character becomes more essential. Examples of the three-dimensional attributes of chrome lines are the vertical edge runout and the depth-dependent optical constants in antireflecting chrome.

5.2.2 Transmission mode optics

Optical dimensional metrology has several accepted advantages such as high-throughput, nondestructive imaging, through-pellicle imaging, and generally a lower cost of ownership. The disadvantages of optical metrology are generally thought to be limited resolution, nonlinear measurements of small features, and proximity effects. These limitations, however, can be partially overcome through good instrument design and well-designed measurement routines which include comprehensive optical scattering simulations. In addition, although conventional optical image resolution may be limited by the measurement wavelength using the Rayleigh criterion, fundamentally the measurement resolution is only moderately wavelength limited when using model-based optical metrology methods. This is shown schematically in Fig. 5.6.

There are several similarities when comparing transmission optical microscopy and exposure techniques using a lithographic stepper. In a

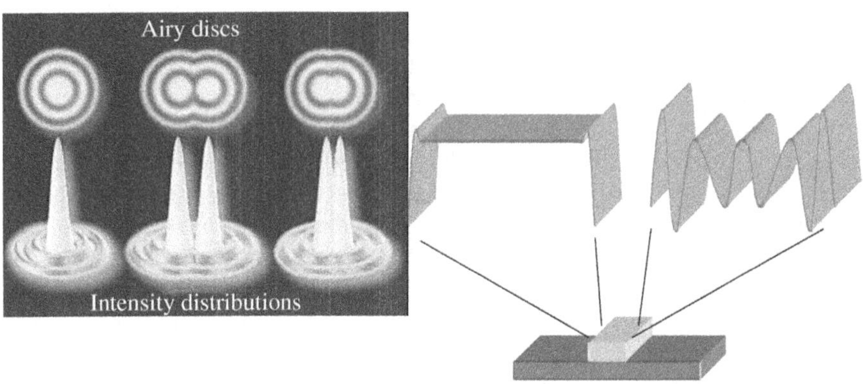

Figure 5.6 The figure on the left shows typical optical Rayleigh limited resolution. This is contrasted with a graphical comparison of threshold-based edge detection and more complicated signature-based model interpreted image analysis on the right.

typical configuration optical mask metrology and lithographic wafer exposure differ in tool parameters such as magnification, wavelength, numerical aperture, and the coherence parameter. However, the relevant mask parameters such as chrome thickness, chrome complex index of refraction n and k, and substrate thickness are the same. These similarities and the parallelism in methodology make optical photomask metrology a particularly useful tool for photomask metrology since an optical measurement tool measures a photomask artifact in a similar fashion to the way it will be used. A good counter example of this is encountered when an e-beam system is used to measure a linewidth. As an example, look at the situation when a thin layer of chrome remains due to improper and incomplete etching near a line edge. At a higher electron beam energy the electron scattering can be insensitive to such a thin layer. The edge detection algorithms will then likely place the edges at slightly different positions resulting in different linewidth determinations for the e-beam and optical tools. The electron beam measurement will be subject to different effects and sensitivities when compared to an optical measurement, and the end result will be a technique-dependent linewidth measurement.

To best understand optical metrology methods it is worthwhile going through the standard key elements of an optical tool. Experimental image profiles are often acquired with a transmitted-light optical microscope or in reflection mode as discussed in the next section. A typical scanning transmission microscope consists of an illumination setup, a three-dimensional scanning stage (x, y, and z/focus), microscope optics including a condenser and objective lenses, a stationary slit, and a photometer. The slit is positioned in the image focal plane of the objective lens, orthogonal to the direction of the x-scanning stage and the photometer is placed to record the intensity of the light passing through the slit. Feature profiles are acquired by scanning the magnified image of the feature past the slit (by scanning the stage) and synchronously recording both the position of the x stage via an interferometer and the transmitted light intensity on the photometer. Typical profiles of some line and space features are shown in Fig. 5.7.

An alternative configuration is to acquire the image with a single "snap shot" using a CCD camera. This approach is in general less time consuming than a scanning system, but prone to a different set of measurement errors. Although this approach is described more thoroughly in the reflection mode section, transmission measurements can and often are made this way. For calibration and standards work, the scanning system allows an interferometer to be used, which directly results in a traceable measurement. In either approach, a Köhler illumination configuration, an illumination technique to generate a uniform intensity distribution across the field of view, is used. The type of illumination and the quality or integrity of the illumination have a substantial

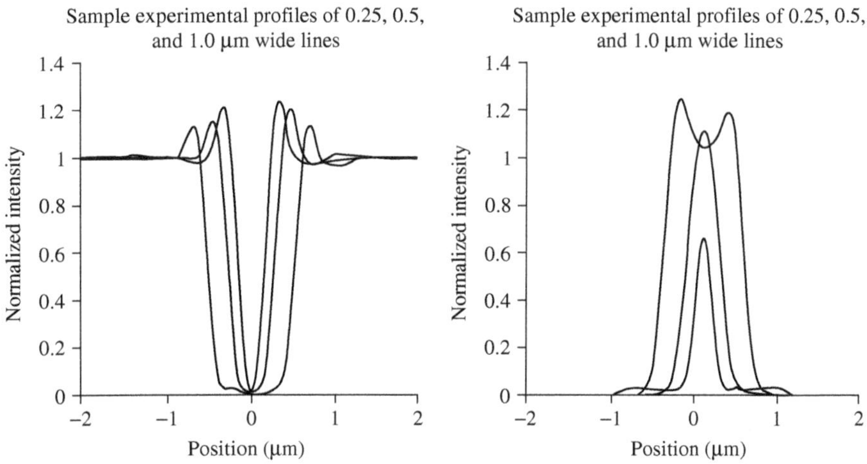

Figure 5.7 Typical profiles of lines (left) and spaces (right) with nominal values of 0.25, 0.5, and 1.0 μm.

effect on the image quality and robustness, and the ability to accurately model an experimental configuration.

A typical illumination system comprises a mercury arc lamp, a wavelength filter, and a system of lenses to provide Köhler illumination.[16] The general procedure to achieve Köhler illumination is to image the source at the back focal plane of the condenser lens. This causes each point on the source to illuminate the entire field of view in the sample plane. A simple ray tracing schematic of this ideal situation is shown in Fig. 5.8.

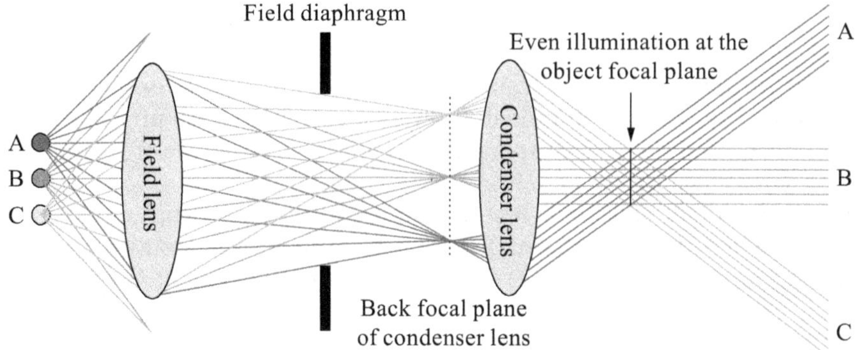

Figure 5.8 A simple ray tracing showing the concept of Köhler illumination. Three point sources (A, B, and C) are imaged into the back focal plane of the condenser lens. The light from each point source illuminates the entire field of view in the object plane giving a uniform illumination intensity, but not necessarily symmetric angular illumination (e.g., if A and C have different intensities).

Although it can be difficult to determine if Köhler illumination has been achieved, there are a few experimental indicators that can be used to determine the degree to which it has been achieved.

Accurate linewidth metrology requires that Köhler illumination be accurately produced. The optical scattering models assume plane wave illumination for the light source. As seen in Fig. 5.8, perfect Köhler illumination results in the sample being illuminated effectively from a continuous distribution of plane waves at different angles as limited by the illumination numerical aperture. The model results are quite sensitive to the details of the illumination such as NA.

Unfortunately, the preceding description oversimplifies the concept of Köhler illumination. For example, if different areas on the extended light source have different intensities, the intensity of the light illuminating the sample will vary with angle. Left-right illumination asymmetry can lead to asymmetric image profiles. On the other hand, if the source is symmetric, but uneven in intensity (e.g., a bright spot in the center of the source), the effective illumination NA will be affected (e.g., The effective NA will be reduced if there is higher intensity light forming the low angle illumination.). Also, since it is not possible to image a three-dimensional object such as the arc or filament onto a plane, there is a lack of depth uniformity to the Köhler illumination if you image a real source to the back focal plane of the condenser lens. One method to help achieve better Köhler illumination is to use a virtual source consisting of an aperture and a diffuser where the diffuser is itself evenly illuminated. Another technique is to randomize the intensity of the source using a randomizing fiber bundle where the positions of the fibers on one end are not correlated to the other end of the fiber. In practice, there are difficulties when implementing either of these methods and often illumination intensity is sacrificed.

Perfect Köhler illumination is represented in the model by a finite set of plane waves approximating the incident cone of illumination. However, as just described, it is difficult to experimentally achieve uniformity of the light incident at the different angles of incidence. It is essential to understand the effects of nonideal illumination on linewidth uncertainty and accuracy and the procedures for determining the effects of nonideal Köhler illumination.

Another optical effect which can complicate the modeling and accuracy of the results is light which enters the imaging system inappropriately. Flare light and glare result from scattered light in the image and object space, respectively. Flare and glare are generally impossible to differentiate and are defined as light that does not contribute to image formation but is indistinguishable from image forming light. In linewidth metrology, significant linewidth errors can be attributed to scattered light if it is not properly minimized or removed. For example, Fig. 5.9 shows modeling results of a 10 percent change in the apparent linewidth when a 5 percent constant offset due to scattered light is

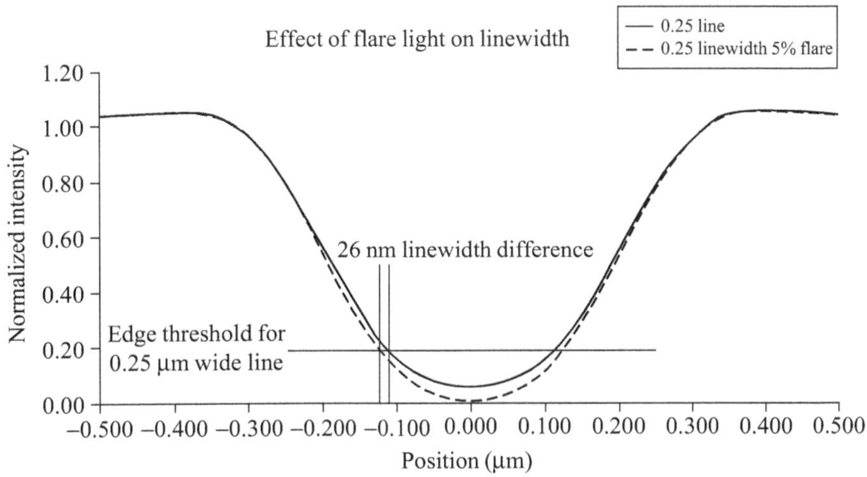

Figure 5.9 Modeled results showing that a 26-nm (10 percent) linewidth difference results from a constant 5 percent flare light being added to the intensity profile of a 0.25-μm-wide line.

added to the profile of a 0.25-μm-wide line. A simple constant offset like this can be corrected by subtracting the measured flare and glare from the profile to match the original model result, although in practice corrections for scattered light are much more complicated. For instance, the amount of scattered light (flare and glare) reaching the detector can change as a wide line is scanned across the field of view—essentially blocking some of the scattered light from reaching the detector. It is essential to reduce the amount of scattered light reaching the detector and at the same time use improved methods for modeling, which correctly compensate for these effects.

It is important to determine the degree of linearity/nonlinearity in the image detection system, whether it is a simple photometer with a sampling aperture or an array detector. There are advantages and disadvantages to using a stationary slit and a photometer versus a CCD-based imaging system. The photometer only has a single response function to quantify, which tends to be fairly linear, while the response of the CCD-based system must be studied in two dimensions. The scanning tools also benefit from on-axis data acquisition in the optical column, which results in smaller optical aberrations. These affects, however, are countered by the lower data acquisition rate of a scanning system as opposed to the quantity of data acquired with a CCD and short acquisition dwell times. The scanning tools are also more susceptible to tool vibration and thermal stability issues which arise when scanning due to their longer measurement times.

In order to directly compare experimental profiles with model profiles, it is essential that the same criterion for focus be used in the model as was used in the measurement. In the modeling software the position of the object focal plane (top of the feature, bottom of the feature, or someplace in between) is well defined and known, where as experimentally the physical relationship between the optical imaging lens and the object is not known, nor is it easily determined. As a result profiles must be acquired through focus, for both the experimental and the simulation data, and the same focus criterion applied to both.

In a typical case an image-acutance-based method of determining the focus is used. After a profile has been acquired the absolute value of the slopes of the profiles at the 50 percent intensity point are averaged. The slopes of various through-focus profiles are compared and the profile with the maximum slope is considered the best focus. Figures 5.10 and 5.11 depict modeling results through focus for a 0.25-μm-wide line and the results of our focusing criterion, respectively. The best focus in this case is not at the top of the feature as one might initially expect, but 40 nm below it.

Achieving Köhler illumination, properly accounting for scattered light, determining detector linearity, and establishing best focus are some of the critical issues that must be addressed when acquiring image profiles that can be compared to model profiles to determine the feature linewidth. Other tool parameters that must be evaluated are listed in Table 5.1, along with object and alignment terms which must be considered in a full uncertainty analysis.[17,18] The two experimental aspects that continue to require

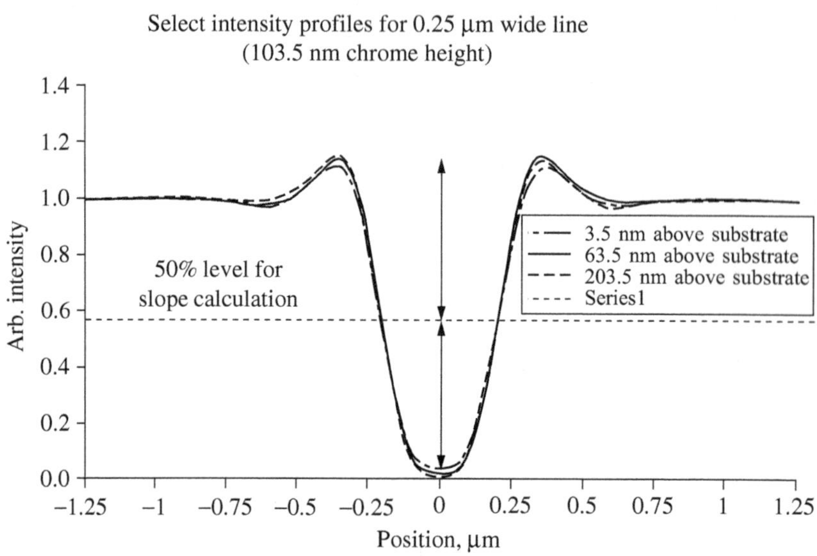

Figure 5.10 Model images for a 0.25-μm-wide line with the object focal plane positioned at $Z = 3.5$, 63.5 (best focus), and 203.5 nm above the substrate. Figure 5.11 shows the results of our focus criterion for these and other Z values.

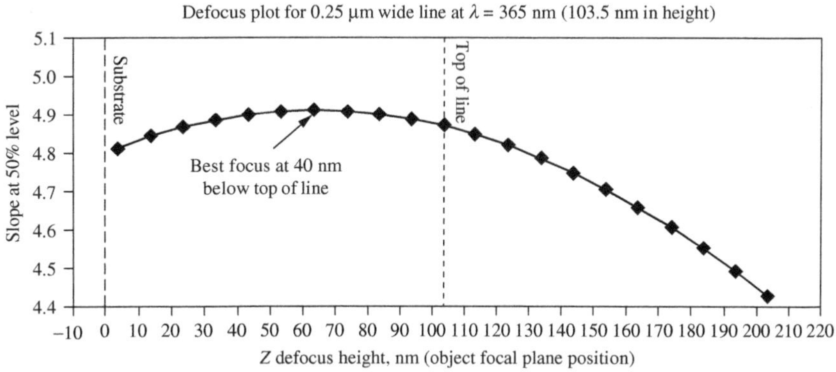

Figure 5.11 Defocus plot showing the results of a focus criterion being applied to a 0.25-μm-wide line. The best focus is found at 40 nm below the top of the feature (63.5 nm above the substrate).

improvement and better characterization are the Köhler illumination setup and the effects of scattered light on the measurements.

5.2.3 Reflection mode optics

Although reflection mode optical methods have been used for years for critical dimension metrology, wafer level metrology is now largely performed

TABLE 5.1 Optical Instrument Parameters Whose Uncertainty must be Evaluated

Object terms	Instrument terms	Alignment terms
Chrome edge runout	Tool-induced shift	Specimen cosine alignment
Chrome n	Scale factor calibration	Defocus
Chrome k	Substrate temperature	Illumination alignment
Substrate thickness	Substrate temperature variation	
Chrome thickness	Air temperature, pressure, RH	
Feature proximity	Illumination wavelength	
	Illumination NA	
	Objective NA	
	Sampling aperture	
	Data noise filter	
	Proximity effects	
	Photometer linearity	
	Image modeling parameters	
	Interferometer resolution	
	Photometer resolution	
	Laser wavelength uncertainty	
	Laser polarization mixing	
	Abbe error	
	Optical image distortion	
	CCD linearity (x, y, intensity)	
	Image processing algorithms	

with SEMs due to the higher resolution required for wafer-sized features. Also, although currently reflection mode tools are largely used for two-dimensional registration metrology purposes and for opaque samples, they are viewed as more complicated than transmission mode optical measurement tools. In fact, this is only partly true, and optical reflection mode, critical dimension metrology tools are commercially available. Furthermore, most of the two-dimensional registration metrology tools have the capability to operate in a CD measurement mode. Most such reflection mode tools typically operate in a conventional bright-field mode with data acquired by a CCD array camera. A complete description and example are described in Ref. 19. An example image from a reflection mode optical tool used in bright field imaging mode with a CCD acquisition system is shown in Fig. 5.12.[20]

Although reflection mode optical microscopes can operate in a scanning mode similar to the approach described earlier for transmission microscopy, they are typically operated in a full-field imaging mode. The details of the optical layout and illumination are again critical to performance in terms of repeatability and reproducibility as is having well-defined data acquisition parameters to achieve accurate modeling results. In a reflection implementation an automated positioning and focusing routine is used to properly place the feature of interest in the field of view. A single objective lens is used for both the illumination and collection optical train, and the appropriate additional relay lenses and apertures are placed in their proper positions. A splitter is used to separate the illumination optical path from the imaging path and different numerical apertures for both the illumination and imaging can be achieved. At this point we will go through the key different aspects of reflection mode CD metrology, which were not covered in the previous sections on transmission optical metrology.

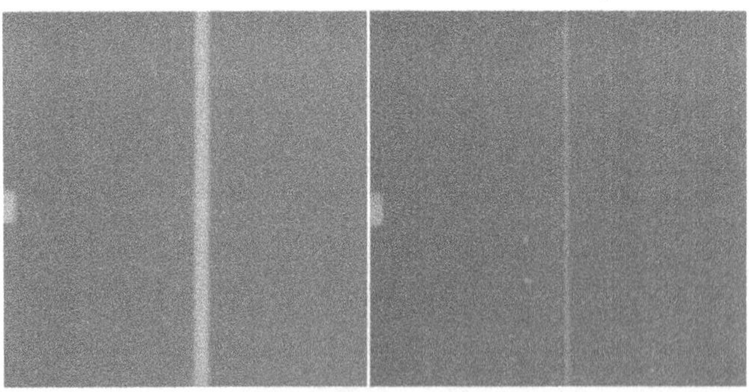

Figure 5.12 Typical images of 1 μm (left) and 0.25-μm (right)-wide lines from the NIST reflection mode tool. (Source: Silver, R. M., J. Jun, S. Fox, and E. Kornegay, "Overlay metrology: recent advances and future solutions," *Future Fab Int.* 11, 2001.)

A CCD array is often used for rapid full-field imaging. A typical CCD array has a 1024 × 1024 array of pixels with on-board digitization. Most metrology CCD arrays are the full-frame transfer type and the data are transferred to a digital frame grabber in a PC-based computer.[21] Although interline transfer cameras and shuttered cameras are used, interline transfer cameras tend to have lower signal-to-noise ratios. Image files and data are then transferred via a network to an image processing computer. In the examples described here, Matlab was used extensively as the programming language as well as for image analysis and graphic presentation.

Other important elements of the optical microscopes, which affect tool performance and accuracy of the results include a fully automated set of positioning and focus control algorithms. These algorithms enable the tool to operate in an optimal repeatability mode and also allow calibration and mapping algorithms to be employed successfully. The tools collect the images using a CCD array which also contributes to excellent repeatability. This allows the tools to operate in a full-field imaging mode, and averaging methods can be effectively applied to the data since there are a significant number of pixels in a CCD image.

Data acquisition. Photomask linewidth data can be collected in 0° and 180° orientations, which allows the use of reversal methods and the determination of the *tool-induced shift* (TIS) error component.[22] The reversal methods are used to evaluate, quantify, and reduce tool asymmetry or identify sample asymmetry. The data acquisition for calibration purposes should be repeated three or more times each day and data from several nonconsecutive days should be acquired. Once a CD tool is calibrated and qualified, it can then be operated in a single repeat mode although data from 0° and 180° orientations can still be taken to test tool symmetry.

Algorithms for metrology and image analysis. A comprehensive and extensive application of algorithms is used in the reflection mode measurement process as well as in other measurement techniques. We will discuss several of the algorithms used in the context of reflection mode microscopy, although many of these are used in transmission microscopy and SEM measurements as well. Several algorithms have been developed and will be discussed here, which can enhance and improve quantitative tool performance and evaluation of outputs and results. The measurement algorithms have a substantial impact on tool performance, accuracy, and tool robustness.

Several algorithms have been analyzed to determine the effects of noise on algorithm performance. We will discuss the effects of horizontal and vertical (digitized signal) noise on the algorithms and on the measurement repeatability. Similarly the effects of noise suppression and filtering methods will be discussed, and the results of various filter widths on data repeatability will also be shown. We further explore the effects of data bias resulting from some data filtering methods and present these results. Lastly, issues regarding subpixel resolution by interpolation and its effects on the edge detection response along with effects from subpixel resolution interpolation and curve fitting with high-order polynomial and spline data fits will be discussed.

Details of the edge detection methods. At this point it is useful to describe some of the different methods used for edge detection. In general there are two classes of edge detection algorithms, which are used, edge threshold methods and those which use the entire profile. And although they may share similarities, linewidth determination and pitch determination in general benefit from a different choice of edge detection. In this section we evaluate four methods of edge detection, centroid, edge threshold, *least squares* (LS) and *correlation* (Corr).[23,24] The centroid method finds the center of a given feature essentially by integrating two portions of the feature intensity pattern with a sliding integration limit. When the integrals from the two sides are equal, that defines the center of the centroid. In the edge threshold method one assigns a value (percentage of total intensity) which is then considered the edge location. Careful evaluation reveals that the centroid and edge threshold methods are not well suited for pitch metrology and are primarily used for linewidth determination. Often, when features are defined largely by significant optical intensity variation at the edge with little change in intensity between the line or feature center and background (contrast), it has been found that the correlation method and least squares method are the most robust.

The least squares method involves calculating the difference between two curves, one from each side of a centerline, which can then be used to determine the pitch. The square of difference between the curves is integrated or summed point by point giving a value which represents the "fit" of the two curves. This method is less sensitive than the correlation method to error when the intensity pattern is a step function as can be encountered in a single edge target with limited intensity change at the edge. The correlation methods used are standard and referenced.[25] The correlation and least squares methods just described work quite well for pitch measurements and also for symmetry analyses such as TIS or quantifying sample asymmetry.

In determining a correlation function, a curve is generated, as described next, which represents the correlation or least squares function. In the two-dimensional analysis method, two masked regions of an image are slid one step at a time on a grid based on the previous interpolation density and a surface is generated. In one dimension, the masked region is stepped again by the interpolation density and a simple curve results. An example of a least squares curve is shown in Fig. 5.13. This figure shows

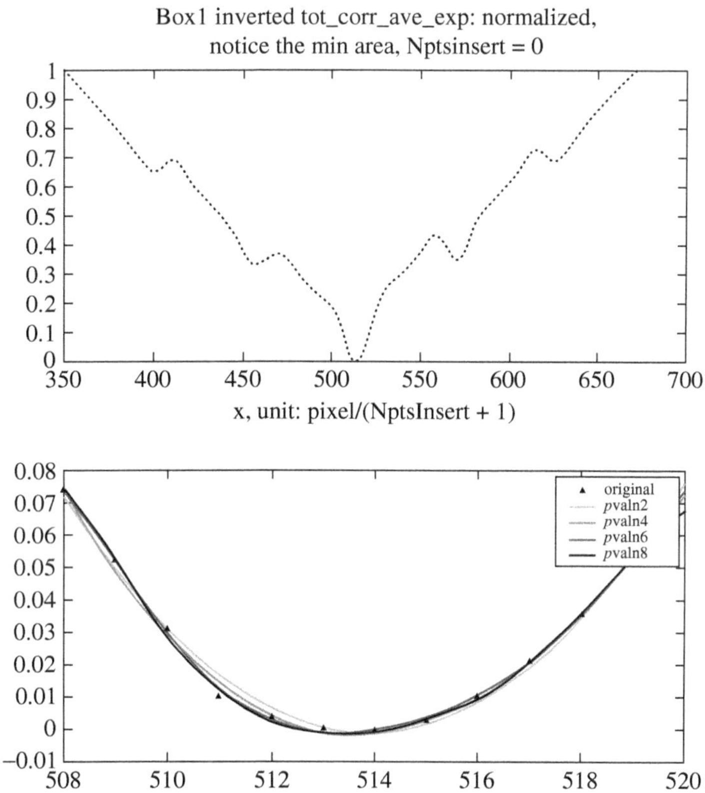

Figure 5.13 Top curve is an example of a least squares curve generated by stepping two masked regions. The lower panel shows the use of curve fitting to further enhance resolution.

a typical curve with pronounced extrema. The maximum or minimum of this curve is then used as the feature centerline. The lower panel of Fig. 5.13 shows an example of the extrema fitting process.

This completes the centerline evaluation process for the determination of pitch. The results for the individual features can be analyzed using the standard assortment of statistical processing tools including standard deviations, mean, TIS, and TIS corrected values. TIS was previously defined as the difference between 0° and 180° measurements and, again, primarily represents the components of tool-induced measurement error.

Filter and interpolation comparisons. Although other filters have been examined in a similar context, a simple moving-window averaging method is discussed next. The goal is to demonstrate and recognize the potential effects of filters on data acquisition and analysis. In the data shown in Fig. 5.14, the resulting correlation or least squares function extrema in response to different filter widths is evaluated. The data

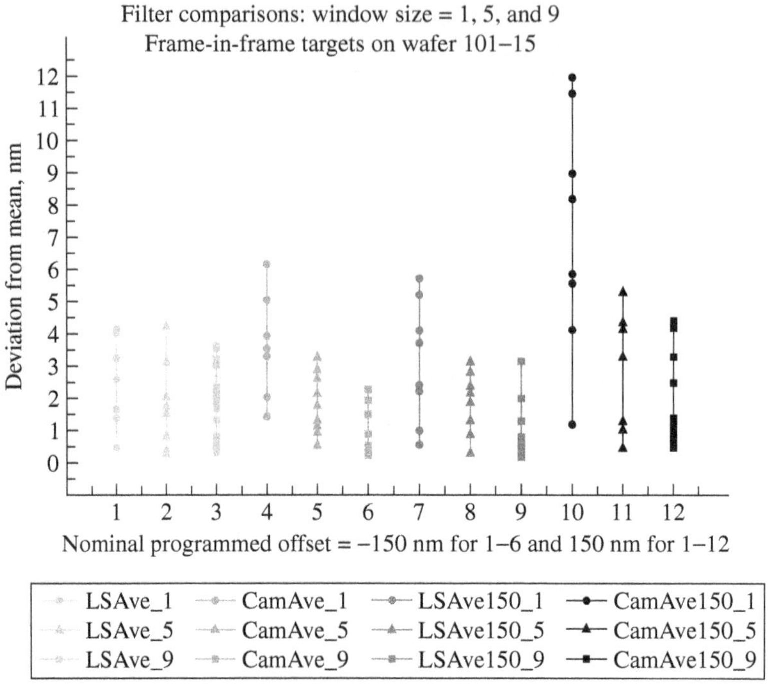

Figure 5.14 These data show the optimal filter size for this data set. Too large a filter introduces a systematic bias and too small a filter causes poor repeatability.

show results from registration measurements of targets from two sites with 150- and –150-nm offsets. These measurements are essentially pitch measurements from overlaid targets. The results show an improvement in reduced 3-sigma repeatability in all four examples between no filtering and a width of nine pixels, which is nominally equivalent to a 70-nm width maximum filter. All of this processing was performed on interpolated data with a density of four points per pixel, as described next. The biggest gains are seen with the application of the first filter of 5 pixels width and with reduced or minimal effects seen beyond this. This figure also shows a comparison of the results for the application of the two different functions, least squares and correlation.

Another of the front-end image processing steps is image interpolation. This is critical to reduced scatter in the repeatability plots. The effect of interpolation at this stage is to allow the stepping process in the generation of the correlation/least squares functions to occur with decreased distances between steps resulting in finer resolution in the Corr/LS curves. This is seen graphically as the apparent pixel spacing in Fig. 5.13. Interpolation adds significant processing time to the calculations, and proceeding beyond the optimal interpolation density does not yield measurement improvements.

Results which demonstrate the effects of different interpolation densities for similar targets and data which were analyzed in Fig. 5.14 are shown in Fig. 5.15. These results show significant improvements in all the examples analyzed. As seen, the improvements can be substantial and are sequential in the application of increasing interpolation densities as shown. The 3-sigma scatter in these measurements was reduced from well over 10 to below 3 nm with the appropriate stepping sizes. In the application of interpolation, there are several methods of fitting, which can be used to place the intrapixel points such as linear, cubic, spline, bilinear, and bicubic.

Background subtraction and gray scale inversion. An important step in image preparation and analysis is background subtraction and the overall choice for the feature gray scale. This also plays an important role in the normalization of profiles for comparison of experiment and theory. We will first discuss the background subtraction issue. This can have a significant effect on outlier generation with some images. This also plays a key role in gray scale inversion if one alters the image zero point. Figure 5.16 shows the application sequences for two background

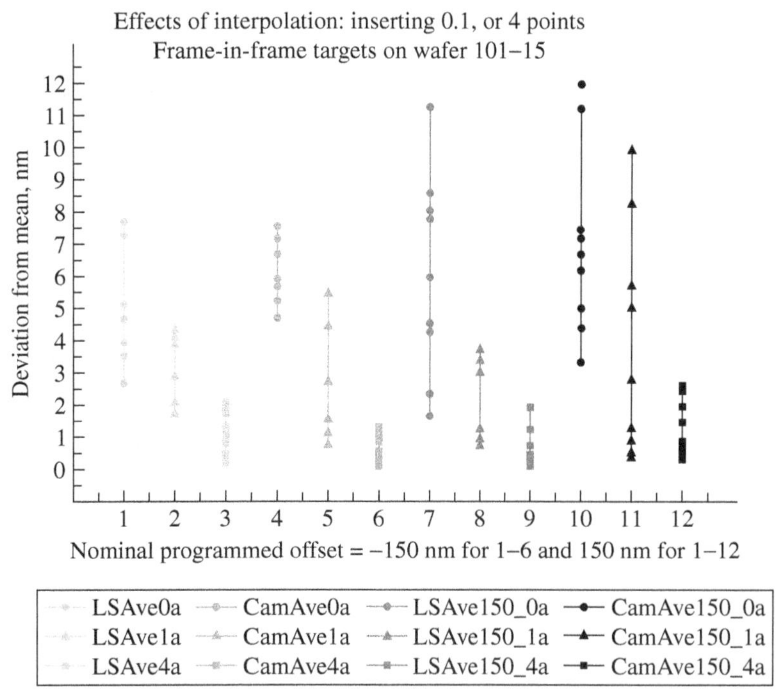

Figure 5.15 This figure shows the effects of increasing resolution by pixel density through interpolation. There is an optimum interpolation density and appropriate utilization of interpolation will improve measurement uncertainties.

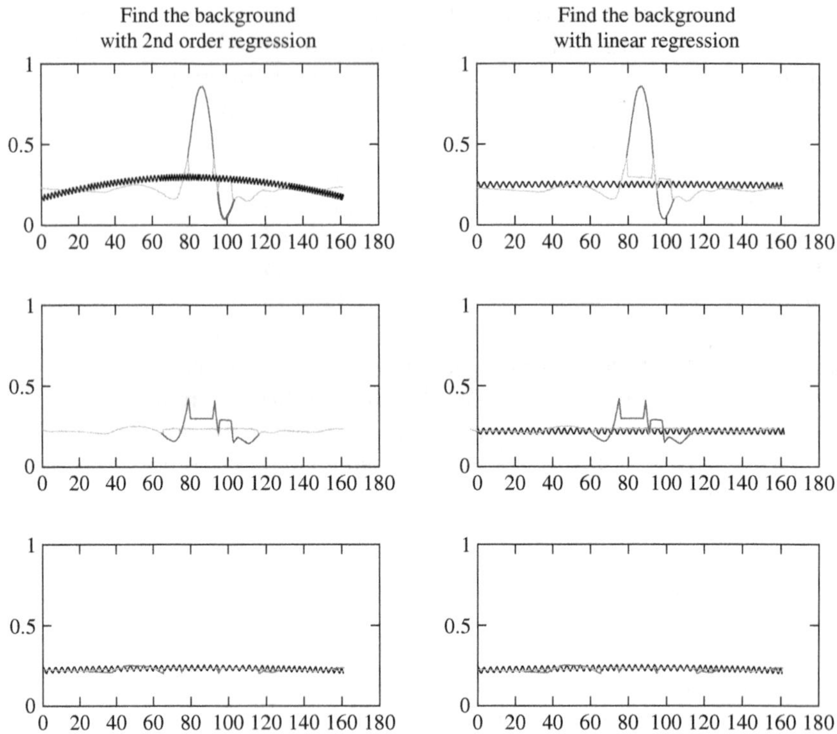

Figure 5.16 Higher order fits are more susceptible to errors. Multiple linear fits match the background well. Applications of background subtraction can significantly reduce outlier data.

subtraction algorithms. The first example is a second order fit to the entire image, followed by the removal of data outside a defined bracket, such as data lying more than three standard deviations from the mean. If this bracket is applied a second time, the resulting fit to background is shown at the bottom of the figure. The right side of Fig. 5.16 shows a similar fit to background except a simple linear regression is used followed by similar bracketing and removal. The linear method was generally chosen since it was less influenced by the actual data peaks. With a well-aligned optical tool, the background is generally flat and well behaved and a linear fit is adequate for removal.

The background can influence the Corr/LS curve results since, as the data are slid relative to one another, different amounts and regions of background are being sampled. Another important method which can be applied is that with good background removal, and the application of an absolute value operator to the entire image, the resulting peaks are all positive in the image. Although the LS method is much less sensitive to this, the Corr method can benefit from the increased number of positive peaks to correlate. This is quite effective with edge intensity patterns which oscillate positive and negative.

For high-contrast images with features that display good optical contrast background subtraction can have minimum effects. Figure 5.17 shows data for five different variations of background subtraction on an image with good intensity and contrast. These results show minimal effects using the different methods. The five variations of the background subtraction methods shown include a zeroed background and an alternative with a full background removal with both methods applied to the LS- and Corr-based curves. Again, for this example, there are no significant differences in the results for the methods applied. However, in several other cases, the outlier and 3-sigma scatter can be significantly improved with the correct method of choice.

Focus algorithm details. Although edge detection and centerline algorithms are an essential part of quantitative image analysis, they also have a central role in auto focus and sample centering hardware control

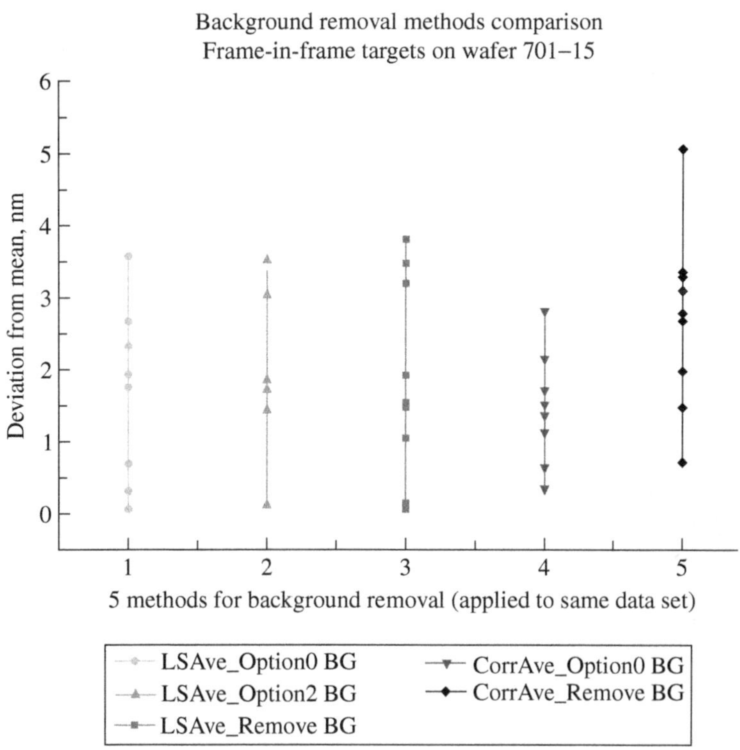

Figure 5.17 This figure shows the effects from background removal using different methods as outlined in the text.

algorithms. These steps are essential for optimized tool performance. The automatic positioning algorithms use much of the image analysis software package just described, although a further determination as to what is the best focus metric needs to be made.

Applying a best focus criterion to the image during acquisition ensures that the tool is always operating in the same focal plane. This is essential to having good symmetry in the edge slopes and intensity patterns, and stable optical distortions for good sample-to-sample and image-to-image repeatability. Although there are a number of different methods and focus criteria which one can apply, different target geometries may respond best to a particular focus algorithm.[26] In this section we evaluate a number of the standard focus metrics—the gradient energy, standard deviation, summed intensity, and contrast of images. These were chosen because of their widespread use in semiconductor manufacturing and other industries.[27, 28] The gradient energy is determined by applying the operator Q on the image intensity where Q is defined by:

$$Q \equiv \sum_x \sum_y \left| \nabla_{xy} \right|^2 \tag{5.1}$$

The discrete gradient may be implemented in numerous ways. One option is to use the Sobel method in order to reduce sensitivity to noise.[29] The standard deviation method is defined as the standard deviation of the image pixel values from the mean of all included pixels. The summed intensity is simply the sum of the pixel values. Contrast is defined as (max − min)/(max + min) intensity. In the examples presented in this section in which algorithms are compared to one another, their responses are normalized to their peak values.

An examination of different algorithm responses to basic chrome-on-quartz features and more complicated metal oxide patterns in bright-field and scanning confocal modes is presented next. As a part of a study of focus algorithms, it is important to investigate the absolute placement of the focal plane by each algorithm, and it has been demonstrated that the placement of the focal plane depends on the material composition of the artifact and more recently the target geometry and optical configuration can both have substantial effects on focus position and focus response.

Figures 5.18 and 5.19 show the response of the various algorithms to a periodic pattern of chrome lines on quartz. In general, the contrast algorithm shows very little variation between out-of-focus and in-focus images for bright-field mode and as a result is not shown for data acquired in the bright-field mode. This particular deficiency can at times be rooted in the performance of the auto-gain circuits in CCD cameras. In bright-field mode, the total intensity of the image is also a poor indicator of image focus. The total intensity data, shown in Fig. 5.18 (bright-field mode), indicate a broad minimum slightly above the nominal focus

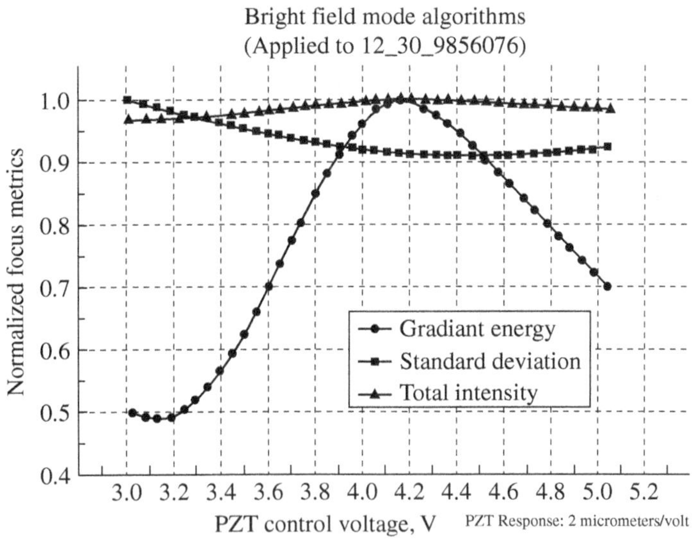

Figure 5.18 Response of focus algorithms to the chrome-on-quartz test patterns in bright-field mode.

position. However, even this behavior is not repeatable among various objects. In the bright-field case, the gradient energy metric provides the sharpest discrimination, although it involves substantial computation time compared to the standard deviation. In reflected light scanning confocal mode, gradient energy, standard deviation, intensity, and

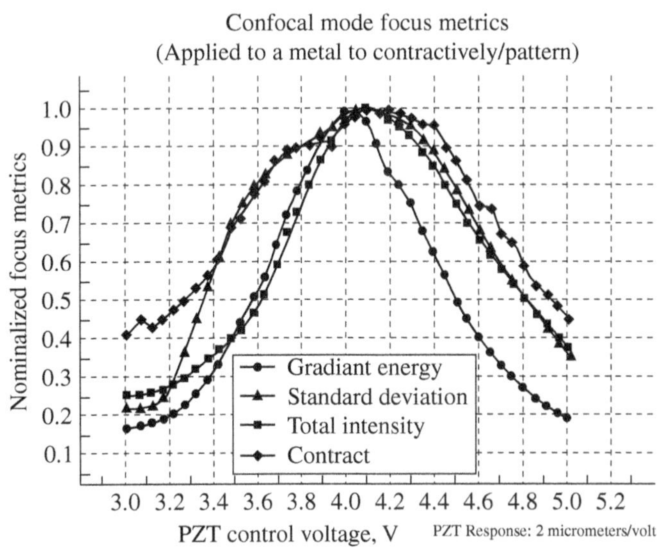

Figure 5.19 Response of the focus algorithms to a metal layer target pattern in confocal mode.

contrast are all suitable focus metrics, although the standard deviation algorithm does discriminate somewhat more sharply than the other methods for this particular image (Fig. 5.19).

Overall, the data on photomasks indicate that for bright-field mode, the gradient energy is the most reliable algorithm (of those tested in this study) and that for scanning confocal mode, either the total intensity or the gradient energy may be used if the issue of absolute focus placement can be resolved. An interesting result to come from these investigations of best focus was that the physical focal position appears different for varying geometries and materials.

Determination of each algorithm's theoretical placement of focus is accomplished by applying each focus algorithm to a set of simulated optical intensity data. Because the feature structure is known in the simulations, the output of the algorithms may be easily compared to a suitably defined focal plane. Comparison of extrema in the data to the feature structure determines the placement of focus. In this study, the algorithms were tested against simulations of the microscope response to chrome lines on quartz, SiO_2 lines on chrome, and SiO_2 lines on silicon. Simulations were developed for both bright-field and confocal modes of microscope operation.

Each algorithm, when applied to chrome lines on a quartz substrate, placed the focus at the top of the feature (in bright-field reflection mode). Similar results were obtained for confocal mode. Figure 5.20 displays

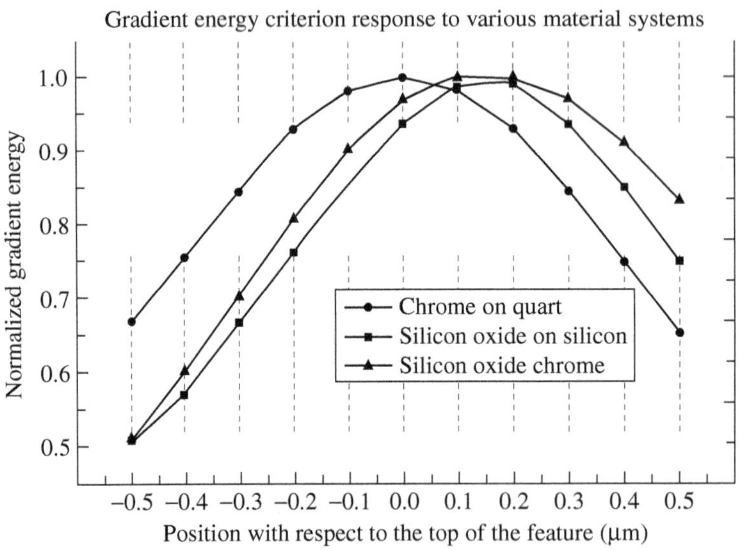

Figure 5.20 Response of focus algorithms to simulated microscope data. The zero point of the abscissa is at the top of the imaged feature.

the results for SiO_2 lines on a chrome substrate. These data demonstrate that the material composition of the artifact *does* affect the theoretical placement of focus. For a dielectric line on a metal substrate (in this case SiO_2 and chrome) it was found that the gradient energy and standard deviation algorithms both placed the focus 200 nm above the top of the feature. In the confocal mode the data show that when applied to lines of SiO_2 on silicon, the algorithms will place the focus approximately 100 nm above the top of the feature. Figure 5.20 indicates that the gradient energy algorithm reaches its extreme value before the total intensity algorithm as the sample is moved toward the objective lens. Although in previous work it was demonstrated that a pitch measurement only weakly depends on focus, the errors associated with pitch measurements can depend on the absolute placement of the optics in the focusing dimension.

For accurate, predictable measurements of three-dimensional structures, the placement of the focal plane and the response of algorithms to particular material systems must be well characterized. This is particularly true for the antireflective chrome materials used in photomask fabrication since they have graded optical properties. The main point is that when comparing theory to experiment or trying to obtain repeatable experimental results, the determination of best focus is very sensitive to the material stack and optical properties.

Linewidth estimation using reflection microscopy. As an example of reflection mode photomask metrology, optical images from SRM 2059 were obtained in reflection mode using 546 nm illumination wavelength, 0.95 collection numerical aperture, and 0.6 illumination numerical aperture.[30] Nominal chrome linewidths of 0.25, 0.5, and 1.0 μm were selected for this investigation. Since AFM analysis resulted in a better estimation of the linewidths selected, as a starting point, model simulations were made using the linewidth values obtained from AFM measurements. It was observed that variations in the optical properties of the chrome used for model inputs resulted in considerable variation in the simulated intensity profiles. Example experimental data and modeled data are shown in Figs. 5.21 and 5.22 for the 0.5- and 0.25-μm lines, respectively. Results from *x-ray photoelectron spectroscopy* (XPS) analysis on similar chrome on quartz features indicated a variation from pure chrome at the bottom of the line to pure chrome oxide at the top of the line, different from the standard n and k values measured by ellipsometry. Lack of reasonably accurate optical properties of the chrome line prompted a study to simulate theoretical profiles using optical properties ranging from pure chrome to pure chrome oxide. Qualitatively, the shape of the simulated profile obtained using pure chrome optical properties was in good agreement with the experimental profile. However, physically, since the

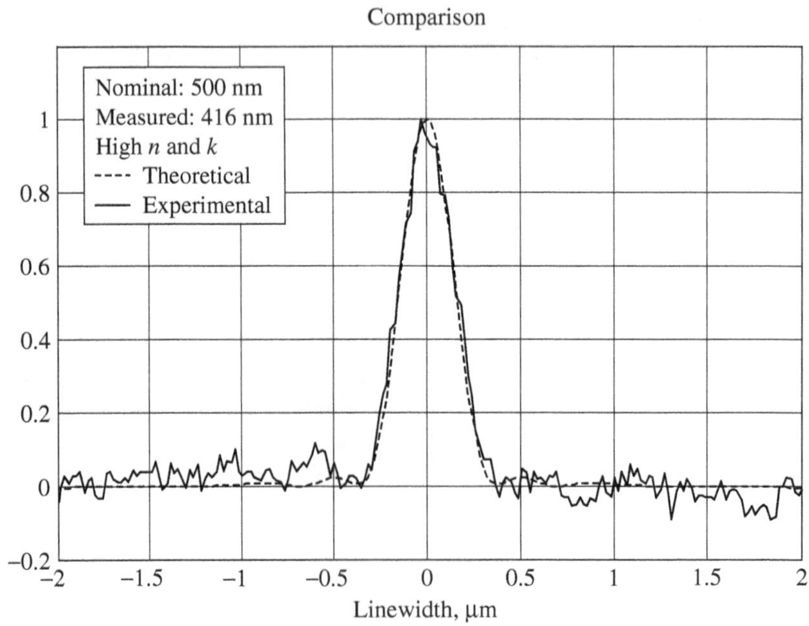

Figure 5.21 Comparisons of experimental and theoretical results for 500-nm-wide line.

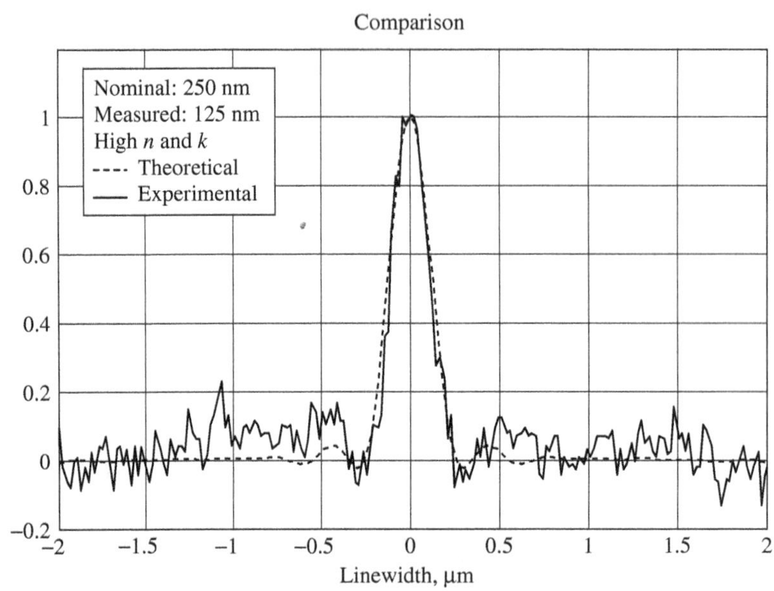

Figure 5.22 Comparisons of experimental and theoretical results for 250-nm-wide line.

linewidth structures are a mixture of chrome and chrome oxide, the linewidths were also estimated using the optical properties of a mixture of pure chrome and chrome oxide.

The procedure for linewidth estimation using reflection microscope optical images for this data is as follows. Profiles were obtained from the optical images and theoretical simulations were appropriately normalized for comparison. Widths at 50 percent of the intensity were obtained for both the theory and the experiment. A known linewidth in the simulation gives a particular measured width at 50 percent intensity. By comparing the width of the experimental profiles at 50 percent with that of the simulation, an estimation of the linewidth was made. Table 5.2 summarizes the estimated values.

5.2.4 Models and simulations in optical microscopy

The importance of modeling has been discussed and alluded to throughout this chapter. It is simply not possible to perform linewidth metrology without modeling to any degree of accuracy in this current generation of manufacturing due to the small feature sizes. The time when simple edge threshold based methods could be applied without modeling has long since passed. This holds true regardless of whether we are discussing scanning electron microscopy, optics, or atomic force microscopy.

There are a few different approaches which can be used to model optical images although all of these models are based on Maxwell's equations. Probably the most widely used methods of the electromagnetic scattering models are the *rigorous coupled wave methods*.[31] Many issues exist with these models. They are used extensively by the industry in scatterometry applications. Extensive evaluation of some models has shown that these models must be thoroughly evaluated for robust numerical convergence and stability. Sensitivity to resonances and interference phenomena is a common problem as well as lack of convergence

TABLE 5.2 Comparison of Experimental Results between Model-Based Reflection Mode Optical Measurements and Modeled AFM Measurements. The Comparison is Shown for Two Different Sets of Optical Material Parameters

		Reflection microscope measurements	
Nominal linewidth, nm	AFM measurements, nm	Considering pure chrome, nm	Considering a mixture of chrome and chrome oxide, nm
250	125	126	124
500	400	422	410
1000	900	928	919

at the nanometer scale. Interference can be a subtle problem when the scattered light is in phase with the sample structure. Theoretical inaccuracies and problems need comprehensive evaluation, especially when accuracy requirements are below the 10-nm uncertainty scale.

Two different theoretical model evaluations and comparisons are presented next. In the preparation of this material it was important to evaluate and use multiple models so that the results and parametric effects in numerous situations not easily obtained in practice could be simulated with confidence. It is important to recognize that we need to rely on these models with full confidence to perform quantitative analysis of linewidth structures with complete expanded uncertainties. As a result, the full testing of two models has given unique insight to several critical issues. The two models are based on an analytic waveguide solution to the electromagnetic problem and a full Maxwell equation solver.[32–34] The general configuration of the sample and electric fields for the calculations is shown in Fig. 5.23. This figure shows the incoming plane wave fields assumed to be Köhler illumination (see Fig. 5.8) with an angular cutoff determined by the illumination numerical aperture. The reflected fields are shown to have both specular and scattered components as shown in this figure. The reflected and scattered fields have appropriately terminated wave vectors so that the NA of the collection optics can be properly accounted for in the calculations.

The full Maxwell equation solver, also referred to as the integral equation method, is considerably more computationally intensive and is therefore run on a main frame computer and generally requires several hours of CPU time. The waveguide calculations can be run on a desktop personal computer in a matter of hours or longer depending on the details of the

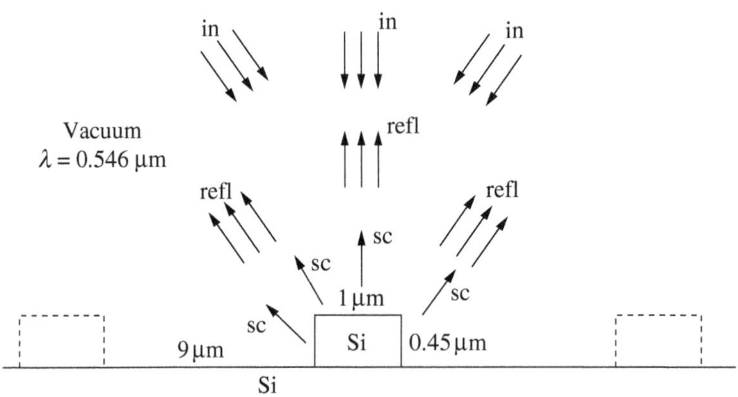

Figure 5.23 General configuration of the sample and electric fields for the calculations and simulations.

calculations. The integral equation method has no specific approximations other than numerical. The analytic waveguide method assumes a repeating structure and is discussed in detail in the references provided. The Maxwell equation solver is a unique, more recently developed set of codes, and we therefore go into some explanatory detail of the theory and code.

The integral equation method. In this model, the scattering object consists of a dielectric or finitely conducting infinite line on a substrate of the same or a different material that is invariant under translations along the line. The general configuration for the simulations is shown in Fig. 5.23. The z axis is chosen along the line and the y axis perpendicular to the substrate. It is assumed that the incident field is a plane monochromatic wave, and the field is decomposed in the region above the substrate into incident field, reflected field, which is the plane wave that would exist in the absence of the line, and scattered field. There also are refracted and scattered fields in the substrate.

Under these conditions, it can be shown that the electromagnetic fields can be expressed in terms of the longitudinal components of the fields, E_z and H_z, and that these fields satisfy the two-dimensional Helmholtz equation,

$$\left(\nabla_{\perp}^2+k_{\perp}^2\right)E_z(x,y)=0 \qquad \left(\nabla_{\perp}^2+k_{\perp}^2\right)H_z(x,y)=0 \tag{5.2}$$

where both field components are required if the incident wave has arbitrary polarization and direction of propagation. These equations are reduced to integral equations for unknown boundary functions, which can then be solved by the point-matching method. Details of these derivations are given in Ref. 32.

Once the boundary functions are determined, the scattered fields in the region above the substrate are given by integrals. These are the exact scattered fields for the infinite line on the substrate, and the only approximations introduced are those of numerical calculations.

To construct an image as obtained by a microscope focused, for instance, at the top of the line, a number of incident plane waves must be defined whose distribution depends on the illumination aperture of the microscope. The fields are calculated at a height where there are essentially only radiation parts of the fields, we then eliminate the evanescent waves and those fields that do not propagate to the microscope (and therefore do not get captured by the microscope) as specified by the imaging aperture, compute the fields on the plane at the top of the line, and add the intensities for each incident wave.

For a function $F(x, y)$ that satisfies the Helmholtz equation, if the function and its normal derivative at a height y_0 are known, the function can be determined at an arbitrary height y above the lines by solving the one-dimensional (1D) Helmholtz equation

$$F(x,y)=\frac{1}{\sqrt{2\pi}}\int_{-\infty}^{\infty} dk_x [\tilde{F}_{0+}(k_x)\exp(ik_y y) + \tilde{F}_{0-}(k_x)\exp(-ik_y y)]\exp(ik_x x) \tag{5.3}$$

where

$$\tilde{F}_{0\pm}(k_x)=\left(\frac{1}{\sqrt{2\pi}}\right)\exp(\pm ik_y y_0)\int_{-\infty}^{\infty} dx [F(x,y_0) \pm (i/k_y)\partial F(x,y_0)/\partial y]\exp(-ik_x x) \tag{5.4}$$

$$k_y = \begin{cases} +\sqrt{k_\perp^2 - k_x^2} & \text{for } |k_x| < k_\perp \\ +i\sqrt{k_x^2 - k_\perp^2} & \text{for } |k_x| > k_\perp \end{cases} \tag{5.5}$$

The contributions to the Fourier transform in Eq. (5.3) with imaginary values of k_y are either exponentially increasing with y, which are set to zero because they are unphysical, or exponentially decreasing with y, which are called evanescent waves. These evanescent waves should disappear as y increases, but errors in numerical calculations and the approximation of the Fourier transform by a discrete Fourier transform do not allow proper recovery of these terms when $F(x, y)$ is computed for $y < y_0$. It is noted that k_y vanishes if $|k_x| = k_\perp$, which means that $\tilde{F}_{0-}$ has to vanish. Numerically, these cancellations may introduce difficulties in integrals such as those in Eq. (5.4).

For the propagating components, one term corresponds to outgoing waves and the other to incoming ones. In the radiation zone, there should be no incoming waves, which means they vanish and the two terms in $\tilde{F}_{0+}$ are equal, so that we do not need to know the normal derivative of the field, whence Eq. (5.4) becomes

$$\tilde{F}_{0-}(k_x)\approx 0 \qquad \tilde{F}_{0+}(k_x)\approx\left(\frac{1}{\sqrt{2\pi}}\right)\exp(ik_y y_0)\int_{-\infty}^{\infty} dx\, F(x,y_0)\exp(-ik_x x) \tag{5.6}$$

The microscope image is then obtained using only those outgoing waves that have a propagation vector that is within the angle subtended by the lens (the collection numerical aperture). We have

$$F_{\text{image}}(x,y)=\left(\frac{1}{\sqrt{2\pi}}\right)\int_{-k_{x,\max}}^{k_{x,\max}} dk_x\, \tilde{F}_{0+}(k_x)\exp(ik_y y)\exp(ik_x x) \tag{5.7}$$

where the range of integration is determined by the numerical aperture, $\sin\vartheta_{\max}$, in

$$k_{x,\max} = \begin{cases} \sqrt{k^2 \sin^2 \vartheta_{\max} - k_z^2} & \text{for } k \sin \vartheta_{\max} > k_z \\ 0 & \text{for } k \sin \vartheta_{\max} \leq k_z \end{cases} \tag{5.8}$$

This is essentially the same calculation used for determining the microscope image with the analytic wave guide model.

As a test of model integrity for the integral equation method, self-consistency tests have been performed. In this analysis, a full scattering calculation is performed at two different heights relative to the sample. Both electric fields are Fourier transformed and terminated with the appropriate cutoff value for the illumination and collection optic angles. One of the fields is then transformed in the Fourier domain to the height at which the other field was calculated. Both fields are then inverse Fourier transformed and the real space electric fields are compared. The results show excellent agreement within the numerical calculation limits. This is an essential test of self-consistency. A consistent, detailed agreement is very strongly indicative of a correct scattering calculation and Fourier transform mathematics.

The analytic waveguide method. This method has been published in the SPIE forum in some detail and therefore only a cursory overview of the analytic waveguide method is presented here.[35,36] This technique uses a modal method called the *rigorous analytic waveguide method* (AWM) applied to gratings. It calculates the eigenmodes in a waveguide decomposition of the scattering essentially exactly. This exactness provides it with the potential for extreme accuracy. The fields are expanded in terms of the eigenmodes of the problem only. These are calculated analytically without use of a Fourier series. The hardest part of the calculation is calculating the eigenvalues.

After calculating the eigenvalues, the method proceeds by calculating a matrix which relates the generalized Fourier series of the fields in one z layer to those in an adjacent z layer by imposing the boundary conditions at the junction. Using these matrices, the coupling matrix between the fields in the substrate and overlaying layer is calculated, and then the scattering matrix is derived by requiring that the upward traveling wave in the substrate (the noncausal wave) vanishes.

For evanescent waves the exponentially growing solution into the substrate must vanish. These conditions close the problem and allow a solution. Fourier optics, as described in the subsection "The integral equation method," is then used to image the outgoing radiation fields or electric fields in the microscope. Intensity images can then be compared at different focal planes or heights relative to the sample.

It is important to recognize specific sensitivities which can be successfully addressed with parametric variation and analysis studies when using rigorous coupled waveguide methods. Errors and issues potentially exist with these models as a result of the infinite repeating structures used in the modeling methods. They are used extensively by the industry in scatterometry applications. Extensive evaluation of some models has shown that these models must be thoroughly evaluated for robust numerical stability. When interested in measurement uncertainties at the nanometer scale, sensitivity to resonances and interference is a problem to be aware of as well as lack of convergence.

An example comparison of theoretical results is shown in Fig. 5.24. This figure shows a 10-μm-wide Cr line, so a detailed comparison of the independent edge profiles can be made. This is in fact a critical test for the agreement of two modeling methods. The independent edges can be examined in detail. Once modeling agreement on these larger features is achieved, then the model can be further exercised by comparing smaller features. Additionally, it is also essential to compare the simulated wide lines with isolated edges to experimental results and ensure accurate agreement between the electromagnetic scattering calculations and the experimental results.

Modeling phase-shifting masks. With the increased use of phase-shifting technology it has become essential to develop the capability to model phase-shifting masks in transmission and reflection mode. For etched quartz masks this requires modifications to the simulation code to allow for asymmetric features and asymmetric feature offsets as input structures. It is now possible to simulate etched quartz phase shifters directly

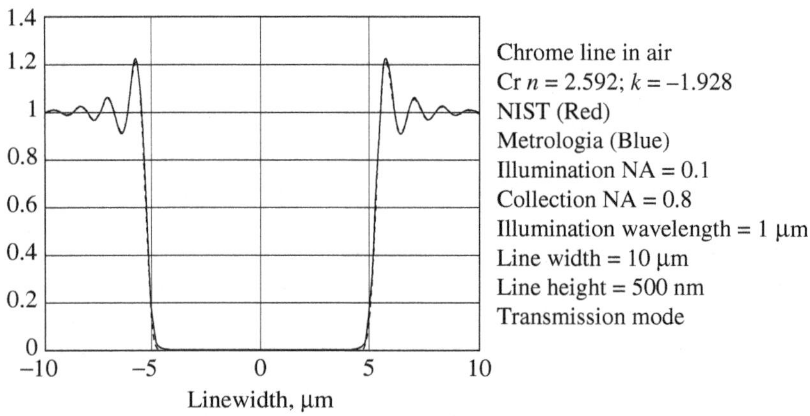

Figure 5.24 Comparison of two theoretical models' results for a wide Cr line.

adjacent to or slightly under chrome features. Results from recent modeling applications and experimental data are compared in this section.

A recent transmitted-light version of the Metrologia simulation program which allows for a structure consisting of a line with an adjacent trench was qualitatively tested and those results are presented here. As a test of model consistency, a nearly opaque chrome line on a quartz substrate that included an adjacent trench in that substrate exhibited no evidence of that trench when the trench was completely covered by the chrome line or when the trench was adjacent to the line, but filled with quartz. However, in this latter case, when the quartz filling of the trench was removed and replaced by a vacuum, the expected asymmetric line profile was obtained. The variation of this simulated line profile, including extending the trench under the line so that it was exposed on both sides of the line, with all of the input parameters of the simulation were explored and found to lead to the expected variations in the line profile (e.g., changes in position, width and depth of the trench, wavelength, illumination and objective numerical apertures).

Modeling results and actual experimental data from a phase-shift chrome-on-quartz photomask fabricated at ISematech were recently qualitatively compared.[37] Three 103-nm-tall *chrome lines* on this mask were optically measured and simulated after SEM measurements widths of 246, 316, and 368 nm were obtained. Each line had a single adjacent trench etched into the quartz substrate that was 171 nm deep and 491, 597, and 667 nm wide, respectively. A "best guess" for the effective complex optical index of refraction of the chrome line was 1.843-i2.195 at a wavelength of 365 nm and 1.475-i2.540 at 547 nm. The quartz substrate was 6350 μm thick and had an optical index of refraction of 1.475 at a wavelength of 365 nm and 1.460 at 547 nm. The medium was considered lossless at both wavelengths. The nominal numerical apertures of the lenses on the measuring microscope were 0.9 and 0.6 for the objective and condenser lens, respectively.

Comparison of simulated transmitted-light line profiles and the corresponding experimental profiles at various wavelengths show the expected qualitative behavior with regard to the various peaks and valleys, but differ somewhat in their quantitative behavior. Two sets of simulation data are shown in Fig. 5.25. These data are interesting as they show the measurement differences between the phase-shifting features at two different wavelengths.

Experimental images of isolated phase-shifted 250- and 500-nm lines are shown in Fig. 5.26, with a schematic on the right. All data were taken at 365-nm wavelength (not the 193-nm design wavelength). The reason for the quantitative differences between the experimental data and the simulations is likely due to the assumption of a single index of refraction and the extremely nonuniform chrome lines.

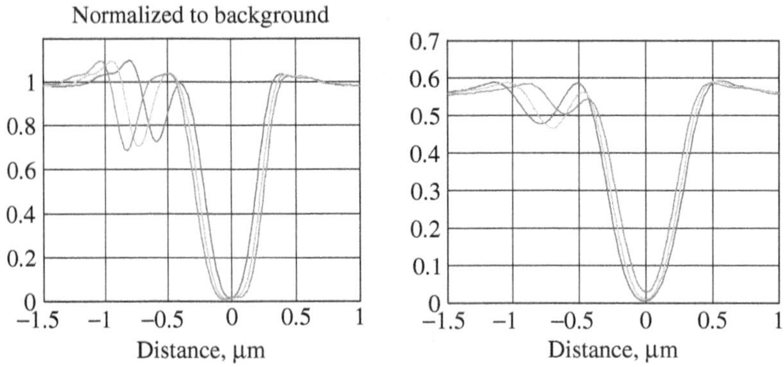

Figure 5.25 Computer simulated transmitted light intensity profiles at 365-nm wavelength on the left and 547-nm wavelength on the right for the three different linewidths described in the text.

Quantitative disagreement between experiment and simulation results can also occur if the experiment does not correspond to what was input to the simulation program. In this respect, the position of the focal plane of the microscope in the experiment is not a directly measurable quantity, and the rated numerical apertures of the objective and condenser lens may not correspond to their effective values when incorporated into the optical system of the microscope and its Köhler illumination system. The model must simulate the actual experimental conditions and parameters accurately.

It is also interesting to compare the reflected light optical images seen in Fig. 5.27 to the transmitted light images in Fig. 5.26. Similar sensitivity and response to the phase-shifting features are seen.

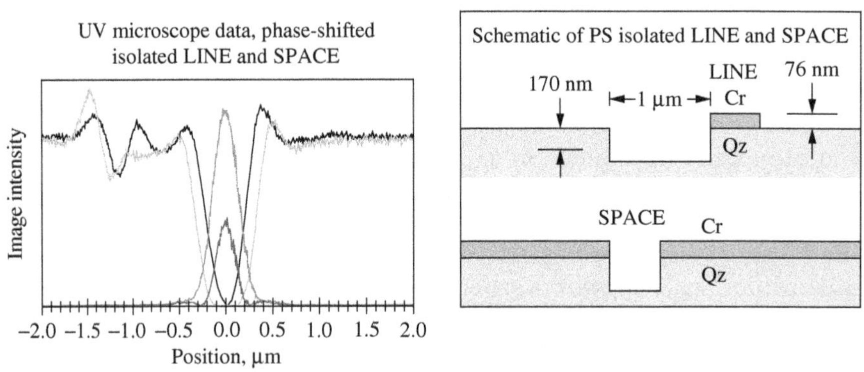

Figures 5.26 Experimental data for lines and trenches are shown on the left and a schematic of the phase-shifting features is shown on the right.

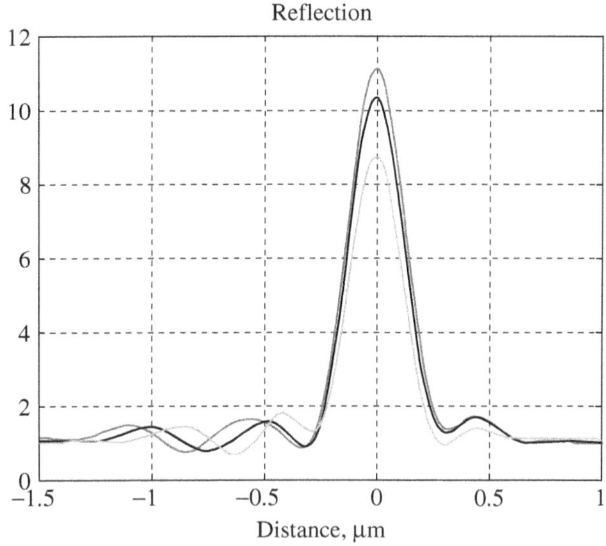

Figure 5.27 Reflection mode data from the same set of phase-shifting features modeled above.

5.2.5 Assessment of the effects of the optical constants *n* and *k*

The optical constants, n and k, have significant effects on optical based measurement outputs and accuracy with similar affects in both reflection and transmission mode measurements. The illumination angle and numerical aperture can increase sensitivity to surface properties. Even when the repeatability is superb, the accuracy may depend strongly on the knowledge of accurate material properties.

The scattering calculations used in conventional optical microscopy show strong dependence on the optical material constants. Solutions to the complete Maxwell's equations for optical scattering from semiconductor surfaces have provided a detailed evaluation of the effects on scattered optical intensities resulting from variations in optical input parameters.[30] These parametric studies have resulted in a systematic evaluation of the sensitivity of the results to errors in the assumed optical constants.

Quantitative evaluation of *n* and *k* effects. The modeling used to simulate conventional optical microscopy configurations has a similar strong dependence on the optical material constants to that observed in models used for optical lithography simulations. The actual scattering calculation methods used, whether full solutions to Maxwell's equations with boundary conditions or rigorous coupled waveguide solutions, are based on the same physical conditions. A comprehensive evaluation of optical parameter effects has been achieved using the complete and more complex model which solves the full optical scattering problem. Detailed evaluation of the effects on linewidth as a result of errors in the optical materials parameters has been studied extensively for photomask features, and a summary of some of the more important effects is given in this section.

The optical parameters found in the literature are quite variable. And although simple process stacks involving just silicon may have accurately known optical constants, more complex materials, and in particular chrome on quartz, have material inhomogeneity and are not as well known. This has been studied in some detail for chrome on quartz and a comparison of model-based ellipsometer determined values were shown to be in disagreement with the observed amount of transmitted light through the Cr layer. An alternative approach is to estimate the composition and resulting optical parameters through a direct measurement using sputtering methods that *etch* through the chrome while taking material composition data using XPS-based methods. This remains an essential area for future research as these optical constants continue to be poorly understood but have significant effect on optical metrology and lithography.

As a quantitative, model-based example of optical parameter effects on light scattering, it is useful to perform a parametric study of linewidth calculations by varying the optical parameters. Linewidth is intrinsic to the object and inherently independent of the method of measurement. Only an object's image can be measured with an instrument, and the image depends on both object and tool parameters $\{P_i\} \equiv P_1, P_2, P_3, \ldots, P_N$ (e.g., chrome thickness, chrome complex index of refraction n and k, edge roughness, illumination wavelength, and objective lens NA, see Table 5.1). The tool parameters are not intrinsic to the object, but they still affect the measured image.

Table 5.3 gives a specific example showing the strong dependence of the resulting measured linewidth and uncertainty in linewidth due to variations and uncertainties in the optical material input parameters.

TABLE 5.3 Some Estimated Parametric Uncertainties at an Illumination Wavelength of 365 nm

Parameter P_i	Nominal	df/dP_i	Uncertainty in P_i	Linewidth uncertainty component
Cr n	1.8	31.50 nm	0.50	16 nm
Cr k	−2.2	5.65 nm	0.50	3 nm
Qz thickness	6350 µm	−0.02 nm/µm	50.00 µm	−1 nm
Cr thickness	103 µm	−0.76 nm/nm	15.00 nm	−11 nm
Wavelength	365 nm	0.16 nm/nm	0.00 nm	0 nm

The instrument intensity image must be modeled (or the difference in images if two measurements are to be compared) in order to identify the locations of the object's edges, and the image scale calibrated if dimensions are to be accurately measured.

The measurement error $\varepsilon(a)$ can usually be expressed in terms of errors in the measurement parameters $\{\delta P_i\}$

$$\varepsilon(a) = f(\{\delta P_i\}) \tag{5.9}$$

where $f(\{P_i\})$ is the measurement process model function and δP_i is an error in the parameter P_i. Table 5.3 shows the effects in the resulting intensity profiles and linewidth estimations due to the uncertainties in some of the relevant optical parameters.

It is important to recognize that the optical parameters n and k vary as a function of wavelength. The multifit models used in estimating optical parameters for complex stacks are very susceptible to errors in estimating the underlying layers which contribute to the scattered fields.

The relative composition difference encountered in antireflective chrome can have significantly diverse local optical properties. The depth and lateral variations in material composition within mask structures can have significant effects on the measured results. As an example, Fig. 5.28

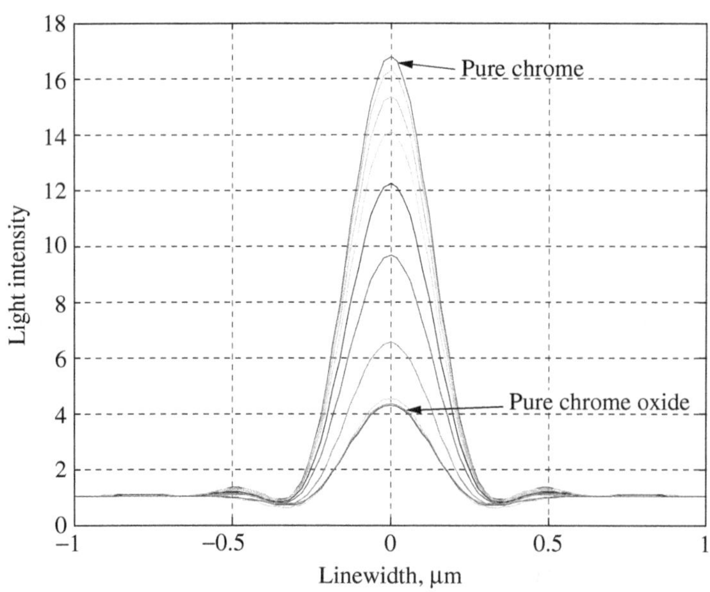

Figure 5.28 Light intensity profile changes due to variations in n and k for an isolated 400-nm-wide line for reflection optical microscope.

shows how the light intensity profile changes due to the variations in n and k for an isolated 400-nm-wide line as the composition of the line goes between pure Cr and Cr oxide (CrO_2). These results were obtained by simulation for a reflection optical microscope. Figure 5.28 demonstrates the importance of using the correct n and k values. These data clearly show the importance of accurately knowing the materials properties and the strong affects on the resulting profiles.

5.2.6 Advanced photomask CD metrology

Although optics is often thought to lose its effectiveness as a metrology tool beyond the Raleigh criterion, there is evidence that optics can be used to image and measure features as small as 10 nm in dimension.[38] Optical imaging and metrology with resolution well below the wavelength is a possibility as a result of increased computational speed, high-resolution CCD array detector technology, improved signal processing, and reliable theoretical light scattering models. The essential new areas of research involve low NA optics, annular illumination schemes, or different illumination schemes from high-angle annular to very low-angle illumination. Since the models are only valid for Köhler illumination conditions, it is strongly desired.

Preliminary scattering calculations and a set of experiments imaging 50-nm-wide silicon lines indicate that enhanced optical methods may be used for manufacturing and metrology at the 65-nm node and beyond. The preliminary calculations show that with an attainable signal-to-noise ratio, detection of features smaller than 10 nm is not only theoretically possible but even potentially practical. Actual optical images have been acquired with good signal-to-noise ratios for silicon features 50 nm in lateral dimension. An example of the potential resolution using these methods is shown in Fig. 5.29. This figure shows three different sets of calculations for repeating structures.

Although conventional scatterometry is routinely used for metrology of features smaller than 100 nm on silicon wafers, this methodology has not yet found its way into photomask metrology. Current investigations of low numerical aperture imaging methods, similar to scatterometry, except using a more conventional optical microscope layout are now underway. This has significant potential to enable full-field imaging of scattered optical fields with sensitivity to high diffraction orders using a library-based look up table.

Introductions to the optical scattering models have been presented in the previous sections. Preliminary simulations indicate that the theoretical limits are well beyond current practice. The challenge will largely be in the experimental apparatus and effectively emulating the theoretical results. These applications of advanced optical metrology methods have not been previously explored, and in part rely on advanced high-speed

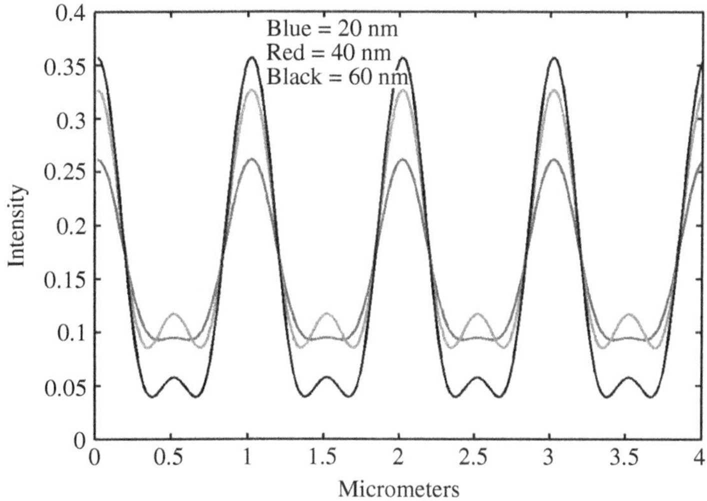

Figure 5.29 This figure shows optical simulations of silicon features as small as 20 nm in size. Although not chrome on quartz, this is representative of the possible limits.

computation of model-based optical profiles in addition to new structured illumination techniques.

The technical approach. There are several key new or enhanced techniques now being employed in next generation, high-resolution optical microscopy. The basic concept is to use specific engineered illumination wavefronts and illuminating fields and collection numerical apertures, which create the scattered field for a given small target dimension. The best parameters will give good signal-to-noise ratios and sensitivity to feature dimension and that aspect of the geometry of interest. These topics are further discussed in Refs. 39 and 40. The entire captured scattered field can then be processed and analyzed for comparison to a simulated profile. It is likely that the entire profile will need to be compared using correlation methods rather than the conventional methodology of using thresholds for linewidth determination. This is only now a possible processing technique using high-quality CCD arrays and significant data acquisition and analysis computation resources.

In the optimization of these methods, numerous simulations and modeling results need to be analyzed to determine the best illumination configuration and collection (objective) NA. Then an optimum sample geometry and optical configuration can be chosen for signal-to-noise enhancement, filtering conditions, and effective signal processing. An example of an experimental optical profile from a set of repeating features of 40 nm/100 nm/40 nm line space line pairs at a 1-μm pitch is shown in Fig. 5.30.

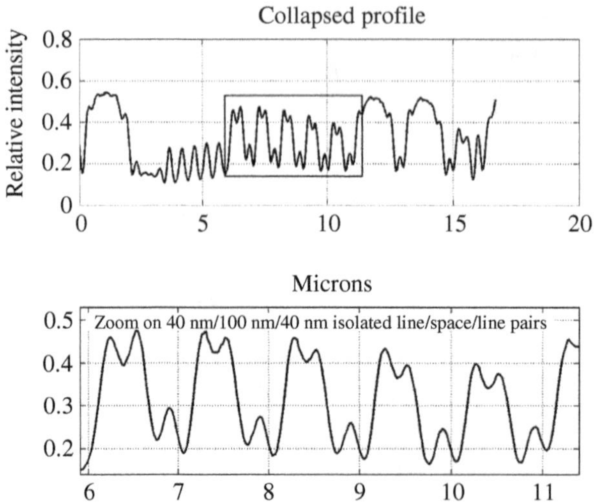

Figure 5.30 This is an experimentally acquired profile. These data show physical results that demonstrate the possibility of future high resolution optical metrology.

5.2.7 CCD mapping and errors

It is essential to fully map the charge-coupled imaging arrays (CCD) used in metrology tools. In this section we demonstrate the methods for CCD array calibration using a self-calibration algorithm applied to a CCD camera. The self-calibration algorithm is applied to the CCD camera to determine a quantitative error map for the correction of measurements made with that camera and is also used in the evaluation of the resulting uncertainty from the camera.[21] Self-calibration is a technique by which an instrument may be calibrated for measurements using an artifact with unknown dimensions.[41,42] Significant theoretical and some experimental work has been accomplished in the development of general self-calibration algorithms and their application to self-calibration of metrology stages, stepper lithography systems, and e-beam writers. In the work presented here we use a 1D, propagating self-calibration algorithm to examine errors in a CCD imaging system, which is comparable to those commonly used in optical metrology. For this CCD analysis method an *x-y* translation/rotation stage was developed, which allows the CCD imaging array errors to be separated from the other optical system errors.

The algorithm is shown schematically in Fig. 5.31. The artifact consists of a line grating with nominally uniform pitch. A single two-line pitch (P_0) is measured in position X_0 at time T_0. The array is then translated and the same feature pitch is measured in position X_1 at time T_1. The ratio between the measured values of P_0 in position X_1 and position X_0

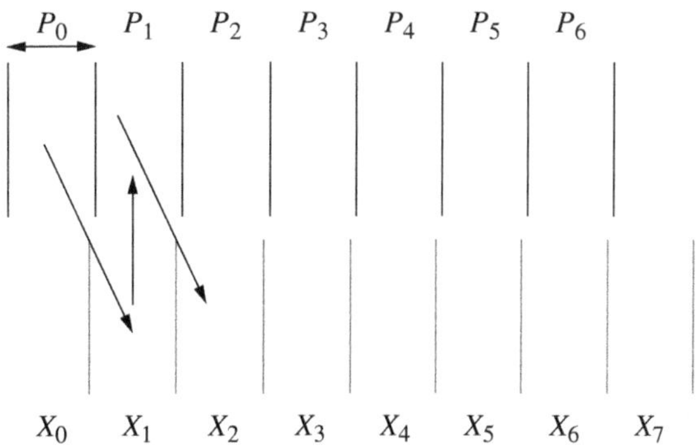

$\gamma_1 = P_0(X_1)/P_0(X_0)$ ... $\gamma_n = P_{n-1}(X_n)/P_{n-1}(X_{n-1})$

$\Gamma_1 = \gamma_1$ ... $\Gamma_n = \gamma_1{}^*\gamma_{n-1}{}^*\gamma_{n-2}\ldots{}^*\gamma_1{}^*\gamma_0$

Figure 5.31 Diagram of the 1D self-calibration algorithm used in this work.

provides a correction factor for position X_1 with respect to position X_0. We denote this factor γ_1. The factor γ_1 is then used to correct the measured pitch P_1 measured in position X_1 at T_0. This process is continued to position X_2. The ratio of P_1 measured at time T_0 in position X_1 to P_1 measured at time T_1 in position X_2 provides the correction factor γ_2. The correction factor appropriate for position X_2 with respect to position X_0 is then $\gamma_1 \times \gamma_2$ which is denoted Γ_2. The correction factor for any position is then—$\Gamma_n = \gamma_n \times \gamma_{n-1} \times \gamma_{n-2} \times, \ldots, \times \gamma_1 \times \gamma_0$, where γ_0 is the known or unknown scale factor corresponding to the feature P_0 in position X_0. It is important to recognize that when using this method, the error associated with the measurement of each feature also propagates (increases) through to γ_n. Although the algorithm is most easily described in terms of position shift steps the same size as the grating pitch, in actual application the artifact is stepped in fractions of the grating pitch of the calibrating artifact. The inter-feature positions are compared to the closest previous integer position. For example, P_0 might be measured in position X_0, $X_0 + \delta$, $X_0 + 2\delta$, $X_0 + 3\delta$, . . . , $X_1 - \delta$, X_1, where δ is a fraction of a grating pitch. The error propagation occurs in integer steps of the pitch. The inner steps allow for analysis of higher spatial frequency distortions.

We now demonstrate the method on a full frame transfer digital camera. We first conducted static precision tests (short term repeats with no movement), and then measured the dynamic precision (repeats with repositioning and refocus) by mapping the distortions imposed by the camera on a particular pitch pattern. In all cases, the measured pitch is the average across the middle third of the pixel array. The static precision test indicated that the static precision of the camera is very good

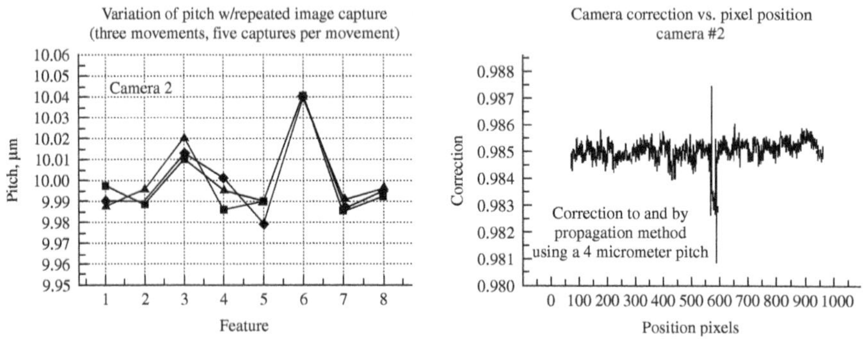

Figure 5.32 Results of the static precision test for Camera 2 are seen on the left and the error map of Camera 2 obtained through application of the 1D self-calibration algorithm is on the right. Correction factors are again shown on the y axis.

and nominally better than 1 nm throughout the field of view. In fact, with this camera the static repeatability is representative of the static precision of the entire optical system.

The variations seen in the dynamic precision test are due to slight changes in position within the frame. Figure 5.32 displays the results of applying the self-calibration algorithm to a series of data taken with this camera. The camera shows random pixel errors on the order of 1 nm for the 4-μm pitch used in this mapping. However, there are localized distortions of magnitude 1.003 to 0.997 of the mean value near pixel (column) 600. An examination of the distortions near the 600th pixel showed that the estimated pitch is first compressed, and then expanded as the feature moves through the area. Based on this result, a CCD camera defect consisting of a column of less responsive pixels was identified and later verified as the cause of this behavior.

5.3 SEM Methods for Mask CD Metrology

In the past the use of traditional optical tools for mask metrology was enough to make the necessary measurements. This has changed with the introduction of advanced masks with assist features that are much smaller than the main features, which are still four times larger than the desired structures on the wafer. Just like in wafer dimensional metrology, as mask feature sizes became too small for traditional optical microscopy, the use of new methods with higher spatial resolution became necessary for quality assessment and dimensional metrology of photomasks. As with wafers, SEMs were the natural choice for measurements of these small structures.

Today, especially for masks with phase-shifting and resolution enhancement structures, SEMs are used in every mask house. The key attributes that SEMs bring to mask metrology are very high spatial resolution, good repeatability and accuracy in delivering averaged and/or localized size and shape information, ease of operation, and high throughput. The ability to produce localized information on essentially any location of the masks is crucial in the development and troubleshooting of mask fabrication procedures. While modern SEMs can easily provide high spatial resolution images, there are also problems concerning SEM-based metrology. Most of these are not insurmountable, the correct choice of instrument and measurement parameters usually presents acceptable to excellent solutions. For some tasks new measurement methods and new types of instruments are needed.

One is the application of model-based metrology that has proven, in the case of wafers, that significant improvements are possible in the repeatability and accuracy of dimensional measurements. The other very promising technology is the use of field-emission variable-pressure SEMs. The resolution of the images obtainable with these microscopes is outstanding; small details of the photomask, even the fine structure of the Cr layer, are clearly visible.

5.3.1 Basics of SEM-based dimensional metrology

High-vacuum, low-landing-energy operation. This is the typical working mode of the current wafer CD SEMs. The low landing energy and fast scanning is essential to control sample charging. Since with this method, sample charging and contamination cannot be fully eliminated, they have to be monitored and controlled for reliable metrology. The image quality depends on the settings but it is usually not as good as with low-vacuum, high-landing-energy operation. Figure 5.33 shows two SEM

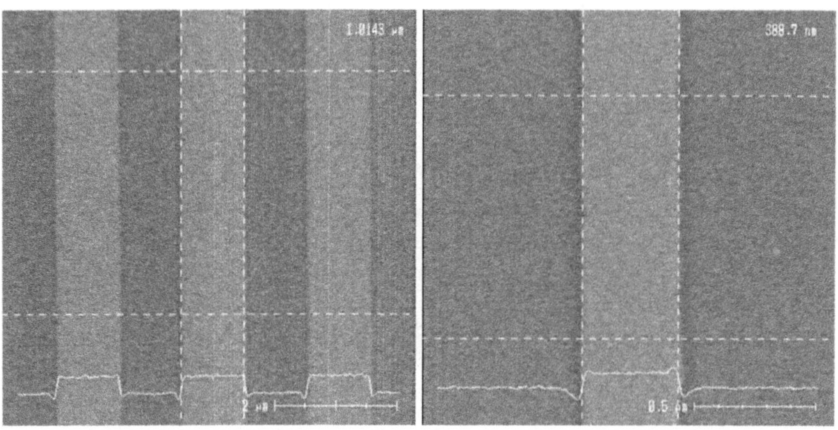

Figure 5.33 High-vacuum, low-landing-energy SEM images of Cr lines of a photomask. (Courtesy of ISMT.)

images of dense and isolated Cr lines on a photomask taken with fast scanning Hitachi S-9200 CD SEM.* The image quality is satisfactory, even some details of the Cr lines are visible; the average of many line scans is certainly good enough for dimensional metrology of linewidths.

The observation conditions were—accelerating voltage 800 V, beam current 8.2 pA, the number of frames averaged 64, magnification 20,000 × (left), 80,000 × (right), and the field of view 6.75 μm (left), 1.7 μm (right).

A typical, optimized mask CD-SEM image and a few parameters related to it are shown in Fig. 5.34. The image quality is quite good; the graininess of the Cr layer is visible. The darker gray lines depict the locations where the electron beam hit the sample for longer times. These could be signs of small degree sample contamination. The single line scan across the center shows good signal, with high peaks at the rim of the contact hole. The quality in general is adequate for reliable SEM dimensional metrology. The number of gray levels automatically set by the instrument is only 126, which could and need to be set close to the possible maximum of 256. The pixels size sets the Nyquist sampling criterion-limited spatial resolution to about 9 nm. The measurement resolution could be about 10 times better, which means with an SEM that works with sufficient spatial resolution structures differing in size by only one nanometer could, in principle, be distinguished. The potential to obtain these results is also limited by the amount and type of noise, stage drift and vibration, and drift and other unintentional motions of the electron beam (due to disturbing external electromagnetic fields). The electron beam in the vicinity of the focus has an "hourglass" shape, so the best focus might be available only in a very short region. In the case of 70 to 100 nm Cr layers this is usually not a problem, but for taller structures this needs to be taken into account.

Low-vacuum, high-landing-energy operation. An alternative technique to low-landing-energy, high-vacuum SEM for photomask metrology and inspection that minimizes, if not eliminates charging problems is environmental or low-vacuum SEM.[43] This technique is also known as variable-pressure SEM (Fig. 5.35).

This form of microscopy was originally proposed early in the development of SEMs, has slowly developed and has been most recently utilized to obtain previously unattainable data in biological, food, and chemical science applications. The application of environmental microscopy to production semiconductor metrology is new because of the need for the technological combination and implementation of high-resolution, high-brightness field emission technology in conjunction with large chamber and sample transfer capabilities to the environmental microscope technology. This combination of technology has not been available until just recently. Figure 5.36 shows an example for the capabilities of this technique.

With the excellent resolution of these field emission variable pressure SEMs, it becomes possible to measure the fine details of Cr or

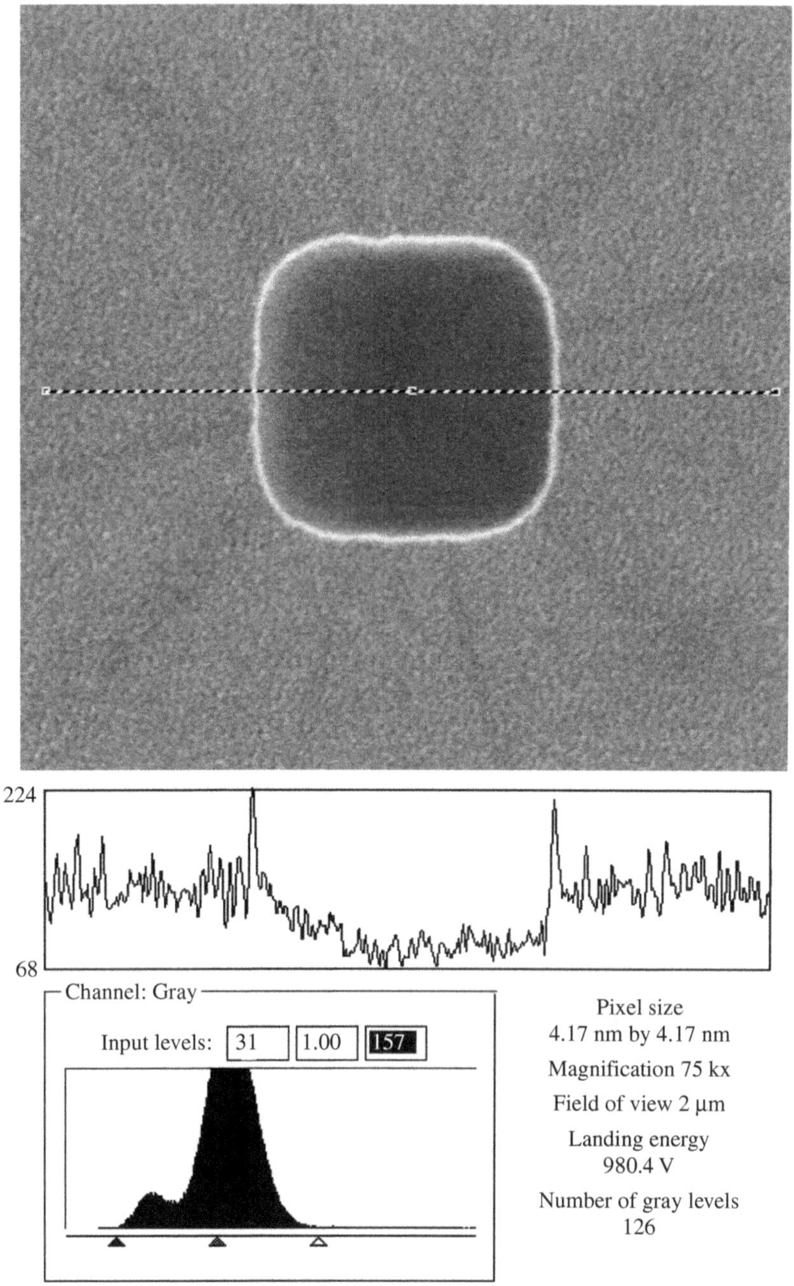

Figure 5.34 A typical mask CD-SEM image and few parameters related to it. The line scan shows the signal intensity distribution across the center line, the histogram is the grayscale distribution across the whole image. (The SEM image is courtesy of Dupont Co.)

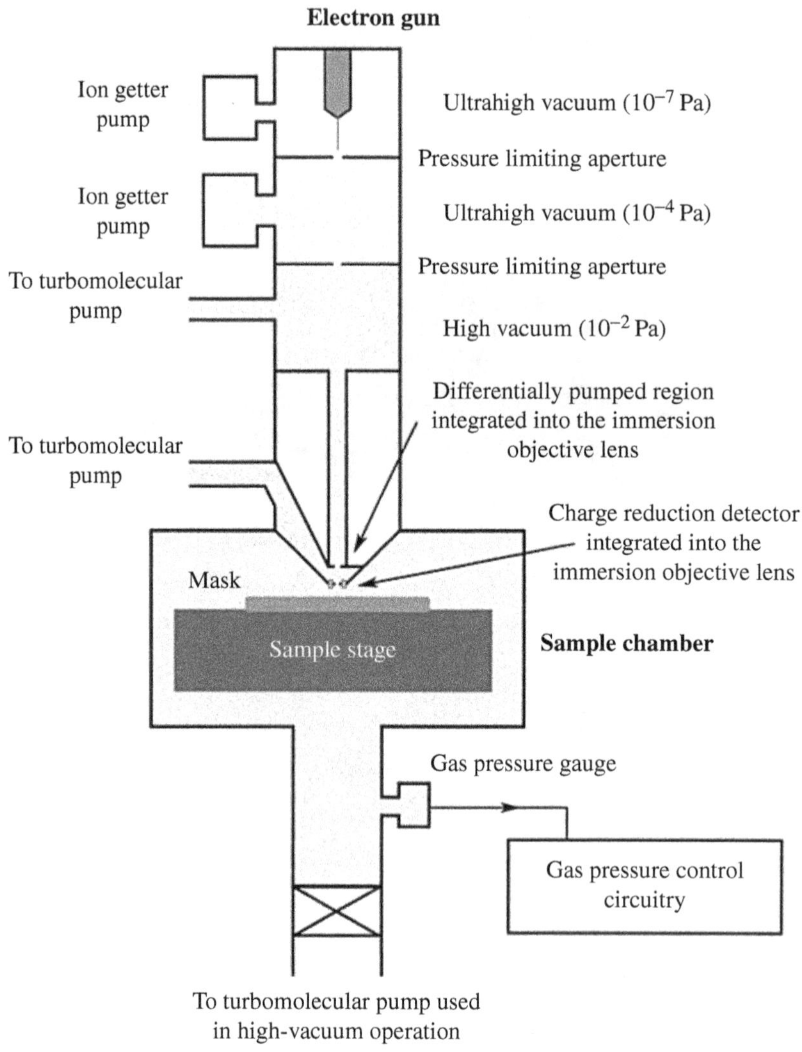

Figure 5.35 Schematic diagram of a field emission variable pressure SEM.

other reticle structures. Figure 5.36 reveals the fine details of the real 3D shape of the rough edge.

5.3.2 Currently available CD SEMs and their performance limits

The currently available mask metrology instruments are mostly modified CD SEMs that were originally developed to fit the requirements of dimensional metrology on wafers. These are low-landing-energy tools,

Figure 5.36 High-resolution image of a photomask taken in a variable pressure Gemini Supra 55VP variable pressure field emission SEM. The field of view is 1002 nm, 46 Pa pressure, 110 k× magnification, 20 kV accelerating voltage. (Courtesy of Carl Zeiss NTS GmbH.)

optimized for the distinct requirements of optical mask metrology. The masks are made on a thick quartz substrate that is an excellent insulator. Therefore one of the biggest problems is sample charging. As with wafers there are several possibilities to try to reduce charging to acceptable levels. With high scanning speed imaging and proper dose and energy control, SEMs can deliver respectable results on most mask samples. High throughput and repeatability are also important parameters. There is a potential of sample contamination under the irradiating electron beam. With rigorous and proper sample handling and clean vacuum, this also can be handled well.

As examples, the main parameters of two industrial instruments are summarized here. All these tools come with sophisticated software support to find the optimum among process control, yield, and device performance through pattern metrology and transfer.

Applied Materials Inc. RETicleSEM is based on a wafer SEM (NanoSEM 3D) platform.[*,44] This automated metrology instrument that can measure reticles and can achieve less than 1 nm precision with high throughput. It can work on a variety of materials like photoresist, glass with chrome, or molybdenum silicide (MoSi) layers. The system has 15° electronic e-beam tilt imaging capability. The good, 2.8 nm resolution and tilt imaging help to get good quality measurement of advanced masks including pre/postetch resist reticles and masks with *resolution enhancement techniques* (RET). There are software algorithms for calculation of corner roundness and contact area.

KLA-Tencor Corp. 8250-R is a critical dimension scanning electron microscope for advanced reticle measurement and control needed for 130 nm device production and 100 nm device development and beyond.[45,*] It is based on their 8200/8400-series wafer CD SEMs. The best measurement precision is better than 2 nm. It has an automatic recipe generation capability to address complex reticle metrology requirements. It can work on a variety of sample materials, including chrome-on-glass and MoSi reticles using the charge equalization technologies. This helps to produce stable images and measurements on reticles. The 8250 series wafer and mask CD SEMs incorporate the same laser stage that provides a capability of measuring the same features on both the reticle and wafer.

The results from the SEM image analysis module can be used in lithography data modeling and analysis software that enhances 8250-R functionality for both mask manufacturing operations and wafer fabs. This allows for comparisons and measurement of reticle features with reticle design data.

It is likely that a great deal of improvement is possible in sample charging, resolution, signal-to-noise ratio, optimization of measurement parameters, and especially in advanced measurement algorithms. Model-based algorithms, which take the physics of the measurement into account, are superior in obtaining the best, most accurate and repeatable results.

Today there are essentially four types of edge algorithms in use in CD SEMs. The maximum derivative type finds the point on the outer slope of the peak where the video signal changes fastest. The regression-to-baseline type finds two fitting lines, one for the base (background) line and one for the outer slope of the peak. The sigmoid type uses the outer half of the peak and background to fit an S-shaped curve. All provide a more or less repeatable way to determine the line-edge parameters, but none of them will point to the true position of the edge. This is done by the metrology engineer who, based on practice, experience, and other, e.g., cross-sectional measurements, will set some correcting factors to arrive at the desirable answer.

There is a significant drawback with these methods—they do not use the whole peak, which contains complex information on the sidewall angle, top and bottom corner radii, height of the wall, shape of the electron beam itself, and the like. This leads to two problems. One is that a large part of the signal is thrown away, and this leads either to a signal-to-noise ratio problem, from which the repeatability and accuracy suffer, or it leads to unnecessarily long acquisition times and to potential loss of throughput. The other problem is that these methods have unpredictable biases, which cannot be accounted for. This is due to the signal differences related to the real shape of the lines, which is varying from one sample to another. In short, none of these methods is able to fully account for the changes of the structure they measure.

The fourth modeled library-based method is aimed at exactly this; it works with modeling that takes all significant sample- and instrument-related parameters into account. For resist and polySi wafer dimensional measurements it is proven to be the best.[46,47]

Line-edge roughness metrology. This is increasingly important as the linewidths shrink and illumination wavelengths shorten. All current optical metrology tools measure the average linewidth over a long line length segment. The scanning electron microscope and atomic force microscope can measure line-edge roughness down to few nanometers. For an example for *line-edge roughness* (LER) measurements, see Fig. 5.37.

The requirements for LER metrology are more stringent; since it is about localized information, better spatial resolution is needed than for linewidth measurements, which can be averaged. LER leads to a measurement bias, which may become a significant problem for small size structures.[48] The line-edge roughness measurement algorithms implemented in current CD SEMs are not based on accurate modeling, therefore the results reported by them are undoubtedly not correct to some extent. It is likely that the solution for accurate mask LER metrology will be similar to LER metrology on wafers.

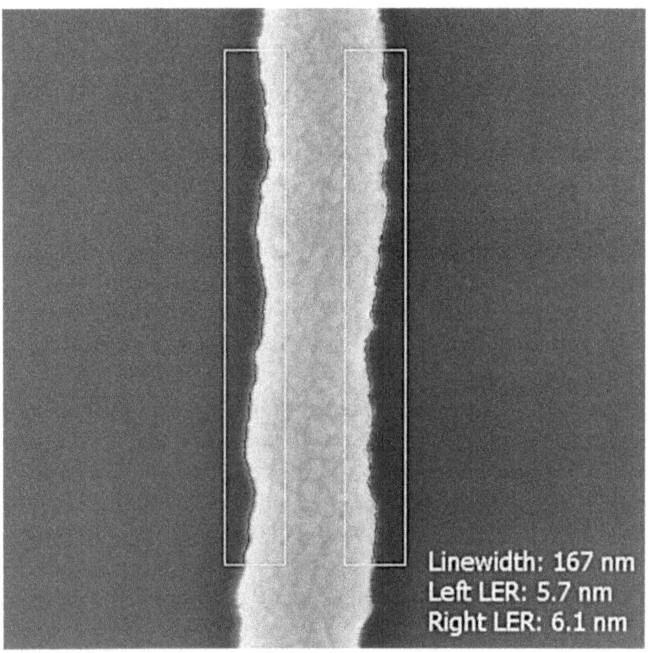

Figure 5.37 Example for line-edge roughness measurements on an image of a Cr line that was taken with an SEM capable of working in low vacuum. 50 k× magnification.

5.3.3 New methods under development

Model-based SEM mask metrology. Modeling is essential to linewidth metrology. All measurement methods generate images or line scans that have a signal peak corresponding to the region near the edge of the line structure on the mask. An accurate model makes it possible to tell where within the peak region the edge is located. To model experimental results and extract accurate feature dimensions from photomask measurements requires robust and accurate models. Currently, there are two 2D SEM models readily available that give good results. The SEM modeling may be performed with MONSEL,[49] a NIST electron trajectory simulator, or with Metrologia of Spectel Co. NIST and Spectel Co. have evaluated by comparisons and made necessary changes to these two software and now they are very close in their capabilities and performance.[50]

New, 3D modeling software is currently under development in collaboration between NIST and a university. This software is designed to work on images and data obtained with variable pressure SEMs.

The *model-based library* (MBL) method is an experimental method that was developed relatively recently.[51] In the past years following the efforts of researchers of Hewlett-Packard and Spectel companies, NIST has been developing methods to match libraries of modeled SEM images to measured images for wafer CD metrology. The model-based SEM metrology on resist and polysilicon structures has shown significant improvements over any other type of SEM-based dimensional metrology. It is possible to customize and apply this method to photomask measurements as well.

In this method a large set of edge shapes is simulated in advance using an electron trajectory simulator (see Fig. 5.38). These shapes and their images or line scans form a database, or library, which may be interpolated. To find the needed sample parameters, including edge position that produce the best match, the measured images are compared to the library (Fig. 5.39). The library entries are peaks and all the data in and near the edge can be included in the fit. This is a significant advantage over other edge finding algorithms.

The precomputed library contains line segments, i.e., peaks and some left and right line regions for the right edge only; these are mirrored for the left edge. The use of separate edges instead of the whole line scan is permitted as long as the curves do not overlap, i.e., the electron beam does not interact with more than one edge. The generation of the library can be done in few hours, because only few thousands of entries need to be computed. The library can continue to be used as long as the same type of mask is used. Finding the best match is very quick, only a few seconds, so there is no penalty in time.

The results of application of MBL to resist and polysilicon wafer samples showed excellent agreement with cross-sectional measurements.[46,47]

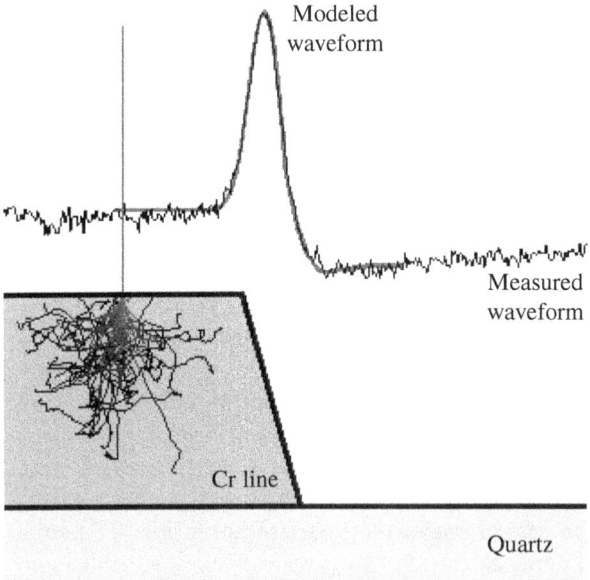

Figure 5.38 Modeled electron trajectories within the sample determine the secondary electron signal as a function of the landing position of the electron beam (smooth curve). The match with the measured waveform determines the edge location accurately.

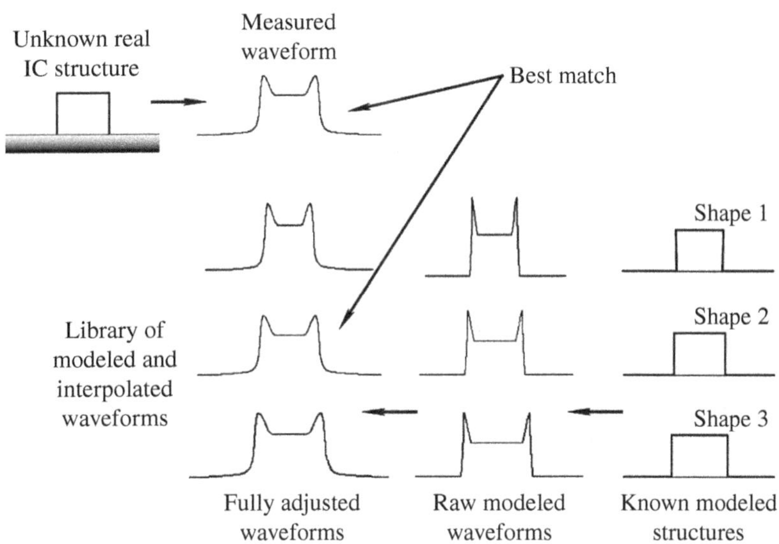

Figure 5.39 Schematic operation of model-based library method. The measured SEM signals from each edge of the unknown sample are compared to signals in the library to determine the parameters that produce the best match.

It is likely that MBL will be better than other methods for masks as well. Capabilities of the existing software therefore need to be improved to do a better job of modeling masks, and the method needs to be validated for application to masks. This work is currently underway.

Variable pressure scanning electron microscopy. This methodology employs a gaseous environment surrounding the sample to help neutralize the charge. Typically the gas used for photomask inspection is water vapor (although other gases can be used). A typical high-vacuum SEM operates with a sample chamber pressure of about 6.7×10^{-3} Pa (5×10^{-5} torr). For high-pressure microscopy work, the chamber pressure is allowed to rise to about 20 to 160 Pa by the injection of the water vapor. High-pressure microscopy offers the advantage and possible application of higher landing energies or accelerating voltages, different contrast mechanisms and better charge neutralization. Higher landing energy means that potentially higher resolution imaging is possible than at the lower accelerating voltages, but, at the cost of increased beam penetration. The consequences of the increased penetration into the various mask samples are currently being assessed.[53]

This methodology employs a gaseous environment to help diminish the charge buildup that occurs under irradiation with the electron beam. Although potentially very desirable for the charge reduction, for various technical reasons, this methodology has not been seriously employed in semiconductor inspection or metrology until just recently. This is a relatively new application of this technology to this area and there is still much to be learned, for example, the best operating conditions for the vacuum and signal collection and detector techniques. But, this technology shows great promise in the inspection, imaging, and metrology of photomasks in a charge-free operational mode. It has been found that even at high accelerating voltage, injection of air of as little as 20 Pa (0.15 torr) into the specimen chamber can reduce the charging potential of an insulator at the surface by as much as an order of magnitude. In addition, this methodology affords a path that minimizes, if not eliminates, the need for charge modeling which is otherwise needed for higher accuracy measurements. The modeling of charging is exceptionally difficult since each sample, instrument, and operating mode can respond to charging in different ways. Further research is currently underway to understand the ways to optimize these operating conditions.

The instrument used for taking the SEM images of Figs. 5.4 to 5.7 was an FEI Sirion NC 600, a thermally assisted field emission, high-pressure microscope. (Images are courtesy of FEI Corp.) This instrument is equipped with a gaseous secondary electron detector and incorporates a large chamber and sample stage that allows for full coverage of a 150-mm photomask, with normal incidence beam and tilt up to 45° for sidewall examination.

Figure 5.40 An image of a severely charging sample observed under high-vacuum inspection conditions (800 V accelerating voltage, slow acquisition speed). The right image was taken on the same sample and location with the high pressure mode (3.4 kV accelerating voltage, slow frame speed, 0.77 torr). The magnifications are 4000×.

One of the major advantages afforded by high-pressure SEM is the potential for the control of specimen charging. Figure 5.40 demonstrates severe charging in an image of a sample observed under standard high-vacuum inspection conditions. The charging at 800 V accelerating potential at slow image acquisition speed results in a distorted and useless image. The right half of Fig. 5.40 is the image taken on approximately the same sample and location with the high-pressure SEM. Note the control of the charging exhibited in this micrograph. Observe also in the image the appearance of contamination deposited during the high-vacuum mode imaging, i.e., the darker center area where, for focus setting, the electron beam hit the sample in slow-scan mode.

Photomasks present a challenging subject for metrology and inspection in the SEM. Figure 5.41 demonstrates high-resolution images of isolated and dense optical mask lines. The roughness of the edges and the grain structure of the chromium lines are readily observable at 50,000 × magnification.

Figure 5.42 presents images that clearly resolve the fine structures of the chromium lines, edge profiles, surface roughness, and residual

Figure 5.41 High-resolution images of isolated and dense photomask lines. 50 k× magnifications, 0.74 torr, 9 kV accelerating voltage.

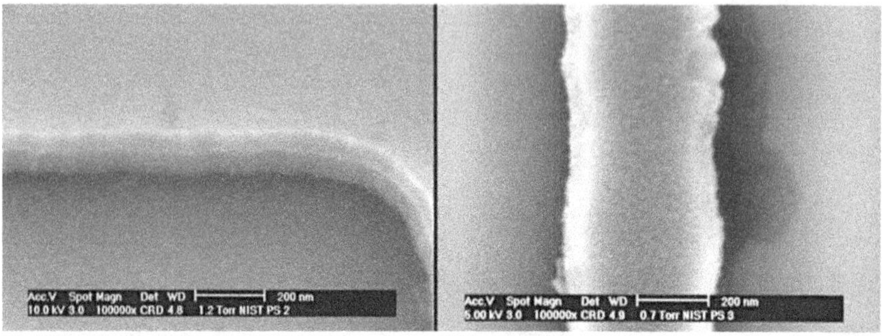

Figure 5.42 Fine structures of the chromium lines, edge profiles, surface roughness, and residual surface contamination from the preceding mask processing. (Left image is a tilted view) 100 k× magnifications.

surface contamination from the preceding mask processing. The left image, which is a tilted view micrograph, clearly shows the fine structures of the edge.

The charge control of the cascade process is thought to remove the local fields that may cause hydrocarbons to be attracted to the scan area. This effect and the very mild plasma of the cascade process are thought to greatly reduce the deposition of carbon contamination on the scanned areas. Figure 5.43 demonstrates a short assessment of electron-beam-induced contamination. This series of three micrographs shows the effects of irradiation of approximately 3 min. Note there is no visible contamination deposited on the sample during the exposure. This is in stark contrast to Fig. 5.34 where brief irradiation under high-vacuum conditions resulted in contamination residue noticeable on the sample. The action of the ions in the plasma that results in a cleaning of the surface may also have a negative effect in damaging sensitive surfaces. This is an area that needs further research and work is ongoing to understand this further.[54]

In conclusion, environmental or high-pressure scanning electron microscopy presents a new approach for accurate metrology and inspection

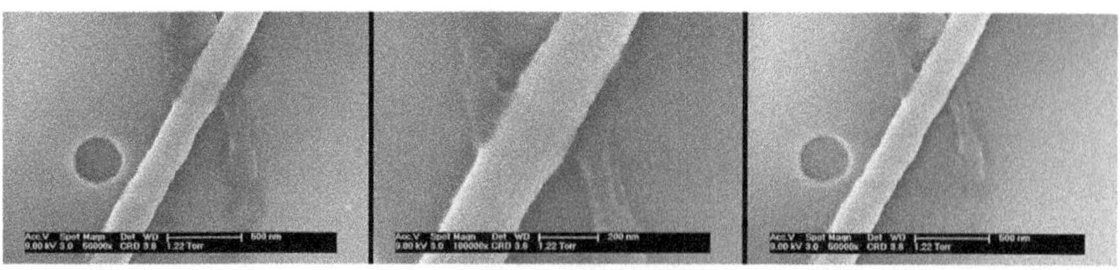

Figure 5.43 Images of a short assessment of electron-beam-induced contamination in an environmental SEM. Starting image on the left taken at 50 k×, irradiation of approximately 3 min at 100 k× (middle) and image taken at 50 k× at the end of the e-beam bombardment. No visible contamination.

of photomask samples. The minimization, if not elimination of sample charging that currently limits the inspection of photomasks in the SEM is a significant step forward. International SEMATECH and NIST are currently investigating the potential this method affords to production photomask metrology.

Metrology of real three-dimensional metrology of mask structures. Ultimately the whole shape of the line needs to be accounted for, including its width, height, sidewall roughness, and bottom edge roughness. The sophistication of models for the design of masks is now at the level that they can deal with real 3D structures, therefore there is a need for the corresponding metrology. In SEMs there is the possibility of taking multiple images at different angles and reconstructing the whole 3D structure of the Cr and phase-shifting features. MBL, with limitations, can also be used to deduce 3D information from top-down view SEM images.

5.4 Atomic Force Microscopy

The atomic force microscope works by scanning a sharp tip over the surface while maintaining a constant small force between the two.[55–56] The AFM may operate in a noncontact or contact mode. In the noncontact mode the cantilever is driven at its resonant high frequency and slight shifts in the resonance occur as a result of the change in the natural frequency of the cantilever due to small changes in the force between the tip and surface. In the contact mode the cantilever actually deflects due to small forces on the tip applied by the sample. A force-sensitive feedback system moves the tip around obstacles presented by the sample topography. The tip motion thus traces an image of the sample's surface, which is recorded and reported by the instrument. By its nature, AFM imaging is generally less composition sensitive than either SEM or optical imaging. In those microscopies the signal depends on scattering or optical properties of the various sample materials. For the AFM in topographic imaging modes, the signal is generated by short-range contact forces. Soft materials may more easily contaminate the tip, causing less stable imaging, but this is unlikely to pose a problem for materials employed in photomasks.

5.4.1 Introduction to atomic force microscopy

The leading AFM manufacturer makes a mask holder for its cross-section three-dimensional instrument,[57] and early on some units were sold to mask manufacturers in Japan. The three-dimensional tool is a successor to the previous two-dimensional cross-section tool. Both instruments have the capability of profiling steep features, including those with vertical or undercut sidewalls. The AFM has not been the traditional tool for mask metrology, partly because optical methods have been good enough and partly because capable AFMs have only recently started becoming available. Robustness of the measurement system may still be a concern. The tip lifetime is a concern and is not predictable, and software for automation reportedly needs work. However, these limitations can probably be overcome. Throughput relative to the other techniques is a more inherent problem. (It is unlikely that tips can ever be scanned as rapidly as electrons or photons used by other techniques.) Measurement times are about on the order of a minute per site. This limits the use of AFM for measurements in wafer production, but throughput may not be as much of a concern in mask applications since far fewer masks than wafers are produced.

In contrast to these disadvantages there are some important advantages with the AFM technique. AFM provides height or depth as well as width information. This is important for some of the new kinds of masks. AFM is a high-resolution imaging technique that can be performed in ambient air and does not charge insulating samples as do techniques requiring charged particles. The accuracy for width measurements on masks is limited by tip shape effects, at about the 5-nm level currently, though this may be improved in the future with better tip characterizers. The technique also has some apparent advantages in that the modeling of the AFM tip-sample interactions and the extraction of the physical sample shape from the resulting images or profiles is viewed as significantly simpler than electron scattering or electromagnetic calculations. Although the physics of the force interactions are quite complicated, at the nanometer scale the problem presents itself in a simpler fashion.

Recently, the AFM provided very useful results as a reference calibration system. In this work, the AFM was successfully employed to measure sub-500-nm-sized chrome-on-quartz features with uncertainties better than reported by the SEM or optical methods. The results from the AFM measurements can then be used to effectively calibrate measurements on a high-throughput optical instrument.

Modeling is required to remove the effect of the tip from raw images. In most cases, the modeling today is rudimentary, consisting of subtracting the tip width from the raw measured value in order to obtain a corrected result. This kind of correction is adequate for determining the gross width values of steep-walled features, but it does not account for the full shape of the tip. For this reason it is unable to account for rounding of corners due to tip roundness. It is also unable to fully account for tip effects on features with intermediate steepness, which make contact with the tip neither on the side nor on the bottom but somewhere in between. Etched polycrystalline silicon samples measured for semiconductor manufacturing tend to have nearly vertical sidewalls with only small areas of rounding near the top or the foot of the line. Chrome-on-glass masks appear to have more significant areas with intermediate slopes. This limitation of the AFM modeling is therefore likely to pose more of an issue on photomasks than it has on the wafer samples that up to now have provided most of the measurement targets for the AFM.

5.4.2 AFM uncertainty evaluation

The uncertainty budget for the AFM data shown in Table 5.4 was developed in collaboration between NIST and ISMT and resulted from work on a reference metrology system based on the Veeco X3D instrument. The combined uncertainty resulting from the table is:

$$\text{The combined standard uncertainty } u_c = [1\ \text{Std}^2 + (5\ \text{nm})^2 + (2.2 \times 10^{-3}\ W)^2]^{1/2}$$

This effort involved a series of mask measurements and comparisons on the same features by various techniques. The uncertainty table (Table 5.4) is meant to summarize and illustrate the AFM measurement capabilities with respect to photomask metrology and its potential use as a reference metrology tool. As can be seen, the source of the largest measurement error is related to the uncertainty in the shape of the tip. This can be improved with better tip characterization.

TABLE 5.4 Explicit Breakdown of the Components Contributing to the Type B Standard Uncertainty of the AFM Photomask Measurements

Type	Source	Standard uncertainty
A	Statistical terms (repeatability, reproducibility, nonuniformity)	1 Std or StdoM
B	Algorithm (edge detection)/measurand definition	*,†
B	Tip width correction (zeroth order)	5 nm†
B	Tip related (higher order—rounding, bending)	*,†
B	Scale calibration (linear term)	$1.0 \times 10^{-3}\ W$‡
B	Nonlinearity	$2.0 \times 10^{-3}\ W$‡
B	Differential Abbé errors	*,†
B	Nonposition dependent motion errors	*,†
B	Cosine errors (in sample plane)	$0.15 \times 10^{-5}\ W$‡
B	Cosine errors (out of sample plane)	$0.15 \times 10^{-3}\ W$‡

*Indicates that contribution has not been thoroughly evaluated or is strongly sample dependent and is usually evaluated for every measurement, but is often negligible.

†Algorithm, measurand-definition, and tip-related components are interrelated and not always easily partitioned.

‡Uncertainty component is proportional to the measured value of width *W*.

ABBREVIATION: Std—standard deviation; StdoM—standard deviation of mean.

5.5 Registration Metrology

The use of two-dimensional measurements is an important element in feature placement metrology. Typically a photomask is qualified and accepted based on a number of specified site measurements. These measurements are usually an assortment of critical dimension measurements and registration measurements. The registration measurements affect within die positioning for a given photolevel and level-to-level alignment from multiple photolithographic process steps. The types of registration errors vary with the type of photomask fabrication tool, but in general there are local errors (within stripe), stripe-to-stripe errors, and lower spatial frequency global errors and distortions throughout the reticle.[58] The main metrology challenge is maintaining a long-range stable measurement environment over 6 in. for nearly an hour. There are both dynamic concerns such as temperature and vibration stability as well as static issues such as metrology stage and interferometry performance.

5.5.1 The basic tools of registration metrology

There are currently no standardized algorithms for the implementation of a full two-dimensional calibration, and several ranges of the metrology tool require calibration. This is a result of the required large measurement field. Unlike a linewidth measurement where the measurement field is on the order of microns, some measurements on the photomask may be nearly 6 in. in length. There has been quite a bit of work over the last 15 years to develop calibration methods appropriate for the field sizes.[59,60]

The challenge of two-dimensional measurements and calibrations can generally be broken into three regimes—(1) the microscale, with measurements typically in the 100-μm length range, useful for calibration of CCD cameras, microscope optical lens systems, fine stage motion systems and the like; (2) die size fields, as encountered on a printed wafer and in stepper lens characterization; and (3) 150 to 200 mm field dimensions as are typical on a photomask, e-beam/optical mask writer, or mask metrology tool, or as encountered in wafer inspection tools doing defect location and review over the entire wafer.

Two-dimensional calibrations generally consist of calibrating the tool scale for both axes as well as orthogonality. However, in addition to this is the detailed mapping of the instrument errors which are effectively a higher-order calibration in two dimensions. The scale calibration is a one-dimensional first-order effect. Self-calibration type methods are very useful in mapping a tool's higher-order errors. Self-calibration does not necessarily require a traceable scale calibration but is intended for general mapping of a metrology tool's distortions, whether they are interferometer mirror errors, stage motion errors, or optical distortions.[61]

Historically, traceable artifacts for calibrating two-dimensional metrology tools were not available in the United States and in fact the calibrations were accomplished at NIST on a one-dimensional metrology system.[62] NIST has measured a number of photomask plates and smaller line scales on our *line scale interferometer* (LSI) for several semiconductor manufacturers and tool vendors. Now, however, fully calibrated two-dimensional standards are available from the NIST Standard Reference Materials office.[63]

Supporting the artifact. How a photomask is supported during a measurement can directly affect the measurement results. This can add a repeatable bias and must be calibrated out as different metrology tools hold the reticles in different ways and then they are held in a stepper in yet another environment. Effects from photomask flexing on measurements at the 100-μm lateral length scale are not a problem. At the full photomask dimension, there is a contribution from sagging and bending directly attributable to the positioning of the mounting points. The problem of mounting and bending of 150-mm photomasks is now well understood and is universally addressable by modeling the photomask flex if one uses the so called Bessel points. Alternatively, using modern static simulation tools one can accurately model more complex fixtures which hold the reticle at the edges and then appropriately compensate for the mounting method in the final data. The end result is that the registration values throughout the reticle can be reported in an as-fixtured format or in a "gravity-free" format.

The artifact support system used in most industrial tools supports the photomask at the outer edges. This is required to protect the inner mask region from scratching or damage since it is typically in the exposure field of the stepper. These tools use a theoretical z algorithm to unbend the data and report the measurements in an effectively zero gravitational field. These bending models are essential since a 150 mm × 150 mm × 6.35 mm photomask held on the extreme edges can have maximum bending in the several micrometers range. This z bending due to gravity can have profound effects on the values for the lateral positions of x-y measurements as large as 40 to 50 nm.[64] The leading algorithms reduce this effect dramatically. This is, of course, an issue as the registration measurements from an x-y metrology tool reported in an unbent fashion must be properly interpreted for use in a stepper or when used for feedback to a mask writing tool.

Previous work shows that these mounting points are somewhat insensitive to their exact locations. A mapping of the bending when the reticle is mounted at the Bessel points shows that the effects of maximum bending on the lateral position values are reduced to less than 9 nm maximum effect over the entire 150-mm photomask. Studies have shown that features measured on an artifact located where the bending effects are at a minimum result in significantly less than 9 nm bending related error.

5.5.2 Two-dimensional calibrations

We now discuss a standard calibration-type sequence for a two-dimensional photomask measurement tool. To accomplish this, a two-dimensional grid artifact has been developed, which can be used on a variety of tools. The calibration artifact is a 150 mm × 150 mm × 6.35 mm quartz blank with chrome features written nominally 100 nm in thickness. An example calibration photomask is shown in Fig. 5.44 and has a series of 27 × 27 unit cells. Each of the "unit cells" consists of three types of features. Two of the features, shown in the lower part of the figure, are each written once per unit cell and are optimized for measurement by photomask metrology tools.

There is a set of fiducial lines 1 mm in length barely visible to the eye without a microscope. These are to assist in alignment of the photomask to the scan axes. The design values for nominal positional accuracy are ±0.1 μm and CD control is also ±0.1 μm. The mask is transmissive except for the opaque printed features. There is an additional set of alignment marks to enable photomask displacements and rotations relative to a lower mask for application in self-calibration algorithms.

The features shown in Fig. 5.44 are 18 μm × 18 μm outer dimension. One is a solid square and the other a frame structure. The frame structure provides edge redundancy and is the current feature being measured in this example. There is also a 13 × 13 microgrid composed of a set of smaller frames measuring 5 μm in outer dimension with 1 μm CD line segments. This array is repeated with the unit cell and is intended for use in calibrating CCD arrays and optical system distortion.

Traceability path. One of the critical elements of this process is defining a step-by-step unbroken traceability path where each step is linked with a complete measurement uncertainty analysis.[9] This discussion is useful in that it covers a number of the essential elements in using and calibrating two-dimensional metrology tools as commonly used in photomask registration metrology.

In this section we outline the sequence of measurements and analysis required to obtain a traceable two-dimensional measurement. The traceability path for this measurement involved the use of the NIST line scale interferometer (LSI) that is a traceable 1D metrology tool. The LSI was used to sample and calibrate a leading edge *x-y* metrology tool through a systematic statistical analysis and calculation of uncertainty.[62] The basic method employed here is based on calibrating a leading edge *two-dimensional metrology tool* (2DMT) with a NIST traceable scale factor in conjunction with verification of the tools performance in repeatability and in tool mapping.

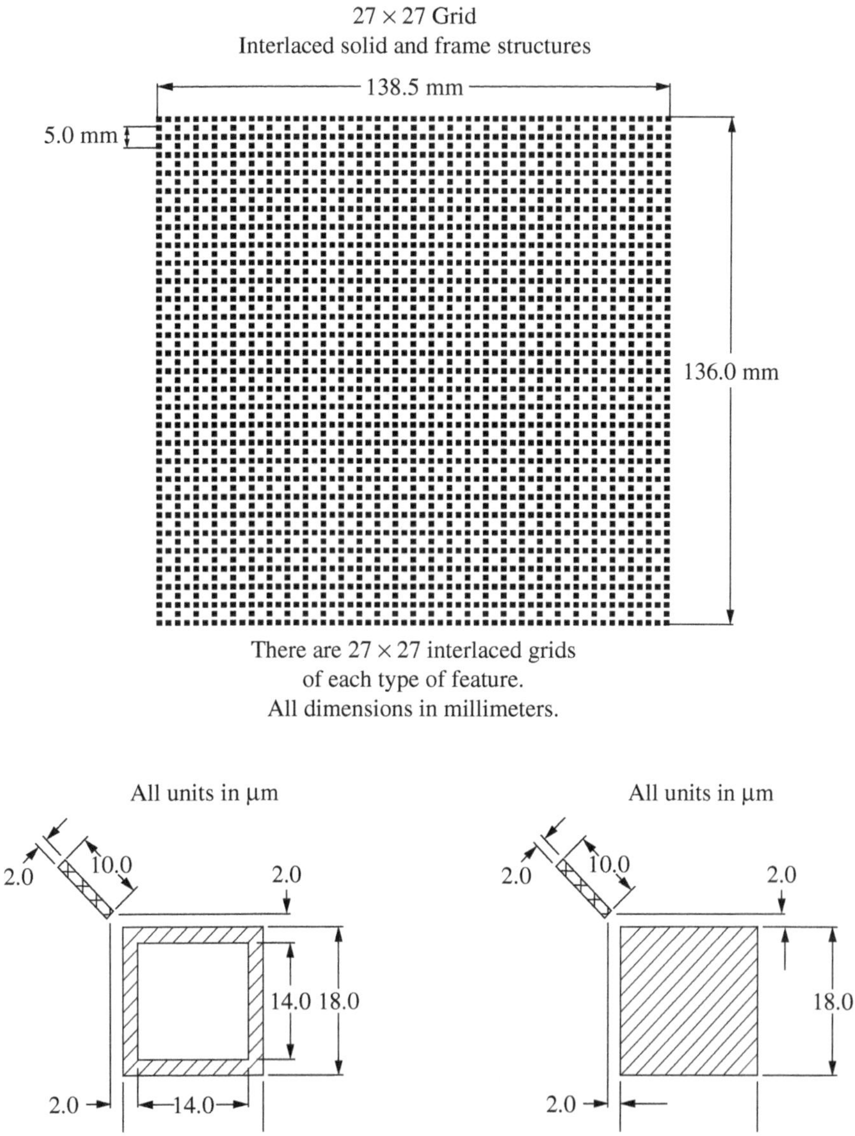

Figure 5.44 2D Grid has 27 × 27 unit cells with a 5000-μm pitch. Each unit cell contains a single frame, a solid box, and a 13 × 13 microarray of frames. Nominal feature dimensions—frame: 2 μm lines, 18 μm edge to edge (square); box: 18 μm edge to edge (square).

The first step is the direct calibration of individual rows by comparing the one-dimensional measurement coordinates of a row to those made on the LSI. This straight forward calibration of uncertainty is the result of random repeatability errors of the industry tool and the total uncertainty of the measurement on the LSI. This step results in the one-dimensional calibration of rows. By comparison of several rows measured on both tools, the one-dimensional scale transformations for the entire field can be determined. In this approach several rows are measured with the LSI, these results are compared to the measured values of the industry tool, and an estimate of uncertainty for the entire field is arrived at. This approach was effective since the manufacturing variations from row to row are small.

Self-calibration methods. The alternative approach is based on self-calibration methods. Here, one data set (first data acquisition) is acquired and the artifact is then physically translated in the tool mounting system by one grid spacing and another data set acquired (second data acquisition). This method was described in more detail as applied to a CCD calibration in Sec. 5.2.7. If one row is measured on the LSI, then that row can effectively calibrate that portion of the industry tool, where it was measured in the first data acquisition. This row, call it machine row 1, measured a second row on the artifact in the second translated data acquisition, call this artifact row 2. Artifact row 2 can now be calibrated and since it was measured in machine row 2 during the first acquisition, can be used to calibrate machine row 2. Machine row 2 can then calibrate artifact row 3 from the second data acquisition. This method can be used to calibrate the entire machine and artifact from a limited number of data acquisitions. The problem here is the accumulation of error becomes too large after about three rows. However, this method can be applied equivalently in both directions resulting in seven calibrated rows if accumulation is acceptable after three calibration translations. This method requires stacking multiple plates so the grid of interest can be translated or rotated on a stable surface with fiducial markings for alignment. This method has been demonstrated on photomask calibrations in detail in Ref. 61.

The preceding discussion yields sets of calibrated rows or columns but does not address orthogonality or straightness of the machine or the artifact. Although one can compare the perpendicular dimensions of rows, which are nominally parallel, this does not address the issue of orthogonality or straightness. Orthogonality errors can be estimated by comparing data from multiple rotations in 90° increments. Analysis of the individual rows at 90° rotation gives both the scale factor errors of the perpendicular axes and the error associated with nonorthogonality of the axes. It should be noted that there are higher-order errors such as fourfold symmetric errors that are not fully captured with this methodology.

The line scale interferometer. We now outline the main elements of a full one-dimensional pitch calibration and examine the different sources of error and measurement uncertainty and then apply this methodology to a full two-dimensional calibration. For one-dimensional high-accuracy measurements NIST uses its line scale interferometer. The LSI consists of a scanning electrooptical line detector, a high-precision one-axis motion system, and a high-accuracy heterodyne interferometer for determining the displacement of the test artifact beneath the line detector. The wavelength of a stabilized helium-neon laser corrected for temperature, humidity, and atmospheric pressure is used as the length standard. The instrument is housed in an environmental chamber in which all the environmental properties are carefully monitored. The complete description and design of the NIST LSI are given in Ref. 62.

The Type A standard uncertainty (often referred to as random uncertainty in this application) for the interferometer readings, u_i is one standard deviation of the mean S_m

$$S_m = \frac{1}{\sqrt{N}}\sqrt{\frac{\Sigma d^2}{N-1}} = u_i \tag{5.10}$$

where d is the deviation of a single measurement from the mean of N measurements. The combined standard uncertainty u_c in each reported length value is

$$u_c = \sqrt{u_i^2 + u_j^2} \tag{5.11}$$

where u_j is the Type B standard uncertainty (often referred to as systematic uncertainty). The expanded uncertainty U in each given length value is $U = ku_c$. The coverage factor $k = 2$ was used, which gives for each reported length value a level of confidence of approximately 95 percent. Table 5.5 gives an explicit breakdown of the components contributing to the Type B standard uncertainty of the LSI measurements reported here. which yields

$$u_j = \sqrt{\left(u_j^2\right)} = \sqrt{(27\times 10^{-16})} = 5.2\times 10^{-8} \tag{5.12}$$

$$u_j = 5.2\times 10^{-8}\times L \tag{5.13}$$

where L is the length measured.

The pitch measurements made on the LSI are from line center to line center using an approximately 7 μm long segment of each side line between the two longitudinal parallel sides of the square frames. Each segment is masked starting from nominally 0.0035 mm below and above the inside corners of the frame structures. The environmental chamber

TABLE 5.5 Explicit Breakdown of the Components Contributing to the Type B Standard Uncertainty of the LSI Measurements

No.	Source	u_j	$(u_j)^2$
1	Laser vacuum wavelength	2×10^{-8}	4×10^{-16}
2	Refractive index equation	3×10^{-8}	9×10^{-16}
3	Atmospheric pressure measurement	2×10^{-8}	4×10^{-16}
4	Relative humidity measurement	1×10^{-8}	1×10^{-16}
5	Determining the amount of CO_2 in laboratory	1×10^{-8}	1×10^{-16}
6	Interferometer alignment	2×10^{-8}	4×10^{-16}
7	Scale temperature measurement	2×10^{-8}	4×10^{-16}
	Sum of $(u_j)^2$		27×10^{-16}

and the artifact temperature were held within ±0.005°C of 20°C during the calibration process. The length is reported at a temperature of 20°C. A coefficient of linear thermal expansion of 0.5×10^{-6}/°C was used in normalizing the lengths to 20°C.

5.5.3 Measurements on a standard production tool

The next step in this sequence is the measurement of the photomask in multiple positions (rotation and/or translation) on a 2DMT. This yields multiple sets of two-dimensional coordinate mappings for the photomask. Comparing data sets from similar orientations following artifact removal and remounting gives a measure of repeatability. Since the leading-edge 2DMTs are well error mapped, back rotating or translating the data sets from different orientations and comparing results yield a measure of the tool's systematic errors. Although this mapping is initially calibrated by an unknown scale factor, it can be rescaled following traceable measurements at NIST.

The same prototype artifact that was measured on the NIST LSI was then measured on an x-y metrology tool. This particular tool is an industry owned metrology tool in a production metrology environment. This is a leading-edge tool. The data shown here do demonstrate the current status of industry tool calibration and the significant progress on error mapping, orthogonality, and overall scale which has been achieved by the industry. The specific tools and companies which performed the individual measurements are not identified.

The 3σ repeatability for this 2DMT is reported by the manufacturer to be approximately 8 nm. This error source is also sampled when we calculate the error map residuals although it is difficult to separate the repeatability from the residual error map uncertainties.

We now discuss the uncertainty components, starting with the scale error. The major error not sampled and corrected in the 2DMT error mapping procedure is scale. For the photomask plates we have compared the 2D measurements with two orthogonal lines on the plate measured with the NIST line scale interferometer. Typical comparisons of the two measurement orientations are shown in Fig. 5.45.

In general the two scales are indistinguishable given the uncertainties of the NIST LSI and the 2DMT. We combine the two axes and apply one scale factor correction (Fig. 5.46) to both axes of the 2DMT measurements.

The uncertainty in this scale factor is calculated by combining the NIST LSI uncertainties and the 2DMT repeatability with the proper weighting.[65] This length dependent uncertainty is calculated to be:

$$u(\text{scale}) = 14.5\text{e}^{-9} \times L\ (\text{nm})$$

Error map residuals. The reproducibility and error map residuals are sampled by measuring the plate in two orientations. The differences in these two measurements contain variability due to the short-term repeatability of the machine and sensor system, day-to-day variations in the environment, and the residual error not compensated in the error map. The data here were analyzed by comparing the distances between each two grid points in the two orientations. Since there are 196 points on each grid, there are 196 × 195/2 or 19,110 different distances in the analysis. The graphs show a scatter plot of all the differences for one sample in Fig. 5.47 and the standard deviations for each separate set of nominal distance measurements for the combination of plates (Fig. 5.48).

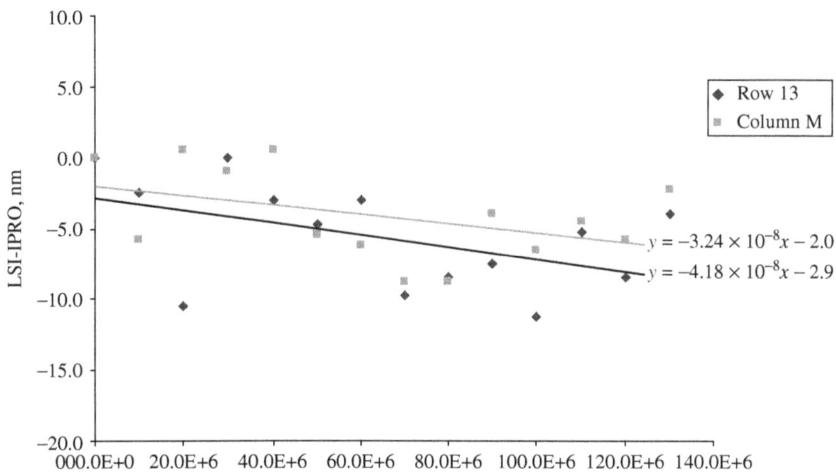

Figure 5.45 Comparison of scale factors calculated from measurements in two orthogonal orientations. The 2DMT results were measured using its two axes while the NIST LSI results were measured by rotating the photomask.

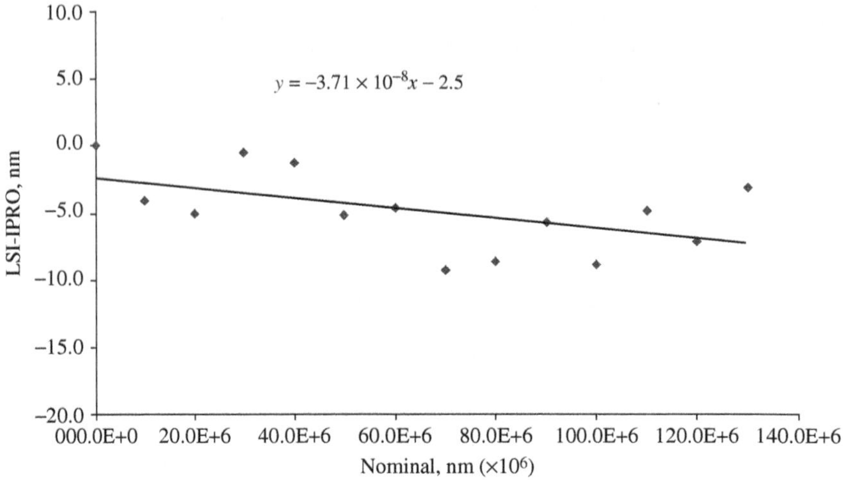

Figure 5.46 Single scale factor applied to 2DMT data from combined orthogonal orientations.

These graphs show no length dependences, and a worst case value as the standard uncertainty due to residual mapping errors is:

$$u(\text{geometry}) = 5 \text{ nm}$$

Thermal expansion. When the grid on the line scale interferometer is measured the effect of thermal expansion is corrected. Since the line scale interferometer thermometer system has an uncertainty of less

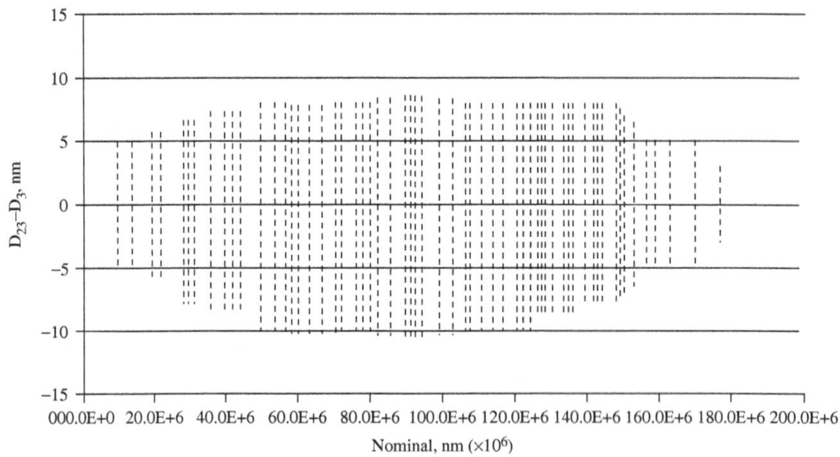

Figure 5.47 Scatter plot showing the differences between distances measured at two orthogonal orientations. For example, each nearest neighbor distance is calculated for each orientation and the difference taken.

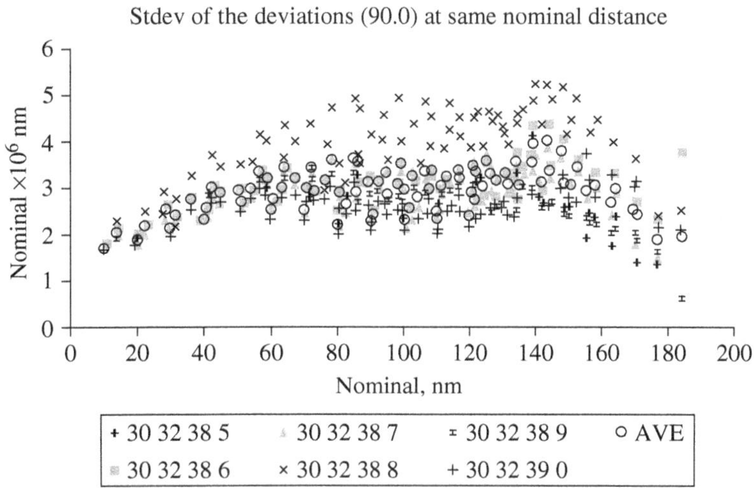

Figure 5.48 The standard deviation of the difference between distances at the two orthogonal orientations is calculated for each nominal distance. This gives an estimate of the reproducibility and error map residuals.

than 1 mK and the coefficient of thermal expansion of fused silica is 0.5 ppm/°C, the uncertainty from the plate temperature is negligible.

Elastic deformation. Since the glass plate bends as described earlier, any bending will change the distance between the grid marks. The plate is supported by three points at the edge of the plate in the 2D machine. The data are then corrected by the software to give the positions of the points for the undeformed plate. For the line scale measurements, we supported the plate at three points that produced negligible bending along the measurement line. Any error in the correction in the 2D data would be sampled adequately in the scale comparison, and therefore does not have a separate estimate of the uncertainty of this effect.

Line geometry effects. The lines are, of course, not perfect. Measurements made using the inside edges of the lines and the outside edges of the lines showed systematic differences in the data. In fact the systematic changes in the grid mark positions caused by changing the edges of the lines were larger in this example than those seen when the same edges were used but the plate rotated 90°. Experiments were performed where the center of the target (frame) was determined using the outside edges of the frame, and then remeasured using the inside edges of the frame. This shows the uncertainty in the frame position caused by variations in the frame linewidths.

When the same edges were used, and the data compared it was found that the grid positions were very repeatable, with a standard deviation of slightly above 2 nm. When the positions found from the outside edges and inside edges were compared, it was found that the standard deviation rose to about 5.5 nm. This difference is caused by the variation in linewidth of the frames. The difference between these standard deviations, about 6 nm, is taken as the standard uncertainty in point positions from linewidth variations.

Plate-to-plate variation. As a part of this exercise one can evaluate the plate-to-plate writing errors and plate-to-plate manufacturing repeatability. These data are summarized in Fig. 5.49. These data give a snap shot summary of a standard state-of-the-art fabrication tool.

Combined uncertainty. The uncertainty components can be combined (RSS) from Fig. 5.49 to obtain the standard uncertainty for each nominal distance on the artifact. These are plotted in Fig. 5.50. Also shown are two approximations to the combined uncertainty—a worst case linear relationship and a more accurate third-order fit.

$$U(\text{combined linear}) = 3.9 \times 10^{-8} \times L\ (\text{nm}) + 8.6\ \text{nm}$$

$$U(\text{combined third order}) = -6 \times 10^{-25} \times L^3\ (\text{nm}^3) + 3 \times 10^{-16} \times L^2\ (\text{nm}^2) + 9 \times 10^{-11} \times L\ (\text{nm}) + 8.6\ \text{nm}$$

The expanded uncertainty ($k = 2$) for the linear relationship is

$$\text{Expanded uncertainty (95 percent confidence level)} = 7.8 \times 10^{-8} \times L\ (\text{nm}) + 17.2\ \text{nm}$$

Data analysis on 150-mm photomask grids. The LSI data show that the uncertainties for the largest distance measurements are on the order of 10 nm and are dominated by the systematic error components. The

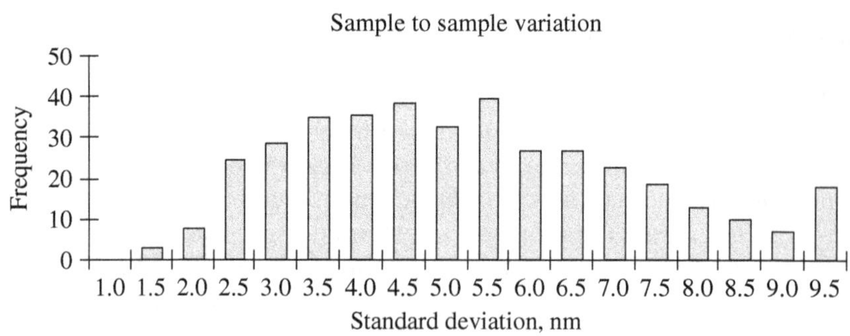

Figure 5.49 The variation in the positions of the grid marks over all measured grid plates from a single run of plates is shown.

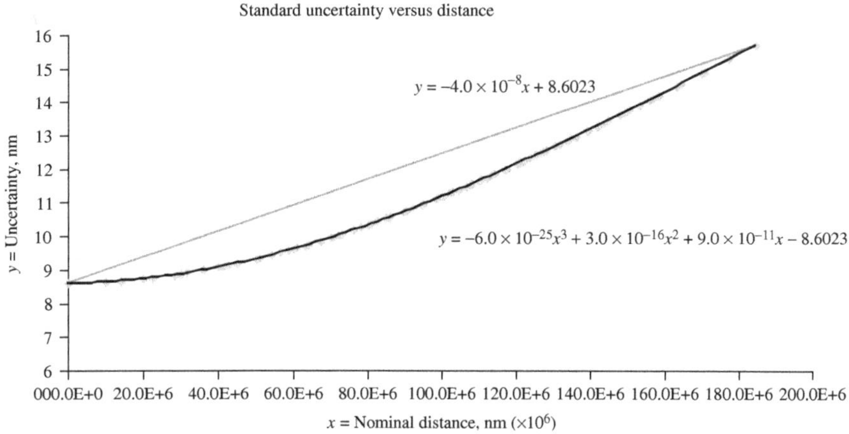

Figure 5.50 Combining the results from Fig. 5.49, the standard uncertainty versus distance as well as two estimates of that uncertainty are shown. A linear, worst case estimate and a more accurate third-order fit.

repeatability measurements for the best 2DMT at the time these measurements were made are at worst in the 10-nm range and the scale errors vary depending on the tool. In this process of developing a calibrated artifact for the calibration of an industry tool, the two primary uncertainty components at this stage are the combined standard uncertainty of the LSI and the random statistical-based errors of the 2DMT. Another error component which needs to be added to this uncertainty estimation is the squareness error, which analysis shows to be of second order. The mask bending error can be minimized by using the Bessel point mounting method and modeled. A detailed analysis shows that these errors are 9 nm maximum before elimination by modeling and can be reduced to an estimated 3 nm.

This two-dimensional calibration sequence just described illustrates most of the key elements in the measurement and calibration of a two-dimensional registration metrology tool. The stage calibration and various environmental contributors to the uncertainty were described and the importance and methods for compensating mask bending were also addressed. Although the references cited go into appropriate depth on the individual components, the essential elements for the calibration sequence and examples applications of the techniques were given. These results also demonstrated the current capabilities of the commercially available tools.

Acknowledgments

The authors would like to thank many individuals who contributed data, results, ideas, and other support to this work. We are grateful to Marc Pedulla and James Potzick for many discussions and contributions to the transmission mode optical section as well as assistance with the registration metrology work. Ravikiran Attota and Michael Stocker provided several contributions throughout the optical section and Michael T. Postek provided support to the SEM section. The authors would also like to acknowledge Ronald Dixson and John Villarubia for their contributions to the AFM and SEM work as well as John Kramar and Robert Larrabbee for their excellent review comments and input. Egon Marx was the lead researcher for the integral equation solver and Mark Davidson had been very helpful with his support and development of the Metrologia model. The authors would like to express their appreciation to the office of Microelectronics Programs (OMP) of the NIST and International SEMATECH for partially funding the work that led to many of the findings presented here.

References

1. "International vocabulary of basic and general terms in metrology," 2nd ed., ISBN 92-67-01075-1 U.S., 1993.
2. Nyyssonen, D., "Spatial coherence: the key to accurate optical etrology," *SPIE: Applications of Optical Coherence* 194, pp. 34–44, 1979.
3. Silver, R. M., R. Attota, M. Stocker, J. Jun, E. Marx, R. Larrabee, B. Russo, and M. Davidson, "Comparison of measured optical image profiles of silicon lines with two different theoretical models," *Metrology, Inspection and Process Control for Microlithography XVI, Proc. SPIE* 4689, p. 409, 2002.
4. Hecht E., and A. Zajac, *Optics*, Addison-Wesley, Reading, Mass., p. 41, 1975.
5. Potzick J., "The neolithography consortium," *Proceedings of SPIE 25th International Symposium on Microlithography* 3998-54, 2000.
6. Habler-Grohne, W., and H. Bosse, "An electron optical metrology system for pattern placement measurements," *Meas. Sci. Tech.* 9, 1998.
7. Lowney, J. R., A. E. Vladár, and M. T. Postek, "High-accuracy critical-dimension metrology using a scanning electron microscope," *Proc. SPIE* 2725, pp. 515–526, 1996.
8. "Guidelines for evaluating and expressing the uncertainty of NIST measurement result," NIST Technical Note 1297, 1994.
9. Taylor, B. N., ed., "The international system of units (SI)," Natl. Inst. Stand. Technol. Spec. Publ. 811, U.S. Government Printing Office, Washington, DC, April 1995.
10. "American Nat'l Stand. for Calibration — U.S. Guide to the Expression of Uncertainty in Measurement," ANSI/NCSL Z540-2-1997, published by NCSL International, ISBN 1-58464-005-7.
11. Devore, J. L., *Probability and Statistics for Engineering and the Sciences*, Duxbury, Belmont, Calif., p. 322, 1995.
12. Banke, B., C. Archie, M. Sendelbach, J. Robert, J. Slinkman, P. Kaszuba, R. Kontra, M. DeVries, and E. Solecky, "Reducing measurement uncertainty drives the use of multiple technologies for supporting metrology," *SPIE* 5373, p. 133, 2004.
13. Davidson, M., "Analytic waveguide solutions and the coherence probe microscope," *Proceedings Microcircuit Engineering* 90, Leuven, Belgium, September 1990.
14. Marx, E., "Integral equation for scattering by a dielectric," *IEEE Trans. Antennas Propagat.* 32, pp. 166–172, 1984.
15. Standard Reference Materials Office of the National Institute of Standards and Technology, Gaithersburg, Md.
16. Evennett P., *Kohler Illumination Centenary*, Royal Microscopial Society, Oxford, UK, 1993.

17. Potzick, J., J. M. Pedulla, and M. Stocker, "Updated NIST photomask linewidth standard," *Proceedings of SPIE 28th International Symposium on Microlithography*. 5038-34, 2003.
18. Silver, R. M., J. Potzick, and Y. C. Hu, "Metrology with the ultraviolet scanning transmission microscope," *Proc. SPIE* 2439, p. 437, 1995.
19. Silver, R. M., J. Potzick, F. Scire, and B. Larrabee, "High accuracy overlay measurements," *SPIE* 2725, p. 320, 1996.
20. Silver, R. M., J. Jun, S. Fox, and E. Kornegay, "Overlay metrology: recent advances and future solutions," *Future Fab Int.*11, 2001.
21. Holst, G. C., CCD arrays, cameras, and displays," 2nd ed., *SPIE*, Bellingham, WA, ISBN 0-81294-2853-1.
22. Silver, R. M., J. Potzick, E, Kornegay, C. Evans, R. Larrabee, and M. McGlauflin, "A method to characterize tool misalignments and distortions," *SPIE* 3050, p. 143, 1997.
23. Silver, R. M., J. Jun, E. Kornegay, and R. Morton, "Comparison of edge detection methods using a standard prototype overlay calibration artifact," *Proc. SPIE* 4344, p. 515, 2001.
24. Nyyssonen, D., and B. Monteverde, "Linewidth edge detection algorithm for coherent image profiles," *SPIE* 1087, p. 146, 1989.
25. Zaslavsky, A., "Overlay measurements and edge detection methods," *SPIE* 3050, p. 418, 1997.
26. Fox, S., E. Kornegay, and R. Silver, "Characterization of imaging optics and CCD cameras for metrology applications," *ULSI Conf. Proc.*, NIST, 2000.
27. Fox, S., R. M. Silver, E. Kornegay et al., "Focus algorithms and edge detection methods," *Proc. SPIE* 3677, March 1999.
28. Lohmann, A., D. Mendlovic, and Z. Zalevsky, "Digital method for measuring the focus error," *Appl. Optics* 38(4), pp. 638–646, 1999.
29. Krotkov, E., "Focusing," *Int. J. of Comput. Vision* 1 p. 223, 1987.
30. Pedulla, J. M., J. Potzick, and R. Silver, "Improving the uncertainty of photomask linewidth measurements," *Proceedings of SPIE 29th International Symposium on Microlithography* 5375-31, 2004.
31. Davidson, M., B. H. Kleemann, and J. Bischoff, "A comparison between rigorous light scattering methods," in *Optical Microlithography X, SPIE* 305, pp. 606–619, 1997.
32. Marx, E., "Electromagnetic scattering by a thick strip on a half-space," Nat. Bur. Stand. (U.S.) Tech. Note 1236, 1987.
33. Wojcik, G. L., J. Mould, Jr., E. Marx, and M. P. Davidson, "Standard reference models for optical metrology simulation," *Proc. of SPIE*. 1673, pp. 70–82, 1992.
34. Nyyssonen, D., "Theory of optical edge detection and imaging of thick layers," *J. Opt. Soc. Am.* 72, pp. 1425–1436, October 1982.
35. Seligson, J., B. Golovanevesky, J. Popawski, M. Adel, and R. Silver, "Overlay metrology simulations—analytical and experimental validations," *Proc. SPIE* 5038, p. 61, 2003.
36. Born, M., and E. Wolf, *Principles of Optics*, 7th ed., Cambridge University Press, Cambridge, UK, 1999.
37. Vladár, A. E., R. Silver, J. Potzick, J. S. Villarrubia, and M. T. Postek, "Development of photomask artifacts and metrology methodology for the 65 nm device generation with extensibility beyond," ISMT Final Report, 2004.
38. Jenkins, F. A. and White, H. E., *Fundamentals of Optics*, McGraw Hill, N. Y., p. 259, 1957.
39. Attota, R., R. M. Silver, M. Bishop, E. Marx, J. Jun, M. Stocker, M. Davidson, and R. Larrabee, "A new method to enhance overlay tool performance," *Metrology, Inspection and Process Control for Microlithography XVIII, Proc. SPIE*5375, p. 395, 2004.
40. Silver, R. M., R. Attota, M. Stocker, M. Bishop, J. Jun, E. Marx, M. Davidson, and R. Larrabee, "High resolution optical overlay metrology," *Metrology, Inspection and Process Control for Microlithography XVIII, Proc. SPIE* 5375, p. 78, 2004.
41. Silver, R. M., T. Doiron, W. Penzes, E. Kornegay, S. Fox, M. Takac, S. Rathjen, and D. Owens, "Two-dimensional calibration artifact and measurement methodology," *Proc. SPIE* 3677, March 1999.
42. Evans, C., R. Hocken, and T. Estler, "Self calibration, reversal, redundancy, error separation, and absolute testing," *Ann. CIRP* 45(2), 1996.
43. Postek, M. T., A. E., Vladár, T. Rice, and R. Knowles, "Potentials for high pressure/environmental SEM microscopy for photomask dimensional metrology," *SPIE* 5038, p. 315–329, 2003.

44. Applied Materials Inc. RETicleSEM, http://www.appliedmaterials.com/products/a_reticlesem.html
45. KLA-Tencor Corp. 8250 R Reticle and Wafer SEM, http://www.kla-tencor.com/j/servlet/Product?prodID=8
46. Villarrubia, J. S., A. E. Vladar, B. D. Bunday, M. Bishop, "Dimensional metrology of resist lines using a SEM model-based library approach," *Proc. SPIE* 5375, p. 199–209, 2004.
47. Villarrubia, J. S., A. E. Vladar, J. R. Lowney, M. T. Postek, Jr.. "Edge determination for polycrystalline silicon lines on gate oxide," *Proc. SPIE* 4344, p. 147–156, 2001.
48. Bunday, B. D., M. Bishop, J. S. Villarrubia, A. E. Vladár, "CD-SEM measurement line edge roughness test patterns for 193 nm lithography," *Proc. SPIE* 5041, p. 127–141, 2003.
49. Lowney, J. R., "Application of Monte Carlo simulations to critical dimension metrology in a scanning electron microscope," *Scanning Microscopy* 10, pp. 667–678, 1996.
50. Dauidson, M., "Metrologia: metrology and lithography simulation software for windows," Spectel, Mountain View, Calif.,.http://www.spectelresearch.com
51. Davidson, M. P., and A. E. Vladár, "Inverse scattering approach to SEM linewidth measurements," *Proc. SPIE* 3677, p. 640–649, 1999.
52. Postek, M. T., A. E. Vladár, M. H. Bennett, "Photomask dimensional metrology in the SEM part I: has anything really changed?" *JM3.* 2003.
53. Postek, M. T., A. E. Vladár, M. H. Bennett, T. Rice, R. Knowles "Photomask dimensional metrology in the SEM part II: high pressure/environmental SEM." *JM3.*, pp. 212–233, 3 (2) 2004.
54. Postek, M. T., and A. E. Vladár, "Application of high pressure/environmental scanning electron microscopy to photomask dimensional metrology," *Characterization and Metrology for ULSI Technology*, D. Seiler et al., eds. AIP Press, N. Y., 2003.
55. Dixson, R., R. Koning, V. Tsai, J. Fu, and T. Vorburger, "Dimensional measurements with the NIST calibrated atomic force microscope." 3677, p.20, 1999.
56. Villarrubia, J. S., "Algorithms for scanning probe microscope image simulation, surface reconstruction, and tip estimation, " *J. Res. Natl. Inst. Stand. Technol.* 102, p. 425, 1997.
57. Veeco Dimension X3D Automated Atomic Force Microscope, http://www.veeco.com/html/product_bymarket_proddetail.asp?MarketID=&ProductID=199.
58. Ye, J., "Errors in high precision mask making metrology," Stanford University Ph.D. Thesis, 1996.
59. Fox, S., "Focus, edge detection and CCD camera characterization for development of an optical calibration standard," University of Maryland, Ph.D. Thesis 2000.
60. Hocken, R. J., and B. R. Borchardt, "On characterizing measuring machine geometry," NBSIR 79-1752, Natl. Bur. Stand. (U.S.), 1979.
61. Takac, M., J. Ye, M. Raugh, R. Pease, C. Berglund, and G. Owen, "Self-calibration in two-dimensions," *Proc. SPIE Microlithography*, 1997.
62. Beers, J. S., Penzes, W. B., "The NIST length scale interferometer," *J. Res. Natl. Inst. Stand. Technol.* 104, p. 225, 1999.
63. Potzick, J., "Re-evaluation of the accuracy of NIST photomask linewidth standards," *Proceedings of the SPIE Symposium on Microlithography* 2439-20, Santa Clara, CA, pp. 232–242, 1995.
64. Emerson, W., "Determination of planeness and bending of optical flats," *J. Res. NBS*, 49, p. 247, 1952.
65. Silver, R., and M. Pedulla, "Two-dimensional grid photomask standard," Certificate of Analysis, Standard Reference Material, 2001.

Disclaimer

Certain commercial equipment is identified in this chapter to adequately describe the experimental procedure. Such identification does not imply recommendation or endorsement by the National Institute of Standards and Technology, nor does it imply that the equipment identified is necessarily the best available for the purpose.

Chapter

6

Process Back End and Defectivity Control

Edita Tejnil, Ted Liang, Florence O. Eschbach, and Alan R. Stivers

6.1 Introduction

The finishing, or back end, section of the photomask fabrication process primarily deals with defects introduced earlier in the process. Unlike silicon wafer processing, where a certain level of defects is acceptable, photomask processing must leave no defect that will significantly affect the resist patterning on the wafer. The exposure field of modern lithographic exposure tools is only a fraction of the wafer size. The photomask pattern must be stepped and repeated as many as 100 times across the wafer to expose the entire wafer. Any flaw in the photomask would be repeated with each exposure. In the extreme case of one integrated circuit die per photomask field, the photomask flaw would be repeated in every integrated circuit, potentially resulting in zero wafer yield. It is therefore advantageous to attempt to repair or clean away defects even though the addition of those extra steps add cost to the process.

Photomask feature sizes decrease to keep up with the similar trend in silicon integrated circuit development. Smaller feature sizes have several consequences. The integrated circuits have a higher transistor count, so the required photomasks become more complex and more costly. Smaller feature sizes make photomasks sensitive to smaller defects. The likelihood of the occurrence of a defect rises at the same time photomask costs rise. The finishing process modules are still key to the overall photomask fabrication process even as great effort is taken to reduce the occurrence of defects.

The finishing section includes four process modules—inspection, repair, clean, and pelliclization. Each of these modules is discussed in detail in this chapter. This introduction will cover the process flows, mask types, and characteristic defect types that the finishing modules address.

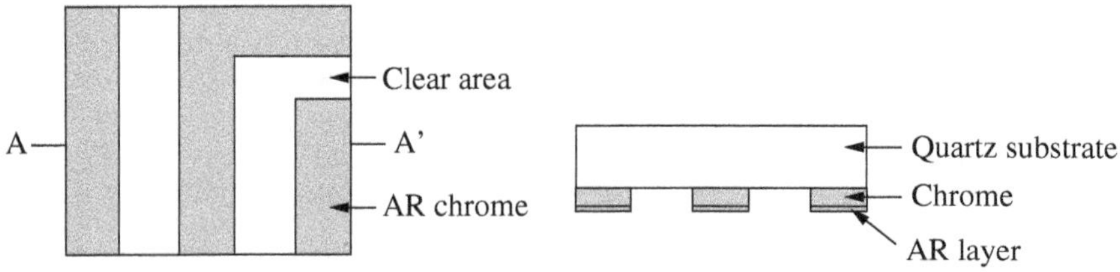

Figure 6.1 Schematic of binary-type photomask. Left: plan view; Right: section A-A'.

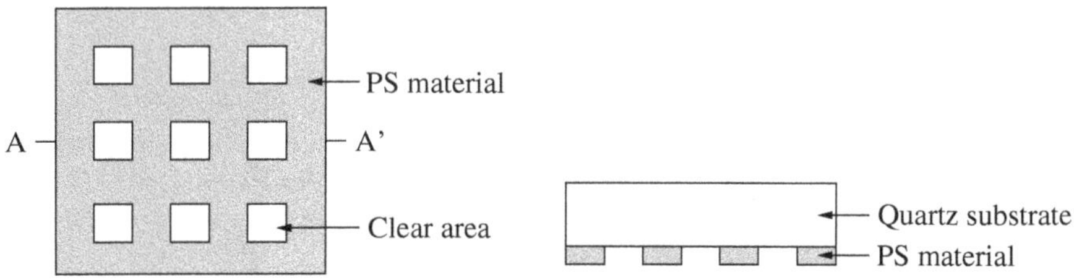

Figure 6.2 Schematic of a halftone PSM-type photomask. Left: plan view; Right: section A-A'. PS Material is the phase-shift material.

6.1.1 Photomask types and their characteristic defects

Photomasks can be categorized by the material and layer that is being patterned. Patterned materials are chrome, a *phase-shift* (PS) material, and quartz. The basic mask types are:

- *Binary:* Patterned chrome layer. See Fig. 6.1.
- *Halftone phase-shift mask (PSM):* Patterned PS material. The halftone material has 6 to 10 percent transmission and shifts the illumination phase by 180° to increase contrast at feature edges. See Fig. 6.2.
- *Tritone PSM:* Patterned PS material and chrome. Similar to halftone PSM but with an additional chrome pattern to compensate for excessive light in dark areas. See Fig. 6.3.
- *Alternating PSM (APSM):* Patterned chrome and quartz. Quartz patterning creates shifters that are etched into the surface to decrease the optical path length of light transmitted through the shifters. See Fig. 6.4.

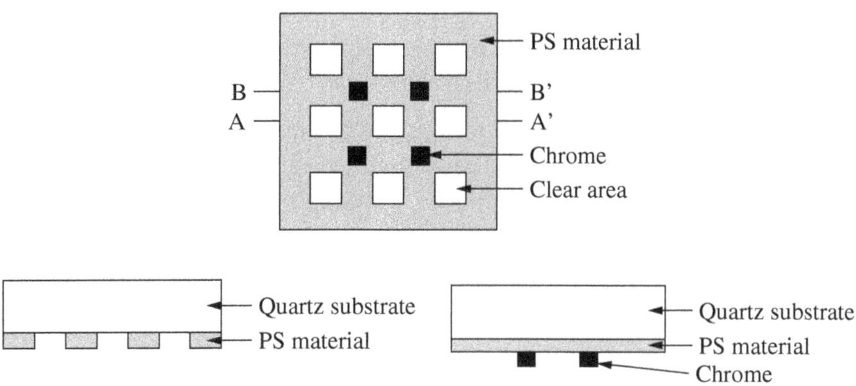

Figure 6.3 Schematic of a tritone PSM-type photomask. Top: plan view; Left: section A-A'; Right: section B-B'.

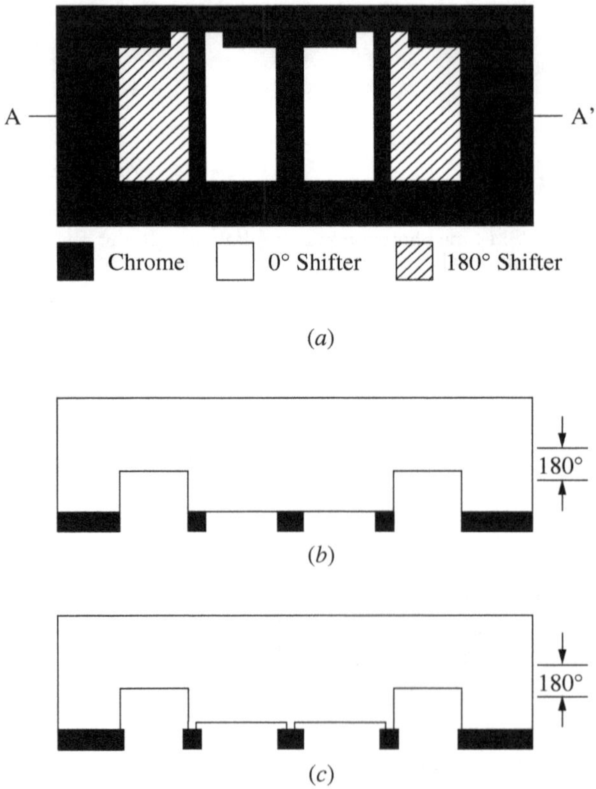

Figure 6.4 Schematic of an alternating-PSM-type photomask: (*a*) plan view; (*b*) section A-A' assuming a dry only quartz etch; and (*c*) section A-A' assuming a dry, then wet quartz etch. The dry-wet process has lithographic advantages but undercuts the chrome making it more susceptible to lift-off during mask cleaning.

- *Chromeless phase lithography (CPL) mask:* Patterned chrome and quartz. Quartz patterning creates structures that diffract the illumination out of the exposure tool entry pupil. See Fig. 6.5.

Each type of photomask is susceptible to defects in the patterned layer. These are caused by defects either in the incoming blank or the patterning process itself. We confine our discussion to localized defects caused and do not consider general process excursions such as feature

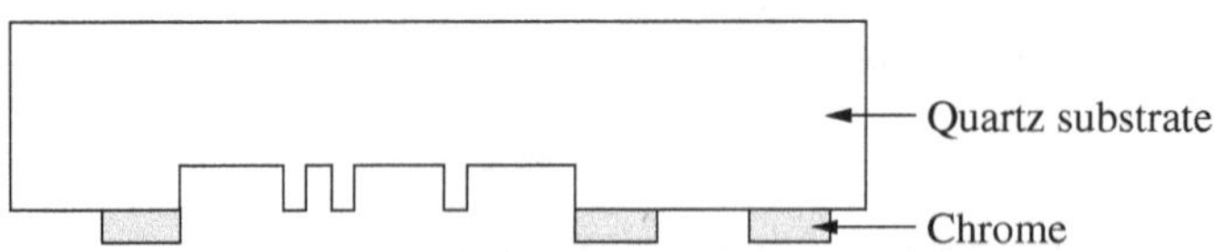

Figure 6.5 Cross-section schematic of a chromeless phase lithography photomask.

dimension errors and incomplete pattern etch. For the purposes of this chapter we define the pattern defects as:

- *Extrusions:* Extra pattern material overlapping into an area that should be clear.
- *Intrusions:* Missing pattern material creating a clear area that should be dark.
- *Pindots:* Extra pattern material in a clear area that does not connect to a dark feature.
- *Pinholes:* Missing pattern material in a dark field not connected to a clear feature.
- *Thin chrome:* These are areas of partial transmission in an otherwise opaque feature.

These defect types are also defined in Fig. 6.6. Pinholes and pindots are usually considered as different defect types than intrusions and extrusions because attachment to a feature has large consequences on defect printability, ease of inspection, and ease of repair.

Alternating PSMs have additional types of defects relating to the quartz patterning. The most common are etched areas in the quartz in the 0° shifter and nonetched areas in the 180° shifters (Fig. 6.7). These can be caused due to defects in the resist layer, particles introduced in the quartz process, or stress cracking in the quartz substrate near shifter corners. The CPL photomask process flow is similar to that of alternating PSM. The defect types are similar.

The term "soft defect" refers to cleanable defects and includes particles, contamination, and crystals/haze. These are common to all mask types. Particles are foreign objects that rest on the mask surface. These are usually 1 μm in size or larger. They are often removable by cleaning. Contamination is a generic term for a material or surface condition that absorbs some of the light in the clear areas. Photomasks used for

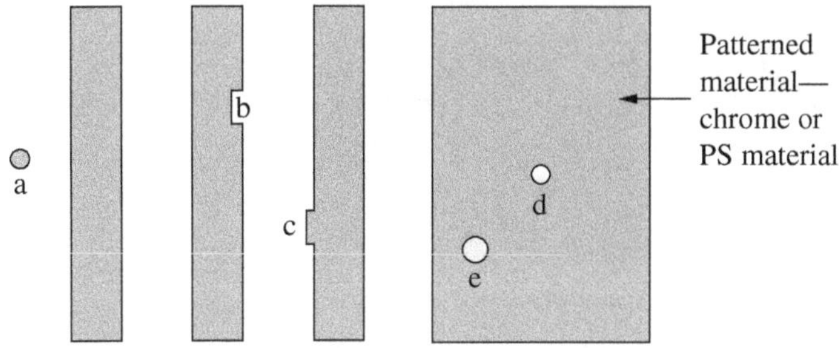

Figure 6.6 Pattern defect definitions as used in this chapter: (*a*) pindot; (*b*) intrusion; (*c*) extrusion; (*d*) pinhole; and (*e*) thin chrome. Dark areas are patternable material: chrome or phase-shift material.

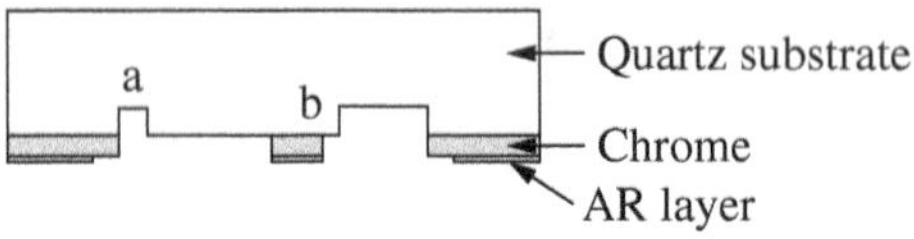

Figure 6.7 Schematic representation of typical quartz defects: (*a*) quartz pit or divot in 0° shifter and (*b*) quartz bump in 180° shifter.

deep ultraviolet (DUV) exposure also can have photoinduced crystals and haze on their front and back surfaces.[1] An example of these defects on the backside of the photomask is shown in Fig. 6.8. These defects are formed from residual ammonium and sulfate ions left over from the cleaning process or from the environment. Their formation is induced by the DUV photon energy. Haze may also be formed from organic species from the environment, pellicles, pellicle adhesives, and plastic shipping and storage boxes.

6.1.2 Blank defects

The chrome layer on photomask blanks is thin, 70 nm or less, and therefore susceptible to pinholes. Chrome pinholes will become pinholes in the pattern if they coincide with opaque areas of the pattern. Contamination in the resist layer can interfere with resist exposure or block the subsequent chrome etch.[3]

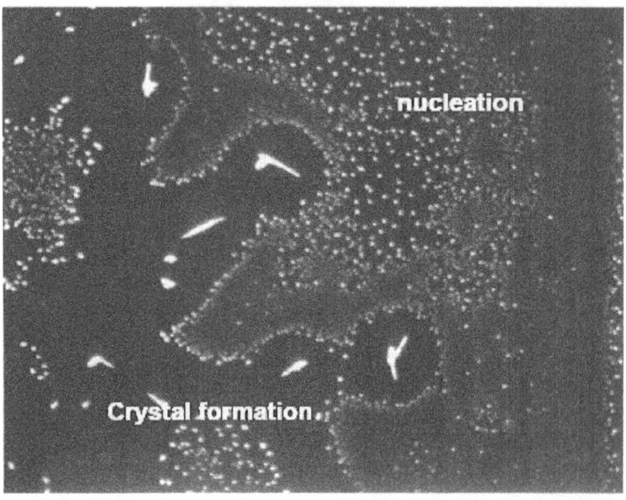

Figure 6.8 Photomicrograph of crystals observed on rear of photomask after several kilojoules of 248-nm laser dose. (*Source:* Eschbach, F., D. Tanzil, M. Kovalchick, U. Dietze , M. Liu, and F. Xu, "Improving photomask surface properties through a combination of dry and wet cleaning steps," *Proc. SPIE 5446*, pp. 209–217, 2004.)

6.1.3 Back-end process flow

The finishing section tasks detect and repair printable pattern defects, detect and clean contamination and particles, mount the protective pellicle, and complete a final inspection.

The process flow is described schematically in Fig. 6.9. The leftmost column in the figure contains the hard defect inspection, disposition, and

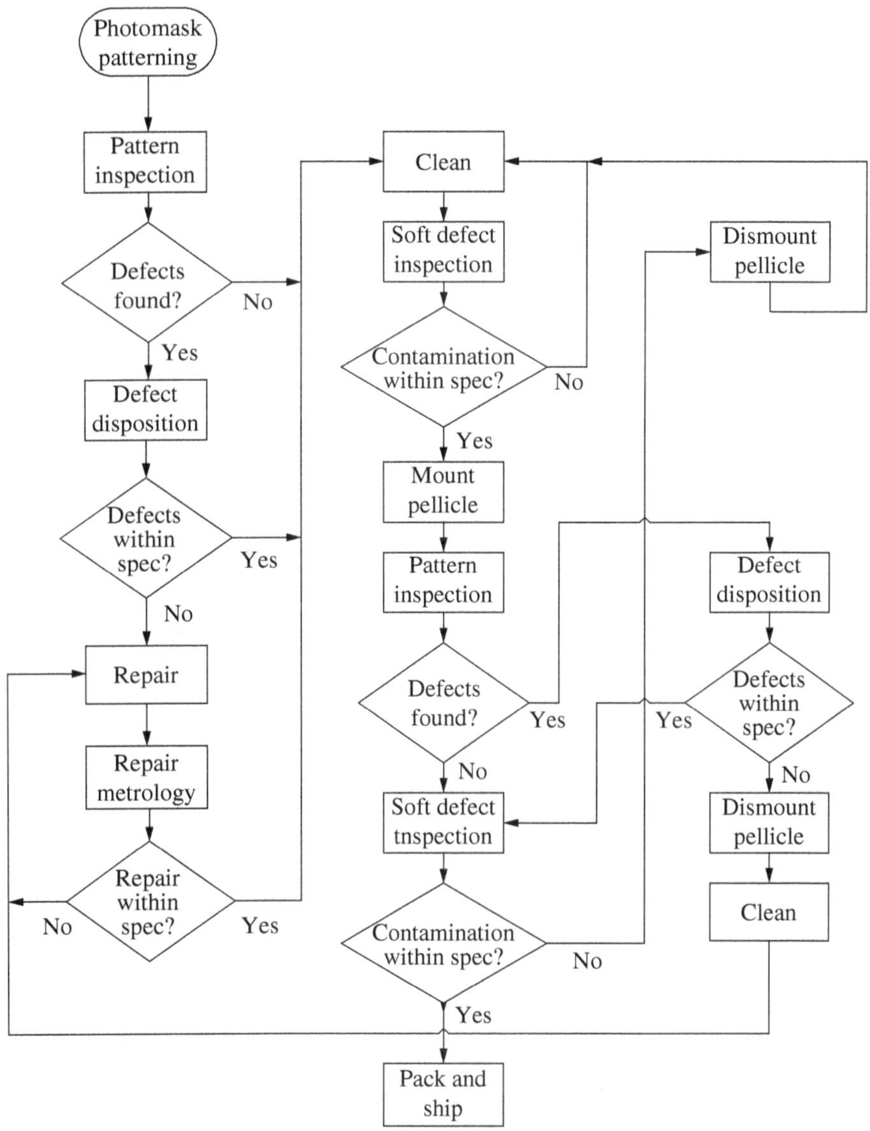

Figure 6.9 Schematic of photomask finishing process flow. Rework loops are included to capture all uses of the inspection and metrology information generated during this section of the photomask fabrication flow.

repair sections. The center column contains the clean, soft defect inspection, and pellicle mount steps. The center column contains pass/fail decision points after inspection steps. The rightmost column defines the processing if there are failures at these decision points. This is given for completeness and flow closure. These can be considered as rework loops and not the main flow.

The flow is composed of the following module elements:

- Pattern or hard defect inspection, including defect disposition

  Purpose: Detect and locate pattern defects and determine which defects need to be repaired.
- Defect repair and repair metrology

  Purpose: Micromachine photomask to correct pattern and measure postrepair pattern against a specification.
- Soft defect inspection

  Purpose: Detect any particles or contamination. This step is generally done at a lower sensitivity than hard defect inspection and can therefore be done quickly.
- Clean

  Purpose: Remove any soft defects (contamination or particles).
- Pellicle mounting

  Purpose: Add the protective pellicle to the photomask.
- Post-pellicle-mount inspection

  Purpose: Outgoing inspection of final product.

6.1.4 Defect dispositioning

After defects are detected during the inspection step, a decision must be made as to what must be done to correct the defect. This is the dispositioning process. Defects are dispositioned according to the photomask's defect specification. The specification is an agreement between the photomask user and photomask fabricator. The photomask user is interested in the effect of photomask defects as printed. The photomask fabricator does not have lithography equipment and therefore cannot duplicate the printing results so important to the photomask user. The defect specification relates the as-printed defect to dimensions that can be measured directly on the mask. These measurements are compared with the specification and used in the dispositioning process.

The photomask defect specification contains separate maximum allowable sizes for each of the defects discussed in Sec. 6.1.1, qualified by their location relative to critical mask features.[4] Table 6.1 shows a simplified

TABLE 6.1 Example Defect Specification for a Technology of an Arbitrary Design Rule (DR)

Defect criticality	Defect Spec. (4×)	Defect examples
Loose	1.4 DR	Isolated pindots and pinholes
Standard	DR	Chrome extensions and intrusions
Critical	0.7 DR	Defects on optical proximity correction (OPC) features, defects in critical spaces (i.e., near adjacent OPC)
Contact	1.1 DR	Missized contact

example of a defect specification. Note that the maximum allowable size for a "standard" defect, measured in photomask dimensions (4×) is often numerically equal to technology main feature size when measured in wafer dimensions (1×).

The photomask user wants all defects that can affect printing to be eliminated. In order to minimize cost, the photomask fabricator wants to eliminate *only* those defects that affect printing. The inspection tool may not be able to distinguish between defects that are slightly larger than specification allows from those that are slightly smaller. Thus, the possibility exists that some of the defects captured by the inspection tool will not exceed the mask defect specification. The defect is first reviewed at the inspection tool to determine whether it is a false or nuisance defect. False defects are detections of defects where there are truly no defects. These can be due to a difficult pattern where the tool's defect recognition algorithms cannot easily distinguish between defects and the pattern. This is most common where the tool is operating near its resolution limit. Nuisance defects are real defects that are judged to be within the size range allowed by specification. The next step in the dispositioning process is to compare measurements of defect size against the defect specification appropriate for the defect. This step is easiest, and most manufacturable, if the specification itself is simple. Defects judged to be in violation of the specification must be repaired. Mask shops have more than one type of repair tool, so the final part of defect dispositioning is defining which repair tool to use.

The rule-based procedure described Table 6.1 compares the defect to an allowable dimension on the photomask. The actual printed defect may depend on factors not comprehended in the specification. The specification often considers worst-case situations, e.g., where the *mask error factor* (MEF) is high, to insure that the all killer defects are eliminated. Some defects that exceed the specified dimension but are in a more forgiving area of the mask may not adversely affect integrated circuit yield. Model-based dispositioning simulates the exposed pattern from the inspection tool image of the defects.[5] Currently these use scalar imaging models to predict the aerial image at the wafer. The photomask fabricator must have knowledge of the exposure conditions that will be used with the mask in order to accurately predict the aerial image. This technique is starting to be used with good success at least at one captive mask shop.

Another dispositioning procedure is using an aerial imaging monitor system, or AIMS.[6,7] The AIMS is a metrology tool that reproduces the illumination and projection conditions that the mask is exposed to in the exposure tool. The AIMS and the exposure tool utilize the same wavelength and partial coherence. The numerical aperture is the same as viewed from the mask in both the AIMS and the exposure tools. Whereas an exposure tool projects a demagnified photomask image onto a wafer, the AIMS projects a magnified photomask image onto a *charge-coupled device* (CCD) imager. In this way the aerial image can be quantitatively measured. A normal analysis includes a process-window evaluation from a through-focus set of images and *critical dimension* (CD) measurements. The dispositioning criterion would then be a maximum allowable CD change through a given focus range. The unique quality of AIMS measurements is that they are actinic, i.e., they use the same wavelength light as the wafer exposure tools. AIMS will measure wavelength-dependent effects in the same way as the wafer exposure tools. One area where this is vitally important is repair qualification. Repair tools can damage the photomask substrate and alter the wavelength dependence of the transmission. Outgoing photomask quality can then only be assured by actinic measurements. The repair application is further discussed in Sec. 6.3.4.

6.1.5 Defect size scaling and implications on requirements

For far-field incoherent imaging through cylindrically symmetric optics the image is given by the expression[8]

$$I_{\text{im}}(\vec{r}) = \int I_{\text{obj}}(\vec{r}')\text{PSF}(\vec{r}-\vec{r}')d\vec{r}' \tag{6.1}$$

where I_{im} = intensity at the image plane
I_{obj} = intensity at the object plane
r and r' = position vectors in the image and object planes, respectively (both referenced to the image plane)
PSF is the point spread function

The point spread function for cylindrically symmetric optics is given by

$$\text{PSF}(\vec{r}-\vec{r}',\lambda/\text{NA}) = \left[2J_1\left(2\pi\frac{|\vec{r}-\vec{r}'|}{\lambda/\text{NA}}\right)\bigg/2\pi\frac{|\vec{r}-\vec{r}'|}{\lambda/\text{NA}}\right]^2 \tag{6.2}$$

where λ is the wavelength, NA is the numerical aperture, and J_1 is the Bessel function of the first kind.[8] Here we have added the λ/NA dependence to the argument of the function. This function's value goes to zero at 0.61 λ/NA. The PSF is essentially a delta function if the feature size of interest is greater than λ/NA and the image and object would be similar. Today, the feature sizes can be on the order of 0.6 λ/NA Photomask features of this size will not be faithfully reproduced in the image.

Dimensional scaling can be described as $r \rightarrow cr$ and $r' \rightarrow cr'$. If we want the image with feature sizes on the order of λ/NA to scale with the object we must have

$$\begin{aligned} I'_{\text{im}}(\vec{r}) = I_{\text{im}}(c\vec{r}) &= \int I_{\text{obj}}(c\vec{r}')\text{PSF}(c(\vec{r}-\vec{r}'),\lambda/\text{NA})d\vec{r}' \\ &= \int I'_{\text{obj}}(\vec{r}')\text{PSF}\left(\vec{r}-\vec{r}',\frac{\lambda/\text{NA}}{c}\right)d\vec{r}' \end{aligned} \tag{6.3}$$

The last equality is due to the functional dependence of Eq. (6.2). The characteristic length of the point spread function must also scale as $\lambda/\text{NA} \rightarrow \lambda/\text{NA}/c$. In the ideal case, the dimensional scaling will be accompanied by a proportional decrease in λ/NA either by decreasing the wavelength or increasing the numerical aperture. In this case the defect specification s would also scale as $s \rightarrow cs$.

This ratio of the characteristic dimensions of the object and point spread function is commonly defined as k_1 factor:

$$k_1 = \frac{d}{\lambda/\text{NA}} \tag{6.4}$$

where d is the characteristic length of the image, i.e., feature size or half-pitch. The image is well resolved at high k_1 and the image features will scale with the object features. At low k_1 (<0.45) the image features scale faster than the object features, leading to a higher sensitivity to photomask feature variation. This effect is quantified by the mask error factor, which is:

$$\text{MEF} = M\frac{\Delta\text{CD}_{\text{image}}}{\Delta\text{CD}_{\text{mask}}} \tag{6.5}$$

where $\Delta\text{CD}_{\text{image}}$ is the feature size change in the image on the wafer, $\Delta\text{CD}_{\text{mask}}$ is the feature size change on the photomask, and M is the optical system magnification.

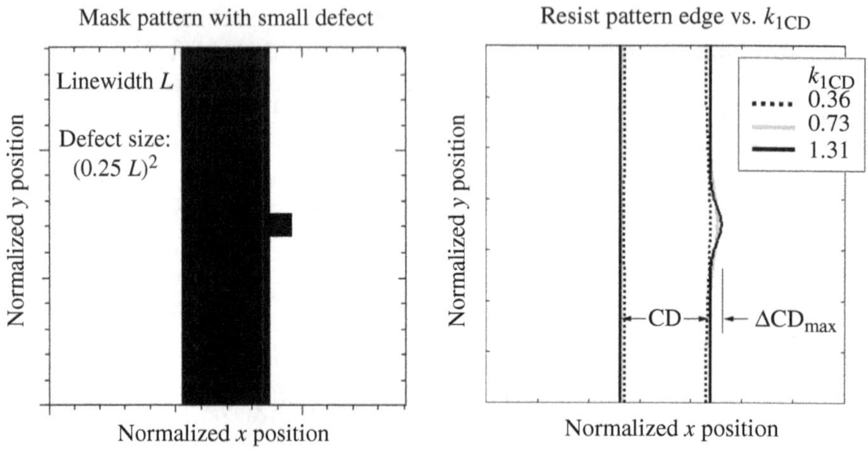

Figure 6.10 Left: mask pattern with semiisolated line of length L and opaque defect of linear size $L/4$. Right: printed resist pattern edge at three values of $k_{1CD} = L/(\lambda/NA)$. CD is the resist width away from the defect. ΔCD_{max} is the width change at the defect at its maximum extent.

The *printability* of a defect is commonly quantified by the ratio $\Delta CD_{max}/CD$, where CD is the feature size of interest and ΔCD_{max} is the change in that size due to the presence of the defect. This is usually measured at the location where the change is maximum. A defect's printability depends strongly on its location in relationship to the photomask pattern.[9] Defects that are not resolvable in isolation can cause a measurable change to a neighboring feature.[10] The effect of k_1 on the aerial image of a defect on the edge of a semiisolated line is shown in Fig. 6.10.[11] It can be seen that the shape and size of the printed defect changes as it becomes more or less resolved. For this case, the printability ratio has a minimum at $k_1 = 0.44$ (Fig. 6.11). In this case of an

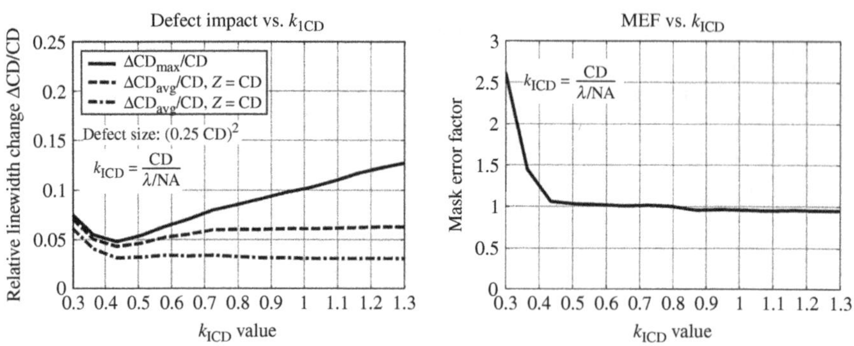

Figure 6.11 Dependence of defect ΔCD (left) and MEF (right) on lithography k_1 for binary masks. ΔCD_{max} is the linear size of the printed defect at its widest point. Two cases of ΔCD_{avg} are plotted: the average of ΔCD over a width of $Z = CD$ and $Z = 2CD$. Note that ΔCD_{max} has a linear k_{1CD} dependence while ΔCD_{avg} has little k_1 dependence. ΔCD_{max} and MEF have opposite trends for $k_1 > 0.4$.

isolated line the printability increases at larger k_1 as the defect is better resolved. The minimum printable defect size can be calculated given a criterion such as $\Delta CD_{max}/CD = K$. 10 percent is a commonly used value for K.

The printability increases at smaller k_1 as the mask error factor increases (Fig. 6.11). This is a general behavior as k_1 decreases. In this case the defect specification would scale faster than $s \to cs$.

6.2 Inspection

6.2.1 Role of mask inspection

Defects on photomasks that affect the image printed on the wafer can influence the functionality of the integrated circuit produced with the lithography process. Mask defect inspection enables detection, analysis, and elimination of defects that print on the wafer and affect the lithography process yield. Since the photomask represents the "master pattern" that is replicated on the wafer by the lithography exposure tool, any "killer defect" on the mask that significantly reduces the wafer yield will be copied to every exposure field. Hence to ensure any wafer process yield, the photomask must have no substantially printable defects. The success in producing photomasks that are effectively *defect free* depends the capability of the mask defect inspection used during the mask fabrication process to detect printable defects.

Mask defect inspection is used extensively during the mask fabrication process, as illustrated in Fig. 6.9. First, the unpatterned masks blank substrates are inspected to ensure their quality. In some cases, the resist-coated blanks are inspected just before the mask is exposed on the pattern generator. Once the resist pattern is exposed and developed, *after develop inspection* (ADI) is often performed to ensure that the exposed pattern is defined correctly. After mask pattern transfer from resist to the underlying layers, the patterned mask is inspected again for different types of defects, including pattern defects and contamination defects. After possible defect repair as well as after final clean and pelliclization, the defect inspection is repeated again to ensure the final photomask is free of defects. Furthermore, once the photomask arrives in the wafer fab it is usually reinspected to ensure that is has not been damaged during the shipment from the mask shop. Numerous inspection approaches are needed to detect the different types of important defects encountered during the different mask fabrication steps.

In addition to ensuring outgoing mask quality during manufacturing using a mature mask process, defect inspection is used during the mask process development to analyze the process capability and evaluate different process improvements.

Importance of high capture rate for mask defects. Before elaborating on mask inspection techniques, it is important to highlight the primary distinction between the defect detection requirements in mask inspection and in wafer inspection. In both the mask shop and the wafer fab, defect inspection plays a vital role in fabrication of the desired product—a wafer with large fraction of functional integrated circuits. During inspection of a photomask, which serves as a master pattern in integrated circuit definition, every substantially printable defect must be found with very high confidence level to ensure any integrated circuit yield. On the other hand, only a fraction of the defects that are present on the wafer must be detected during wafer inspection to ensure some integrated circuit yield.[12] A larger fraction of defects that are detected and eliminated on the wafer through process improvement corresponds to better integrated circuit yield. Hence to ensure wafer circuit yield, defects on the photomask must be detected with near perfect effectiveness (>99.9 percent) while wafer process defects can be detected with only partial effectiveness (typically 50 to 95 percent).

The defect detection is described by the probability that a particular defect is captured during an inspection, which is often referred to as defect capture rate. A typical defect capture rate is schematically depicted in Fig. 6.12. For a particular type of defect, the probability of defect detection usually, but not always, increases with the size of the defect, going from no detection to partial detection to reliable detection. Since a significantly higher capture rate is necessary in mask inspection, for a given critical defect size requirement, the mask inspection capability needs to be significantly better than wafer inspection capability as illustrated in Fig. 6.12.

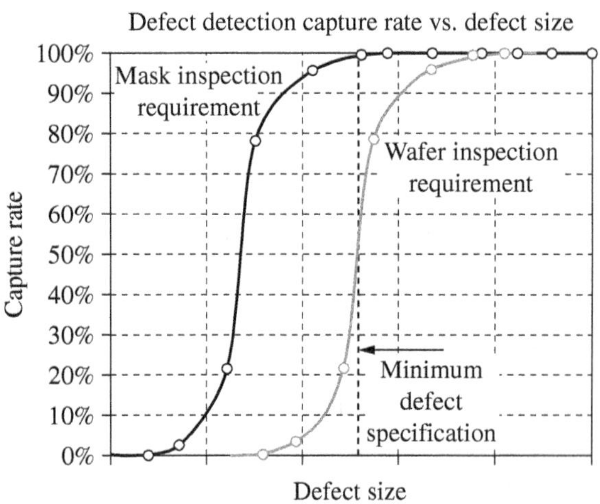

Figure 6.12 Schematic representation of dependence of defect detection probability on defect size.

6.2.2 Inspection system overview

Mask inspection tools represent some of the most complex systems used in the semiconductor industry, on par in cost and complexity with lithography exposure tools and mask pattern generators. The complexity of mask inspection tools is driven by the need for high resolution and high speed in detection of defects. Mask inspection systems employ microscopy and light scattering techniques to probe the mask surface. The key components of an inspection tool include (1) an energy source that generates the probe beam, (2) a mask stage that controls the position of the mask and enables scanning of its entire surface, (3) illumination, collection, and detection systems that generate a signal from the mask surface, (4) a data processor that runs defect detection algorithms during the inspection, and (5) a data storage device that can archive the inspection data for analysis and retrieval.[13] For pattern inspection, additional data processing modules are usually needed to manipulate the mask pattern database.

General description of defect detection. The defect detection in an inspection system can be represented schematically as shown in Fig. 6.13. Though actual inspection systems are significantly be more complex, Fig. 6.13 illustrates the basic principle of defect detection, where the detection signal is collected as a function of position along the mask surface. For unpatterned mask blanks, the inspection signal may be proportional to the raw detected signal. For patterned masks, the inspection signal in Fig. 6.13 may correspond to the difference signal between the test die and a reference die, as discussed in detail in Sec. 6.2.5. Although Fig. 6.13 shows the inspection signal in one dimension, in general, the inspection signal is obtained along both directions along the mask surface. In most inspection systems, the signal is obtained (sampled) at regular spatial intervals along the surface,

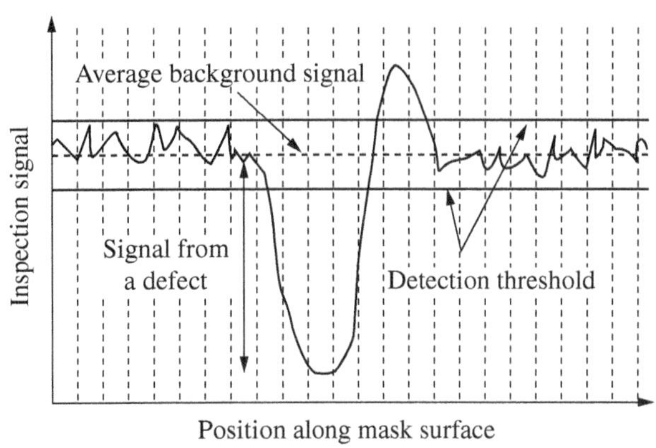

Figure 6.13 Diagram of defect inspection signal.

often referred to as pixels. The pixel size depends on the inspection system design. To achieve high-speed inspection, it is desirable to have as large a pixel size as possible to minimize the detection algorithm processing time. On the other hand, the detection signal must be sampled at sufficiently small intervals, determined by the Nyquist sampling theorem, to resolve small mask features and defects. Most inspection systems operate at pixel sizes near the Nyquist sampling size appropriate for the inspection tool wavelength and optical design. Typically, advanced defect inspection tools operate with a pixel size that is about 1.2 to 1.6 times the size of the smallest defect that can be reliably detected. In the vicinity of a defect, the detected signal changes and, if the signal from the defect is larger than a detection threshold, the defect is detected. Most inspection systems report the defect location and the number of pixels that are above the detection threshold. This apparent defect size given by the number of detection pixels is usually an imperfect measure of the actual defect size. The defect size may be measured by careful analysis of the inspection tool image of the defect or by other defect metrology tools.

As the surface is scanned, the background signal fluctuates owing to noise arising from the sample under test and from the inspection system. Sometimes, the background noise may exceed the detection threshold even when a real defect is not present on the sample. Such detections are usually referred to as false defects. To achieve high sensitivity, the detection threshold is usually set as close to the noise floor as possible while avoiding too many false defects.

The ability of an inspection system to reliably detect defects depends on the signal contrast of the defect and on the system noise. The inspection sensitivity may be improved either by improving the defect contrast or by reducing the noise. The defect contrast C is usually taken to be the absolute value of the difference between the intensity produced by the defect I_{defect} and the average background signal $I_{\text{background}}$, normalized by the average background signal, or $C = |I_{\text{defect}} - I_{\text{background}}|/I_{\text{background}}$. The defect contrast depends on the size and shape of the defect, properties of the defect and the sample surface, and the design and properties of the inspection tool. The detection noise depends on many factors that may be categorized as those related to the inspection system and those related to the sample. The system noise depends on the stability of the environment, the stability of the signal acquisition, and the inherent fluctuations of the light or particle source. If the system is operated in a shot-noise-limited regime, as may be the case when fast sample scans are desired or when the reflected signal from the sample is low, the system intensity noise may depend on inspection speed. The system noise may be traded off against inspection speed in this regime. Fluctuations in the properties of the sample surface, such as those arising from surface roughness or pattern variations, also contribute to the detection noise. An increase in the noise generally necessitates a looser detection threshold to avoid false defects, leading to lower detection sensitivity overall.

Mask inspection sequence. The sequence of steps used during mask inspection is illustrated in Fig. 6.14.

After the mask is loaded into the inspection system, it is necessary to perform a calibration to determine and adjust the strength of the inspection signal at the detector with the mask on the stage. For a patterned mask, the calibration regions need to include the different pattern types on the mask. For instance for a chrome-on-glass mask, the calibration areas should include both glass and chrome regions. If the mask is inspected both in transmission and in reflection, the calibration is necessary for each of the two modes to establish the proper inspection signal operating range.

For a patterned mask, an alignment step is needed to establish the defect coordinate system, usually defined with respect to a set of alignment mark patterns that are placed in well-defined locations on the mask, usually outside the circuit device area. The alignment also compensates for any mask load/unload translation and rotation variations. For mask blank inspection with no pattern and no alignment marks, the alignment step is not performed. The defect coordinate system is established through the accuracy of the mask stage and the repeatability

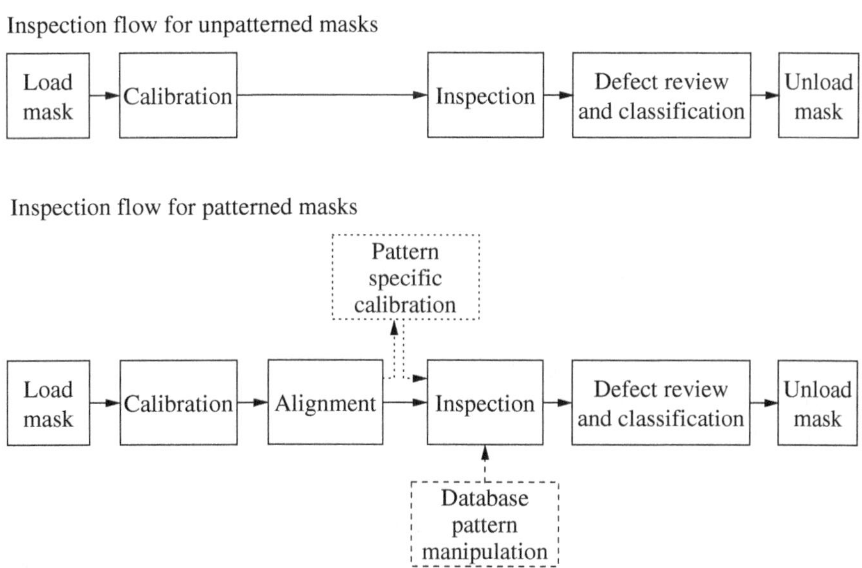

Figure 6.14 Mask inspection flow diagrams.

of loading and unloading the mask. For patterned mask inspection based on the STARlight approach [see the subsection "Soft-defect and contamination inspection (STARlight)"] that searches for contamination defects, an additional calibration on the pattern usually follows the alignment step.[14]

Following calibration (and alignment, if needed), the entire mask area of interest is inspected. The inspection generally reports the locations and approximate size of the detected defects. In some cases, the inspection system may also capture the defect images and report certain properties of the mask pattern, such as feature size uniformity across the inspection area.[15] In pattern inspection, a map of the mask features and their properties may be retrieved from the pattern database and used by the defect detection algorithms. Following inspection, a defect review is used to understand the properties of the detected defects. In many systems, the review involves capturing images of the detected defects by driving the mask stage to the locations identified during inspection. Some inspection systems are capable of capturing the defect images during the inspection. The acquired defect images are subsequently classified either by an operator or automatically.[5,16] The classification determines the defect type and its criticality in terms of its effect on the printed wafer. The classification also identifies (a) whether the defect should be ignored, if it is not printable, (b) which procedure should be used to eliminate it if it is printable (i.e., type of repair or clean), and (c) whether the mask should be rejected, if the defect cannot be repaired.

The entire inspection process needs to be performed in a relatively short period of time in order to maximize the inspection tool throughput and enable a cost-effective mask manufacturing process. Typical inspection times vary from about 30 min for mask blank inspection of a large mask area (~200 cm^2) to about 2 h for pattern inspection of the active device area (~130 cm^2). For certain advanced critical layer masks, the inspection times may be several hours. The rate with which data must be acquired during inspection is dictated by the desired inspection time and by the number of pixels on the mask. The data acquisition rate is inversely proportional to the inspection time and to the pixel size squared because the number of pixels scales with the mask area. For example, for the 90-nm technology generation, the data acquisition rates needed node for the most advanced mask inspections are several hundred megapixels per second. Future technologies will require proportionally larger data acquisition and processing rates. For each pixel, the detection algorithm must process the raw inspection signals in real time without significantly affecting the overall inspection time. To accommodate the demand for these extreme processing speeds in real time, mask inspection data processing systems are some of the fastest computer systems anywhere.

6.2.3 Inspection techniques based on optical microscopy

Most mask inspection systems utilize optical microscopy to detect defects on the mask surface. The inspection approaches include either scanning or imaging optical systems, usually employing monochromatic radiation. To meet the demands of semiconductor device feature scaling, i.e., Moore's law, the mask inspection optical systems have evolved to meet the demand for detection of ever-smaller critical defects among ever-smaller mask pattern features over the last two decades. During this time frame, the optical resolution of the mask inspection systems has been improved through use of shorter probe wavelengths and large optical system numerical apertures. Lithography wafer exposure systems have similarly relied on wavelength scaling for improving resolution.

Optical resolution. The resolution of an optical system can be described by in terms of the operational wavelength λ and the numerical aperture NA of the optics at the mask. In general, the minimum distance S between two features that can be resolved is given by

$$S = \frac{k\lambda}{\text{NA}} \tag{6.6}$$

where k is a proportionality constant that depends on many factors, including the optical system design, optical aberrations, and feature type.[17–19] For diffraction-limited optical systems, including most commercially available mask inspection tools, the minimum k is on the order of 0.5. The numerical apertures used in mask inspection systems typically ranges from 0.7 to 0.95. The minimum detectable defect size d also scales with the optical resolution scaling factor λ/NA and can be written as

$$d = \frac{q\lambda}{\text{NA}} \tag{6.7}$$

The defect detection proportionality factor q depends on the defect type, the properties of the mask near the defect, the detection threshold level, and the detection algorithm. In practical defect inspection systems, the proportionality constant q may be about 0.2 and possibly as small as 0.1 for certain types of defects or types of inspection. At a given inspection wavelength, the minimum detectable defect size will generally be smaller when the background noise can be minimized. For example, the minimum detectable defect size is likely to be smaller at a given wavelength for mask blank inspection than for patterned mask inspection owing to fewer sources of noise in unpatterned mask inspection than in patterned mask inspection.

Historically, the demands on mask inspection tool resolution have not been as stringent as those for wafer exposure tools because the mask features used to be four or five times larger than wafer features owing to the demagnification of the exposure tool optics. Thus for a given semiconductor technology generation, most mask inspection tools have employed longer wavelength radiation than the wafer exposure tools. However, with the implementation of optical proximity correction and resolution enhanced lithography, the optical enhancement features on the mask, such as serifs, cutouts, and assist bars have become comparable to the smallest features on wafers that are printed with 4× or 5× demagnification. The primary mask features remain larger than the wafer features but are decorated with the small resolution enhancement features. Thus in recent years, the demands on mask inspection tool resolution have become comparable to those for wafer lithography exposure tools. However, meeting the resolution requirements is compounded by the historical trend where the wavelength scaling on the mask inspection side lags behind the wavelength scaling on the wafer exposure side. For example, in the time frame when critical layer lithography was performed using 248 nm wavelength, the most advanced mask inspection systems operated at 365 nm wavelength. Currently with the 90-nm wafer technology node in production, the most advanced high-resolution mask inspection tools employ 257nm radiation for masks used at the 193-nm exposure wavelength.

With deep ultraviolet lithography masks used at 248-and 193-nm wafer exposure wavelengths, certain defects pose an additional challenge for mask inspection systems. Specifically, contamination defects that degrade the mask transmission significantly at the deep ultraviolet wavelengths (248 nm and especially 193 nm) may reduce the transmission only slightly at the 365-nm inspection wavelength. Such defects are thus strongly printable on the wafer but very difficult to detect owing to poor contrast at the inspection wavelength. Ideally, mask defects with optical properties strongly dependent on wavelength should be inspected at the wavelength of light being used during mask exposure. Recently an alternative mask inspection approach based on mimicking the wafer exposure tool optics (see the subsection "Aerial image simulation inspection") that employs the same wavelength as the wafer exposure tools has been developed.[20,21]

Light sources for inspection. One primary difference between mask inspection tools and wafer lithography systems is the type of light source. With the transition of wafer exposure lithography from g-line (436 nm) and i-line (365 nm) wavelengths from mercury arc lamps to deep ultraviolet region at 248 and 193 nm, the lithography light sources have become excimer lasers that produce radiation in a series of short pulses.

By contrast, most inspection tools for masks (and wafers) employ continuous wave radiation. Since in most systems, the mask area is covered during inspection by continuously scanning the illuminated field of view across the mask surface, a light source producing radiation continuously (or at a repetition rate enabling coverage of all points on the mask surface) is necessary. Recently, some inspection systems have employed pulsed sources as well.[20] The light sources employed by various inspection systems are listed in Table 6.2.

The average light source power required for mask inspection depends on the system noise, the pixel size, and the desired inspection time. The number of photons per pixel needed to distinguish the defect signal from the background noise depends on the probability that the noise exceeds the detection threshold level. The detection threshold determines the detection sensitivity and in general depends on the type of defect and detection algorithm. For pattern inspection, the best sensitivity corresponds to detection threshold (and defect contrasts) of 5 to 20 percent relative to the background bright-field detection signal. For blank inspection, the best sensitivity may correspond to detection thresholds of 1 to 2 percent relative to the bright-field background.

To first order, the useable detection threshold level is just above the noise floor. In cases where the photon shot noise is the primary noise source, the noise variance is proportional to the average number of detected photons. The useable detection threshold level N scales as the square root of the mean number of detected photons, which is proportional to the square root of the background clear field intensity ($N \propto \sqrt{I_{\text{background}}}$). In this case, the relative threshold level T, corresponding to the threshold N normalized by the clear field background $I_{\text{background}}$, is proportional to the inverse square root of the background clear field intensity ($T = N/I_{\text{background}} \propto 1/\sqrt{I_{\text{background}}}$). As a result, the required number of photons per pixel scales with the squared inverse of the relative minimum

TABLE 6.2 Inspection Light Sources

Wavelength, nm	Source type	Continuous wave
633	Helium-neon laser	Yes
546	Mercury lamp	Yes
488	Argon-ion laser	Yes
365	Mercury lamp	Yes
364	Argon-ion laser	Yes
266	Frequency-quadrupled Nd: YAG laser	Yes
257	Frequency-doubled argon-ion laser	Yes
198	Frequency-doubled argon-ion laser at 244 nm mixed with Nd:YAG laser at 1064 nm	Yes
193	ArF excimer laser	No

detection threshold level ($I_{background} \propto 1/T^2$).[22] In general, a shot-noise-limited inspection using a threshold just above the noise floor requires more photons per pixel to detect lower contrast defects using tighter detection threshold than to detect defects with larger contrast using a looser detection threshold. Overall, mask inspection systems typically require light sources with average powers of less than about 1 W.

Optical microscopy for inspection. The majority of mask inspection systems are based on either scanning microscopy, scanning confocal microscopy, or imaging microscopy.[18,19,23] These approaches employ *bright-field defect detection* where the primary beam transmitted or reflected by the mask surface is collected by the detector and analyzed by the detection algorithm. Simplified schematic representations of scanning microscopy, scanning confocal microscopy, and imaging microscopy are shown in Fig. 6.15. The schematics are shown for the case

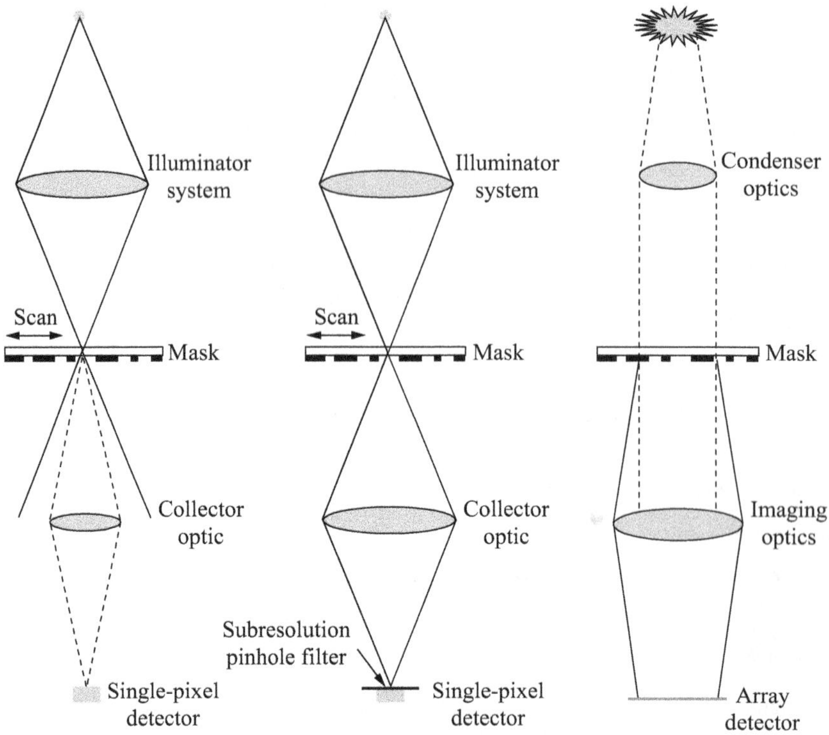

(*a*) Scanning microscopy light from spatially coherent source (*b*) Scanning confocal microscopy light from spatially coherent source (*c*) Imaging microscopy light from spatially incoherent extended source

Figure 6.15 Optical microscopy approaches used for inspection.

where the mask inspection signal is collected in transmission mode. For mask inspection performed in reflection mode, the system schematics would be analogous.

Comparison of scanning and imaging optical systems reveals that the respective roles of the illuminator and the light collecting optics are effectively reversed in the formation of the image. While in imaging systems, the image-forming objective influences the image properties most critically, in scanning systems it is the beam profile function produced by the illuminator system that affects the image quality. For example, while in an imaging system defocus corresponds to a change in the amplitude transmission function of the image forming lens, in a scanning system it is related to changes in the beam profile properties due to defocus of the illuminator. In confocal microscopy, both the illumination and light collection optics contribute to the image quality.[18]

In both scanning microscopy and scanning confocal microscopy, the beam is focused tightly to a small spot and the inspection signal data are acquired for each point (pixel) on the surface as the mask stage is scanned relative to the optical system. An illuminator system focuses light from a source to a small beam spot in the plane of the mask under inspection. The light transmitted through the mask is captured by a collector lens and measured with a detector. The image of the mask pattern is constructed by recording the inspection signal as the focused beam is scanned over the mask. Historically, mask inspection tools have used primarily scanning spot approaches. However, the demands for increased resolution call for reduced inspection spot size and increased number of pixels on the mask surface that has to be inspected. Combined with the requirement that the overall mask inspection time does not change from one semiconductor technology generation to the next, scanning spot inspection does not scale favorably in terms of scanning speed and illumination intensity. Given these constraints, beam spot intensities could eventually exceed the mask damage thresholds with improved resolution and increase in mask feature density.[24] To alleviate this problem, the inspection data acquisition schemes are being changed from single-pixel serial data collection to a more parallel data collection. This includes using either multiple scanning-spot inspection beams simultaneously[25] or implementing imaging microscopy inspection.[23,26]

In imaging microscopy, a region of the mask is illuminated and optically imaged onto an array detector to obtain a signal from multiple pixels simultaneously. This technique has historically been used in semiconductor manufacturing for wafer inspection but has recently been adopted for mask inspection.[23,26] Simultaneous acquisition of the inspection signal from multiple pixels on the mask enables defect inspection that is sensitive to small defects, does not damage the mask, and can be performed with reasonable speed. The image-forming optics usually magnifies the mask region sufficiently to avoid pixel size limitations of the available array detectors, such as *charge-coupled devices* (CCDs) for instance.[27] The light detecting material used in inspection systems is usually silicon. To implement the parallel pixel data acquisition approach with imaging optics, uniform illumination of the inspection field of view is needed to minimize the system noise. The detected signal is usually carefully calibrated to compensate for any illumination and detector sensitivity nonuniformities across the array.

There are two main types of CCD arrays with different acquisition and readout schemes that are suitable for inspection applications. With the full-frame readout, a two-dimensional image can be extracted from a CCD matrix after exposure by serial transfer of the pixel signals along the rows and columns of the array. Multiple exposures and inspection field stitching is required to inspect the entire mask. With the alternative line-scan readout, a one-dimensional array of CCD pixels is used to construct the two-dimensional mask image by detecting the image of an illuminated line field scanned across the surface. In some applications, where relatively low intensity illumination may be preferred or available, the necessary signal may be collected using *time delay integration* (TDI) line-scan sensor with multiple adjacent photoelement lines effectively performing multiple line scans in parallel.[28] In TDI, each line on the object is exposed and detected multiple times in the different detector lines while the mask is scanned synchronously with the transfer of charge from line to line. Integration of the signal from the multiple exposures has an averaging effect that also reduces signal noise. If a pulsed light source were used with a TDI sensor, the averaging process would also reduce the pulse-to-pulse noise allowing reduced pulse repetition rate. TDI is successfully used in many current wafer defect inspection tools and has also been applied to mask defect inspection.[23,28]

6.2.4 Other inspection techniques

Light scattering techniques. Although most mask inspection systems employ bright-field optical microscopy, light scattering approaches are suitable for certain mask inspection applications, such as mask blank inspection. The principle of light scattering inspection is illustrated schematically in Fig. 6.16. The mask surface is illuminated with a focused beam of light. The light that may be scattered by a defect on the surface away from the primary illumination beam is collected by one or more detectors and analyzed as the illumination beam scans the mask surface. Such *dark-field scanning-spot inspection* has traditionally been used to detect defects and particles on wafers, mainly unpatterned.[12,29–31] Detection of the scattered signal at large angles with respect to the reflected (or transmitted) beam may be sensitive to certain types of

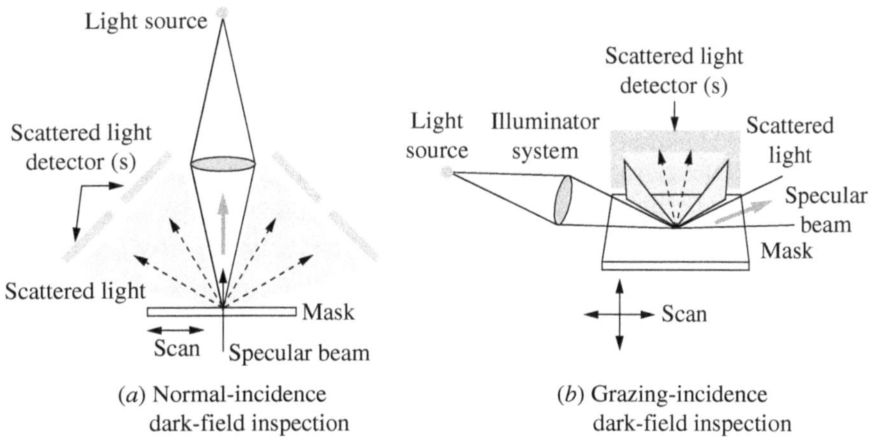

Figure 6.16 Schematic of light scattering dark-field microscopy.

defects, such as particles, small surface bumps, or pits that scatter light into high angles relative to the main beam. One drawback of the dark-field approach is that it is difficult to implement for patterned surfaces, where the radiation scattered by the defect must be separated from the radiation scattered by the background pattern. Furthermore, modeling scattering from a pattern is very challenging, making die-to-database inspection (see the subsection "Die-to-die inspection") relatively impractical with dark-field inspection. As a result, dark-field scattering inspection is primarily used for unpatterned mask surfaces.

SEM inspection. Inspection based on *secondary electron microscopy* (SEM) represents a possible option for mask inspection. The advantage of electron beam imaging is its high resolution that can potentially be significantly better than traditional approaches based on optical imaging at deep ultraviolet wavelengths. An inspection system based on electron beam imaging, called SEMspec, has been developed in the early 1990s.[22,32,33] The electron beam image of a patterned mask appears similar to that shown in Fig. 6.23. Pattern inspection using electron beam is challenging for several reasons. First, the image is dominated by the signals from feature edges, making pattern-to-pattern comparison difficult in the presence of small edge variations, such as pattern line-edge roughness. To avoid detection of such edge imperfections, which do not significantly affect the wafer image, the sensitivity to real printable defects usually has to be detuned. Second, the different materials on the mask surface, such as chrome and glass, may produce very similar secondary electron signals. The resulting poor relative contrast may make it difficult to distinguish the different types of mask pattern features in the image. Third, since mask substrates are not conductive, it is possible that surface charging may significantly affect the image quality and stability across the mask surface. Additionally, as some features are charged by the scan beam, their image contrast may change owing to the induced localized voltage change. Furthermore, the SEM image properties listed here, make it difficult to implement die-to-database inspection with the electron beam. The SEMspec system operates only in die-to-die mode and has been used primarily for wafer inspection and to a limited extent for mask inspection.[34]

6.2.5 Patterned mask inspection

During inspection of patterned masks, the ability of the inspection system to distinguish the defects from the pattern is imperative. The most commonly used technique for detecting defects within a pattern compares the inspection tool signal under test with a reference signal. The reference signal can either be measured on another nominally identical mask pattern or be calculated using the pattern database. Another pattern inspection technique, called STARlight inspection, finds defects by combining mask signals obtained from the pattern in both transmission and reflection modes.

Die-to-die inspection. One of the most straightforward pattern inspection approaches is called *die-to-die inspection* because the defects are searched for by comparing signals from two patterns on the mask that are nominally identical. The die-to-die approach has the advantage of being relatively simple. Its primary disadvantage is the requirement that the photomask has two or more identical copies of the pattern under inspection. Furthermore, most inspection systems require that the two identical copies of the pattern are aligned with one another in one direction, i.e., they cannot be in random locations on the photomask. Having multiple copies of the same pattern on the mask may be impractical when a relatively large die sizes are desired on the wafer and only one die can fit within the mask field. This may be the case in the early phase of an integrated circuit development cycle when the die sizes are large or in certain applications with large die sizes, such as microprocessor applications. However, many applications do not require the full mask field size per chip and can employ die-to-die mask inspection.

The method of collecting the inspection signal depends on the design of the inspection system. In some systems, the scans from both dies on the photomask are obtained using only one optical system and temporarily stored in the data processor before analysis by the defect detection algorithm.[35] Other systems used two imaging objectives to collect the inspection signals from both dies simultaneously.[36]

The detection of defects using the comparison of a test pattern to a defect-free reference pattern, used in the of the die-to-die inspection mode is illustrated in Fig. 6.17. The signal S produced by each defect is determined from the difference between the test image with the defect I_{test} and a defect-free reference image $I_{reference}$, or $S = I_{test} - I_{reference}$. With this approach, the average background of the defect inspection signal is zero, except in the vicinity of the defect, where the signal can either be negative or positive, depending on defect type. In addition to the information about the defect, the difference image contains noise contributions from the inspection tool signal acquisition system, pattern alignment errors, and mask imperfections. The mask imperfections include both the negligible pattern variations that do not affect the wafer pattern and the larger printable defects of interest. In actual inspection systems capable of reliable defect detection that is not limited by excessive false defect counts, the inspection signal from the defect of interest needs to be above the overall image noise level.

Cell-to-cell inspection. One approach to minimizing the inspection signal noise is to compare two patterns that are very close together on the mask during the defect detection. This especially reduces the noise contributions from the variation in the mask properties across the patterned field.[36] The technique is appropriate for periodic patterns, where the pattern-to-pattern comparison can be made between adjacent periodic cells. This cell-to-cell inspection approach is useful for memory circuits that contain many repeating units. However, virtually all integrated circuits contain some areas where the patterns are not repeated. Other inspection methods must be used to ensure that these nonrepeating areas are free of defects.

Die-to-database inspection. In cases where the photomask does not have two or more copies of the pattern, the reference image has to be calculated from the pattern database. (Mask-to-mask inspection is another

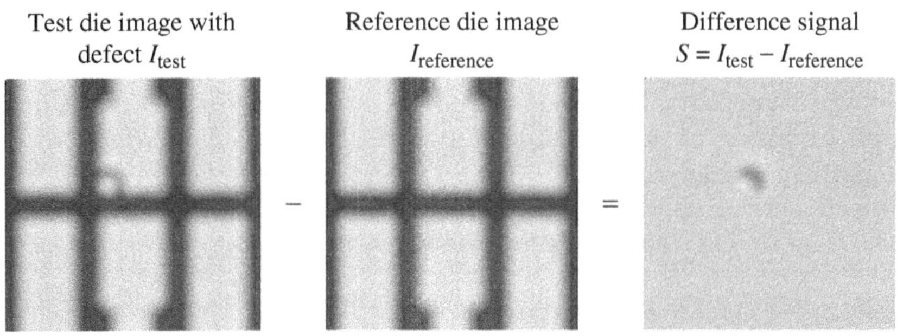

Figure 6.17 Defect inspection principle where the test pattern is compared to a defect-free reference pattern.

costly and impractical option.) In *die-to-database inspection*, the measured signal is compared to a computer rendered image that is calculated from the pattern database using information about the mask and the inspection tool.[35] The advantage of die-to-database inspection is that it effectively verifies the integrity of the pattern on the mask, i.e., ensures that the actual photomask pattern matches the intended pattern design. The illustration of the defect detection principle in Fig. 6.17 applies to die-to-database as well as to die-to-die and cell-to-cell inspection modes.

The die-to-database inspection system generates the reference image from the pattern database using computational engines and algorithms specifically designed for this purpose. Since the inspection tool must verify that the fabricated pattern on the mask matches the intended design, the pattern data used in the database rendering are the same data that are used in the mask pattern generator. In order for the computationally intensive database rendering process to not slow down the inspection, the pattern database is usually partially processed on a separate designated computational engine before the actual mask inspection. This step is often called *database preparation* or *data prep*. During the mask inspection, these partially processed data are used to calculate a representation of the pattern in real time. The rendered database mimics the expected inspection tool signal for the pattern of interest. Referring to the mask inspection flow diagram in Fig. 6.14, in addition to data preparation and real-time database rendering, an additional pattern calibration step is usually needed in die-to-database inspection to account and compensate for small pattern fidelity variations within each mask and differences between masks fabricated using different mask making processes. The calibration areas may be selected either manually by the tool operator or automatically during the database preparation step. Owing to the complexity of the die-to-database inspection system, only a handful of inspection tool suppliers offer this capability.[23,26,37,38]

On a given inspection system, die-to-database inspection is usually slightly less sensitive than die-to-die or cell-to-cell inspection. This sensitivity reduction results from higher overall noise level in the die-to-database inspection signal compared to die-to-die inspection. The increased noise is caused by subtle imperfections in modeling of the mask pattern reference signal. Since defects are only detected when the defect-induced change in the signal exceeds the system noise, the increased noise leads to reduced defect sensitivity.

Inspection of resolution enhancement masks. With the advent of resolution enhancement techniques in photolithography and the concomitant mask pattern complexity, mask inspection and especially die-to-database inspection has become more challenging. For conventional chrome-on-glass photomasks that have only two types of pattern, i.e., bright and dark, only one level of pattern data is needed to model the database pattern. However, for phase-shift masks having multiple pattern types that are defined in multiple mask patterning steps, several levels of data are needed to correctly render the inspection tool image from the database. For instance, in addition to chrome regions, alternating phase-shift masks and chromeless phase-shift masks generally have glass regions producing several different phase shifts. Embedded phase-shift masks and ternary embedded phase-shift masks have clear regions of exposed glass, partially opaque phase-shifting regions, as well as chrome regions. In general, at least one pattern database layer is needed for each mask pattern defining step, such as chrome patterning and glass patterning. Masks with assist features represent another resolution enhancement approach that poses a challenge for mask inspection. Such masks may have only bright and dark features but they have both primary features that will be printed on the wafer and assist features, such as scattering bars, that improve the patterning performance of the primary features. In this case, different data layers may be used for the primary and assist features and die-to-database inspection needs to combine the feature information to correctly model the pattern.

Ideally, the inspection system should be capable of handling a variety of masks used in different semiconductor lithography applications, which is quite challenging given the multitude of possible resolution enhancement techniques. One of the difficulties for resolution enhancement mask inspection is poor inspection tool image contrast of certain types of defects, such as glass defects on alternating phase-shift masks,[39] and of certain features, such as optical proximity correction assist bars. The poor contrast is compounded by the difference in the exposure wavelengths of the mask inspection tool and the wafer exposure tool. Traditionally, the mask inspection tools have employed longer wavelengths than the wafer exposure tools because the primary mask features are generally four to five times larger than wafer features, owing to the demagnification of the wafer exposure tool optics (see the subsection "Light sources for inspection"). Although with optical proximity correction and resolution enhancement features on the mask, the mask and wafer features have become comparable in size, most inspection systems still employ longer wavelengths than exposure tools. On the other hand, some resolution enhancement techniques have been adopted on mask inspection systems to improve contrast and detection sensitivity for certain types of defects.[40–42]

One option to enhance sensitivity of different defects, different versions of the defect image can be compared. For instance, to exploit the well-known differential defocus phase defect contrast mechanism,[41,42] two images of the same phase defect taken at two different focus conditions $(I_{test})_A$ and $(I_{test})_B$ of the inspection tool can be combined to produce the image difference signal $S = (I_{test})_A - (I_{test})_B$. This concept is illustrated in Fig. 6.18 for glass phase defect. To perform such a comparison in practice, the inspection tool hardware needs to be modified from the existing configurations to acquire images of the same defect under different conditions. Furthermore, in this type of comparison, the image difference may be nonzero even when no defects are present due to differences in the images of the primary mask features under the two different conditions.

Aerial image simulation inspection. Most mask inspection systems acquire the mask image at the highest resolution possible to obtain the best defect and feature contrast. Since the main purpose of the photomask is to pattern wafer features, an alternative inspection approach is based on mimicking the optical system of the wafer exposure tool. This so-called *aerial image simulation* (also known as aerial image measurement system or AIMS) has been used extensively in mask shops for mask defect metrology and dispositioning[43,44] and has recently been extended for defect inspection with scanning of the entire mask field of view.[21] With the aerial image simulation scheme, the mask image recorded by the inspection tool matches the image produced by the wafer exposure tool. This requires the inspection tool wavelength to match the wafer exposure tool wavelength. Since the wafer exposure tool is a demagnifying imaging system, the numerical aperture at the mask side NA_m is scaled by the demagnification ratio M compared to the numerical aperture at the wafer side NA_w ($NA_m = NA_w/M$). To model the light collection from the mask in the exposure tool, the aerial image simulation system uses the same mask illumination and the same mask side numerical aperture NA_m as the wafer exposure system to form the mask image. The inspection signal obtained with the aerial image simulation

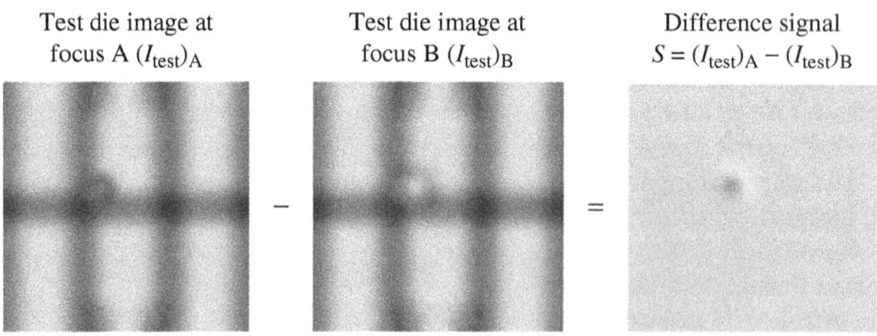

Figure 6.18 Defect inspection principle where the test pattern is compared under different conditions.

represents a highly filtered image of the mask pattern, like that formed on the wafer. The AIMS-based inspection can be employed both in die-to-die and die-to-database inspection modes. It has been demonstrated in die-to-die mode.[20,45]

Observing the mask during defect detection in the same way optically and at the same illumination wavelength as the way the mask is imaged during wafer exposures enables the AIMS-based inspection to detect only defects that truly affect the wafer pattern. For example, organic contamination defects, which may be difficult to detect by conventional approaches using long wavelengths, can be easily detected by AIMS inspection at the 193-nm wafer exposure wavelengths. Nonprintable defects that do not affect the wafer, but may cause nuisance detections during conventional inspection tools should not be detected. Furthermore, like on the wafer, resolution enhancement features on the mask are not resolved in aerial-image-based inspection. The inspection not only detects localized defects but also simultaneously verifies the capability of the resolution enhancement features, such as phase-shifting areas, assist features, and optical proximity correction features to correctly form the wafer image. In principle with complex mask patterns, the AIMS-based inspection can potentially employ simpler defect detection algorithms, compared to conventional inspection because the detected pattern matches the relatively simple image formed on the wafer rather than the complex mask features.

Soft-defect and contamination inspection (STARlight). Pattern-to-pattern comparison inspections efficiently detect pattern defects, such as clear and opaque patterning errors, that have good image contrast relative to the primary pattern features. However, pattern-to-pattern inspection may not be sensitive to contamination defects that often have relatively poor inspection image contrast. Contamination defects, often called soft defects because they can usually be removed by photomask cleaning, may be particles from mask handling or residues from resist patterning, etching, and cleaning. An interesting approach based on Simultaneous Transmitted And Reflected Light (STARlight) inspection has been developed to detect contamination defects.[46,47] The STARlight pattern inspection technique finds defects by combining mask signals obtained from the pattern in both transmission and reflection modes. The concept of STARlight inspection is illustrated in Fig. 6.19. Inspection signals from each point on the mask are mapped onto a plane with the transmitted light signal on one axis and reflected light signal on the other axis. The signals are normalized by the intensity of the brightest primary features on the mask in each of the two imaging modes. If the signals fall within a region that is associated with the mask pattern, then no defect is detected. However, if the signals fall outside the region associated with

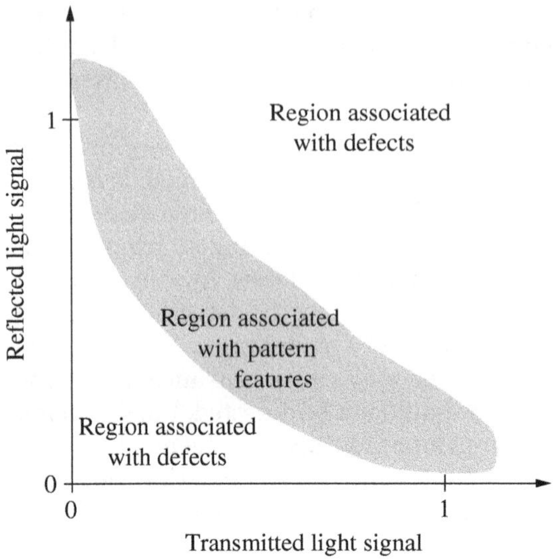

Figure 6.19 Principle of STARlight inspection for detection of contamination defects.

the mask pattern, then a defect is flagged. The region in Fig. 6.19 associated with the pattern features is determined during a calibration procedure performed before the inspection, as illustrated in the inspection flow diagram of Fig. 6.14. It is imperative that the calibration images used during the calibration include representative pattern features but not include any defects of interest. If a defective image were used during the calibration, the defect signature would be considered part of the pattern and the detection sensitivity for that type of defect would be significantly reduced. STARlight inspection is suitable in detection of defects that are optically different from the mask pattern, such as particles and organic residue defects. However, STARlight inspection is not sensitive to defects that have similar appearance as the pattern. Also the STARlight detection capture rate is generally not a monotonic and increasing function of the defect size (such as that in Fig. 6.12 for conventional inspection). In fact for a given type of defect, STARlight inspection may be very sensitive to defects in a certain size range but not detect any defects that are smaller or larger.

After-develop resist inspection. Following mask resist exposure and development in the mask fabrication sequence, an inspection step is often used to verify the fidelity of the resist pattern before etching. Since this step enables reworking of the mask exposure step, it can reduce mask costs. After-develop (aka postdevelop) resist inspection is especially valuable when the mask process requires multiple patterning steps, such as that for alternating phase-shift masks and embedded phase-shift masks. Thus in many cases, the postdevelop resist inspection involves defect detection within a resist pattern formed over patterned mask features, such as chrome or embedded phase shifter. Owing to complexity of the pattern under test in this case, either die-to-die or STARlight inspections are usually used in resist inspection. In some cases, an integrated mask fabrication process can be developed whereby the resist thickness is tuned to optimize the sensitivity of resist inspection and to simultaneously maintain good patterning and etching performance of the resist. One difficulty with resist inspection is possible resist outgassing under the high-power inspection tool probe beam. As a result, inspection systems specifically dedicated for resist inspection may be necessary to prevent mask cross-contamination in an integrated mask manufacturing line.

6.2.6 Inspection of unpatterned mask blanks

Among the numerous defect inspections performed during the mask fabrication process, inspection of unpatterned photomask blanks is necessary to minimize mask yield loss and need for defect repair after patterning. Substrate and blank inspection are especially important in fabrication of extreme ultraviolet (EUV) lithography masks because it may not be possible to repair certain defects after mask patterning using the available repair methods. In some circumstances, unpatterned resist-coated mask blanks may be inspected before patterning to verify the quality of the resist coating. However, since the inspection tool probe beam can partially expose the mask resist and degrade its performance, after-coat resist inspection is seldom used. Another unpatterned inspection step must be performed on the backside of the photomask as well as on the pellicle. Generally, the specifications for defects on the backside of the photomask and on the pellicle are significantly looser than those for the patterned side of the mask.

Mask blank inspection, mask backside inspection, and pellicle inspection can utilize the same inspection techniques. The most commonly used systems for unpatterned inspection include confocal microscopy and dark-field light scattering approaches. Die-to-die systems can also be used in the absence of the mask pattern. Since unpatterned inspection is generally considered only a quality check, as opposed to a step that significantly modifies the mask or directs the processing sequence, high throughput is desirable.

6.3 Mask Repair

Mask repair is the process of restoring a defective mask to its good lithographic condition. It refers to the editing of localized pattern structures by adding and removing of materials, and, in most cases, this structural editing step after the mask is fully patterned enables the yield of useful masks.

Generally, mask repair refers to the elimination of so-called "hard" pattern defects in contrast to the removal of "soft" defects by cleaning. The hard defects are described and categorized in Sec. 6.1.1. The goal of mask repair is to fix all the defects that are printable, meaning that the defects affect the resist CD on the wafer if they were not repaired. The reader may refer to Sec. 6.1.4 for a more detailed discussion of defect printability and its effect on device performance.

There are four different repair tool platforms based on laser ablation, *focused ion beam* (FIB), electron beam, and mechanical nanomachining. Laser ablation is widely used for chrome removal. FIB using *gallium* (Ga) *liquid metal ion source* (LMIS) and electron beam are generally used for both deposition and etching. Mechanical nanomachining based on *atomic force microscopy* (AFM) works best for removal only.

With the introduction of aggressive OPC structures, new materials, and numerous resolution enhancement techniques, and the demands for stringent lateral as well as vertical dimensional control, advanced optical masks have become increasingly complex. All of these add to the high cost of mask making and make it nearly impossible to manufacture defect-free masks. And it is no longer practical to just restart a new mask. Therefore, mask shops have put ever-increasing emphasis and efforts in mask repair at the back end of the production line to fix every defect possible in order to rescue an otherwise useless mask (scrap). Mask repair plays an increasingly important role in yielding advanced masks.

The mask repair module started to appear in mask shops in early 1970s when it became economical to fix the defects rather than remaking a new mask. Since then different types of repair technologies have been developed using photons, ions, electrons, and mechanical tips.

In this section, we will provide a general description of mask repair followed by a discussion of repair requirements put forth by lithographic specifications. Based on these requirements, we review and discuss various repair technologies and their respective applications and limitations. For each technology, the basic working principle and generic repair procedure are described. Therefore, at the end of this section the readers will have a general and practical understanding of the repair processes, and also be able to decide on the types of tools to choose for their particular mask repair applications.

6.3.1 General description of mask repair

After a mask is inspected, the captured defects are reviewed and classified either immediately on the inspection tool (online review) by a factory technician or offline with some types of automatic defect review software. These review and classification steps determine the defects to be repaired. Before repair, characterization is often required as to provide additional information about the physical size and lithography effect of a defect. This information can determine the best repair process to use.

The physical size or shape of a defect can either be measured with the repair tool's imaging capability in situ or an ex situ AFM. AFM measurements are mostly needed for quartz bump defects on an APSM mask. The bump height and profile is used for setting up repair parameters and etch time in the case of FIB- and electron-beam-based repairs since there is no material interface for *end-point detection*. Defect shape information can be obtained with the AFM function of the nanomachining tool. AIMS metrology is used to measure the effect of a defect on lithographic printability, i.e., CD change with respect to the reference CD (see Fig. 6.10). Data from such physical and optical characterization are employed as a guide to determine the process or tool to use for the repair of different defects.

As depicted in Fig. 6.20, the mask is then loaded into the repair tool and the repair process proceeds with mask alignment, defect survey, defect imaging, repair parameter setup, and postrepair imaging. Defect coordinates from pattern inspection files can often be transferred to the repair tool via network connections in the factory or manually entered during the setup. More process details for each repair tools will be described in Sec. 6.3.3.

Postrepair images obtained by the repair tool are usually not sufficient to qualify the repair. Rather, repair qualification is performed with AIMS optical imaging measurements. Details of repair qualification and disposition are given in Sec. 6.3.4. In some cases, iterations are needed to complete the repair based on measurements and qualifications.

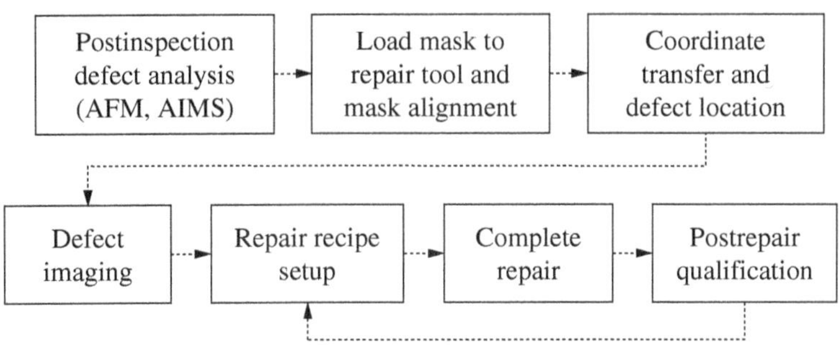

Figure 6.20 Typical steps involved in a repair process. In some cases, iterations may be needed.

6.3.2 Repair requirements

Photomask repair requirements are derived directly from defect specifications based on resist printability on silicon. Defect printability depends on the device design and specific lithography processes, both of which are proprietary information that semiconductor manufactures do not always disclose. Nevertheless, a printable defect is normally defined as one that causes a maximum allowable CD change at particular resist pattern geometries. The printability criteria are often represented by the ratio ΔCD/CD. 5 to 10 percent are ratios commonly used for these criteria. Any defects larger than this maximum allowable size must then be repaired.

The International Technology Roadmap for Semiconductors (ITRS) provides a source for defect specifications that both the mask and equipment makers can refer to as the common goal in their respective development efforts.[48] As a rule of thumb, for lithography processes with MEF around 1, the size of a printable pattern defect on the mask is roughly about 20 percent of the line width. For example, for 45-nm technology node (65-nm half-pitch), the printable defect size is 52 nm (4×, as measured on the mask).

We can select a repair tool or process for a particular application once the defect specifications are understood. Traditionally, the repair tool and repair process are characterized by two key parameters—repair edge placement accuracy and precision, and substrate optical transmission loss. While these two parameters are useful and easily quantifiable, other parameters must be considered with increasing importance for the characterization of a technology for repairing masks with aggressive *resolution enhancement technique* (RET) designs. A capable repair technique should be able to repair defects under these constraints.

1. *Spatial resolution:* The tool must have adequate imaging resolution to "see" the defects. Then, the removal or deposition must be confined to the minimum pattern (e.g., line or space) on the mask in order to avoid mask damage or postrepair steps. The important figure of merit is the minimum feature size that the repair tool can correct and not necessarily the tool's probe size. The minimum feature size is defined as the minimum etched or deposited features (lines or dots) made by scanning the beam in a one pixel-wide line.[49] The minimum effective beam size depends on the probe size, any beam-gas scattering, the materials to be etched, the process gas, and the beam scan time needed to achieve the etch depth or deposition height as required by the repair. The time required to repair a defect is generally longer than that of acquiring an image with a metrology tool (e.g., CD SEM). The minimum is also affected more by system drift. The minimum effective beam size can be many times larger than the imaging resolution (which can be a few

nanometers) due to the aforementioned considerations (see the subsection "FIB").

2. *Edge placement:* This refers to positioning of the removal or deposition of a repair with respect to a line edge of defect-free reference pattern (Fig. 6.21). The placement accuracy is characterized by the remaining CD error, often measured by AIMS (see Sec. 6.3.4). Figure 6.21 uses a simple clear (mouse bite) repair as an example to illustrate edge placement. The repair box is placed to match the clear defect and the resulting electron-beam-induced deposition is shown on the right.[52] Repair edge placement is affected by imaging resolution, edge recognition, initial edge placement, system stability, and endpoint detection. In some cases, repair edge location must be shifted with respect to the main feature edge to produce the desired printed resist CD.

3. *Damage:* Transmission (or reflectivity for EUV mask) loss and phase error quantify substrate damage caused by the repair process. Common requirements are a transmission loss of less than 5 percent and phase error less than 5°. At the 193-nm exposure wavelength a 5° phase error corresponds to a quartz depth control of better than 5 nm. Each repair tool can have its own damage mechanisms. Contamination added to the mask or incomplete material removal will cause transmission loss. Examples of contamination are gallium implantation in FIB repair, chrome redeposition in laser repair, and embedded materials in AFM nanomachining. Another type of unwanted damage from repair is the destruction of pattern structures around the defect. For example, overetch due to lack of end-point detection, river bedding, and poor vertical profile control are few side effects from nonideal repair processes that often require postrepair treatment or compensation to remedy the damage.

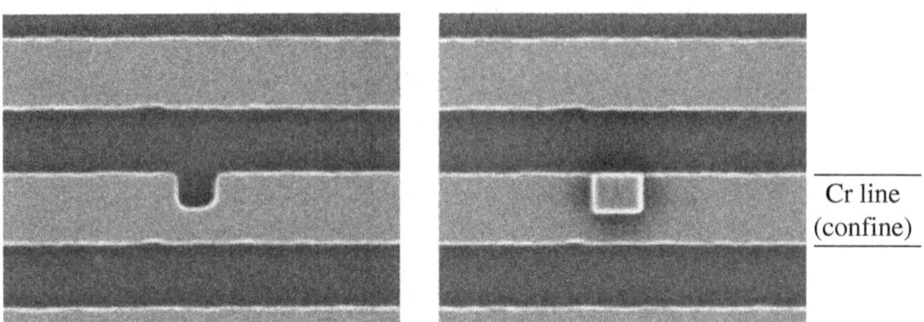

Figure 6.21 SEM micrographs of a simple edge defect (mouse bite on a chrome line) illustrating repair edge placement. The physical edge placement of the repair is qualified optically using AIMS. (*Source:* Liang, T., E. Frendberg, D. Bald, M. Penn, and A. Stivers, *Proc. SPIE 5567*, pp. 456–466, 2004.)

4. *Removal selectivity:* Material removal requires the probe (beam or tip) to scan over an area slightly larger than the defect. This exposes substrate material immediately adjacent to the defect to the same process that removes the defect. The defect removal process mechanism must be selective with respect to the substrate in order to avoid damaging the surrounding substrate.

5. *Durability:* Deposited material must have sufficient durability and adhesion to the photomask to prevent it from being removed during wet cleaning. The material must not erode or decompose during exposure to lithographic illumination. Materials deposited from organic or metalorganic precursors using FIB or electron beam generally attain good adhesion to mask surfaces. Depositions from organic precursors can be unstable under DUV illumination.

The increased complexity of the mask features has pushed mask repair of critical defects to the tools' capability limits. All these five parameters need to be included as figures of merit for determining the capability of a technology to repair these critical defects.

Clear defects. A clear defect is simply the partially missing chrome from a pattern. Many names are used to describe these types of defects such as mouse bites, intrusion, and pinholes. This type of defect is relatively easy to repair since most deposited materials are opaque enough. The critical issue is film adhesion to the surface and integrity against mask clean and optical exposure as previously discussed. In current mask production, FIB-induced carbon deposition is commonly employed using precursors such as pyrene and styrene.[50] In general, deposition is less sensitive to the beam scan parameters. Recently, it has been reported that electron-beam-induced deposition can also be used for this application.[51–53]

Opaque defects. This class of defects refers to the extra chrome to a pattern, often called chrome extensions or pin dots. Defect removal can be done with laser ablation, gas-assisted etch using FIB, electron beam, or nanomachining. Depending on the defect nature (size, shape, and location), one technique is preferred over the other. For example, laser ablation is most efficient and fast for removing large defects.

Phase defects. This category includes all phase-related defects in APSM and embedded PSM (EPSM), and requires special consideration. These defects are the most difficult to repair because one must meet the additional specification of phase error. In APSM, phase defects can be quartz bumps or pits (divots). Shifter defects in EPSM can be extra or missing shifter.

For APSM, most phase defects are quartz bumps, and they can be removed by mechanical cut or etch with FIB or electron beam. Since there is no interfacial signal for end-point detection, one must rely on bump height (obtained from AFM measurements) to determine repair time in order to control the depth of removal, thus the phase control.

Pattern reconstruction. Photomask patterns have become more ornate with ever more aggressive OPC and the application of subresolution assist features. Repairing defects on these masks require special software that can reconstruct or duplicate a complicated pattern from a reference.

In pattern reconstruction, attaining accurate placement and feature fidelity is important. The reference pattern is generally represented by an image taken by the repair tool such as a secondary electron or ion image from FIB. This reference image is then placed to the desired location and used as the template to reconstruct the missing pattern. Such reconstruction can either be additive via deposition or subtractive by removal.

6.3.3 Repair techniques and their applications

Many techniques have been explored for mask repair involving the use of photons, ions, electrons, and mechanical nanomachining for material removal or addition. Commercial mask repair tools installed in mask shops today are based on four platforms, and they are, in the order of their emergence—laser ablation and deposition, FIB-based etch and deposition, AFM-based nanomachining, and electron-beam-based etch and deposition.

Laser ablation for chrome removal as a repair technology started in the early 1970s and still remains as the workhorse for opaque defect repair today.[54,55] Lasers with shorter wavelengths (UV to DUV) and pulses (nanosecond to femtosecond) are used in today's advanced laser repair tools in order to offer higher spatial resolution and better overall performance (see the subsection "Laser-based repair" for details). The ablation mechanism is thermal melting and evaporation with long-pulse-duration (nanosecond and picosecond) lasers to nonthermal evaporation for short-pulse-duration (femtosecond) lasers. Nonthermal chrome removal using femtosecond lasers emerged in mid-1980s developed by IBM in-house,[56–58] driven by the need to reduce chrome splatter and thermal diffusion-caused dimensional control. Even with DUV lasers and high NA optics, however, the relatively large spot size (about 100 nm) of the laser beam obviously sets the limit for extending laser repair technology to masks with narrow spaces.

FIB emerged in the early 1980s as the primary technology for repairing x-ray lithography masks.[59] X-ray proximity lithography was researched in the early 1970s as a demonstration of high resolution replication process,[60] and the efforts were intensified in the 1980s as a leading candidate for *next-generation lithography* (NGL), a potential successor to optical lithography starting with the quarter-micron technology generation. Laser ablation is not effective on the heavy metal pattern absorber on thin nitrides membrane of x-ray masks. FIB-based tools using gallium liquid metal ion sources had matured into a microscopic beam technology by that time, and they became an obvious choice for the 1:1 x-ray mask repair. FIB was then applied for optical mask repair. The main advantage of FIB over laser is its high spatial resolution, about 10 to 20 nm under normal repair conditions. This high spatial resolution makes FIB *gas-assisted etch* (GAE) appropriate for repairing critical masks. The fundamental drawback of FIB is the damage caused by gallium ion sputtering of and implantation into the substrates, which limits the extendibility of FIB repair technology, particularly for etch of phase-shifting masks.

AFM-based mechanical nanomachining was developed for mask repair starting in 1998 and it became a commercial product in 2000. It is now widely used for a variety of applications,[61] including the removal of quartz phase bump in APSM and shifter material in EPSM masks, and the trimming of carbon depositions from FIB on binary masks. Due to the precision control afforded by the AFM platform, this repair technology has inherently high spatial resolution and superior edge placement control. Unlike tips designed for metrology, tips for machining require enough mechanical strength to act as a cutting tool. This requirement limits their shape and size. Tip shape and size, therefore, has posed a limit to the repair of mask patterns with certain geometries and aspect ratio.

The emerging electron beam repair technology offers damage-free repair processes best for deposition and phase defect removal. Electron beams possess a higher spatial resolution compared to photon and ion beams. Electron-beam-induced deposition of a variety of materials has long been widely demonstrated as a way for direct generation of nanostructures eliminating the need for the resist imaging layer used by projection lithography.[62,63] Etching using electron-beam-induced chemical reactions has also been reported on a number of surfaces.[52,64–66] Earlier exploratory work of electron- beam-induced deposition was done for x-ray mask repair.[67,68] By using electrons instead of ions for deposition and etching, one avoids the ion damage from physical sputtering and implantation since electrons do not cause change to the substrate and the processes are purely chemical in nature. It has recently been demonstrated that electron beam repair produced superior resolution for both deposition and etch and good depth profile for quartz etch.[52]

In the remainder of this section, each of the four repair technologies are described in detail, including the fundamental principles based on which the tools as well as the processes are developed. We then describe for each technology the repair procedures and the process parameters that affect the repair performance. We discuss each repair technology's advantage, limitation, and extendibility to support the lithography roadmap in future generations.

Laser-based repair. Laser beam illuminating a material can result in many physical and photochemical processes on the surface and in the bulk.[69,70] Both laser-induced deposition and laser ablation were explored for mask repair,[55–58,71,72] with the latter still being widely used today.

In the presence of a precursor gas, laser radiation can cause the gas molecules to decompose and thus result in the localized deposition on a surface. The decomposition takes place by photolysis and pyrolysis. The deposition size is limited by laser spot diameter, which is much larger than those of FIB and electron beam. So laser-induced deposition is resolution limited for today's mask repair.

For laser ablation, depending on the light pulse duration, there are two fundamentally different mechanisms in material removal.[73,74] For lasers with pulses in the nano- to picosecond regime, the ablation is via thermal evaporation. For laser with subpico- to femtosecond pulses, the ablation is a nonthermal process that provides unique advantages over the traditional thermal ablation for mask repair. The sudden application of laser power results in an explosive and nearly instantaneous removal. Therefore, laser "zapping" is commonly used to describe such ultrafast speed of the removal process. Today, laser mask repair tools come mainly from three suppliers—Quantronix, NEC, and IBM with models such as DRS M855[75] with femtosecond DUV laser, LM700A[71] with picosecond UV laser, and MARS2[57] with femtosecond DUV laser, respectively.

Lasers of wide range of wavelengths from near IR to DUV have been utilized, ranging from *yttrium aluminum garnet* (YAG) lasers to excimer lasers. DUV lasers are necessary for producing smaller beam spots to achieve higher spatial resolution. The optical components of a laser repair tool are typical of those of an optical microscope, but with more advanced laser source, illumination systems, and objectives. As with other repair platforms, laser repair tool is also equipped with precision stage and defect navigation software.

A schematic description of the MARS2 laser mask repair tool is shown in a block diagram in Fig. 6.22. The tool consists of a ~100-fs pulsed laser at about 1-kHz repetition rate with deep ultraviolet imaging optics, a precision air bearing stage, and a computer control system. Laser repair tools normally have multiple objectives operating at UV and DUV wavelengths to provide a range of imaging magnifications with a large range

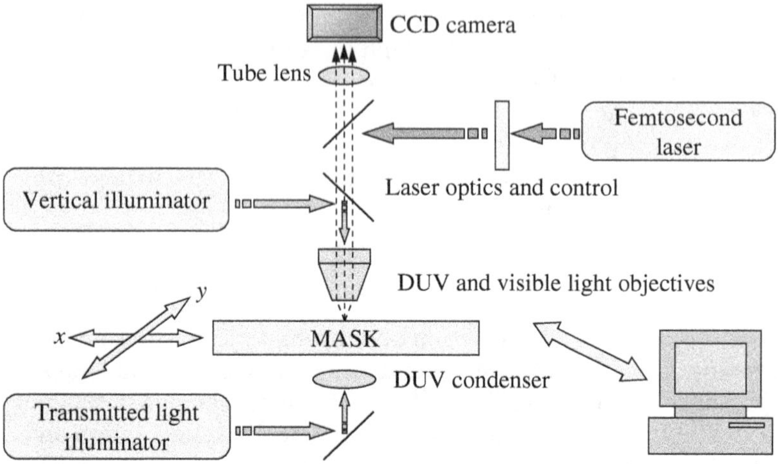

Figure 6.22 Schematic description of the IBM MARS2 femtosecond laser repair system. (Reprinted with permission from Wagner, A., R. Haight, and P. Longo, "MARS2: an advanced femtosecond laser mask repair tool," *Proc. SPIE 4889*, p. 457, 2002.)

of fields of view. The DUV objective is used for ablation as well as imaging with about 100-nm resolution.[57]

Laser-induced thermal ablation. In thermal ablation, the energy in a laser pulse is absorbed by the target (e.g., chrome mask absorber), causes the target to melt and evaporate.[73,74]

When a pulse of laser light is absorbed by the chrome film, it excites the electrons to higher energy states. The subsequent decay of these excited electrons transfer energy to the lattice. Heat generated from phonon emission (lattice vibrations) results in a quick temperature rise. Since the electron coupling to phonon emission process is much shorter (~1 ps) as compared to the laser pulse width,[74] the heat will accumulate within the chrome layer during the laser pulse, allowing thermal diffusion in the chrome layer. The thermal equilibrium process inevitably requires higher threshold laser fluence than nonthermal ablation with femtosecond lasers described later. This mechanism determines the characteristics of laser-induced thermal ablation for mask repair. The removal resolution is controlled by thermal diffusion rather than the laser spot size and is, therefore, quite poor. The ablation causes splatter or roll-up at the edges due to molten materials in the mixture of liquid and vapor.[73,112] It can also result in quartz damage.[76] These phenomena limit the use of nanosecond to picosecond laser ablation to chrome defects in large mask geometries.

The threshold laser fluence for vaporization decreases at shorter pulses.[73,74] This in turn results in less damage and better repair resolution with ultrashort pulse lasers.

Laser-induced nonthermal ablation. When using subpico to femtosecond pulses, laser ablation takes a different mechanism. Here, the energy pulse is much shorter than the ~1-ps cooling time for excited electrons through phonon emission compared to laser-induced thermal ablation, and the heat conduction inside the chrome layer is essentially negligible. Because of the low electronic heat capacity, the excited electrons attain very high transient temperatures and form a plasma that expels the atoms. This fast nonthermal process causes direct vaporization of the material.[74]

The main benefits of nonthermal femtosecond laser ablation are as follows. First, since there is not thermal conduction, the removal resolution is no longer limited by diffusion. Instead, it is controlled by the laser beam size, which can be reduced to about 100 nm with DUV wavelengths and a high NA objective.[57] And the threshold laser fluence is lower. Second, the ablation can be precisely controlled because of no molten material involved to cause splatter. This allows high level of edge placement precision. Third, the removal depth can be precisely controlled by the number of pulses delivered to the target. Therefore, one can remove materials layer by layer to avoid damage to the quartz substrate. This is a result of the fact that ablation depth is a logarithmic function of laser fluence.[74,77]

Figure 6.23 shows an example of chrome removal with femtosecond pulse laser—(a) the clean edges are in clear contrast to the thermal ablation, (b) where the splattered and rolled up chrome residues are readily evident.

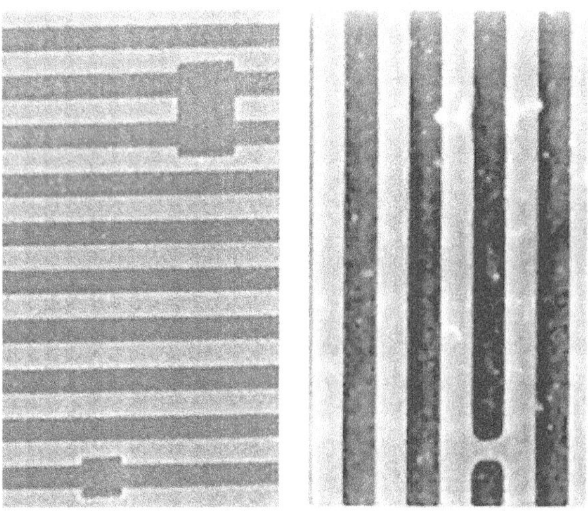

(*a*) (*b*)

Figure 6.23 An example of femtosecond laser ablation of a chrome absorber. The nonthermal ablation creates clean chrome removal with sharp edges, in clear contrast to the thermal ablation with a nanosecond pulsed laser; (Reprinted with permission from Haight, R., D. Hayden, P. Longo, T. Neary, and A. Wagner, *J. Vac. Sci. Technol.* B17(6), p. 3137, 1999.Copyright 1999, AVS The Science & Technology Society.).

FIB. Among all the repair technologies, FIB is the most widely studied for both fundamental physics and applications.[59,78–81] The use of gallium liquid metal as a high brightness ion source enables a rather simple and stable FIB machine, which is extremely important for a mask repair tool in a manufacturing environment. Focused ion beam milling with GAE and ion-induced deposition are two processes that allow the use of FIB for localized materials removal and addition, respectively, for mask repair. Notably, SII NanoTechnology, Inc., and FEI Corporation are two major suppliers of FIB-based mask repair tools.

A FIB mask repair tool consists of an electrostatic ion column with Ga LMIS in a Taylor cone design, a process gas inlet, a laser interferometer controlled x-y stage, flood gun for charge control, detector, and sophisticated GUI-based computer control. The high vacuum chamber usually maintains a base pressure around 10^{-6} torr under normal operations. Figure 6.24 is a schematic description of a FIB system showing only the main components of the hardware.

Unlike electron columns, ion columns use electrostatic lenses to focus the much heavier ions. The focused ion beam arriving at the sample exhibits a gaussian-shaped distribution with spot size limited mostly by chromatic aberration, resulting from ion energy spread due to Coulomb repulsion. As will be seen later, the beam tails cause the so-called "overspray" observed in FIB deposition.[49] To achieve smaller beam size, FIB can be operated at higher acceleration voltage and lower beam current. Beam with higher energies will result in more damage to the mask due to more ion sputtering and deeper ion implantation. On the other hand, lower energy ion beams give poor resolution and milling performance.

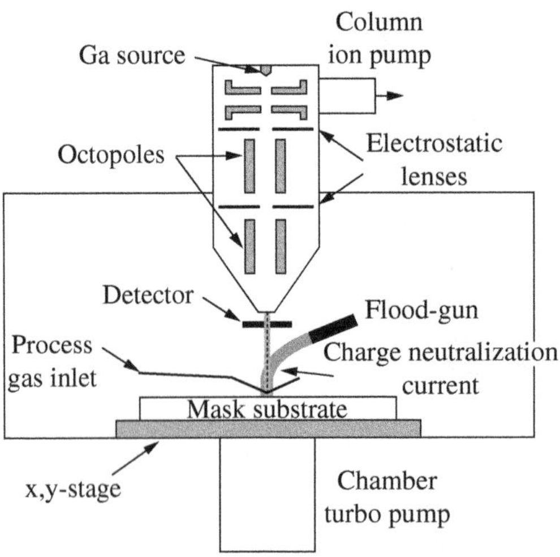

Figure 6.24 A schematic of FIB mask repair system. (Courtesy of C. Marotta of FEI Company.)

As a balance, low current in the range of 5 to 10 pA is recommended at 20- to 30-keV beam energies for mask repair. Under these conditions, the beam diameter is around 10 to 30 nm.

The molecules employed for enhancing sputtering and FIB-induced deposition are delivered to where the beam impinges on the surface of the mask with a gas injection system. The system consists of the source, delivery control and monitors, and the injectors placed very close to the surface for ensuring a uniform and highly localized flow. For easy handling and control, the sources of etch gases (e.g., Br_2, XeF_2) and deposition precursors (e.g., hydrocarbons or organometallics) are usually in solid or liquid phases contained in reservoirs. Temperature then controls the gas pressure by heating or cooling the reservoirs.

As with any charged particles, there is a surface charging problem for nonconducting quartz substrate during FIB imaging and repair. Few side effects take place when charging occurs. When charges build up at the mask surface, the surface potential affects the landing voltage of the ion beam and causes the beam focus and magnification to vary. Charging also causes the beam to drift resulting in more placement errors. Two techniques are generally utilized concurrently to reduce and mitigate charging. One is inject electrons to the surface from a flood gun to neutralize the surface and the other is a dynamic drift correction to bring the beam back to the original position with the help of a reference mark, which is often a milled hole (filled later on) or a deposited feature on the chrome absorber. With such controls, an *edge placement deviation* (EPD) about 10 to 15 nm (3σ) is achievable for advanced FIB systems.

Ion milling or sputtering of materials has been comprehensively studied and the fundamental mechanisms are well understood.[82,83] In simplified terms, ions transfer enough energy and momentum to the target atoms and knock out the bonded atoms from the surface. The sputter yield of 30-keV Ga FIB for chrome is about 2.5 (1 gallium ion releases 2.5 chromium atoms).[84] With the presence of particular gas on the surface, the yield of physical sputtering can increase for certain materials, thus enhancing the removal rate via a chemical process of FIB-induced reactive species (ions and radicals) reacting to the target atoms.[49] This is commonly referred as GAE–gas assisted etching. XeF_2 has been shown to enhance MoSiON and quartz removal to 5 to 20 percent depending on the GAE conditions such as gas pressure and beam scan parameters.[85,86] Bromine enhances chrome removal rate by about 20 percent and search for better enhancement chemistries have proven not fruitful.[87] The GAE by-products must be volatile enough to leave the surface in order for such process to work.

For FIB-induced deposition, the precursor molecules are decomposed by the primary ions and the resultant secondary electrons, leaving the atoms or fragments on the surface along with embedded gallium ions. Therefore, deposits are in the form of a compound (polycrystalline structures can form if metal-containing precursors are used) always consisting of gallium and carbon (from the precursor or contaminations in the chamber or on the surface). In commercial mask repair tools with precursor gases such as pyrene and styrene, the deposited opaque carbonaceous film is partially cross-linked polymers with high gallium content (>20 percent).[50,88]

Figure 6.25 is a schematic description of beam scan parameters in a typical FIB GAE etch or deposition process. The circles represent the ion beam spot on the mask surface. The beam can scan pixel by pixel over the defined repair area (frame), as outlined by the dashed rectangle, in a raster or serpentine mode. Dwell time t_D specifies the single beam exposure time. The figure shows a raster scan where the beam is scanned from left (point A) to right (point B) and blanked off in x- or y-direction before starting again from C. Line refresh time defines t_L how long the beam is blanked off at the end of the line scan before returned to the next line. In a serpentine mode, the beam scans continuously in both directions without being blanked off. In either mode, the beam will be blanked off at the end of the frame (point D) with s specified frame refresh time t_F before returning to point A. Another parameter, as shown in Fig. 6.25, is pixel spacing d that describes the separation between two adjacent pixels. The beam overlaps during scan if d is less than the beam diameter. GAE etch and deposition process development for mask repair of different types of defects (size and materials) often involves the optimization of these parameters in engineering *design of*

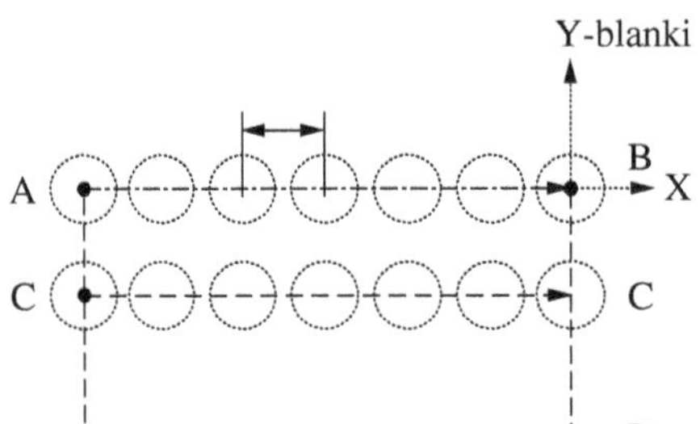

Figure 6.25 Description of beam scan parameters in a FIB (or Electron beam) setup for creating a rectangular pattern. The circles represent beam spots. The beam scans pixel by pixel with spacing d and single exposure dwell time t_D. The beam can be blanked off at the end of line and frame, in either x or y directions, to allow process optimization and avoid deposition or etch in unwanted areas.

experiments (DOEs). Especially for etching, there must be enough time for the gas molecules to adsorb on the surface. In general, a monolayer of adsorbates shall be established before the ion beam (same for electron beam) impinges on the surface. For most surfaces, one Langmuir (10^{-6} torr•s) of gas exposure will cover the surface with one monolayer.

Some studies have been reported about the effects of gas types, gas flow, beam scan conditions on etch and deposition performance for mask repair.[49,78,85,87–92] In general, the high deposition and etch rate is achieved at shorter beam dwell time and longer refresh time. This allows enough time for the gas molecules to adsorb on the surface. Other variables that affect the repair process are the mask cleanliness and the adsorption properties of gas molecules on the mask surface. In actual mask repair, engineers have to develop many recipes to suit wide variations of defect size, shape, locations, and many types of materials. Mask repair is the most manual operation in a mask shop.

There have been extensive article publications on FIB mask repair studies,[53,85,86,93–95] and they cover, in the order of use and difficulty, deposition, GAE removal of chrome, GAE removal of shifter in EPSM, and quartz bump in APSM. There has been no solution for depositing dielectric films with sufficient transmission for repairing quartz divot since the film is essential opaque due to gallium and C content.[96]

One of the important attributes in controlling FIB GAE process is end-point detection for knowing when to stop the etch process. For chrome removal, secondary electron or ion signal can be used to detect the chrome-quartz interface. For FIB GAE quartz removal (same as electron-beam-induced etch), one must rely on AFM-measured height as an integral part of the repair process, where the prerepair bump height determines the FIB dose (or scan time) and postrepair AFM measurements checks for removal status. Here, a repeatable and stable quartz etch rate is critical to ensure consistent control. The nonregular shape of the defects inevitably make the etch process more challenging.

Damage from ion implantation and sputtering. When ion beam impinges on the mask surface whether during imaging or repair, some gallium ions will be implanted into quartz substrate resulting in a dark area commonly referred as gallium stains. The penetration depth is described by the mean projected range (R_P) and straggle (ΔR_P). The ion distribution in SiO_2 as simulated with TRIM is shown in Fig. 6.26 along with the tabulated values of total ion penetration depth ($R_P + 2\Delta R_P$) for few ion energies.[82]

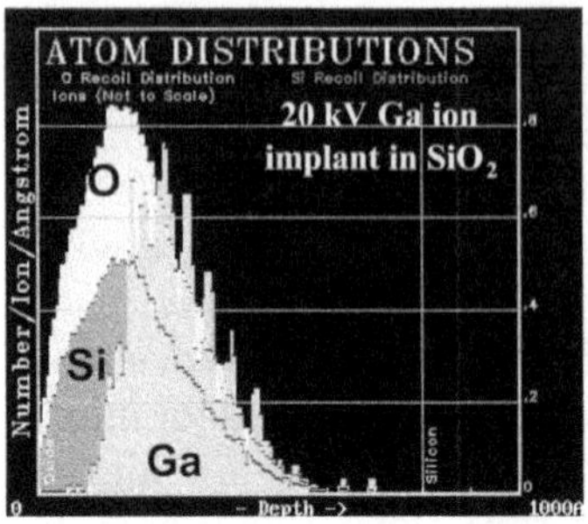

Ga ion implantation depth into quartz

Ion energy, keV	R_p, nm	ΔR_p, nm	$R_p + 2\Delta R_p$, nm
10	11.9	3.9	19.7
20	19.1	5.8	30.7
30	25.6	7.5	40.6

Figure 6.26 Gallium ion distribution, recoiled silicon and oxygen atoms, and ion projected range in SiO_2. (*Source:* Liang, T., A. Stivers, R. Livengood, P.-Y Yan, G. Zhang, and F. Lo, *J. Vac. Sci. Technol.* B18(6), p. 3216, 2000.)

For 30-kV FIB commonly used in repair tool, the ion depth is 40 nm, and this set the limit for any postrepair treatment such as wet and/or dry etch which removes a layer of quartz to rid of the gallium ions along the way. One of the implementations for APSM is a 40-nm undercut for image balance and, accordingly, quartz bump removal using FIB could precede prior such undercut process.[97]

The effect of gallium implantation on optical transmission loss was first measured by Prewett and Heard.[84] A more comprehensive study[98] was conducted more recently to evaluate transmission loss of fluorine (F)-doped fused silica for 157-nm lithography[99] on exposure to gallium and krypton (Kr) ions and electrons. One of the conclusions is the same as that of Prewett and Heard that at the same gallium ion dose the transmission loss becomes worse at shorter wavelengths, rendering the incapability of Ga FIB for 157-nm mask repair. Since the transmission loss is proportional to the gallium ion density inside the quartz, one must limit the ion dose in repair process development including using the lowest possible during defect imaging.

Two options have been proposed as alternatives to gallium ions for mask repair to reduce and eliminate transmission loss from substrate damage. One option is to use inert gas ion source such as Kr. Experimental results have shown that implantation of gas ions on quartz substrate causes essentially no transmission loss at 193 nm[100] and about 8 percent for 157-nm mask substrate.[98] Tremendous difficulties in developing FIB with gas ion sources, either using the "supertip" for *gas field ion source* (GFIS)[101] or the *multicusp plasma* (MP) ion source,[102] for a production worthy tool. Hence, there has no active effort today for alternative FIB development for mask repair applications.

Another quartz damage issue encountered in FIB mask repair is the so-called "river-bed" formation, which is a direct result of the physical ion milling of materials with similar sputtering yields and the milling rate dependence on incident angle. Overetching in a river bed causes phase errors as well as transmission loss from gallium stains. Figure 6.27

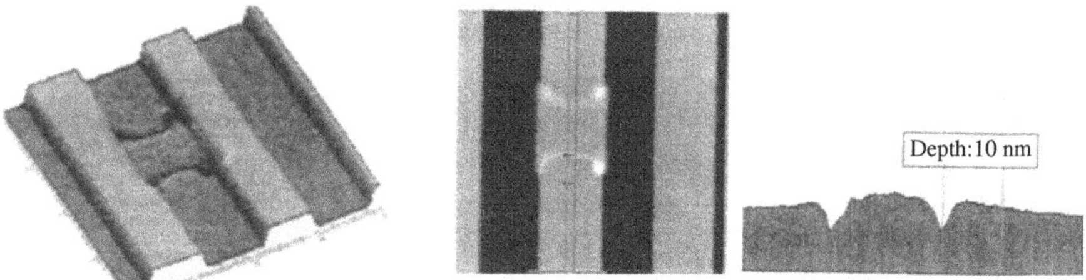

Top view Cross-sectional view

Figure 6.27 AFM scanned images of river bed formation after a chrome bridge is removed with FIB. The image at the right depicts the shape of the river bed. (Reprinted with permission from Hiruta, K., S. Kubo, H. Morimoto, A. Yasaka, R. Hagiwara, T. Adachi, Y. Morikawa, K. Iwase, and N. Hayashi, "Advanced FIB mask repair technology for ArF Lithography," *Proc. SPIE* 4066, p. 523, 2000.)

displays AFM images of FIB removed chrome bridge showing the river bed. Some relief can be achieved by carefully placing the repair box with respect to the defect and instituting a multipass GAE process where the repair box is adjusted according to the reduced defect size.[103] GAE with higher enhancement is also desired to reduce river bed.[85,86,104] Nanomachining and electron-beam-induced chemical etching has been proven to perform better for quartz removal than FIB.[105]

AFM nanomachining. Based on the AFM platform, materials can be removed by sharp mechanical nanotip scratching on the surface. Such nanomachining provides a quite straightforward technique for high resolution and precision mask repair. Today, RAVE LLC is the sole supplier for AFM-based repair systems and have been applied for repairing of many types of masks.[61,106–109] This technique is most suited for PSM repair.[110] The high accuracy of nanomachining comes at the cost of long repair time.

Because this is a physical process, most debris from the mechanical cut remains on the surface near the repair site. In practice, the debris is cleaned off the mask surface with a combination of sweep and cryogenic cleaning using CO_2. The sweep is to loosen the debris with the same nanotip used for cutting. Now the debris particles are bounded on the surface only with van der Waals forces. A plume of CO_2 in a mixture of solid particles and gas is directed to the mask surface. The debris is removed from the surface by momentum transfer from high-speed CO_2 particles and carried away by the CO_2 gas. The CO_2 solid and gas mixture is created by expansion of liquid CO_2 aerosol to free space from a nozzle positioned close to the mask surface. This "Eco-Snow"[111] process was developed by Eco-Snow Systems Inc.[112]

Figure 6.28 is a schematic description of the repair process. The defect is first scanned to provide size information for recipe setup. Then the defect is machined off and cleaned away with the Eco-Snow process. After cleaning, the site is scanned again for edge placement and depth (end-point) measurements.

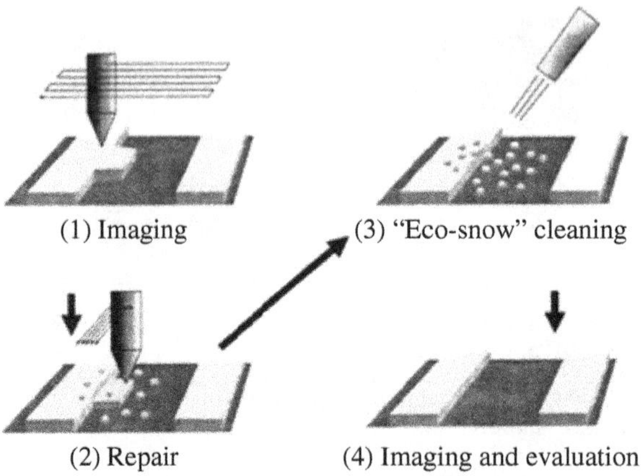

Figure 6.28 A simplified description of the major steps in an AFM-based nanomachining repair. (Reprinted with permission from Nishiguchi, M., T. Amano, S. Sasaki, Y. Morikawa, and N. Hayashi, "Photomask repair performance of the SiON/Ta-Hf attenuating PSM," *Proc. SPIE* 5256, p. 1200, 2003.)

Nanomachining for mask repair gives a high degree of control for edge placement and cutting depth, often achieved by iterative cut and scan. It has been found that an overcut into the substrate is needed to completely remove the targeted material (e.g., remove chrome on quartz). In general, for chrome and MoSiON, the overcut depths are about 10 and 5 nm, respectively, to ensure successful repairs. Overcuts are not needed for quartz removal in APSM mask repair, where the material removed is the same as the substrate material.

The shape of the nanotip determines the maximum aspect ratio of the repair as well as the mechanical strength of the tip. The cutting tip must therefore maintain a certain shape to be useful. Tip wear determines its usable life therefore influences the process cost of ownership. For line and space features, asymmetric tip or the tip can be positioned to provide one vertical face so as to maintain a vertical wall during repair.

Electron-beam-based repair. There are advantages in the use of electrons for mask repair.[51–53,113] An electron beam can induce deposition and chemical etching of materials of interest to mask repair while causing no damage to the mask substrate. An electron beam mask repair tool is similar to a FIB system, where the ion column is replaced with an electron column. Electron optics, with their magnetic lenses, generally have higher resolution than ion optics. Therefore, mask repair tools based on electron beams have a fundamental advantage in both resolution and damage. These advantages cannot be exploited until the difficulties associated with the technology are resolved. These are susceptibility to surface voltages, a high dose requirement, and no sputter yield. Beam deflection due to electric fields is proportional to the charge-to-mass ratio and the electron charge-to-mass ratio is over 100,000 times greater than that of a singly charged gallium ion. Surface charging has a higher potential to degrade repair position accuracy of electron beams compared to ion beams. A higher electron dose is needed for a repair compared to the ion case, so the potential for surface charging is higher. Finally, the damage-free property of electrons means they have no sputter yield. All material removal and deposition depends on chemistry alone. The technique trades off sputter yield for freedom from damage and the potential for high etch selectivities. These limitations have only been recently overcome by a combination of tool and process development.

Oxide surface charging can be reduced and high resolution retained by operating at an electron landing energy of about 1 keV. The total electron yield of SiO_2 at that energy is close to unity ($\delta + \eta = 1$, where δ is the *secondary electron* (SE) yield and η, the backscattering coefficient). Surface charging is dynamically balanced when the total electron yield is unity. Mask repair is not directly affected by collection of SE or backscattered electrons by a detector. The well-known effects of surface voltages on electron beam metrology techniques are not all pertinent to mask repair. The effect of surface charging can be compensated with low pressure gases in the chamber or other proprietary solutions.[114]

It is generally believed that the SEs are most responsible for the electron-beam-induced deposition and etching since one would expect a higher interaction cross section of low energy SE with the adsorbed species on the surface than that of the primary electrons. Some reports did correlate deposition rate with SE yield[115,116] which is higher at lower primary electron energy. Other experiments show no direct correlation.[116]

Figure 6.29 contains SEM micrographs of patterns deposited by a FIB and an electron beam. The diffuse tail of the focused ion beam spot is revealed in ion-beam-induced depositions. The term generally applied to this characteristic is *overspray*. By contrast, the corners of the electron

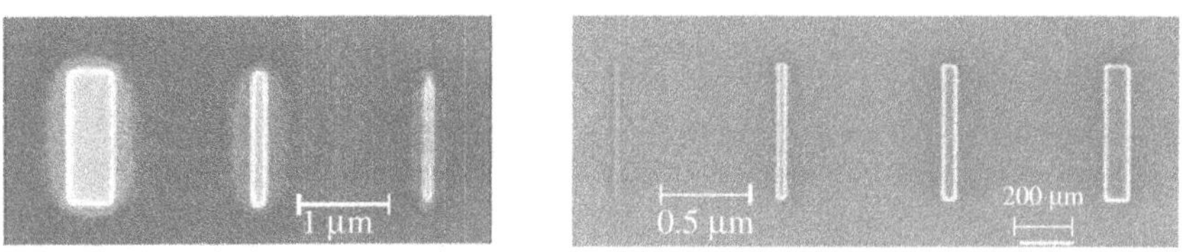

FIB Electron beam

Figure 6.29 Deposition comparison of resolution and overspray.

beam deposited features are sharp and free of overspray, reflecting the high resolution of the technique. Additional critical criteria for deposition are adhesion and durability against mask clean and exposure to DUV light. Both electron beam deposited carbon[95] and Pt/C mixture[51,52] are stable under multiple standard acid wet cleans.

Electron beam processes can remove quartz in high aspect ratio trenches with excellent profile control. Figure 6.30 presents an example showing an etch of 115-nm depth at the bottom of π-phase trench of about 350-nm deep from the chrome surface. Electron-beam-induced etching makes the quartz bump removal possible on a fully processed APSM mask. Electron beam techniques have demonstrated etching of 25-nm-wide lines into quartz.[52]

6.3.4 Repair qualification

There are two aspects in qualifying a repair process. One is to determine if the repair meets the resist print CD requirements, encompassing edge placement, transmission, and phase errors. The other is the repair durability to cleaning and exposure to illumination during lithography. These are also measured by their effect on resist printing. AIMS is generally utilized to measure the CD instead of actual resist print tests.

The mask is imaged under the exact illumination conditions (wavelength, partial coherence, and numerical aperture) that emulate the specific lithography tool in which the mask will be used. Figure 6.31 shows two such AIMS-captured images for a clear defect on a binary photomask before (Fig. 6.31*a*) and after (Fig. 6.31*b*) repair. A series of images are captured through the lithography tool's depth- of-focus range. Plotted in Fig. 6.31*c* are a series of five through-focus intensity profiles at the location indicated in Fig. 6.31*a*. Defect and repair qualification are normally defined as the CD change through focus. This is particularly true for high MEEF and phase defects.

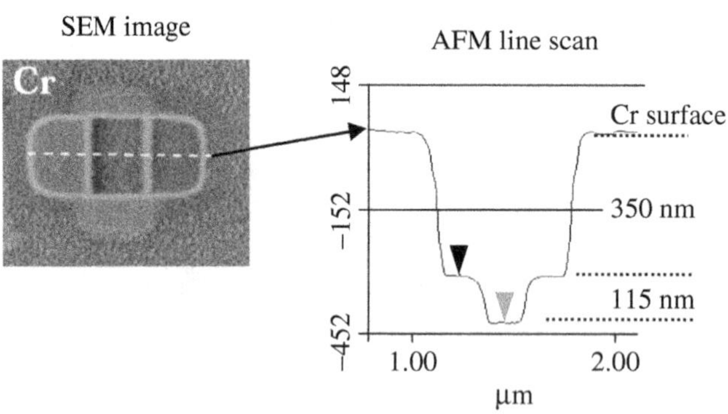

Figure 6.30 Quartz etch with XeF_2 results.

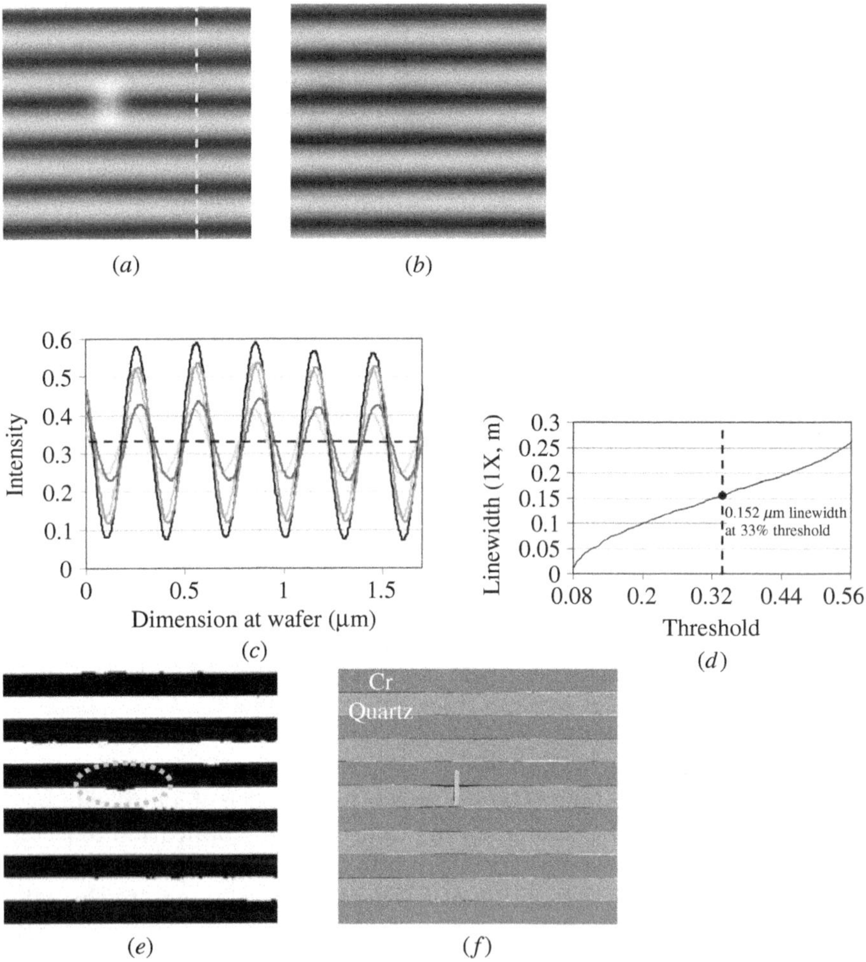

Figure 6.31 General description of AIMS analysis procedure for defect dispositioning and repair qualification: (*a*) pre-repair image (*b*) post-repair image (*c*) intensity profile through focus (*d*) plot of the linewidth versus intensity threshold (*e*) threshold image and (*f*) post-repair versus reference difference image.

An intensity threshold value must be chosen in order to measure the CD from the AIMS image. A plot of line width versus threshold as shown in Fig. 6.31*d* is used to determine the threshold value that produces the targeted line CD. In this example, the patterns are equal lines and spaces with a pitch of 0.305 μm (1×). From the plot, one finds that a 33 percent threshold gives a 0.152-μm linewidth. 33 percent is the isofocal point as shown by the dotted line in Fig. 6.31*c*.

To qualify the repair, one can compare the line CD at the defect with a reference line CD, all extracted from the intensity plot at 33 percent threshold at a chosen focus point. One can also use the difference image as show in Fig. 6.31*f* to evaluate CD changes. The difference image is produced by subtracting the reference image from the defect image Fig. 6.31*e*. The CD difference is then automatically read out and visually exhibited by the dark (extra chrome) or lighter (missing chrome) pixels as shown. Such procedure presents a two-dimensional view of the defect impact on wafer CD.

6.3.5 NGL mask repair

Mask repair for next generation lithography requires special considerations since both the materials and architectures are different from the transmissive optical masks. Brief description will be given in this section only to highlight the most promising repair techniques that meet the unique requirements of three unique types of masks—reflective mask for *extreme ultraviolet lithography* (EUVL), membrane mask for *x-ray proximity lithography* (XPL), and stencil mask for *electron and ion projection lithography* (IPL and EPL).

Reflective EUV mask. EUV mask repair must avoid damage to the delicate Mo-Si multilayer (ML) reflective blank. A commonly used absorber is TaN (or TaBN) about 70 nm thick. High performance EUV masks that use Ru capping layers (2- to 4-nm thick on top of ML) do not require SiO_2 buffer layers. Since ML absorbs UV and DUV light, we cannot use laser ablation. The overcut (~10 nm) required by AFM nanomachining of TaN will either destroy the top ML stack or require the use of a buffer layer. Therefore, mechanical nanomachining finds limited application to EUV mask repair. Gallium ion implantation into the ML will result in reflectivity loss via gallium absorption of EUV light and damage to the ML interface.[49] To avoid this, a buffer layer must be added to the mask. FIB may find limited use for deposition in clear defect repair.

The nondamaging nature of electrons makes electron-beam-based repair a feasible technology for EUV mask repair. The authors have demonstrated the promising capability of electron-beam-based deposition and etching for EUV mask repair,[51,52] which includes high resolution and high etch selectivity of TaN against Ru (~70:1). Therefore, electron beam mask repair enabled the new EUV mask architecture using Ru capping layer. Figure 6.32 shows an example of programmed pattern defect removal on EUV mask with 70-nm TaBN absorber and 2.5-nm Ru capping layer.[52,117] The left defect is removed as shown in Fig. 6.32*b* with 1-keV electron- beam-induced etching using XeF_2. The AFM scan indicated no obvious Ru loss after repair (as shown by the circled red line). For comparison, the gray line is a scan across the defect on the right that was not repaired.

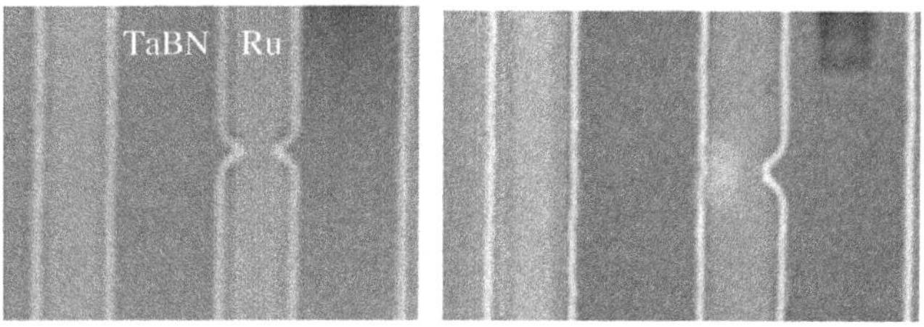

(*a*) SEM images of defects before and after repair

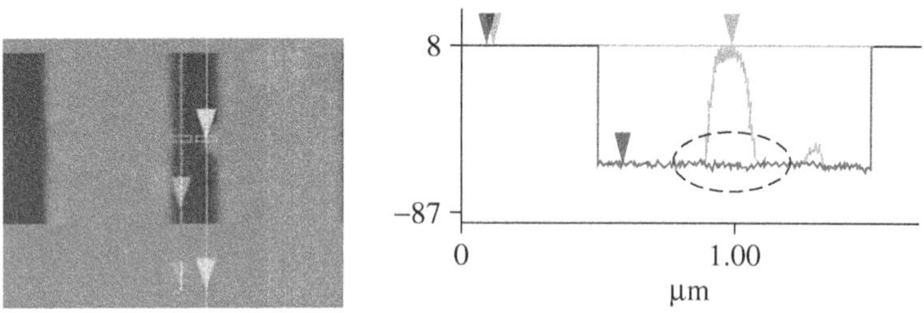

(*b*) AFM image and line scans

Figure 6.32 Programmed absorber defect removal on TaBN EUV mask with electron-beam-induced etching using XeF_2.

Membrane mask. Such type of mask is used for x-ray proximity lithography. The absorber of high-Z metals is supported by thin nitride or carbide membranes. Obviously, laser ablation and nanomachining cannot be utilized to repair membrane mask due to absorption and membrane break, respectively.

For etch and deposition, both FIB- and electron-beam-based methods are capable techniques, and the processes are similar to optical mask repair, except the materials to etch and deposit (high-Z metal is needed) are different. Since the high transmissivity of x-rays, gallium implantation is less a problem than for optical mask. Most research and development work was carried out in the late 1980s and early 1990s.[59]

Stencil mask. In principle, laser ablation could work for removal. FIB- and electron-beam-based methods will work for both etch and deposition. Since there is no supporting structure (except distributed struts) for imaging patterns and, therefore, there is nothing for the deposition to take place in a normal scan mode. So, clear defect repair requires special beam scan control such that the deposition is first anchored to the patterns and from where the deposition can proceed. Repairing a stencil mask truly requires real three-dimensional nanostructuring and both FIB and electron beam have proven to be capable of making such structures.

6.4 Photomask Cleaning

6.4.1 Introduction

Photomask clean requirements are driven by different criteria than wafers; however, photomask cleaning processes face similar challenges as wafer cleaning even though effects at the photomask and wafer levels are different. There are several analogies between photomask and wafer cleaning. Due to the 4× reduction during exposure, photomasks have traditionally had looser soft defect specifications than wafers. With the advent of embedded phase-shift masks soft defect specifications on photomasks are becoming tighter reaching the technology node (design rule) for the most critical layers.

Table 6.3 lists common wafer and photomask cleaning challenges highlighting differences and final effect on semiconductor manufacturing processes.

6.4.2 Photomask clean processes

There are several photomask cleaning steps in a typical photomask manufacturing process. A binary mask requires three basic cleaning steps—initial clean (includes resist strip or postresist strip), postmetrology-repair clean, and final clean. An additional pellicle adhesive clean can be required if the photomask is to be reworked postpellicle mount (see Fig. 6.8). Phase-shift masks requiring two or more levels of patterning steps have several additional cleaning operations—precoat clean (resist level 2), initial clean (postresist level 2 strip), postmetrology-repair clean (level 2), final clean (level 2), and rework (pellicle adhesive) clean (Fig. 6.33). Photomask clean becomes critical since any defects remaining after the first patterning step can affect the hard defect levels of the subsequent patterning steps.

A photomask clean process consists of five major steps:

Step 1: Acid step for organic removal. Nitric acid with peroxide oxidizer or traditional piranha chemistry (H_2SO_4:H_2O_2) are used to remove organics and in some processes can be used for resist strip. The chemical concentration can range from concentrated H_2SO_4 (97 percent) and H_2O_2 (30 percent) at ratios of 2:1 to 3:1 in single reticle cleaning tools to more dilute solutions (>10× deionized water dilution) in immersion photomask cleaning tools. The temperature of the H_2SO_4:H_2O_2 ranges from 60 to 120°C. Higher temperatures have stronger organics dissolution capability and are used for photomask clean processes relying on acid clean for stripping resist.

Step 2: Rinse. The acid organic removal step is followed by a rinse step using *deionized water* (DI). In single photomask processing tools DI water is dispensed on the photomask directly on the spin station, while

TABLE 6.3 Photomask and Wafer Cleaning Common Challenges and Effect Comparison

Challenge	Photomask	Wafer	Comments
Soft defect (SD) specification (technology node, design rule)[48]	Defect-free mask: zero remaining light absorbing defects in active area at soft defect spec size.	Defect density: <0.07 defects/cm^2 at half minimum feature size for technology node.	One photomask killer defect will print on every wafer die resulting in 0 percent wafer/lot yield.
Soft defect dispositioning size (technology node)[48]	Photomask soft defect dispositioning size follows wafer technology node (minimum feature size on wafer, design rule).	Particles 1/3 wafer minimum feature will affect function of wafer die. Metrology capability is limiting wafer particle size dispositioning to 1/2.	4× reduction from photomask to wafer allows for looser photomask soft defect size specifications.
Backside defects[48]	Photomasks are transmissive. Glass side defects are dispositioned at looser SD specification than front side defects (>10× looser spec).	Backside defects, if large >5× design rule, can affect wafer focus plane. Defects introduced through wafer process (chucking) can be source of cross contamination during wafer clean process.	Due to photomask transmissive nature, photomask handling (edge) is less prone to backside defects generation and/or cross contamination.
Anions, cations, transition metals[118–123]	Residual SO_4^{2-} from piranha clean and ammonium silicate from NH_4OH clean act as seed for ammonium sulfate crystals formation on stepper exposure. SO_4^{2-} levels as low as 2 ng/cm^2 can result in crystals formation[119]	Alkali metals and heavy metals damage electrical characteristics of transistor device (increase in junction leakage current, degradation in breakdown voltage of oxide film) if present at >10^{10} atoms/cm^2.	Cumulated stepper exposure results in photoinduced defects formation on photomasks. (crystals, contamination). Photomasks can receive >10 kJ/cm^2 exposure dose over product lifetime.
Substrate etching with NH_4OH[124–128]	EPSM MoSi chemical etching results in EPSM phase loss affecting photomask optical properties.	Etching results in microroughening of silicon surface (affecting dielectric breakdown field) and generates crystal pits.	Photomask clean process parameters must combine cleaning efficiency, phase loss reduction (EPSM), and minimal residual anions/cations.

(Continued)

TABLE 6.3 Photomask and Wafer Cleaning Common Challenges and Effect Comparison (*Continued*)

Challenge	Photomask	Wafer	Comments
Mechanical aspect of clean[129–141]	Megasonic-full- jet-induced damage: chrome lifting, chrome thinning (APSM, OPC features)	High aspect ratio poly lines sensitive to mechanical aspect of clean (megasonic, full jet)	Photomask aspect ratios are lower (<2.5) than wafer aggressive aspect ratios. Subresolution features unique to photomasks are prone to clean-induced mechanical damage.
Environmental inorganic (NH_3, SO_2) and organic vapors in wafer fab[118–123]	Inorganic vapors (photomask storage) react with ammonium silicate and residual SO_4^{2-} to form ammonium sulfate crystals.	Environmental NH_3 vapors and organic bases (amines) affect resolution of wafer chemically amplified photoresists.	Both organic and inorganic vapors in wafer fabs contribute to photoinduced defects on photomasks.

NOTE: Soft defects refer to cleanable defects and include particles, contaminations, and crystals/haze.

in immersion stations the photomask is immersed in a rinse tank overflowing with deionized water.

Step 3: Particle removal. A particle removal process follows the organic removal step and involves NH_4OH or SC1 ($NH_4OH : H_2O_2$) chemistries coupled with megasonic energy providing mechanical removal of dislodged particles. Particle removal relies on the etching action of NH_4OH base solution. Due to the high pH of the solution (pH >10), the particles, dislodged from the substrate through etching, have negative zeta potential. This results in repulsive forces between the particles and the photomask substrates (SiO_2, chrome, MoSi) that prevent the dislodged particles from redepositing onto the substrate. Megasonic energy is applied simultaneously and allows for lift off of the particles which are subsequently transported away by diffusion and convection in both single reticle and immersion type photomask clean platforms. High frequency (>1 MHz) is used to target submicron particle removal. Optical proximity correction features on binary photomasks as well as features with chrome undercut on APSM 180° features (see Fig. 6.4*c*) are prone to mechanical damage through chrome lifting. Megasonic process parameters (frequency, power, and flow dynamics) have to be optimized in order to achieve optimal balance between particle removal and photomask pattern damage minimization.

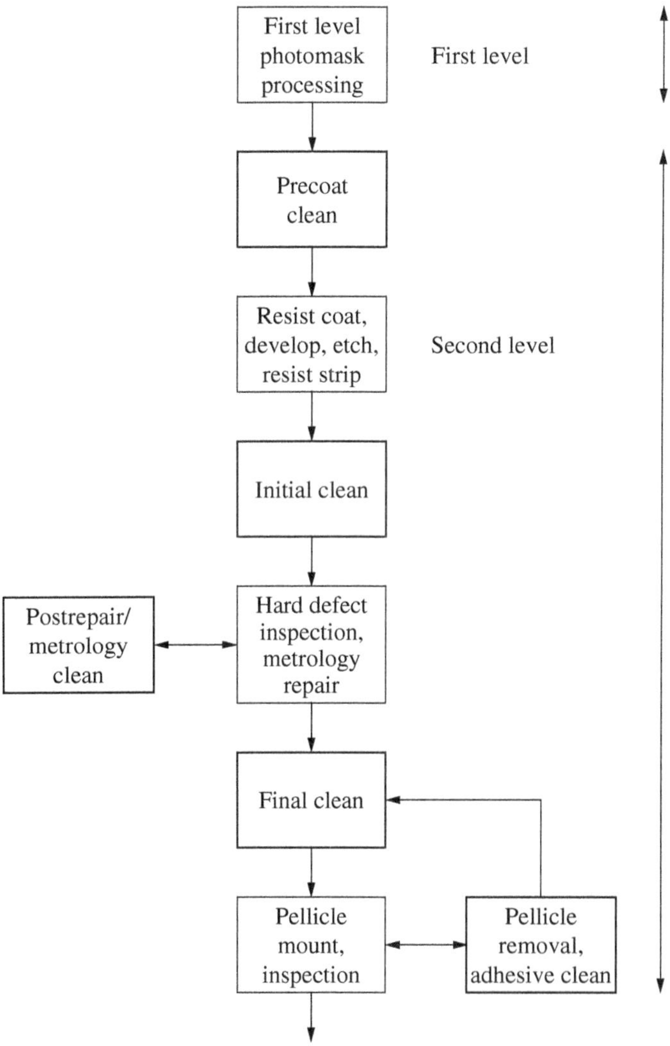

Figure 6.33 Cleaning steps in two level photomask patterning process (EPSM, APSM).

Step 4: Rinsing/neutralization. Rinsing using deionized water coupled with or without megasonic energy follows the particle removal step. High temperature deionized water is recommended to rinse off any sulfate ions residual from the piranha clean process. Residual NH_4^+ ions from the particle removal react with SiO_2 quartz substrate to form ammonium silicate which can also act as seed for photoinduced defects forming on cumulated 248- or 193-nm stepper photomask exposure.[118–123]

Step 5: Drying. Spin rinse dry processes are compatible with single reticle photomask clean platform while *isopropyl alcohol* (IPA) dry processes are implemented on immersion-type platforms. The challenge of spin rinse dry is to overcome static buildup during high-speed spin process which can lead to subsequent static damage. Eliminating static charge on the photomask surface also prevents particle redeposition. IPA drying relies on IPA concentration gradients that lead to surface tension gradient created at the IPA/substrate interface. The Marangoni flow goes from low to high tension gradients leaving the substrate free of water droplets. The surface gradient drying IPA technique is traditionally used to dry hydrophobic substrates for which spin rinse drying technique may not be adequate.

Strategies to avoid particle redeposition during and postmask cleaning are discussed in Sec. 6.4.4.

6.4.3 Photomask substrates

Photomask types and substrates have been mentioned in Sec. 6.1.1. Binary and alternating phase-shift mask materials include chrome and quartz. Phase-shift materials used in halftone phase-shift masks (PSM) and Tritone PSM include MoSi, TiSiON, ZrSiO, TiSiO/Cr, and TaSiO for 193-and 157-nm lithography.

Quartz. Due to the transmissive nature of photomasks, residual absorbing soft defects on quartz have the potential, depending on the location on the die, to print at the wafer level resulting in zero wafer yield. Soft defects on quartz in critical device areas are dispositioned according to specifications that become tighter with the technology node.

Traditional quartz cleaning chemistries include piranha (H_2SO_4:H_2O_2) for organic removal followed by ammonium hydroxide containing chemistries for particle removal. Exposure of SiO_2 to NH_4^+ during the NH_4OH step results in ammonium silicate Si-O-NH_4 formation.[142,143] Diffusion of SO_4^{2-} in the porous quartz matrix during the piranha step has been reported as major cause for photoinduced ammonium sulfate crystals formation on cumulated exposure in 193-nm and 248-nm lithography.[118–121,142,143] During stepper exposure residual SO_4^{2-} diffuses out off the quartz matrix to react with environmental NH_3 vapors or with ammonium silicate present on the surface of the photomask to form ammonium sulfate crystals [$(NH_4)_2SO_4$]. UV 172-nm flood exposure preceding a hot DI water rinse introduced as final photomask cleaning steps has been reported as an efficient way to simulate laser photoexposure. This causes diffusion of embedded SO_4^{2-} reaction which then reacts with sources of NH_4^+. The photoinduced $(NH_4)_2SO_4$ is then eliminated during the subsequent hot DI-water rinse step.[122–124]

Chrome. Chrome on a binary mask consists in a 70- to 120-nm-thick film containing chromium, nitrogen, and oxygen. An antireflective layer deposited on top of the chrome film reduces mask reflectivity and therefore stray light in the exposure tool. Soft defects on chrome are typically organic photoresist material residues from the photomask patterning step. Even though contamination on chrome has no wafer printability effect, soft defect on chrome must be eliminated as residual contamination can dislodge during subsequent cleaning steps and redeposit over quartz and/or shifter material. Residual contamination on chrome can also become nuisance defects complicating soft defect inspection. Residual particles under pellicle located on chrome have to be eliminated through cleaning due to the potential of the particle to dislodge during photomask handling, and redeposit over a transmissive area creating absorbing killer defects on quartz or shifter.

Chrome damage (lifting, thinning, faded antireflective coating) can occur during photomask cleaning and is mostly induced by the mechanical portion of the clean process, i.e., megasonic, full jet, brush). Chrome undercut areas located at the quartz 180° shifter/Cr interface of APSM photomasks are prone to chrome lifting during photomask clean (Fig. 6.4). Serifs and other small features added by aggressive optical proximity correction are also prone to chrome damage. In addition, extensive cleaning and exposure to H_2SO_4 and NH_4OH chemistries can lead, with cumulated cleans, to the formation of ammonium sulfate resulting in etching of antireflective coating.

Phase-shift material. The partially transmissive phase-shift material in PSM shifts the phase of the light. Thickness and roughness control in PSM is crucial as changes in these parameters can effect the phase of the light. Clean process improvements for MoSi shifter materials used in 193-nm lithography are reported.[122,124–128] Due to the aggressive nature of traditional piranha/NH_4OH photomask clean processes, alternative chemistries are being evaluated. 172-nm ultraviolet cleans (UV172 nm) and ozonated water are being evaluated as an alternative to H_2SO_4:H_2O_2 for the organic removal step[122,124–126] and hydrolysis water as an alternative to NH_4OH for the particle removal step.[127] In addition, temperature reduction has proven effective in reducing phase loss induced by MoSi shifter etching during the neutralization step with NH_4OH-based chemistry.[127] The introduction of new shifter materials for future lithography generations presents a challenge to combine cleaning efficiency, minimize phase loss, and eliminate residual anions and cations acting as seeds for photoinduced defects.

6.4.4 Photomask cleaning platforms

Photomask cleaning equipment has traditionally followed wafer cleaning equipment development. Two basic platforms are used for photomask processing. Single reticle processing (Convergent, Pretech, STEAG-Hamatech, Shibatec) and immersion cleaning platforms(Akrion, WACOM, Convergent). Single reticle process tools include one spin station in which all photomask clean steps take place (acid, rinse, neutralization with megasonic, rinse, spin rinse dry). Different chemistries are distributed through dedicated nozzles and chemical mixing takes place at point of use prior to nozzle dispense. The immersion cleaning platform is similar to traditional wafer bench systems and includes cleaning tanks with dedicated chemistries. A typical immersion photomask system includes piranha clean, ozonated water clean (with megasonic), SC1 (with megasonic), and IPA dry.

New platforms must support the increasing need for chemical purity and DI water quality. All sources of contamination must be minimized or eliminated. Rinsing tanks, in addition to chemical tanks, are also dedicated to specific cleaning chemistries to prevent cross-contamination. Strategies for preventing contamination include single pass chemistry, chemical and DI water filtering, bacteria elimination strategies for DI water (dead leg removal, UV ozone), single photomask clean process, high-quality microenvironments, usually with *standard mechanical interfaces* (SMIFs), and ionization of cleaning tool microenvironment for *electrostatic discharge* (ESD) prevention. Practical factors such as use of clean room space, energy consumption, and chemical and DI water consumption are forcing the redesign of photomask cleaning equipment.

6.4.5 Novel photomask cleaning processes for nanoparticle removal

Wet cleaning chemistries have been used as traditional processes for global particles/contamination removal. The shrinking of semiconductor devices requires removal of sub-100-nm soft defects for 90-nm and future technology nodes. Fundamental understanding of particle adhesion and removal mechanisms should drive the development of global and/or local cleaning processes for damage-free removal of sub-100-nm particles on photomasks.

Particle adhesion fundamentals. Particle adhesion forces in a liquid media can be approximated as the sum of van der Walls forces, deformation forces, and electrostatic double layer forces.[144–148]

Adhesion force. The adhesion of a submicron particle on a substrate is dominated by the van der Waals forces, which is given by

$$F_{vdW} = \frac{AR}{6z_0^2} \tag{6.8}$$

The adhesion force can cause particle or substrate deformation for soft particles on hard surfaces and the adhesion force increases with increasing deformation at the interface. The total adhesion force consists of two additive components, namely, the force acting between the adherents before deformation at the instant of the first contact, F_{vdW}, and the force acting on the contact area due to the deformation. The total adhesion force including van der Waals force and adhesion-induced deformation:[149–156]

$$F_a = F_{vdw} + F_{\text{deformation}} = \frac{AR}{6z_0^2}\left(1 + \frac{a^2}{Rz_0}\right) \tag{6.9}$$

where A = Hamaker constant, for two dissimilar materials "1" and "2" in the medium "3" $A_{132} = (\sqrt{A_{11}} - \sqrt{A_{33}})(\sqrt{A_{22}} - \sqrt{A_{33}})$

R = particle radius

z_0 = separation distance between the particle and the substrate (for smooth surfaces, it is taken as 4 Å)

a = contact radius between the deformed particle and the surface

which is given by Eq. (6.10):

$$a = \sqrt{\frac{2W_a R}{H}} \tag{6.10}$$

where W_a is work function of adhesion between particles and surface, $W = 2\sqrt{\gamma_1\gamma_2}$, γ_1 and γ_2 are surface free energy of the two contact materials, and H is the hardness of the deforming material (softer material).

All materials immersed in an aqueous environment acquire a surface charge due to preferential adsorption of ions present in the liquid phase or due to dissociation of surface groups. The charge on the surface of the particle or any other surface immersed in the liquid is balanced by an equal but oppositely charged layer in the adjacent liquid, resulting in a so-called electrical double layer. The Deryagin-Landau-Verway-Overbeek (DLVO) theory[157–161] estimates the repulsive and attractive forces resulting the overlap of electric double layers and the London-van der Waals force, as a function of interparticle distance. The total interaction (F_T) occurring between the particles and the photomask surface in the solution is the sum of van der Waals force (F_a) as introduced in Eq. (6.9) and the electrostatic force (F_R), as shown in Eq. (6.11).

$$F_T = F_a + F_R \tag{6.11}$$

The calculation of the electrostatic force is complex due to the overlapping of the diffused double layers between two surfaces and must rely on numerical solutions or various approximations. Overbeek's approximation can be used for the calculation of the electrostatic force between the wafer and particle due to larger than ±25-mV zeta potentials on most surfaces, as shown in Eq. (6.12):

In the case of a charged particle near a charged surface, the electrostatic double layer force is given by

$$F_R^{\Psi} = \pi\varepsilon_r\varepsilon_0 a\left(\Psi_{01}^2 + \Psi_{02}^2\right)\frac{\kappa e^{-\kappa H}}{1 - e^{-2\kappa H}}\left[\frac{2\Psi_{01}\Psi_{02}}{\Psi_{01}^2 + \Psi_{02}^2} - e^{-\kappa H}\right] \tag{6.12}$$

where F_R^{ψ} = electrostatic double layer force interacting between a sphere of radius R and a plate with constant potential

ψ_{01} = zeta potential of the particle of radius R

ψ_{02} = zeta potential of the substrate

ε_r = dielectric constant of the medium

ε_0 = dielectric permittivity of a vacuum

κ = Debye-Huckel parameter of the electrolyte solution

Drag force. Flow past a particle leads to a drag force acting on the particle. The drag force on a spherical particle in a newtonian fluid can be expressed by Eq. (6.13)

$$F_d = \frac{C_D}{C_C}\frac{\pi}{8}\rho_l d_p^2 u^2 \tag{6.13}$$

where C_D = drag coefficient

ρ_l = density of the cleaning liquid

d_p = diameter of the particle

u = streaming velocity

C_C = Stokes-Cunningham slip correction

In water, $C_C \cong 1$. However, in the air it cannot be neglected and is given by

$$C_C = 1 + 2\text{Kn}(1.257 + 0.400e^{-0.550\text{Kn}}) \tag{6.14}$$

where the Knudsen number Kn is given by

$$\text{Kn} = \frac{2\lambda'}{d_p} \tag{6.15}$$

where λ' is the fluid's mean free path.

Removal mechanism. Three possibilities exist when a submicron particle detaches from a surface in liquid media. The particle may either slide, roll, or be lifted away.[162–164] The drag force introduced by the cleaning mechanism, the electrostatic double layer, and the adhesion force becomes crucial to particle removal mechanism. When the removal force overcomes the adhesion force, the particle is lifted by sliding. The force required to remove a particle by rolling is less than that required by sliding or lifting. When the removal moment overcomes the adhesion resisting moment, the particle is removed by rolling. The ratio of the removal moment to the adhesion resisting moment is given by Busnaina et al. as shown in Fig. 6.34.[164]

Both adhesion and removal forces decrease as the particle size decreases. However, the removal force decreases much faster than the adhesion force because of the quadratic relationship with particle compared to a linear dependence for the adhesion force. Therefore for fixed process parameters the removal moment/adhesion moment (RM) ratio decreases with decreasing particle size indicating that certain removal mechanisms may reach a threshold for efficient particle removal at small particle sizes. Particle size thresholds of 100 nm have been reported by Olim[165] and Zhang[166] for megasonic cleaning process on silicon substrates.

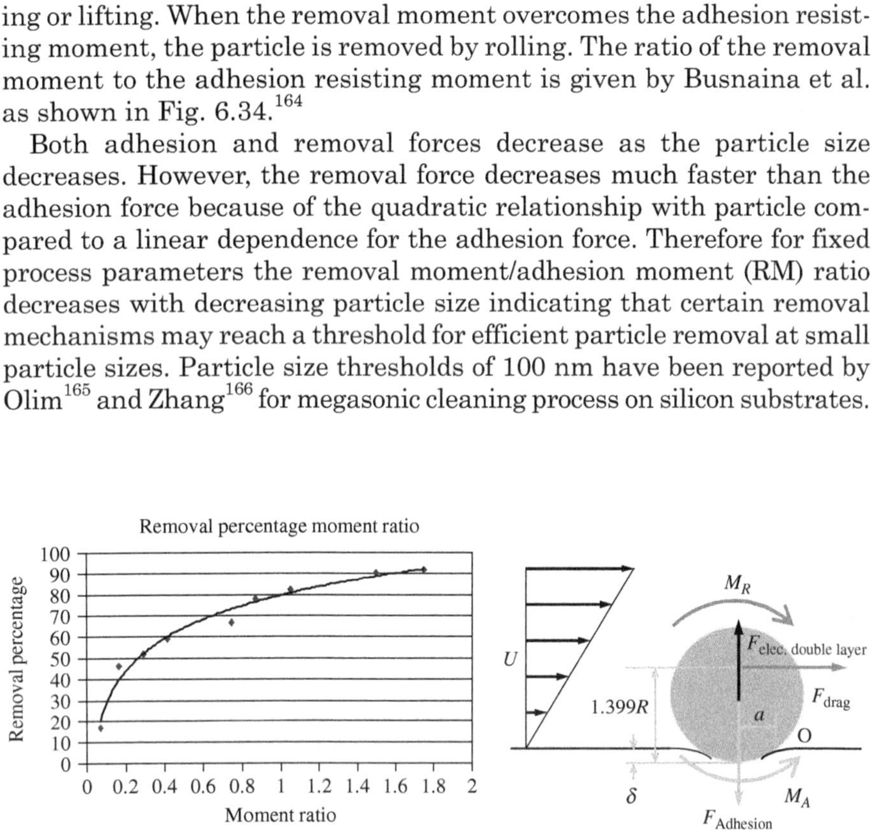

- Removal percentage vs. moment ratio (silica removal experiment, Ref x).
- The figure shows when RM > 1, 80 percent of particles are removed.

Rolling removal mechanism

$$\text{RM} = \frac{\text{Removal moment}}{\text{Adhesion resisting moment}}$$

$$\text{RM} = \frac{F_d(1.399R - \delta) + F_{\text{dl}} \cdot a}{F_a \cdot a}$$

Figure 6.34 Removal moment for particles on a silicon substrate δ is defined as follows: $\delta = R - (R^2 - a^2)^{1/2}$. (*Source:* Busnaina A. A., H. Lin , N. Moumen, J. W. Feng, and J. Taylor, "Particle adhesion and removal mechanisms in post-CMP cleaning processes," *J. of Semiconductor Manufacturing, IEEE Trans.* 15, pp. 374–82, 2002.)

Megasonic energy, high pressure jets, and spinning of the substrate to provide centrifugal force and fluid flow across the substrate are techniques used in photomask cleaning to force particles to slide or/and roll off. Brush cleaning is also an efficient rolling particle removal mechanism, however, due to concerns with scratching, brush cleaning should only be used on the backside and is finding limited use on the front side or active area of photomasks.

Megasonic cleaning. Megasonic cleaning is the most widely used noncontact cleaning technique.[129–135] The dominant mechanism of particle removal in megasonic includes cavitation, microstreaming, and Schlichting streaming In megasonic cleaning a thin boundary layer is generated on semiconductor surfaces. Within this boundary layer, also referred to as acoustic boundary layer, (~0.6 μm for silicon at 1 MHz), fluid motion with vortices (Schlichting streaming) and micron size bubbles are generated (microstreaming and cavitation). Particle removal in megasonic cleaning relies heavily on acoustic streaming and the reduction of the boundary layer thickness on the substrates. The removal drag force induced by acoustic streaming is larger than that induced by hydrodynamic flow at the same velocity. For acoustic streaming flow with constant megasonic intensity, higher frequency leads to a larger drag force because of a larger streaming velocity and a thinner acoustic boundary layer.[130–135] Megasonic energy has also been reported to induce pattern damage due to implosion of cavitation bubbles resulting in transfer of cavitation energy to the surrounding media.[139,140,141] Pressure gradients generated by the megasonic waves propagating into the fluid can also exert damaging forces. In addition, the oscillatory nature of the acoustic field can result in fatigue damage. The cavitation threshold in fluids increases with frequency and is also reduced with gaseous species dissolved in the fluid. As a cavity collapses, the fluid around the bubble can be subjected to extreme temperatures and pressures causing excitation of the molecules present in the gas or fluid and leading to the emission of light on relaxation also called sonoluminescence.[167] Sonochemical reactions in water decompose water molecules into hydrogen atoms and hydroxyl radicals and dissolved gases can lead to the formation of other chemical species, i.e., nitrogen gas leading to formation of nitrogen oxides.[168,169] Experimental cleaning studies on wafers have confirmed that gas dissolution can improve cleaning efficiencies for nanoparticle removal and provide higher bubble stability resulting in less violent cavitation.[170] The development of megasonic-based damage-free sub-100-nm particle removal processes on photomask substrates will require optimization of megasonic parameters (frequency, power, gas dissolution) and fundamental understanding of cavitation and sonoluminesce effects using both experimental and modeling approaches.

Dry cleaning techniques—particle removal. Even though van der Walls forces are smaller in liquid than in air, the impulse forces resulting from cleaning processes providing high velocities (>1 km/s) exceed by several orders of magnitude than the van der Walls adhesion forces attracting particles to substrates. Novel dry cleaning processes such as laser plasma shock,[171–177] cryokinetic aerosol cleaning, and electrohydrodynamic cleaning[178] can provide impact velocities >1 km/s. Concerns with pattern damage due to mechanical and localized heating stresses and plasma effect have to be fully addressed.

Nevertheless, the development of dry, nondestructive cleaning techniques shows promise for localized defect removal and the achievement of defect-free photomasks for sub-65-nm nodes. The introduction of dry cleaning techniques also addresses the concerns of chemical waste, chemical purity, and DI water consumption associated with traditional wet cleaning processes.

Laser plasma shock. Extensive literature exists on the use of short pulsed lasers for direct pulsed irradiation processes.[171–177] Inertial forces generated induce rapid thermal expansion resulting in high surface acceleration adequate for particle removal. A pulsed laser beam converged to the focal point of a convex lens can cause a local increase in temperature and energy density leading to dielectric breakdown of air, forming a microplasma. The expansion of the plasma and the hot core of air surrounding the plasma create an expanding spherical shock wave with velocities ranging from 1 up to 10 km/s. The pressure at the shock wave can be directed to a particle on a surface. The force results in particle debonding through rolling and/or sliding. Damage-free removal of 60-nm particles on silicon wafers has been reported using a 370-mJ pulsed laser at 1064-nm wavelength.[171] Interaction between the plasma and substrate material can induce surface roughening and pattern damage through mechanical and thermal effects. Additional work is required to develop damage-free laser plasma shock cleaning processes integrated with inspection for local particle removal applicable to future generation photomasks.

Dry cleaning techniques—contamination removal. Surface cleaning with ultraviolet radiation at 253- and 172-nm wavelengths is an established cleaning technique for organic removal.[179] Ozone generation occurs during UV exposure and forms a three way reaction of atomic with molecular oxygen. In the presence of water vapor, highly-reactive hydroxyl radicals are formed through a hydrogen abstraction reaction of atomic singlet oxygen with water. This process initiates oxidation and decomposition of organic compounds. In addition, photochemical decomposition of organic compounds is also initiated with free radicals formed by the dissociation of oxidative additives introduced in the UV chamber.[179–181] UV 172-nm processes have been implemented in photomask cleaning manufacturing for both organic removal and anion/cation reduction through diffusion-induced photoreactions.[122,124]

Future photomask cleaning technologies. Additional novel cleaning techniques such as cryokinetic aerosol,[182,183] CO_2 snow,[184] supercritical fluid,[185,186] vacuum cavitational streaming,[187] and electrohydrodynamic[178] are being evaluated for damage-free nanoparticle removal on both wafers and photomasks. The photomask industry should leverage on wafer clean process optimization learning and breakthrough developments in wafer cleaning technologies.

6.5 Pellicalization

6.5.1 History of pellicles and their requirements

Pellicles or protective covers applied to photomasks were introduced in the late 1970s to increase wafer yields in the semiconductor industry.[188–195] Pellicles act as particle protection for photomask/pellicle assemblies. The separation between the active photomask area and the pellicle film is kept at a distance (5 to 10 mm) sufficient to ensure that particles falling onto the pellicle film during reticle handling be kept out off the focus range of the photolithography stepper therefore preventing for landed particles to create a printable shadow at the wafer level (Fig. 6.35), Photomasks of dimensions 5 in. × 5 in. × 1/8 in. require pellicles on both sides of the photomask. Pellicles are only required to protect the active photomask area on 6 × 6 × 1/4 in. photomasks. The thicker substrate allows for looser specifications to be applied to glass side defects. Pellicle defect size specifications are orders of magnitude looser than that applying for the active photomask area. Photomasks are typically dispositioned for submicron defects on the active photomask area while photomasks with pellicle defects larger than 10 μm have no wafer printability effect.

6.5.2 Pellicle specifications

Pellicles are composed of four basic components—metallic frame, pellicle membrane, adhesives, and pressure relief valve located on the frame (Fig. 6.36). The metallic frame is typically made out of anodized aluminum, and its critical parameters are height and flatness. The height must be at least 5 mm to keep particles on the pellicle surface out stepper focus range. The flatness must be better than 50 μm for easy adhesion to the mask. The pellicle membranes first introduced for i- and g-line lithography were made out of nitrocellulose. Fluoropolymers were introduced for DUV lithography (248- and 193-nm wavelengths) pellicle

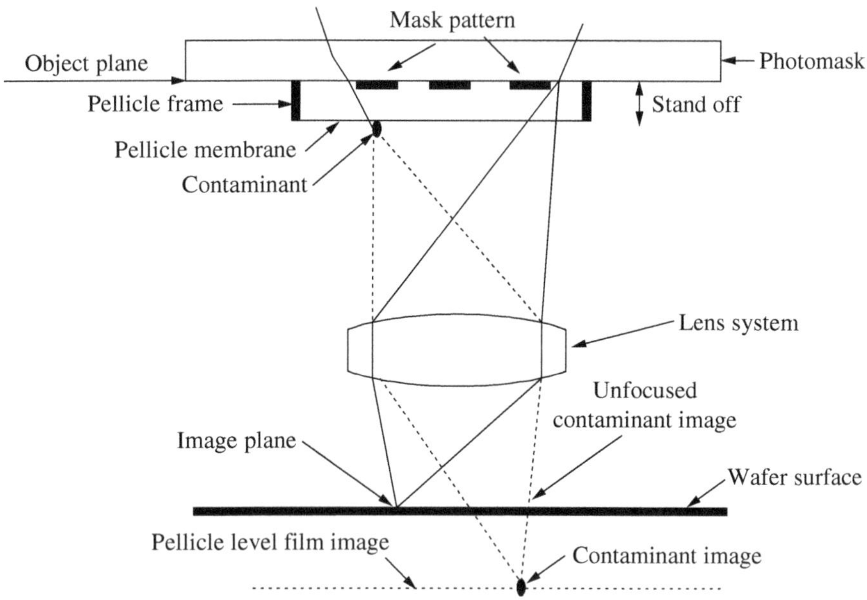

Figure 6.35 Pellicle-photomask assembly in optical lithography tool.

membranes due to optical transparency requirements (>99 percent transmission at exposure wavelength). Fluoropolymers used for DUV pellicles are either Cytop (Asahi Glass)[195] or Teflon AF (Dupont Photomask)[196] (Fig. 6.37).

The pressure relief valve releases pressure buildup in the pellicle-reticle interspace occurring during air shipment. Pressure is relieved through a hole in the aluminum frame covered with a microfilter preventing particles from entering the pellicle interspace area. Three types of adhesives are used in the pellicle assembly—the membrane adhesive bonding the pellicle membrane to the frame, the mask adhesive

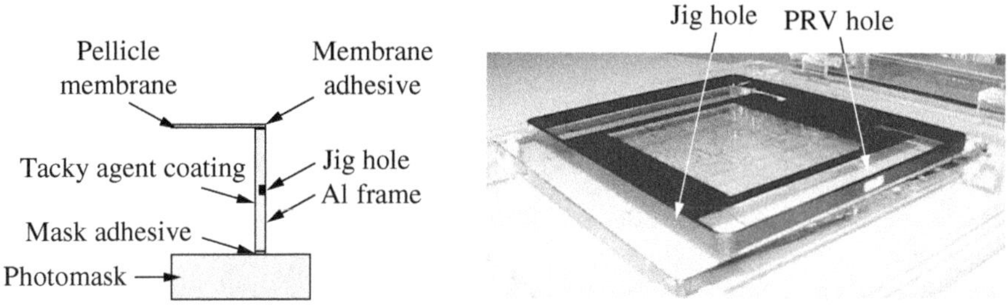

Figure 6.36 Photomask-pellicle assembly. Left: schematic detail; Right: photograph of mounted pellicle.

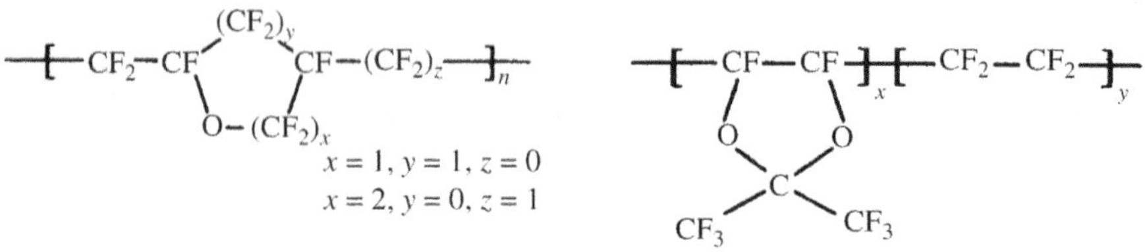

Figure 6.37 Polymer structure of DUV pellicle membranes.

bonding the frame/membrane assembly to the photomask, and the tacky agent lining the internal portion of the frame to trap any particles moving in the pellicle-photomask interspace (Fig. 6.36). The key criteria for pellicle adhesives are—(1) to provide adequate seal to prevent particles from migrating within the pellicle-photomask interspace and (2) low outgassing to prevent photoinduced defect formation prevalent in DUV lithography. Pellicle adhesives used by Mitsui, FST, Shin Etsu, MLI, and Inko manufacturers are varied and include fluoropolymers, polyurethane, acrylic, silicone, and epoxy materials. Pellicle critical specifications are summarized in Table 6.4. Pellicle manufacturing takes place in Class 1 to Class 10 clean room environments, and pellicles are required to meet particle specifications of zero particles >0.5 μm for DUV lithography. The application of the pellicle to the photomask can be performed in horizontal or vertical configuration using manual or fully automated tools. The pellicle mounting process requires precise control for mounting pressure and pellicle placement accuracy (±0.3 mm). Figure 6.38 illustrates a horizontal semiautomated pellicle mounter

TABLE 6.4 Specifications for 193 nm Pellicle and Frame

	Attribute	Specification
Membrane	Transmission	>99 percent at 193 nm
	Maximum particle count	None of size >0.5 μm
	Lifetime	Retains >99 percent transmission and zero photoinduced defects with >4 kJ/cm^2 exposure
	Strength	No breaks or distortion by 30 psi air jet
	Material	Aluminum
Frame	Size	Depends on application
	Flatness	<50 μm peak to valley
	Inner wall adhesive	
	Material	Tacky agent
	Particle	No particles > 100 μm size
	Vent hole	PTFE membrane with 0.3 μm filter

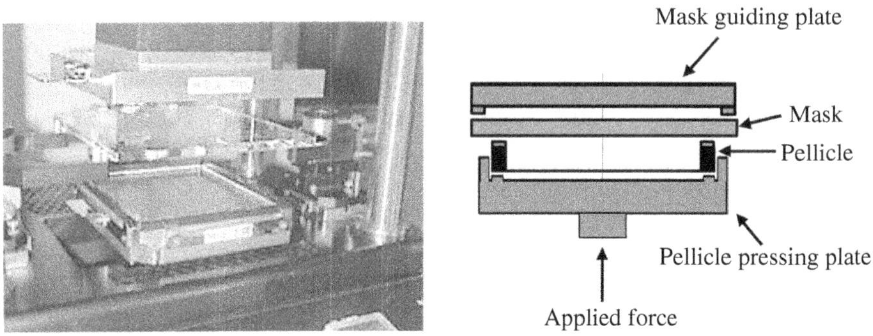

Figure 6.38 Semiautomated pellicle mounter.(*Source:* Cotte, E. P., R. L. Engelstad, E. G. Lovell, F. Eschbach, D. Tanzil, "Effects of soft pellicle frame curvature and mounting process on pellicle-induced distortions in advanced photomasks," *Proc. SPIE* 5040, pp. 1044–1054, 2003.)

the critical components of which are the reticle guiding plate (for placement accuracy) and the pellicle pressing plate (for adhesive seal). Mounting pressure conditions are adjusted based on the adhesive compliance. Pellicle-induced distortion studies were conducted on photomask-pellicle assemblies. The flatness of the pellicle frame and the mechanical properties of the mask adhesive were determined the parameters to have the most effect on the photomask registration after pellicle assembly.[197–206]

6.5.3 Pellicle lifetime requirements and performance (193 nm)

Pellicles are expected to maintain high transmission (>98 percent) at the stepper exposure wavelength for the lifetime use of the photomask in high-volume manufacturing environments. Depending on the device lifetime this requirement can translate into >1000 wafer lots exposures corresponding to 10,000 J equivalent exposure dose received by the pellicle. The required photomask exposure dose can be calculated from the wafer level exposure dose using Eq. (6.16).

$$D_M = \frac{N_w N_f D_R}{M^2 T_{PO}} \tag{6.16}$$

where D_M = reticle dose requirements
N_w = number of wafers
N_f = number of fields per wafer
D_R = resist dose
M = exposure tool demagnification
T_{PO} = optical transmission of the projection optics

With the introduction of 193-nm lithography, photoinduced reactions are found to limit the lifetime of photomasks due to either crystal or submicron contamination formation. The main root cause for crystal formation has been linked to residual sulfate from the photomask clean process leading to the ammonium sulfate.[207,208] However, the defect signature and composition of photoinduced submicron formation has been correlated to pellicle adhesive outgassing (Fig. 6.39).[207] Due to the variety of adhesives types used and the contribution of the fab environment in submicron defects formation, several compounds were identified as photoinduced contaminants from pellicle outgassing ranging from organic to inorganic in composition.[207,208]

Pellicle membrane lifetime expectancy can also be affected by organic environmental vapors. Mitsui published Cytop membrane lifetime expectancy in various organic saturated vapor environments (Fig. 6.40).[209] Aromatic hydrocarbons and amines were determined to be detrimental to pellicle lifetime leading to pellicle membrane thinning with accumulated 193-nm exposure dose.

As shorter lithography wavelengths are introduced to meet ITRS roadmap of shrinking device dimensions, the synthesis of an organic polymer membrane capable of sustaining the energy of shorter wavelength

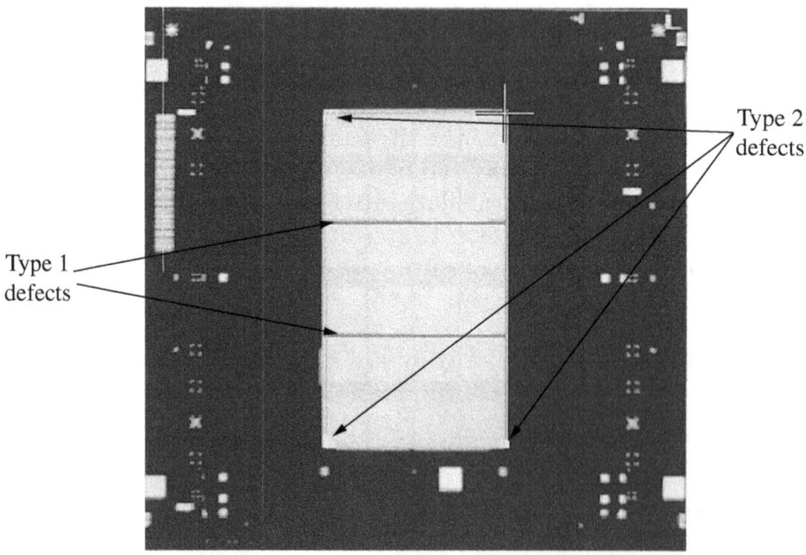

Figure 6.39 Ammonium sulfate crystals (Type I) and pellicle outgassing submicron defects (Type 2) detected by STARlight inspection on 248 nm photomask. (Reprinted with permission of Junsik L., L. Dongwook, K. Munsik, J. Hoyong, H. Oscar, Hynix, "Investigation of sub-pellicle defect formation at KrF Lithography," *Proc. SPIE* 5446, pp. 231–237, 2004.)

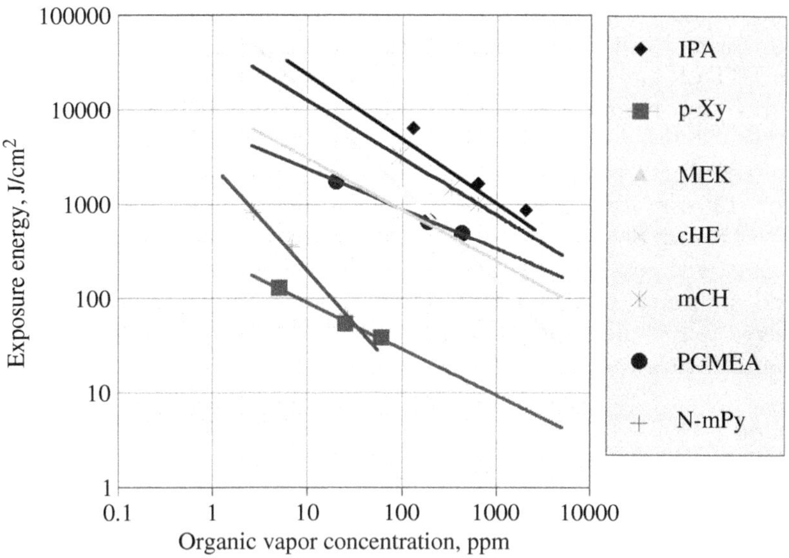

Figure 6.40 Cytop membrane exposure dose at 1 percent transmission drop in various organic vapor environments. The membrane was exposed to isopropyl alcohol (IPA), para-xylene (p-Xy), methyl ethyl ketone (MEK), cis-hydroxyethylene (cHE), methylcyclohexane (mCH), propylene glycol monomethyl ether acetate (PGMEA), and nitrosomethylaminopyridines (N-mPy). (Courtesy of Hiroaki Nakagawa of Mitsui Chemicals; Source: Kozeki, T. (Mitsui), *Proc. SPIE* 4409, pp. 438–46, 2001.)

stepper lasers (157 nm, EUV 13.4 nm) becomes a challenge. There is extensive literature describing fluoropolymer photodegradation on 193- and/or 157-nm exposure.[210–219]

Massachusetts Institute of Technology Lincoln Laboratories (MIT-LL) conducted intensive durability studies on Cytop.[207,220] MIT-LL work also determined prolonged pellicle lifetime in oxygen-containing environments. MIT-LL reports that the free radicals formed on irradiation were created in the CF_2 backbone. Irradiation of the free radical formed resulted in formation of peroxy radical in air $-CF_2OO$ and unsaturated –C=C– groups in nitrogen.

According to MIT-LL irradiation under nitrogen causes formation of radicals, which exhibit long lifetimes in the absence of any quenching reagent such as oxygen or water (hydrogen). As a result, the radiation-induced radical concentration builds until radical-radical interactions begin to occur, resulting in the formation of unsaturated –CF=CF– bonds. The π electrons accompanying these bonds give rise to the increased absorbance in the far ultraviolet. During irradiation in oxygen, the formed radicals are quickly converted to peroxy groups. When the concentration of peroxy groups becomes large, peroxy-peroxy or peroxy-fluorocarbon radical interactions occur, resulting in the formation of organic peroxides, which exhibit lower absorbance than the perfluoroalkenes formed in the absence of oxygen or hydrogen. The data collected by MIT-LL and Mitsui suggest strong influence of environmental factors on Cytop membrane 193-nm photodegradation pathways and kinetics.[209,220–223]

Extensive efforts were placed to develop an organic pellicle membrane for 157-nm lithography.[210–217] Organic polymer membranes exposed to 157-nm or shorter wavelengths experience significant transmission drop with less than 10 J of cumulated exposure dose due to photoinduced polymer chain break down. Inorganic thin films and alternatives to organic pellicle membranes are being investigated for 157-nm and future lithography technologies.[203,217]

References

1. Grenon, B. J., C. Peters, K. Battacharyya, and W. Volk, "Formation and detection of sub-pellicle defects by exposure to DUV system illumination," *19th Annual BACUS Symposium on Photomask Technology* 3873, pp. 162–176, 1999.
2. Eschbach, F., D. Tanzil, M. Kovalchick, U. Dietze, M. Liu, and F. Xu, "Improving photomask surface properties through a combination of dry and wet cleaning steps," *Proc. SPIE* 5446, pp. 209–217, 2004.
3. Howard, W. B., and K. Krause, "Improving reticle quality through reticle blank inspection,"*Proc. SPIE* 4066, pp. 401–408, 2000.
4. Mansfield, S., R. Ferguson, L. Liebman, A. Molless, and A. Wong, "Method to determine printability of photomask defects and its use in phase-shift mask evaluations," *18th Annual BACUS Symposium on Photomask Technology and Management, Proc. SPIE* 3546, pp. 651–660, 1998.
5. Bald, D., S. Munir, B. Lieberman, W. Howard, and C. Mack, "PRIMADONNA: a system for automated defect disposition of production masks using wafer lithography simulation," *22d Annual BACUS Symposium on Photomask Technology*, B. Grenon and K., eds., Proc. SPIE 4889, pp. 263–270, 2002.
6. AIMS™ may be claimed as a trademark and service mark of IBM Corporation.
7. Martino, R., R. Ferguson, R. Budd, and J. Staples, "Application of the aerial image measurement system (AIMS) to the analysis of binary mask imaging and resolution enhancement techniques," *Proc. SPIE* 2197, pp. 573–584, 1994.
8. Reynolds, G. O., J. B. DeVelis, G. B. Parrent, Jr., B. J. Thompson, *Physical Optics Notebook: Tutorials in Fourier Optics*, SPIE Optical Engineering Press, Bellingham, Wash., p. 30, 1989.
9. Rudzinski, M., L. Zurbrick, D. Pettibone, and M. Ananth, "Investigation of reticle OPC defect printability and detectability for 180 nm technology," *SPIE Symposium on Photomask and X-Ray Mask Technology VI, Proc. SPIE* 3748, pp. 535–545, 1999.
10. Neureuther, A. R., P. Flanner III, and S. Shen, "Coherence of defect interactions with features in optical imaging," *J. Vac. Sci. Technol.* B5, pp. 308–312, 1987.
11. Stivers, A. R., and E. Tejnil, "Dependence of mask-defect printability and printability criteria on lithography process resolution," *21st Annual BACUS Symposium on Photomask Technology*, G. T. Dao and B. J. Grenon, eds., *Proc. SPIE* 4562, pp. 122–129, 2001.
12. Gise, P., "Surface particle detection technology," in *Handbook of Contamination Control in Microelectronics—Principles, Applications, and Technology*, D. L. Tolliver, ed., Noyes Publications, Westwood, N.J., 1988, Chap. 12.
13. Matsuda, T., T. Tojo, and S. Yabumoto, "Metrology, defect inspection, and repair," *Sub-Half-Micron Lithography for ULSIs*, K. Suzuki, S. Matsui, and Y. Ochiai, eds., Cambridge University Press, Cambridge, UK, 2000, Chap. 7.

14. STARlight™ may be claimed as a trademark of KLA-Tencor Corporation.
15. Eran, Y., G. Greenberg, and A. Joseph, "Mask inspection and real-time line width measurements," *Proc. SPIE* 2884, pp. 138–48, 1996.
16. Munir, S., D. Bald, V. Tolani, and F. Ghadiali, "DIVAS: an integrated networked system for mask dispositioning and defect management," *Proc. SPIE* 5043, pp. 114–122, 2003.
17. Born, M. and E. Wolf, *Principles of Optics*, 6th ed., Pergamon Press, Oxford, 1980, Chap. 8.
18. Wilson, T., and C. Sheppard, *Theory and Practice of Scanning Optical Microscopy*, Academic Press, London, 1984, Chap. 3.
19. Kiess, H., "Microscopy: light," *Encyclopedia of Chemical Physics and Physical Chemistry, Volume II. Methods*, J. H. Moore, and N. D. Spencer, eds., Institute of Physics Publishing, Bristol, UK, 2001, Chap. B1.18.
20. Liebe, R., H. Haffner, S. Hemar, A. Rosenbusch, J. Chen, and F. Kalk, "Aerial image based mask inspection—a development effort to detect what might impact printing image quality on wafers," *Proc. SPIE*. 5038, part 1, pp. 177–84, 2003.
21. Kenan, B., Y. Eran, A. Karpol, E. Elyasaf, and E. Tirosh, "Method for reticle inspection using aerial imaging," *US Patent*, US 6,268,093 B1, July 31, 2001.
22. Meisburger, W. D., A. A. Desai, and A. D. Brodie, "Requirements and performance of an electron-beam column designed for x-ray mask inspection," *J. Vac. Sci. Technol.* B9(6), pp. 3010–3014, 1991.
23. Alles, D., P. Ter Beek, S. Juang, J. Wiley, and K. Hsia, "Reticle inspection system using DUV wavelength and new algorithm platform for advanced reticle inspection for 0.13 micron technology node," *Proc. SPIE* 4066, pp. 462–71, 2000.
24. Tejnil, E., and A. R. Stivers, "Options for at-wavelength inspection of patterned extreme ultraviolet lithography masks," *Proc. SPIE* 3873, pt. 2, pp. 792–803, 1999.
25. Stivers, A. R., T. Liang, M. Penn, B. Lieberman, G. Shelden, J. Folta, C. Larson, P. Mirkarimi, C. Walton, E. M. Gullikson, and M. Yi, "Evaluation of the capability of a multilbeam confocal inspection system for inspection of EUVL mask blanks," *Proc. SPIE* 4889, pp. 408–417, 2002.
26. Oohashi, K., H. Inoue, T. Nomura, A. Ono, M. Tabata, and H. Suzuki, "Newly developed mask inspection system with DUV laser illumination," *Proc. SPIE* 4066, pp. 452–61, 2000.
27. Keefer, M., R. Pinto, C. Dennison, and J. Turlo, "The role of metrology and inspection in semiconductor processing," in *Handbook of Thin-Film Deposition Processes and Techniques—Principles, Methods, Equipment and Applications*, 2d ed., K. Seshan, ed., William Andrew Publishing, Norwich, NY, 2002, Chap. 6.
28. Inoue, H., K. Okuda, T. Nomura, H. Tsuchiya, M. Tabata, "Image sensing method and defect detection algorithm for a 256 Mbit and 1 Gbit DRAM mask inspection system," in *Photomask and X-Ray Mask Technology IV*, N. Aizaki, ed., *Proc. SPIE* 3096, pp. 462–469, 1997.
29. Lee, F., "Lithography process monitoring and defect detection" in *Handbook of VLSI Microlithography—Principles, Tools, Technology and Application*, 2d ed., J. N. Helbert, ed., William Andrew Publishing, Norwich, New York, 2001, Chap. 3.
30. Smith, M. L., "Analysis of surface defects," in *Surface Inspection Techniques—Using the Integration of Innovative Machine Vision and Graphical Modeling Techniques*, Professional Engineering Publishing Limited, London, 2001, Chap. 7.
31. Buckner, B. D., L. Suresh, and E. D. Hirleman, "Surface particle detection for the 0.07 mm generation and beyond," *Proc. SPIE* 3275, pp. 90–101, 1998.
32. Sandland, P., W. D. Meisburger, D. J. Clark, R. R. Simmons, D. E. A. Smith, L. H. Veneklasen, B. G. Becker, et al., "An electron-beam inspection system for x-ray mask production," *J. Vac. Sci. Technol.* B9(6), pp. 3005–3009, 1991.
33. Brodie, A. D., Z.-W. Chen, J. Jau, and D. Meisburger, "Inspection of optical phase-shifting masks with an automated electron-beam system," *J. Vac. Sci. Technol.* B12(6), pp. 3595–3599, 1994.
34. Garvin, J. F., R. L. Guldi, N. Sridhar, M. Tinker, R. Cappel, T. Cass, J. Roberts, "Fast yield learning using e-beam wafer inspection," *Proc. SPIE*4229, pp. 85–91, 2000.
35. Tojo, T., M. Tabata, K. Yamashita, H. Tsuchiya, T. Watanabe, C. Itoh, A. Ono, H. Inoue, K. Okuda, and H. Yoshino, "Mask defect inspection method by database comparison

with 0.25-0.35 μm sensitivity," *Jpn. J. Appl. Phys.* 33, pt. 1, No. 12B, pp. 7156–7162, 1994.

36. Morikawa, Y., T. Ogawa, K. Tsuchiya, S. Noguchi, and K. Nakashima, "Performance of cell-shift defect inspection technique," *Proc. SPIE* 3096, pp. 404–414, 1997.
37. Tateno, M., N. Takayama, S. Murakami, K. Hatta, S. Akima, F. Matsuo, M. Otaki, B. G. Kim, K. Tanaka, and N. Yoshioka, "Inspection capability of high-transmittance HTPSM and OPC masks for ArF lithography," *Proc. SPIE* 5130, pp. 446–53, 2003.
38. Rosenbusch, A., M. Har-zvi, and G. Gottlib, "Inspection of production alternating PSM reticles using UV-based 365-nm reticle inspection tool," *Proc. SPIE* 4562, pp. 130–137, 2002.
39. Tejnil, E., A. R. Stivers, R. S. Schenker, and L. S. Zurbrick, "Evaluation of printability and inspection of phase defects on hidden-shifter alternating phase-shift masks," *Proc. SPIE* 4066, pp. 376–387, 2000.
40. Kincade, K., "Mask inspection rises to the subwavelength challenge," *Laser Focus World* 40(7), pp. 92–98, 2004.
41. Zurbrick, L., J. Heumann, M. Rudzinski, S. Stokowski, J.-P. Urbach, and L. Wang, "Alternating phase shift mask inspection using multiple simultaneous illumination techniques," *Proc. SPIE* 4754, pp. 511–516, 2002.
42. Zurbrick, L., M. Rudzinski, S. Stokowski, and L. He, "Alternating phase shift mask inspection through the use of phase contrast enhancement techniques," *Proc. SPIE* 4889, pp. 241–246, 2002.
43. Ferguson, R. A., R. Martino, R. A. Budd, and G. Hughes, "Etched-quartz fabrication issues for a phase-shifted 256M DRAM pattern," *J. Vac. Sci. Technol.* B11(6), pp. 2645–2650, 1993.
44. Budd, R. A., D. B. Dove, J. L. Staples, H. Nasse, and W. Ulrich, "A new mask evaluation tool, the microlithography simulation microscope aerial image measurement system," *Proc. SPIE* 2197, pp. 530–540, 1994.
45. Hemar, S., A. Rosenbusch, and R. Falah, "CD uniformity control suing aerial image-based mask inspection," *Proc. SPIE* 5038, pt. 2, pp. 929–934, 2003.
46. Emery, D. G., Z. K. Saidin, M. J. Wihl, T.-Y. Fu, M. Zywno, D. F. Kvamme, and M. E. Fein, "Automated photomask inspection apparatus and method," *US Patent*, US 5,737,072, April 7, 1998.
47. Dayal, A., N. Bergmann, and P. Sanchez, "Implementation of high resolution reticle inspection in wafer fabs," *Proc. SPIE* 5038, pt. 2, pp. 1153–1160, 2003.
48. 2003 International Technology Roadmap for Semiconductors, See http://public.itrs.net/Files/2003ITRS/Home2003.htm
49. Liang, T., A. Stivers, R. Livengood, P.-Y Yan, G. Zhang, and F. Lo, *J. Vac. Sci. Technol.* B18(6), p. 3216, 2000.
50. Harriott, L. R., and M. J. Vasile, *J. Vac. Sci. Technol.* B6(3), p. 1035, 1988.
51. Liang, T., A. Stivers, M. Penn, D. Bald, C. Sethi, V. Boegli, M. Budach, K. Edinger, and P. Spies, *Proc. SPIE* 5446, pp. 291–300, 2004.
52. Liang, T., E. Frendberg, D. Bald, M. Penn, and A. Stivers, *Proc. SPIE* 5567, pp. 456–466, 2004.
53. Itou, Y., et al., *Proc. SPIE* 5446, p. 301, 2004.
54. Kaplan, R., *Solid State Technol.* p. 74, April 1976.
55. Tison, J. and M. Cohen, "Lasers in mask repair," *Solid State Technol.* 30, p. 113, 1987; *Proc. SPIE* 1809, p. 182, 1992.
56. Haight, R., D. Hayden, P. Longo, T. Neary, and A. Wagner, *J. Vac. Sci. Technol.* B17(6), p. 3137, 1999.
57. Wagner, A., R. Haight, and P. Longo, "MARS2: an advanced femtosecond laser mask repair tool," *Proc. SPIE* 4889, p. 457, 2002.
58. Schmidt, M., P. Planigan, and D. Thibault, *Proc. SPIE* 4889, p. 1023, 2002.
59. Blauner, P. G. and J. Mauer, "X-ray mask repair," *IBM J. Res. Develop.* 37(3), p. 421, 1993.
60. Spears, D. L., and H. I. Smith, "X-ray lithography: a new high resolution replication process," *Solid State Technol.* 15(7), pp. 21–26, 1972.
61. White, R., M. Verbeek, R. Bozak, and M. Klos, *Proc. SPIE* 4562, p. 213, 2002.

62. Koops, H. W. P., J. Kretz, M. Rodolph, M. Weber, G. Dahm, and K. Lee, *Jpn. J. Appl. Phys.* 33, p. 7099, 1994.
63. Matsui, S., T. Ichihashi, and M. Mito, *J. Vac. Sci. Technol.*, B7(5), p. 1182, 1989.
64. Coburn, J. W., and H. F. Whinters, *J. Appl. Phys.* 50(5), p. 3189, 1979.
65. Winkler, D., H. Zimmermann, M. Mangerich, and B, Traunner, *Microelectron. Eng.* 31, p. 141, 1996.
66. Nakamae, K., H. Tanimoto, T. Takase, H. Fujioka, and K. Ura, *J. Phys. D: Appl. Phys.* 25, p. 1681, 1992.
67. Koops, H. W. P., R. Weiel, D. P. Kern, and T. H. Baum, *J. Vac. Sci. Technol.* B6(1), p. 477, 1988.
68. Kohlmann-von Platen, K. T., L.-M. Buchmann, H. C. Petzold, and W. H. Brunger, *J. Vac. Sci. Technol.* B10(6), p. 2690, 1992.
69. Ehrlich, D. J., *J. Vac. Sci. Technol.* B19(4), p. 969, 1983.
70. Elliott, D. J., *Ultraviolet Laser Technology and Applications*, Academic Press, Burlington, Mass., 1995.
71. Yoshino, Y., Y. Morishige, S. Watanabe, Y. Kyusho, A. Ueda, T. Haneda, and M. Oomiya, *Proc. SPIE* 4186, p. 663, 2001.
72. Oprysko, M. M., and M. W. Beranek, *J. Vac. Sci. Technol.* B5(2), p. 496, 1987.
73. Pronko, P. P., S. K. Dutta, D. Du, and R. K. Singh, *J. Appl. Phys.* 78(10), p. 6233, 1995.
74. Chichkov, B. N., C. Momma, S. Nolte, F. von Alvensleben, and A. TÅnnermann, *Appl. Phys.* A63, p. 109, 1996
75. http://www.quantron.com/IND_PRS_PRS.htm
76. Yan, P., Q. Qian, J. McCall, J. Langston, Y. Ger, J. Cho, and B. Hainsey, *Proc. SPIE* 2621, p. 158, 1995.
77. Preuss, S., A. Demchuk, and M. Stuke, *Appl. Phys.* A61, p. 33, 1995.
78. Orloff, J., M. Utlaut, and L. Swanson, *High Resolution Focused Ion Beams: FIB and its Applications*, Springer Verlag, New York, 2003.
79. Melngailis, J., *J. Vac. Sci. Technol.* B5(2), p. 469, 1987; *Proc. SPIE* 1465, p. 36, 1991.
80. Orloff, J., *Rev. Sci. Instrum.* 64(5), p. 1105, 1993.
81. Wagner, A., "Applications of focused ion beams," *Nucl. Instr. Meth.* 218, p. 355, 1983.
82. http://www.srim.org "Particle interactions with matter" and references herein. From this website, users also can download the widely popular simulation package.
83. Many books and articles have been published on this topic. See for example, (a) D. P., Townsend, J. C. Kelly and N. E. W. Hartley, *Ion implantation, Sputtering and their Applications*, Academic Press, Burlington, Mass., 1976; (b) F. J., Ziegler, J. P, Biersack, U. Littmark, *The Stopping and Range of Ions in Solids*, Vol. 1 of series *Stopping and Ranges of Ions in Matter*, Pergamon Press, New York, 1984.
84. Prewett, P. D., and P. J. Heard, *J. Phys. D: Appl. Phys.* 20, p. 1207, 1987.
85. Lessing, J., D. Ferranti, G. Sundaram, L. Nagal, and M. Verbeek, *Proc. SPIE* 3748, p. 609, 1999.
86. Lessing, J., T. Robinson, T. Morrison, and T. Holtermann, *Proc. SPIE* 5256, p. 1208, 2003.
87. Stanishenvsky, A., K. Edinger, J. Orloff, J. Melngailis, D. Stewart, A. Williams, and R. Clark, *J. Vac. Sci. Technol.* B21(6), p. 3067, 2003.
88. Saitoh, K., H. Onoda, H. Horimoto, T. Katayama, T. Watakabe, and T. Kato, *J. Vac. Sci. Technol.* B6(3), p. 1032, 1988.
89. Petzold, H.-C., and P. J. Heard, *J. Vac. Sci. Technol.* B9(5), p. 2664, 1991.
90. Young, R. J., J. R. A. Cleaver, and H. Ahmed, *J. Vac. Sci. Technol.* B11(2), p. 234, 1993.
91. Harriott, L. R., *J. Vac. Sci. Technol.* B11(6), pp. 2102, 2200, 1993.92. Edinger, K., and T. Kraus, *J. Vac. Sci. Technol.* B18(6), p. 3190, 2000.
92. Edinger, K., and T. Kraus, *J. Vac. Sci. Technol.* B18(6), p. 3190, 2000.
93. Most practical repair work has been reported in SPIE and BACUS conferences. Due to fast advances of optical lithography, many previous results become irrelevant for the next generation masks as the requirements changes. Therefore, it becomes meaningful that only the most recent and fundamental studies are referenced in this book when some practical examples are shown.
94. Ferrantti, D., A. Graupera, J. Marshman, D. Stewart, and S. Szelag, *Proc. SPIE* 5256, p. 546, 2003.

95. Itou, Y. et al., *Proc. SPIE* 5446, p. 301, 2004.
96. Wanzenboeck, H. D., M. Verbeek, W. Maurer, and E. Bertagnolli, *Proc. SPIE* 4186, p. 148, 2001.
97. Lin, B., S. Hsu, I. H. Huang, K. Chen, F. Hsieh, T. Hsu, H. Liu, A. Kroyan, F. Hsu, and J. Huang, *Proc. SPIE* 5256, p. 103, 2003.
98. Liang, T., A. Stivers, G. Liu, G. Dao, V. Liberman, M. Rothschild, S. T. Palmacci, and L. Scipioni, *2d Int'l. Symp. on 157nm Lithography*, Dana Point, Calif., May 2001.
99. Liberman, V. V., T. M. Bloomstein, M. Rothschild, J. H. C. Sedlacek, R. S. Uttaro, A. K. Bates, C. Van Peski, and K. Orvek, *J. Vac. Sci. Technol.* B17(6), 3273, 1999.
100. Scipioni, L., D. Stewart, D. Ferranti, and A. Saxonis, *J. Vac. Sci. Technol.* B18(6), p. 3194, 2000.
101. Börret, R., K. Jousten, K. Böhringer, and S. Kalbitzer, *J. Phys. D: Appl. Phys.* 21, p. 1835, 1988.
102. Leung, K. N., *J. Vac. Sci. Technol.* B17, p. 2776, 1999.
103. Fan, S., M. Hsu, A. Tseng, J. Chen, D. van den Broeke, H. Lei. S. Hsu, and X. Shi, *Proc. SPIE.* 4889, p. 221, 2002.
104. Robinson, T., A. Graupera, T. Morrision, and M. Ramstein, *Proc. SPIE*5446, p. 384, 2002.
105. Hiruta, K., S. Kubo, H. Morimoto, A. Yasaka, R. Hagiwara, T. Adachi, Y. Morikawa, K. Iwase, and N. Hayashi, "Advanced FIB mask repair technology for ArF Lithography," *Proc. SPIE* 4066, p. 523, 2000.
106. Laurance, M., *Proc. SPIE* 4186, p. 670, 2001.
107. Brinkley, D., R. Bozak, B. Chiu, C. Ly, V. Tolani, and R. White, *Proc. SPIE* 4889, p. 232, 2002.
108. LoBianco, B., R. White, and T. Nawrocki, *Proc. SPIE* 4889, p. 909, 2002.
109. Nishiguchi, M., T. Amano, S. Sasaki, Y. Morikawa, and N. Hayashi, "Photomask repair performance of the SiON/Ta-Hf attenuating PSM," *Proc. SPIE* 5256, p. 1200, 2003.
110. Chen, J. et al., *Proc. SPIE* 4562, p. 786, 2002.
111. Eco-Snow® may be a registered trademark of Eco-Snow Systems, Inc.
112. Brandt, W., *Proc. SPIE* 4562, p. 600, 2001.
113. Edinger, K. et al., *Proc. SPIE* 5256, p. 1222, 2003.
114. Edinger, K. et al., "Electron-beam-based photomask repair," *J. Vac. Sci. Technol.* B22(6), 292, 2004.
115. Lipp, S., L. Frey, C. Lehere, E. Demm, S. Pauthner, and H. Ryssel, *Microelectron. Reliab.* 56(11/12), p. 1779, 1996.
116. Kunz, R. R., and T. M. Mayer, *J. Vac. Sci. Technol.* B5(1), p. 427, 1987.
117. Liang, T., E. Frendberg, D. Bald, M. Penn, and A. Stivers, *Presentation at 2004 BACUS conference*, Monterey, Calif.
118. Johnstone, E. V., L. Dieu, C. Chovino, J. Reyes, D. Hong, P. Krishnan, D. Coburn, and C. Capella, "193nm Haze Contamination: A close Relationship between Mask and its Environment," *23d Annual BACUS Symposium on Photomask Technology, Proc. SPIE* 5256, 2003
119. Ishii, H., A. Tobita, Y. Shoji, H. Tanaka, A. Naito, H. Miyashita, "Root cause analysis of crystal growth at ArF excimer laser lithography," *Proc. SPIE* 5446, pp. 218–24, 2004.
120. Lee, J., D. Lee, M. Kim, H. Jung, and O. Han, "Investigation of sub-pellicle defect formation at KrF Lithography," *Proc. SPIE* 5446, pp. 231–237, 2004.
121. Shiao, C., C. Tsai, T. Hsu, S. Tuan, D. Chang, R. Chen, and F. Hsieh, "Evaluation, reduction and monitoring of progressive defects on 193 nm reticles with low-k_1 process," *Proc. SPIE* 5446, pp. 225–30, 2004.
122. Eschbach, F., D. Tanzil, M. Kovalchick, U. Dietze, M. Liu, and F. Xu, "Improving photomask surface properties through a combination of dry and wet cleaning steps," *Proc. SPIE* 5446, pp. 209–217, 2004.
123. Eschbach, F., D. Selassie, D. Tanzil, B. Grrenebaum, H. Liu, P. Sanchez, M. Toofan, M. Murray, and R. Villacorta, "ArF lithography reticle crystal growth contributing factors," *Proc. SPIE* 5567, pp. 497–505, 2004.
124. Lee, D. W. et al., "Effect of UV/O3 cleaning process for reducing sulfuric residue on mask process," *Proc. SPIE* 5567, pp. 521–528, 2004.

125. Jeong, W. et al., "The characteristics of residues and optical change of HT PSM during stepwise wet cleaning and optimization of HT PSM cleaning process," *Proc. SPIE* 4754, pp. 597–605 , 2002.
126. Tange, K., Y. Nagamoura, K. Hosono, Y. Oomasa, K. Kido, A. Hayashi, Y. Kikuchi, I. Imagawa, Y. Matsuzawa, and H. Usui, "The development of refined cleaning technique focusing on ecological viewpoint," *Proc. SPIE* 4409, pp. 418–429, 2001.
127. Chen, G., J. Reyes, J. Wood, I. Kashkoush, L. Dieu, and R. Novak, "Immersion system process optimization for 248 nm and 193 nm photomasks- Binary and EAPSM," *Proc. SPIE* 5256, pp. 518–519, 2003.
128. Jeong, W., D. Park, E. Park, Y. Cho, S. Choi, H. Kwon, J. Kim, and S. Choi, "A comparative evaluation of mask cleaning performance," *Proc. SPIE* 5256, pp. 510–517, 2003.
129. Busnaina, A. A., Kashkoush, I. I., and Gale, G. W., "An experimental study of megasonic cleaning of silicon wafers," *J. Electrochem. Soc.* 142, pp. 2812–2817, 1995.
130. Gale, G. W., and A. A. Busnaina, "Removal of particulate contaminants using ultrasonics and megasonics: a review," *J. Particul. Sci. and Technol.* 13, pp. 197–211, 1995.
131. Busnaina, A. A., and G. W. Gale, "Removal of silica particles from silicon substrates using megasonics cleaning," *J. Particul. Sci. and Technol.* 15, 1997.
132. Schwartzman, S., A. Mayer, and W. Kern, *RCA Review* 46, p. 81, 1985.
133. Busnaina, A. A., and G. W. Gale, "Roles of cavitation and acoustic streaming in megasonic cleaning," *J. Particul. Sci. and Technol.* 17, pp. 229–238, 1999.
134. Busnaina, A. A., and F. J. Dai, *Adhesion* 67, pp. 181–93, 1997.
135. Busnaina, A. A., and T. M. Elsawy, "Post-CMP cleaning using acoustic streaming," *J. Electron. Mater.* 27, pp. 1095–1098, 1998.
136. Eckart, C., *Phys. Rev.* 73, p. 68, 1948.
137. Schlicting, H., *Boundary Layer Theory*, 6th ed., McGraw-Hill, Berlin, Germany 1968.
138. Elder, S. A., *J. Acoust. Soc. Amer.* 31, p. 54, 1959.
139. Deymier, P. A., A. Khelif, B. Djafari-Rouhani, J. O. Vasseur, and S. Raghavan, *J. Appl. Phys.* 88, p. 2423, 2000.
140. Deymier, P. A., J. O. Vasseur, A. Khelif, B. Djafari-Rouhani, L. Dobrzynski, and S. Raghavan, "Streaming and removal forces due to second-order sound field during megasonic cleaning of silicon wafers," *J. Appl. Phys.* 88, p. 6821, 2000.
141. Khelif, A., J. O. Vasseur, Ph. Lambin, B. Djafari-Rouhani, and P. A. Deymier, "Theory of acoustic scattering by supported ridges at a solid/liquid interface," *Phys. Rev. E.* 65, 036601, pp. 1–7, 2002.
142. Grenon, B. J., C. Peters, K. Battacharyya, and W. Volk, "Formation and detection of sub-pellicle defects by exposure to DUV system illumination," *19th Annual BACUS Symposium on Photomask Technology, Proc. SPIE* 3873, pp. 162–176, 1999.
143. Battacharyya, K., W. Volk, D. Brown, J. Ayala, and B. J. Grenon, "Investigation of reticle defect formation at DUV lithography," *22d Annual BACUS Symposium on Photomask Technology, Proc. SPIE* 4889, pp. 478–87, 2002.
144. Rimai, D. S., and Busnaina, A. A., "The adhesion and removal of particles from surfaces," *J. Particul. Sci. and Technol.* 13, pp. 249–70, 1995.
145. Rimai, D. S., D. J. Quesnel, and A. A. Busnaina, "The adhesion of dry particles in the nanometer to micrometer-size Range," *Int. J. Colloid Surface A* 165, pp. 3–10, 2000.
146. Krishnan, S., A. A. Busnaina, D. S. Rimai, and D. P. DeMejo, "The adhesion-induced deformation and the removal of submicrometer particles," *J. Adhes. Sci. Technol.* 8, pp. 1357–1370, 1994.
147. R. Bowling, "A theoretical review of particle adhesion," *Particles on Surfaces*, K. L. Mittal, ed., 7, Plenum Press, New York, 1988.
148. Bhattacharya S., and K. L. Mittal, *Surf. Technol.* 7, p. 413, 1978.
149. Rimai, D. S., and A. A. Busnaina, "The adhesion and removal of particles from surfaces," *J. Particul. Sci. Technol.* 13, pp. 249–270, 1995.
150. Rimai, D. S., D. J. Quesnel, and A. A. Busnaina, "The adhesion of dry particles in the nanometer to micrometer-size range," *Int. J. Colloid Surfaces A* 165, pp. 3–10, 2000.

151. Krishnan, S., A. A. Busnaina, D. S. Rimai, and D. P. DeMejo, "The adhesion-induced deformation and the removal of submicrometer particles," *J. Adhes. Sci. Technol.* 8, pp. 1357–1370, 1994.
152. Johnson, K., K. Kendall, and A. D. Robers, *J. R. Soc. London A* 324, p. 301, 1971.
153. Derjaguin, B. V., V. M. Muller, and P. Toporov, *J. Colloid Interface Sci.* 67, p. 378, 1978.
154. Muller, V. M., V. S, Yushchenko, and B. V. Derjaguin, *J. Colloid Interface Sci.* 77, p. 91, 1980.
155. Muller, V. M., V. S. Yushchenko, and B. V. Derjaguin, *Colloid Surface* 7, p. 251, 1983.
156. Maugis, D., and H. M. Pollock, *Acta Metall.* 32, p. 1323, 1984.
157. Johnson, K., K. Kendall, and A. D. Robers, *J. R. Soc. London A*, 324, p. 301, 1971.
158. Derjaguin, B. V., V. M. Muller, and P. J. Toporov, *Colloid Interface Sci.* 67, p. 378, 1978.
159. Muller, V. M., V. S. Yushchenko, and B. V. Derjaguin, *J. Colloid Interface Sci.*, 77, p. 91, 1980.
160. Muller, V. M., V. S. Yushchenko, and B. V. Derjaguin, *Colloid Surface*, 7, p. 251, 1983.
161. Maugis, D., and H. M. Pollock, *Acta Metall.* 32, p. 1323, 1984.
162. Wang, P., and D. Bell, *Third International Symposium on Cleaning Technology in Semiconductor Device Manufacturing, Proc. Electrochem. Soc. PV94-7*, p. 132, 1994.
163. Lin, H., A. A. Busnaina, and I. I. Suni, "Physical modeling of rinsing and cleaning of submicron trenches," *Proceedings, IITC, 3d Annual International Interconnect Technology Conference*, San Francisco, Calif., 2000.
164. Busnaina A. A., H. Lin, N. Moumen, J. W. Feng, and J. Taylor, "Particle adhesion and removal mechanisms in post-CMP cleaning processes," *J. of Semiconductor Manufacturing, IEEE Trans.* 15, pp. 374–82, 2002.
165. Olim, M., J. *Electrochem. Soc.* 144, p. 3657, 1997.
166. Zhang, D., "Fundamental study of megasonic cleaning," Ph.D. dissertation, Department of Electrical Engineering, University of Minnesota, Minneapolis, Minn., 1993.
167. Suslick, K. S., *Sonoluminescence*, Kluwer Academic Publishers, Norwell, Mass., 1999.
168. Ohmi, T., M. Toda, M. Katoh, K. Kawada, and H. Morita, *Mat. Res. Soc. Symp. Proc., Vol. 477*, Materials Research Society, p. 4, 1997.
169. Makino, K., M. M. Mossoba, and P. Reisz, *J. Amer. Chem. Soc.* 104, p. 3537, 1982.
170. Schmidt, M. O., G. Vereecke, R. Vos, F. Holsteyns, M. Baeyens, and P. Mertens, *Solid State Phenom.* 92, pp. 147–150, 2003.
171. Cetinkaya, C., and M. Peri, "Non-contact nanoparticle removal with laser induced plasma pulses," *Nanotechnology* 15, pp. 435–440, 2004.
172. Cetinkaya, C., R. Vanderwood, and M. Rowell, "Nanoparticle removal from flat substrates with pulsed-laser induced plasma and shock waves," *J.Adhes. Sci. Technol.* 16, pp. 1201–1214, 2002.
173. Vanderwood, R., and C. Cetinkaya, "Nanoparticle removal from trenches and pinholes with pulsed-laser induced plasma and shock waves," *J.Adhes. Sci. Technol.* 17, pp. 129–147, 2003.
174. Hooper, T., and C. Cetinkaya, "Efficiency studies of particle removal with pulsed-laser induced plasma," *J. Adhes. Sci. Technol.* 17, pp. 763–776, 2003.
175. Lee, J. M., C. Curran, and K. G. Watkins, *J. Appl. Phys. A* 73, p. 219, 2001.
176. Lee, J. M., and K. G. Watkins, *J. Appl. Phys.* 89, p. 6496, 2001.
177. Lee, J. M., K. G. Watkins, and W. M. Steen, *J. Laser Appl.* 13, p. 154, 2001.
178. Mahoney, J., C. Sujo, J. Perel, P. Kopalidis, and R. Becker, "Precision surface cleaning using microcluster beam," *Electrochem. Soc. Proc.* 99-36, pp. 429–36, 1999.
179. Falkenstein, Z., "Surface cleaning mechanisms utilizing VUV radiation in oxygen containing gaseous environments," *Proc. SPIE* 4440, pp. 246–55, 2001.
180. Gericke, H., and J. Comes, "Energy partitioning in the reaction O(1D) + $H_2O \rightarrow$ OH + OH. The influence of O(1D) translational energy on the reaction rate constant," *Chem. Phys. Lett.* 81(2), pp. 218–22, 1981.
181. Falkenstein, Z., "Ozone formation with (V)UV-Enhanced Dielectric barrier discharges in dry and humid gas mixtures of O_2, N_2/O_2 and Ar/O_2," *Ozone-Sci. Eng.* 21, pp. 583–603, 1999

182. Narayanswami, "A theoretical analysis of wafer cleaning using cryogenic aerosol," *J. Electrochem. Soc.*146, pp. 767–774, 1999.
183. Wagener, T., and K. Kawaguchi, "Improved yields for the nano-technology era using cryogenic aerosols," *IEEE/SEMI Advanced Semiconductor Manufacturing Conference*, pp. 467–471, 2004.
184. Banerjee, S., and A. Campbell, "Principles and mechanisms of sub-micron particle removal by CO_2 cryogenic technology," *Particles on Surfaces Conference*, June 2004, Philadelphia, Pa. (in press).
185. Weibel, G., and C. Ober, "An overview of supercritical CO_2 applications in microelectronics processing," *Microelectron. Eng.* 65 pp. 145–152, 2003.
186. Rothman, L., R. Robey, M. Ali, and D. Mount, "Supercritical fluid processes for semiconductor device fabrication," *IEEE/SEMI Advanced Manufacturing Conference*, pp. 372–375, 2002.
187. Fredrick, C., and D. Gray, "Submicron cleaning using vacuum cavitational streaming," *Particles on Surfaces Conference*, June 2004, Philadelphia, Pa. (in press).
188. Shea, V., and W. J. Wojcik, "Pellicle cover for projection printing system," *US Patent* 4.131,363, Dec 26, 1978 (IBM patent).
189. Johnson, R., "Automatic cleaning technique for pellicles," *Semicond. int.*, p. 164, April 1986.
190. Iscoff, R., "Pellicles 1985: an update," *Semicond. Int.* 8(4), p. 110, April 1985.
191. Augason, G. C., "A device for applying plastic film antireflection coatings to optical elements," *Infrared Phys.*, 24(6), pp. 567–569, November 1984.
192. Bruner, T. A., C. D. Ausschmidtt, and D. L. Duly, "Pellicle mask protection for 1: 1 projection lithography," *Solid State Technol.*, Vol. 26, p. 135, May 1983.
193. Hershey, T. L., "Pellicles on wafer steppers with lenticular optics," *Solid State Technol.* Vol. 26, p. 89, July 1983,
194. Iscoff, R., "Pellicles—a means to increase die yield," *Semicond. Int.* Vol. 5, p. 95, September 1982; *Microelectron. Reliab.* 23(3), p. 595, 1983.
195. Cytop™ may be claimed as a trademark of Asahi Glass Company, Limited.
196. Teflon® may be claimed as a registered trademark of E.I. du Pont de Nemours and Company.
197. Kondou, M. et al., *Proc. SPIE* 3095, pp. 212–228, 1997.
198. Chen, W. et al., *Proc. SPIE* 3546, pp. 167–172, 1998.
199. Roeth, K. D. et al., *Proc. SPIE* 3412, pp. 440–446, 1998.
200. Cotte, E. P., R. L. Engelstad, E. G. Lovell, Y. M. Shkel, F. O. Eschbach, E. Y. Shu, D. Tanzil, and R. M. Calhoun, "Numerical and experimental studies of pellicle-induced photomasks distortions," *Proc. SPIE* 4562, pp. 641–651, 2001.
201. Cotte, E., R. Engelstad, E. Lovell, Y. E. Shkel, F. Eschbach, E. Shu, D. Tanzil, and R. Calhoun, "Distortions in advanced photomasks from soft pellicles," *Microelectron. Eng.*, Vol. 61-62, pp. 187–192, 2001.
202. Cotte, E. P., R. L. Engelstad, E. G. Lovell, F. Eschbach, and D. Tanzil, "Experimental and numerical studies of the effects of materials and attachment conditions on pellicle-induced distortions in advanced photomasks," *Proc. SPIE* 4754, pp. 579–588, 2002.
203. Shu, E. Y., F.-C. Lo, F. O. Eschbach, and E. P. Cotte, "Hard pellicle study for 157-nm lithography," *Proc. SPIE* 4754, pp. 558–569, 2002.
204. Cotte, E. P., R. L. Engelstad, E. G. Lovell, and Y. M. Shkel, "Numerical and experimental studies of pellicle-induced photomask distortions," *Proc. SPIE*.4562, pp. 641–651, 2002.
205. Cotte, E. P., R. L. Engelstad, E. G. Lovell, Y. M. Shkel, Eschbach, "Distortions in advanced photomasks from soft pellicles," *Microelectron. Eng.* 61-62, pp. 187–92, 2002.
206. Cotte, E. P., R. L. Engelstad, E. G. Lovell, F. Eschbach, and D. Tanzil, "Effects of soft pellicle frame curvature and mounting process on pellicle-induced distortions in advanced photomasks," *Proc. SPIE* 5040, pp. 1044–1054, 2003.
207. Junsik L., L. Dongwook, K. Munsik, J. Hoyong, H. Oscar, Hynix, "Investigation of sub-pellicle defect formation at KrF Lithography," *Proc. SPIE* 5446, pp. 231–237, 2004.

208. Bhattacharyya, K., W. Volk, B. Grenon, D. Brown, and J. Ayala, "Investigation of reticle defect formation at DUV lithography," *Proc. SPIE* 4889, 478–487, 2002.
209. Kozeki, T. (Mitsui), *Proc. SPIE* 4409, pp. 438–46, 2001.
210. French, R. et al., *Proc. SPIE.* 4000, pp. 1491–1502, 2000.
211. French, R. et al., *Proc. SPIE* 4346, 2001.
212. French, R., J. Gordon et al., "Fundamentals of transparency in polymers for use as 157 nm soft pellicles," *3d International Symposium on 157 nm Lithography*, 2002.
213. Matsukura, I., *3d International Symposium on 157 nm Lithography*, 2002.
214. Tregub, A., F. Eschbach, J. Powers, F. C. Lo, S. Shigematsu, and H. Nakagawa, "Development of fluoropolymer membranes transparent and resistant to 157 nm exposure," *4th 157 nm Symposium*, Yokohama, Japan, 2003.
215. Eschbach, F., A. Tregub, K. Orvek, J. Powers, and F. C. Lo, "157 nm photo degradation of fluoropolymer model compounds," *4th 157 nm Symposium*, Yokohama, Japan, 2003.
216. Zimmerman, P. A., and A. Whittacker, "Understanding degradation mechanism of fluoropolymers used as soft pellicles during exposure at 157 nm," *4th 157 nm Symposium*, Yokohama, Japan, 2003.
217. Kuo, Y., J. Lu, and P. Zimmerman, "Passivation of the 157 nm pellicle with nanometer thin films," *4th 157 nm Symposium*, Yokohama, Japan, 2003.
218. Liberman, V. et al., "Improved pellicles for 157nm lithography, a nanocomposite approach, *4th 157 nm Symposium*, Yokohama, Japan, 2003.
219. Kuo, Y., J. Lu, and P. Zimmerman, "Passivation of the 157 nm pellicle with nanometer thin films," *Proc. SPIE* 5256, pp. 195–198, 2003.
220. Grenville, A., and V. Libermann, *Proc. SPIE* 3236, 1997.
221. Miyazaki (Mitsui), *Proc. SPIE* 4066, pp. 295–299, 2000.
222. Shigematsu, S. (Mitsui), *Proc. SPIE* 3748, pp. 166–176, 1999.
223. Liberman, V. et al., "Damage testing of pellicles for 193nm lithography," *Proc. SPIE* 3334, pp. 480–495.

Chapter

7

Resolution Enhancement Techniques

Robert J. Socha

7.1 Introduction

As lithography continues to resolve ever-decreasing feature sizes, the lithography community developed a number of resolution enhancement techniques (RETs) to resolve these smaller features. These RETs are discussed in this chapter. Since the mask allows the most degrees of freedom, this chapter concentrates on many of the mask enhancement techniques developed by the lithography community. However, imaging in lithography is a coupled problem, i.e., the image on the wafer depends on the illumination optics, the mask, and projection optics. Therefore, each part of the coupled problem is discussed in this chapter. For additional information on general RETs, consult Refs. 1, 2, and 3.

In the beginning of the chapter, basic Fourier optics is discussed because an understanding of Fourier optics provides the foundation for understanding the various RETs. The concepts of Fourier optics are expanded to include partial coherence of the source by using scalar imaging methods. These scalar imaging methods, Abbe formulation and Hopkins formulation, are described in this chapter. These scalar imaging methods are used in commercially available simulation software. Scalar imaging methods give the necessary theory needed for understanding the various RETs. After discussing the theory needed to understand the RET, the various RETs are presented beginning with *optical proximity correction* (OPC) and *phase-shift mask* (PSM). These mask enhancement techniques can be further optimized by understanding the effect of the illumination on forming the image. Therefore, a section on various illumination and mask optimization techniques is also discussed in this chapter. This chapter concludes with a section on advanced topics including immersion lithography and the effect of polarization.

The RETs presented in this chapter can be organized by their strengths, which are related to the k_1 factor. The k_1 factor is given by Eq. (7.1),

$$\mathrm{CD} = k_1 \frac{\lambda}{\mathrm{NA}} \tag{7.1}$$

where CD is the critical dimension of the feature, λ is the wavelength, and NA is the numerical aperture of the lithography system. The minimum possible k_1 is 0.25 for printing a CD that is half the pitch. In Fig. 7.1, a diagram showing the strengths of the various RETs is presented. In Fig. 7.1, the diagram is divided into five types of RETs; from left to right these types are illumination, mask, optical proximity correction method, resist, and scanner. The techniques with the lowest k_1 are at the top of the pyramid. However, as k_1 decreases, the complexity of the technique also increases. Complex techniques require more innovated

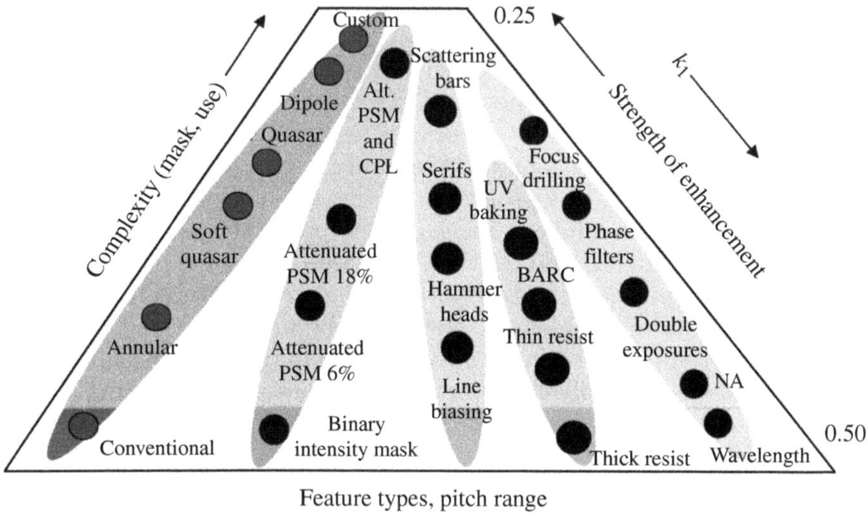

Figure 7.1 A pyramid showing the different types of RETs. The RETs are divided into five groups—illumination, mask type, OPC, resist, and scanner. The top of the mask is for the most aggressive RET and the lowest k_1. Many of these RETs can be combined to achieve the lowest k_1 possible. (Courtesy of Jo Finders, ASML.)

software to implement the technique. Furthermore, the complex techniques are not general and may not be applicable to every design. For example, all designs cannot be phase shifted with alternating PSM or aggressive off-axis illumination (OAI) leads to forbidden pitches. The consequences of each of these techniques are discussed in more detail in the following sections.

7.2 Fourier Optics

The theory of imaging to describe a lithography system has been in existence for many years. The lithography system is similar to a microscope. Consequently, theory developed to describe imaging from the microscope can be applied to the lithography system. The theory of imaging with a microscope begins with Fourier optics, and an understanding of Fourier optical theory is essential to understanding resolution enhancement techniques. The Fourier optical theory begins with describing diffraction from the mask. The diffraction from the mask creates a wave propagating to the projection lens. This diffracted wave represents the spatial frequencies of the mask. The spatial frequencies are represented by the diffraction orders. Therefore this section starts with Sec. 7.2.1 on diffraction orders. Furthermore, since resolution enhancement techniques often use off-axis illumination, theory to describe imaging with off-axis illumination is discussed in Sec. 7.2.2. Imaging of a mask can also be described with system theory as discussed in Sec. 7.2.3. Finally, the effect of focus on imaging is described in Sec. 7.2.4.

The lithography system is made of several distinct parts—the illumination optics, the mask, the projection optics, and the wafer. As shown in Fig. 7.2, RETs can be applied to each of these parts. However, this chapter focuses on the mask and on the illumination. In a typical optical lithography system, the user can increase the resolution by increasing NA or decreasing the wavelength. However, the projection lens pupil is not accessible, so pupil filtering is not described in this chapter. Also, the photoresist has a large impact on improving resolution and making some low resolution enhancement techniques possible. Photoresist, however, is outside the scope of this text and is not described in this chapter.

The lithography system is shown in Fig. 7.2 with the four distinct parts—the illumination optics, the mask, the projection optics, and the wafer. In current lithography systems, the source is nearly monochromatic as a laser creates the illumination. Furthermore, the illumination

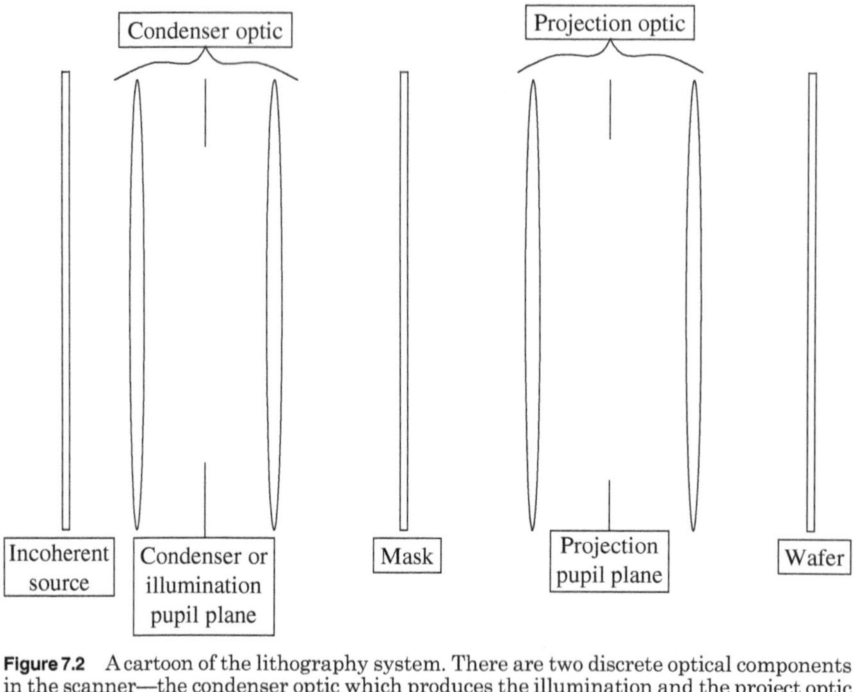

Figure 7.2 A cartoon of the lithography system. There are two discrete optical components in the scanner—the condenser optic which produces the illumination and the project optic which collects the mask diffraction orders and produces the mask image on the wafer.

optics is incoherent in the spatial domain. The coherence of the illuminator is described in more detail in Sec. 7.3. The incoherence is accomplished by using diffusers and optical elements. These elements essentially create a random phase in the ray traveling from the source to the mask. In addition to the illumination, many RETs modify the mask to increase the resolution and process window. In order to understand how modifying the mask improves resolution, the effect of the mask on diffraction must be understood through Fourier optics. Therefore, in this section Fourier optics is described, and the information presented in this section is used in the later RET sections.

Fourier optics was first developed by Abbe in order to describe the image formed by a microscope.[4] Prior to Abbe's description of the microscope, Airy described the diffraction from a circular disk.[5] Diffraction from a circular disk is widely used in imaging because lens elements are circular. The resolution limit of a lens is set by the extent that the lens can collect diffraction from the object. The resolution limit of the lens was reported by Rayleigh.[6] The Rayleigh criterion for the resolution limit is widely used in lithography to describe the minimum resolution of a lithography system. These references include historical references, which first mention Fourier optics, and textbooks, which describe Fourier optics in more detail than described in this section.[7,8]

7.2.1 Diffraction orders

Imaging in optical lithography is limited due to diffraction from the mask. As a wave passes through the mask the light is diffracted. For an infinite grating the light is diffracted into a set of discrete diffraction orders. These diffraction orders represent the spatial frequencies of the mask. In Fourier optics, the diffraction orders are calculated by making several assumptions. The first assumption is that the electric and magnetic fields are uncoupled, i.e., the electric field does not influence the magnetic field and vice versa. This assumption implies that only the diffraction orders of the electric field are needed to calculate the image of the mask. The second assumption is that the vector effects of diffraction from the mask are negligible. The third assumption is that diffraction from the mask is calculated by assuming that the mask is infinitely thin, and the mask topography does not influence the diffraction orders. With this assumption, the mask is represented by a scalar electric field transmission, and this representation is often called a scalar mask. These three assumptions are valid when the mask features are larger than a wavelength. As resolution is pushed and k_1 is lowered, these assumptions break down. However, Fourier optics gives insight into how the lithographic image is formed, and Fourier optics is an adequate approximation to image formation. In the regime where these Fourier optic assumptions break down, rigorous *electromagnetic field* (EMF) theory is used to calculate the diffraction orders from the mask. EMF theory is necessary for phase-shift masks and for hyper-NA (NA > 1) as discussed in Secs. 7.5 and 7.7.1, respectively. For phase-shift mask, the mask topography invalidates the assumption that the electric and magnetic fields are uncoupled. For hyper-NA, the mask begins to polarize the diffraction orders, and vector effects are no longer negligible.

In Fourier optics, the spatial frequencies representing the mask are calculated by determining the diffraction orders of the mask. Each diffraction order represents a ray light traveling in a specific direction. The direction of this ray is given through a condition developed by Bragg. This is known as the Bragg condition and is given in Eq. (7.2):

$$n\lambda = P \sin \theta_n \tag{7.2}$$

In Eq. (7.2), the wave of light impinging on a grating of pitch P is diffracted into a number of discrete diffraction orders. These diffraction orders are represented by the diffraction order number n and each order travels at an angle θ_n normal to the mask. This is diagramed in Fig. 7.3. In Fig. 7.3, an opaque grating of pitch P is on aclear quartz blank. The zero diffraction order travels straight through the grating. The +1 order, however, is diffracted into an angle of θ_{+1} from the normal of the grating. As the diffraction order number n increases, the angle traversed by this order increases.

As well as having a direction specified by the diffraction angle, each diffraction order has a magnitude and phase. The magnitude and phase

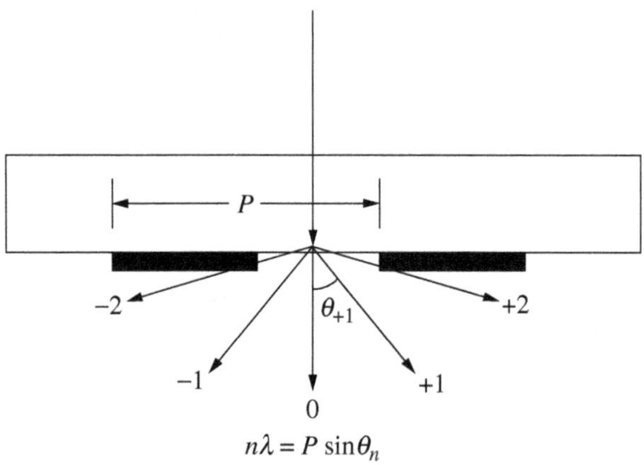

Figure 7.3 Diffraction orders from a periodic structure. For a periodic structure, discrete diffraction orders are formed. The diffraction angle θ_n is calculated with Bragg condition.

are calculated through the Fourier transform of the object. The Fourier transform of the object is first calculated by converting the object into the electric field transmission. The electric field transmission of the object f at position (u, v) is given by Eq. (7.3).

$$f(u,v) = \sqrt{T(u,v)}e^{i\phi(u,v)} \tag{7.3}$$

In Eq. (7.3), the intensity transmittance of the object is given by T and the phase of the object in radians is given by ϕ. For example, a 6 percent attenuating PSM has transmittance of 0.06 and phase of π. The Fourier transform of this electric field object is given in Eq. (7.4) along with the inverse Fourier transform in Eq. (7.5).

$$F(m,n) = \frac{1}{P_x P_y}\int_0^{P_x}\int_0^{P_y} f(u,v)e^{-iu\left(\frac{2\pi}{P_x}m\right)}e^{-iv\left(\frac{2\pi}{P_y}n\right)}du\,dv \tag{7.4}$$

$$f(u,v) = \sum_{-\infty}^{\infty}\sum_{-\infty}^{\infty} F(m,n)e^{+iu\left(\frac{2\pi}{P_x}m\right)}e^{+iv\left(\frac{2\pi}{P_y}n\right)} \tag{7.5}$$

In Eqs. (7.4) and (7.5), $F(m,n)$ represents the diffraction order magnitude and phase. The m represents the diffraction order number in the x direction and the n represents the diffraction order number in the y direction. Equations (7.4) and (7.5) assume that the object is periodic with period P_x in the x direction and with period P_y in the y direction. Diffraction orders exist only for a periodic structure as a nonperiodic object has a continuous spectrum of spatial frequencies. Although most device structures are not periodic, the structure is forced to be periodic in simulation and OPC software. Placing sufficient empty space around the structure creates an isolated structure such that the next period does not influence the image of the isolated structure. The sufficient empty space is often called a guard band and typically is on the order of $5\lambda/\text{NA}$.

Using Eqs. (7.3) and (7.4), the diffraction orders can be calculated for a 1D space of arbitrary transmittance. For a space of width a on a 1D period P_x, the diffraction orders are given by Eq. (7.6).

$$F(m) = \begin{cases} \dfrac{a + \sqrt{T}e^{i\phi}(P_x - a)}{P_x} & m = 0 \\[2ex] \dfrac{\left(1 - \sqrt{T}e^{i\phi}\right)}{\pi m}\sin\left(\dfrac{\pi a m}{P_x}\right) & m \neq 0 \end{cases} \tag{7.6}$$

For a binary mask with a 180-nm opaque chrome line on a 540-nm pitch ($a = 540 - 180 = 360$ nm), the diffraction orders were calculated and are shown in Fig. 7.4. In Fig. 7.4, the zero order has the largest magnitude followed by the ±1 orders. It is important to note that the diffraction order magnitude decreases as m increases, $F(m) \propto 1/m$ in Eq. (7.6). This implies that lower spatial frequencies carry most of the energy and do most of the image formation. Because of this, imaging with only a few diffraction orders makes low k_1 lithography possible.

The effect of each diffraction order on forming the aerial image can also be envisioned pictorially. In Fig. 7.5 an isolated opaque line is imaged onto the wafer. The image with the zero order represents a lithographic system that is only capable of collecting the zero order. The image with only the zero order has no image modulation. The image with the zero and first orders shows that there is modulation in the image. If the modulation contrast of image is sufficient, the photoresist, which acts as a nonlinear filter, is capable of converting this image into a discrete line. If more orders are collected by the projection lens, the contrast of the image increases. This is shown in Fig. 7.5 in which the image from collecting five orders, the 0, ±1, and ±2 orders, has higher contrast than the image from collecting three orders.

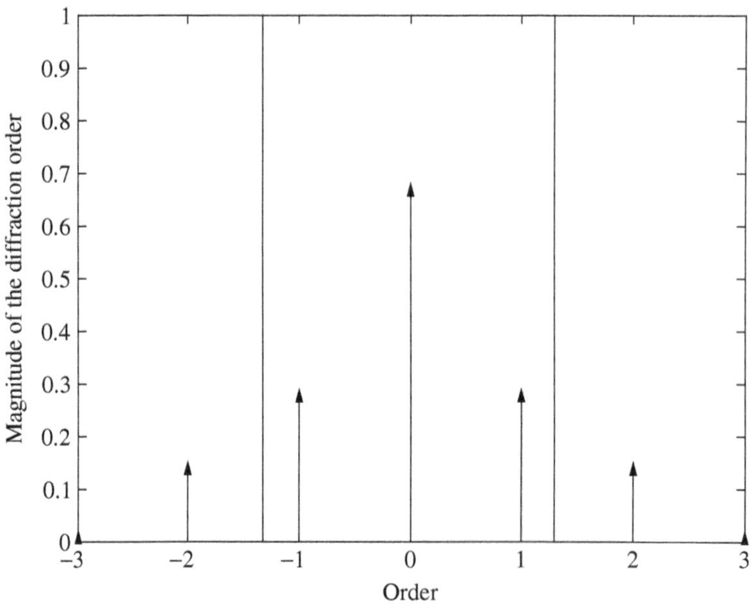

Figure 7.4 Magnitude of the diffraction orders for a 180-nm opaque line on a 540-nm pitch. For a scanner with NA = 0.7 and λ = 248 nm, only the 0 and ±1 orders are collected by the projection lens.

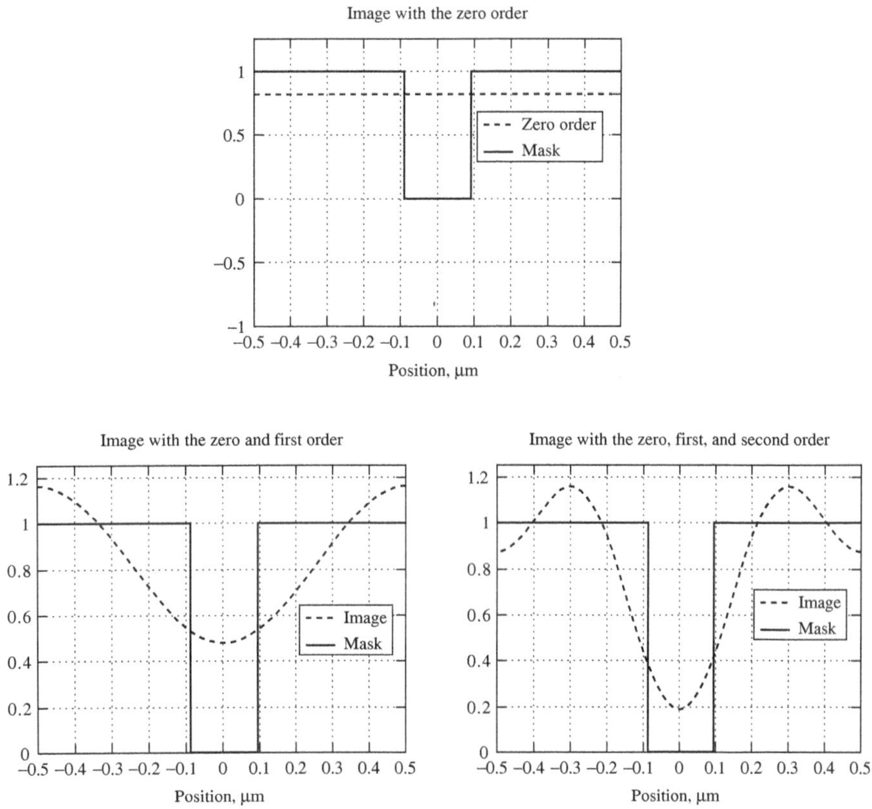

Figure 7.5 Reconstruction of an electric field transmission for a line. Note as more orders are used in the reconstruction the edges are defined.

For a two-dimensional (2D) structure, the effect of collecting the lower spatial frequencies is shown in a photo of John Petersen in Fig. 7.6. In Fig. 7.6, the Fourier transform of John shows that most of the diffraction orders are centered in lower diffraction orders (lower spatial frequencies). If we place a low pass filter that is only capable of collecting 1.5 percent of all the spatial frequencies, the image of a blurred image of a person can be reconstructed. If 5.9 percent of the orders are collected with another low pass filter, the image resembles John. However, if 13.2 percent spatial frequencies are collected, the image is clearly John. As the number of spatial frequencies increase, the contrast and sharpness of the image of John improves. However, device patterns are not as complex as the image of the human face. Therefore, device patterns do not require the same number of frequencies needed to recognize a human face. Consequently, fewer spatial frequencies or diffraction orders are needed in lithography.

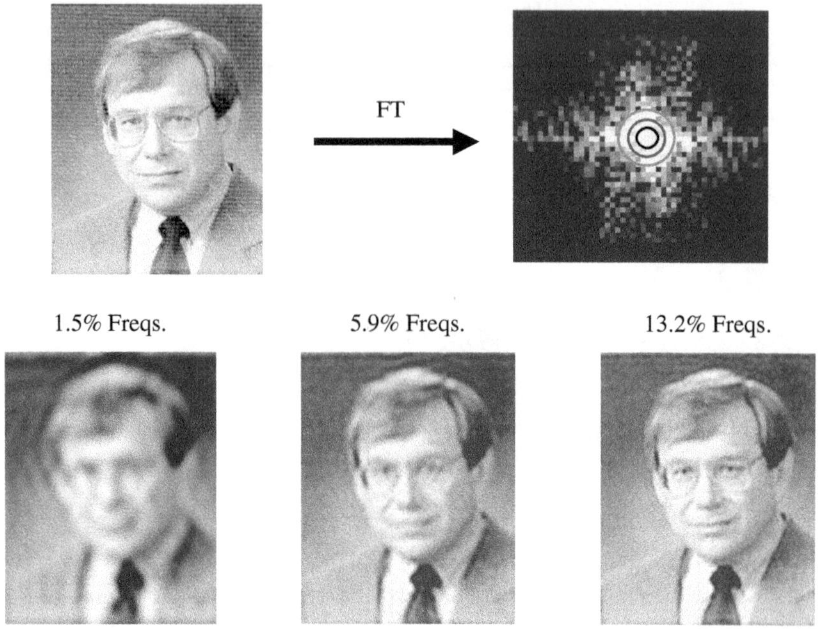

Figure 7.6 Images of John Petersen after being low pass filtered. As high frequencies are collected the resolution of John increases.

The projection lens is capable of only collecting the lower spatial frequency diffraction orders. The maximum angle that the lens can collect is determined by the numerical aperture of the projection lens as given in Eq. (7.7). By substituting Eq. (7.7) into Eq. (7.2), the maximum diffraction order that the projection lens can collect can be determined. The maximum diffraction order is shown in Eq. (7.8). In Eq. (7.8), floor is the lowest integer part.

$$\sin\theta_{\text{max}} = \text{NA} \tag{7.7}$$

$$n_{\text{max}} = \text{floor}\left(\frac{P\cdot \text{NA}}{\lambda}\right) \tag{7.8}$$

Equation (7.8) is the maximum collected diffraction order for a single beam impinging onto the mask along the optical axis.

Since the projection lens can collect the lower diffraction orders, the projection lens forms an image on the wafer that is the low passed filtered image of the mask pattern. In Fig. 7.7, the projection lens is only capable of collecting the 0 and ±1 diffraction orders. However, by

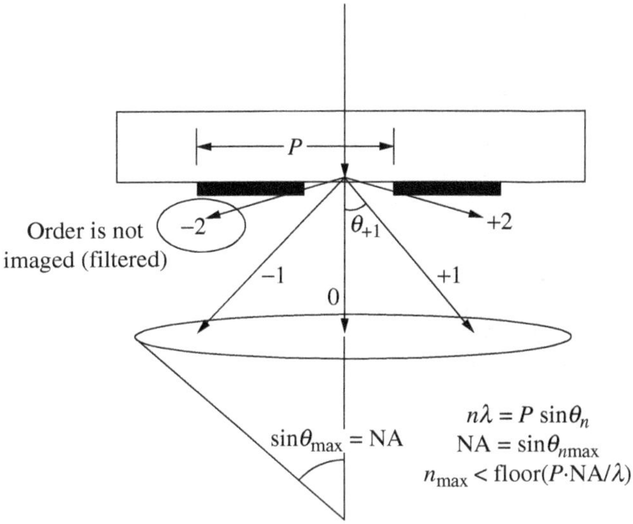

Figure 7.7 The angular spread of the diffraction orders from the mask, and the orders that the lens can collect. The lens acts as a low pass filter and the maximum angle it can collect is given by the floor (PNA/λ).

increasing the NA or decreasing the wavelength, more of the higher diffraction orders could be collected. Likewise as indicated by Eq. (7.8), if the 0 and ±1 orders are only collected, a smaller pitch is resolvable by increasing the NA or decreasing wavelength. As shown in Fig. 7.5, the projection lens must collect the first order in order to achieve modulation at the wafer. By using Eq. (7.8), the smallest resolvable pitch using on-axis illumination can be calculated as shown in Eq. (7.9).

$$P_{\min} = \frac{\lambda}{\mathrm{NA}} \tag{7.9}$$

Figure 7.7 shows how the diffraction orders are collected and imaged onto the wafer as the optical system is sliced parallel to the optical axis. It is also instructive to slice through the pupil plane of projection lens perpendicular to the optical axis. For on-axis illumination, the location of the diffraction orders in the projection pupil is shown in Fig. 7.8. In Fig. 7.8, the projection pupil is plotted in the spatial frequency domain. The largest spatial frequency that the projection pupil can collect is given by NA/λ, and the diffraction orders are located n/P in the spatial frequency domain. As the pitch increases, the distance between the diffraction orders in the spatial frequency domain decreases. This implies

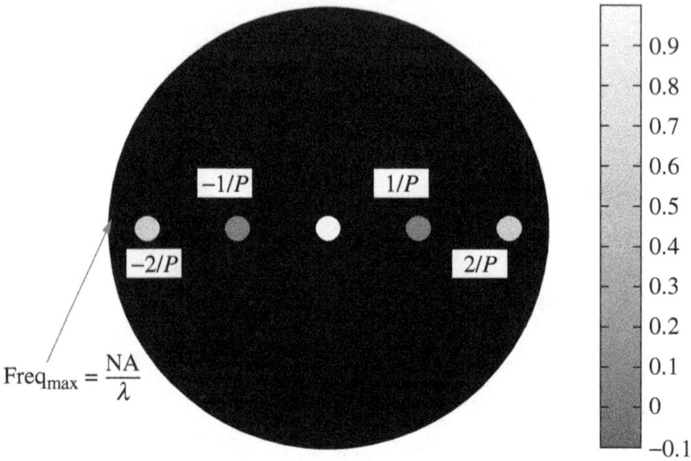

Figure 7.8 Location of the diffraction orders in the projection pupil. The diffraction orders are located in multiples of $1/P$. The maximum frequency the projection pupil can collect is NA/λ.

that the lens can capture more diffraction orders for a larger pitch as compared to a smaller pitch.

Figures 7.7 and 7.8 illustrate the location of the diffraction orders in the projection lens, and Eq. (7.2) describes the angle of the diffraction orders mathematically. As Eq. (7.2) shows, the angle of the diffraction orders increases as the pitch decreases. The NA determines the maximum angle collected by the projection lens. As the NA increases, the projection lens collects higher angle diffraction orders. As the pitch decreases, the angle of the diffraction orders increases. Depending on the pitch, a low NA projection lens is incapable of collecting the ±1 orders; however, a high NA projection lens is capable of collecting the ±1 orders. Therefore, in order to resolve a smaller pitch, the NA must continue to increase.

Using the equations and concepts presented earlier, the electric field at the wafer (the image plane) can be calculated. The electric field at the wafer is calculated by taking the inverse Fourier transform of the collected orders. Consider a 100-nm opaque chrome (T = 0) line in a 250-nm pitch is imaged with on-axis illumination. Using Eq. (7.8), the projection lens only collects the 0 and ±1 diffraction orders. The inverse Fourier transform of these three diffraction orders is calculated by using Eq. (7.5) as shown in Eq. (7.10).

$$e(x) = \sum_{m=-1}^{+1} F(m) e^{ix\left(\frac{2\pi}{P_x} m\right)} \tag{7.10}$$

The magnitude and phase of the diffraction orders are calculated using Eq. (7.6). This equation is substituted into Eq. (7.10) resulting in Eq. (7.11).

$$e(x) = \frac{a}{P_x} + \frac{2}{\pi}\sin\left(\frac{\pi a}{P_x}\right)\cos\left(x\frac{2\pi}{P_x}\right) \tag{7.11}$$

The electric field for this example is plotted in Fig. 7.9. As expected the electric field magnitude decreases under the chrome line and increases in the clear areas.

For a two-dimensional image, Eq. (7.10) can be modified by taking the inverse Fourier transform over two dimensions as shown in Eq. (7.12). In Eq. (7.12), the electric field at the wafer is calculated by taking the inverse Fourier of the diffraction orders collected by the projection lens.

$$e(x,y) = \sum_{m=-\infty}^{\infty}\sum_{n=-\infty}^{\infty} F(m,n)e^{ix\left(\frac{2\pi}{P_x}m\right)}e^{iy\left(\frac{2\pi}{P_y}n\right)} \tag{7.12}$$

From the electric field, the energy of the light impinging on the wafer is calculated. In lithography, the energy absorbed on the wafer creates

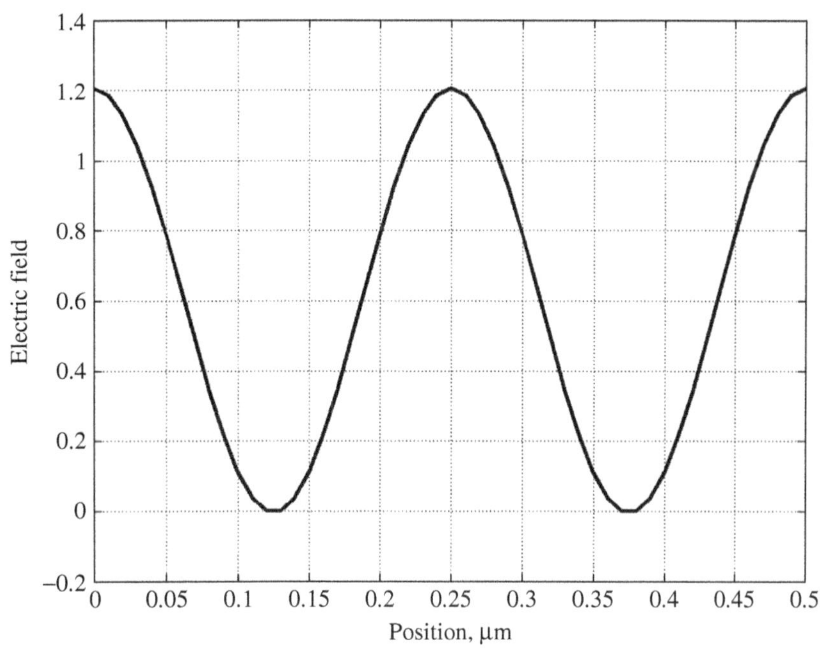

Figure 7.9 Electric field of a 100-nm chrome line with a 250 nm pitch when the 0 and ±1 orders only are collected.

the reaction in the photoresist to define the pattern on the wafer. The energy absorbed on the wafer is proportional to the intensity of the aerial image. The intensity is equivalent to the magnitude of the electric field squared as shown in Eq. (7.13).

$$I(x,y) = e(x,y) \cdot e^{*}(x,y) \tag{7.13}$$

After the energy is absorbed on the wafer, the wafer must be developed, which defines the pattern to be etched into the wafer. An optical image parameter is needed that relates to the ability of the pattern to be developed. The gradient of the development rate log is a parameter that is directly related to the capability of defining a feature in the photoresist. The gradient of the development rate log r is proportional to the gradient of the intensity log as shown in Eq. (7.14). Therefore, the gradient of the intensity log is a parameter that is directly related to the development of the photoresist. Because of this direct relationship, the gradient of the intensity log is a parameter that is often quoted in lithography, and the gradient of the intensity log has been given a special name called the *image log slope* (ILS).

$$\hat{n} \cdot \nabla \ln[r(x,y)] = \gamma \hat{n} \cdot \nabla \ln[I(x,y)] \tag{7.14}$$

In addition to ILS, the *normalized image log slope* (NILS) is a parameter that is commonly used in lithography. The ILS has units of inverse length while the NILS is unitless. The NILS is defined in Eq. (7.15). In Eq. (7.15), the variable w has units of length, and the target critical dimension is most often used for w. When using the target CD, an NILS above 1 is needed at the feature edge in order to resolve a feature in currently available photoresists. In Fig. 7.10, the intensity is plotted for the binary mask example used in Fig. 7.9. The NILS in Fig. 7.10 is 1.22, which has sufficient NILS to begin to form a line into the photoresist.

$$\text{NILS}(x,y) = w\hat{n} \cdot \nabla \ln[I(x,y)] \tag{7.15}$$

In addition to determining if the feature can be developed, the NILS is also directly proportional to the *exposure latitude* (EL) of the process. By increasing the NILS, the EL increases. The EL is the amount of exposure dose variation the process can tolerate. The dose variation typically comes from the scanner or from the laser in the optical lithography system. In addition to maximizing the EL, the *depth of focus* (DOF) is another parameter that needs to be maximized. The DOF is the amount of focus variation the process can tolerate. Since the photoresist has a finite thickness, the DOF needs to be larger than the

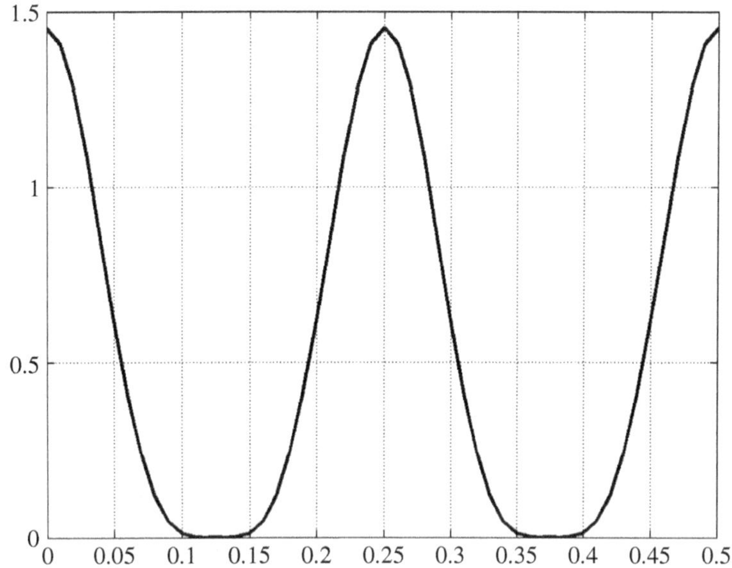

Figure 7.10 The intensity of the electric field of Fig. 7.9 for a 100-nm line with a 250-nm pitch. The NILS is 1.22.

thickness of the photoresist in order to maintain a sufficient image throughout the photoresist. In addition to the photoresist thickness, the DOF needs to be large enough to compensate for photoresist thickness variations due to wafer topography, and the DOF needs to be large enough to compensate for scanner focus variation. Unfortunately, the DOF and EL have an inverse relationship, i.e., the DOF can be improved at the expense of EL and vice versa. Therefore, in lithography, the DOF at some percentage of EL is often quoted. The goal of all the RETs presented in this chapter is to find ways to maximize the DOF and EL. Consequently, the success of the RET is determined by calculating the process window is a plot of EL as a function of DOF.

Figure 7.5 showed that contrast of the image increases as more diffraction orders are collected by the projection lens. The NILS is directly proportional to the contrast, i.e., as the contrast increases the NILS increases. Consequently, in order to improve the NILS, all RETs try to increase the number of diffraction orders that are collected by the lens. Therefore, by increasing the number of orders, the NILS and the EL improve, and techniques to increase the number of orders collected are needed. One technique to increase the number of orders collected by the projection lens involves using off-axis illumination.

7.2.2 Off-axis illumination

In lithography, off-axis illumination is used in order to decrease the minimum resolvable pitch. In this section, the effect of off-axis illumination on image formation is described by extending the theory in Sec. 7.2.1 for on-axis illumination. For off-axis illumination, the Bragg condition in Eq. (7.2) must be modified by the incident angle θ_i, as shown in Eq. (7.16). In Eq. (7.16), the zero order ray travels through the grating and the diffraction angle of the zero order ray, θ_0, is equivalent to the incident angle ray θ_i. This implies that the other orders are offset by the incident angle. By offsetting the diffraction orders by the incident angle, higher diffraction orders can be collected by the projection lens as shown in Fig. 7.11. In Fig. 7.11, the 0, −1, and −2 orders are now collected for off-axis illumination. Figure 7.11 should be compared to Fig. 7.7. In Fig. 7.7, only the −1, 0 and +1 orders are collected; however, Fig. 7.11 shows that the second order is collected when using off-axis illumination. Since higher diffraction orders improve the NILS, this implies that one can improve the resolution by using off-axis illumination.

$$n\lambda = P(\sin\theta_n + \sin\theta_i) \tag{7.16}$$

In addition to improving the resolution, off-axis illumination allows one to print a smaller pitch. The NA as shown in Eq. (7.7) is the largest angle the projection lens can collect. The largest angle that the illumination optics can create is given by the NA of the illumination, NA_i. The NA of the illumination NA_i is defined in Eq. (7.17) where σ is quoted more often rather than NA_i. With Eqs. (7.16) and (7.17), the maximum order collected by the projection lens is derived for off-axis illumination

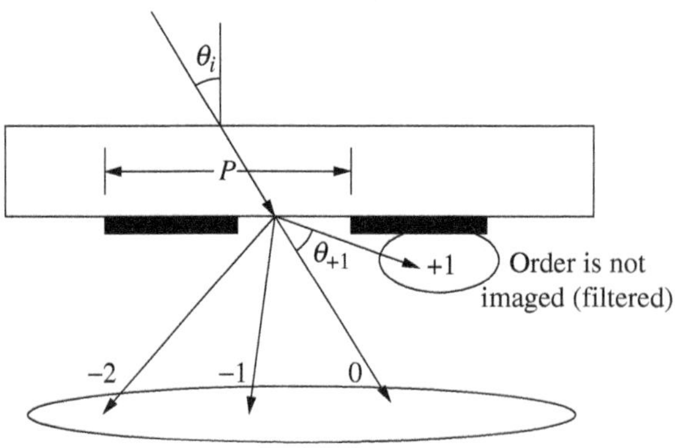

Figure 7.11 Diffraction orders from a periodic structure when illuminated with off-axis illumination.

as shown in Eq. (7.18). Equation (7.18) states that with off-axis illumination the largest order collected by the projection lens increases by a factor of 1 + σ. Equation (7.18) can also be rearranged in order to calculate the smallest resolvable pitch. The projection lens must collect at least the first order to achieve modulation in the aerial image. Thus, by setting $n = 1$ in Eq. (7.18), the minimum resolvable pitch is calculated as show in Eq. (7.19).

$$\sigma = \frac{\mathrm{NA}_i}{\mathrm{NA}} \tag{7.17}$$

$$n_{\max} = \mathrm{floor}\left[\frac{P \cdot \mathrm{NA}(1+\sigma)}{\lambda}\right] \tag{7.18}$$

$$P_{\min} = \frac{\lambda}{\mathrm{NA}(1+\sigma)} \tag{7.19}$$

Equation (7.19) indicates that the smallest resolvable pitch is $\lambda/2\mathrm{NA}$ for an illuminator with $\sigma = 1$. From this expression the k_1 factor can be calculated. In Eq. (7.20), the minimum half pitch k_1 is stated. The minimum half-pitch k_1 is 0.25. It is important to note that a k_1 of 0.25 is only achievable when the lithography system has an illuminator with $\sigma = 1$. Also at k_1 of 0.25 the first order is just collected by the projection lens and the feature is resolvable. However, this does not result in a robust process, and a k_1 of at least 0.27 is probably the minimum k_1 with sufficient process window.

$$\begin{aligned} 2\mathrm{CD}_{\text{half-pitch}} &= P_{\min} = \frac{\lambda}{2\mathrm{NA}} \\ \mathrm{CD}_{\text{half-pitch}} &= \frac{\lambda}{4\mathrm{NA}} = k_1 \frac{\lambda}{\mathrm{NA}} \end{aligned} \tag{7.20}$$

In Fig. 7.12, the projection pupils of a dense line are plotted for three different conventional illumination sigmas. In Fig. 7.12, the pupils are created for $\lambda = 248$ nm and for NA = 0.6. The 360-nm pitch cannot be resolved for $\sigma = 0.05$ as $n_{\max}$ in Eq. (7.18) is 0. This implies that only the zero order can be collected as shown by the pupil on the left in Fig. 7.12. Using Eq. (7.19), a σ of at least 0.15 is needed to resolve the 360-nm pitch. In Fig. 7.12, using $\sigma = 0.5$ the ± 1 orders are inside the projection pupil, and there is adequate modulation in the intensity resulting in an NILS of 2.08. As σ increases, the NILS also increases for a dense line. A $\sigma = 0.75$ has an NILS of 2.29. This implies that higher σ is needed to improve the process window of dense lines.

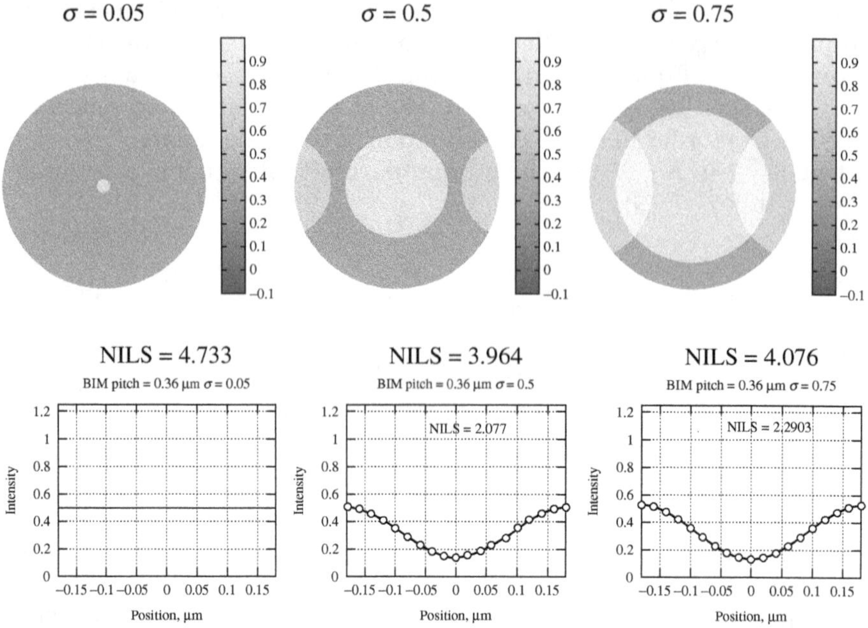

Figure 7.12 Projection pupil filling for various conventional illuminators with $\lambda = 248$ nm. The projection pupil has an NA of 0.6. The feature is a 180-nm chrome line on a 360-nm pitch. The intensity for each illuminator is also plotted. Note that the highest sigma has the largest contrast.

In the presence of off-axis illumination, Eq. (7.12) for forming the electric field using on-axis illumination must be modified.

$$e(x,y) = \sum_{m=-\infty}^{\infty} \sum_{n=-\infty}^{\infty} F(m,n) \exp\left[ix \frac{2\pi}{P_x}\left(m + \frac{P_x \alpha}{\lambda}\right)\right] \exp\left[iy \frac{2\pi}{P_y}\left(m + \frac{P_y \beta}{\lambda}\right)\right] \tag{7.21}$$

where

$$\sqrt{\alpha^2 + \beta^2} \le \sigma_{\max} \text{NA}$$

In Eq. (7.21), α and β represent the location of the off-axis illumination point on the horizontal and vertical axes of the illumination pupil. It is important to note that a single off-axis illumination point causes a shifted image. In order to get an unshifted image, the illuminator must be symmetric with respect to the α and β axes.

The intensity of the image can be calculated when there are multiple off-axis illumination points. For a spatially incoherent source used in optical lithography systems, the intensities from each point in the illumination source are independent of each other. Therefore, the total intensity at the wafer is calculated by adding the intensities from each illumination source when there are multiple off-axis illumination points. This method is known as Abbe imaging formulation and is discussed further in Sec. 7.3.1.

7.2.3 Effect of the projection pupil on the image

The theory described in Sec. 7.2.1 can also be described through system theory. In system theory, an input is fed into a system to produce an output. The system is represented by a transfer function that determines how the output is produced from a given input. The output is calculated by convoluting the input with the transfer function of the system. Instead of using convolution, the Fourier transform can be used to calculate the output from the input. The Fourier transform of the output is given by multiplying the Fourier transform of the input with the Fourier transform of the transfer function. In lithography, the projection lens represents the transfer function. The Fourier transform of the input is the diffraction orders from the mask. The diffraction orders are the spatial frequencies of the mask object. The spatial frequencies at the wafer are then calculated by multiplying the projection lens transfer function with the diffraction orders from the mask. The electric field at the wafer is calculated by taking the inverse Fourier transform of the spatial frequencies at the wafer. This process is described in the block diagram in Fig. 7.13 and is described mathematically in the next paragraph.

The transfer function of the lithography system is the projection pupil function K which can be described mathematically by Eq. (7.22). Equation (7.22) is the transfer function of a projection lens without defocus.

$$K(\xi,\psi) = \begin{cases} 1 & \sqrt{\xi^2 - \psi^2} \leq \text{NA} \\ 0 & \text{otherwise} \end{cases} \tag{7.22}$$

In Eq. (7.22), ξ and ψ are the spatial frequency variables for the horizontal and vertical axes of the projection lens. K is 1 when the diffraction order is inside the projection lens and is zero when the diffraction order is outside the projection lens. Because the diffraction orders from

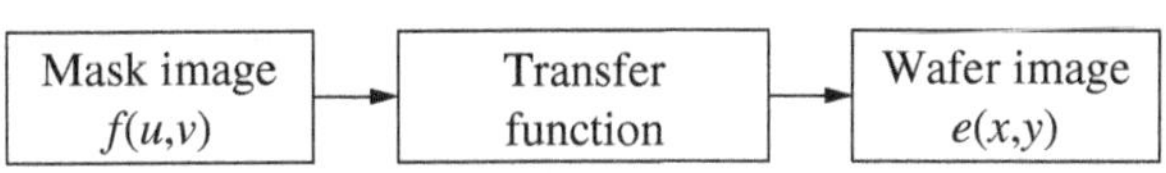

Figure 7.13 System theory block diagram representation of the lithography system.

the mask are discrete, $K(\xi, \psi)$ in Eq (7.22) can be replaced with a discrete representation of the projection pupil, $K(m, n)$, as shown in Eq. (7.23).

$$K(m,n) = \begin{cases} 1 & \sqrt{\left(\dfrac{m\lambda}{P_x}\right)^2 - \left(\dfrac{n\lambda}{P_y}\right)^2} \leq \text{NA} \\ 0 & \text{otherwise} \end{cases} \tag{7.23}$$

The spatial frequencies at the wafer, $E(m, n)$, are calculated by multiplying the projection pupil function K by the diffraction orders from the mask, F, as shown in Eq. (7.24). The electric field at the wafer, $e(x, y)$, is calculated by taking the inverse Fourier transform of the wafer spatial frequencies, $E(m, n)$, as shown in Eq. (7.25).

$$E(m,n) = K(m,n)F(m,n) \tag{7.24}$$

$$\begin{aligned} e(x,y) &= \sum_{m=-\infty}^{\infty} \sum_{n=-\infty}^{\infty} E(m,n) e^{ix\left(\frac{2\pi}{P_x} m\right)} e^{iy\left(\frac{2\pi}{P_y} n\right)} \\ &= \sum_{m=-\infty}^{\infty} \sum_{n=-\infty}^{\infty} K(m,n)F(m,n) e^{ix\left(\frac{2\pi}{P_x} m\right)} e^{iy\left(\frac{2\pi}{P_y} n\right)} \end{aligned} \tag{7.25}$$

Another way of creating the image on the wafer uses the concept of convolution from system theory. In system theory, the output is the convolution of the input with the transfer function. The transfer function of the projection pupil is the inverse Fourier transform of K. The transfer function of the projection pupil, k, is known as *the point spread function* (PSF) or the Airy function of a circular aperture. The point spread function is given in Eq. (7.26) and is plotted in Fig. 7.14. The mask pattern can be convoluted with the projection pupil transfer function to create the electric field at the wafer as shown in Eq. (7.27).

$$k\left(\rho = \sqrt{x^2 + y^2}\right) = 2\frac{J_1\left(\frac{2\pi}{\lambda}\text{NA}\rho\right)}{\frac{2\pi}{\lambda}\text{NA}\rho} \tag{7.26}$$

$$e(x,y) = f(x,y) \otimes k(x,y) = \iint e(x',y')k(x-x',y-y')dx'dy' \tag{7.27}$$

In Fig. 7.14, the point spread function is plotted for two wavelengths KrF (248 nm) and ArF (193 nm). As the wavelength decreases, the central lobe of the PSF is tighter, which implies that a lower wavelength lithography system can resolve smaller resolution features. The PSF is also plotted for two numerical apertures. As the NA increases, the central lobe of the PSF is tighter, which implies that a higher NA lithography system can resolve smaller resolution features.

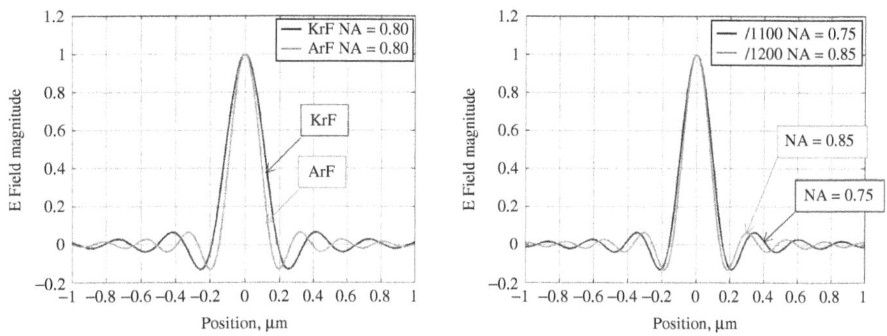

Figure 7.14 Point spread functions for different wavelengths (KrF $\lambda = 248$ nm and ArF $\lambda = 193$ nm) and for different NA (0.75 and 0.85).

The point spread function was used by Lord Rayleigh to describe the resolution limit. In the resolution limit of a telescope, Rayleigh calculated the angular separation between two stars that is required to distinguish each star. Rayleigh stated that the angular separation is limited by the size of the telescope aperture, the NA, and by the wavelength of the light emanating from the star. The resolution limit developed by Rayleigh is identical to the resolution limit used in lithography, $k_{1\lambda}$/NA, as given in the introduction.

7.2.4 Image depth of focus

The previous sections discussed the theory needed to calculate images and to calculate the minimum resolution. In addition to understanding how to create an image, the effect of imaging through focus is often calculated because the image must be maintained through a DOF range to be considered a robust process. The projection pupil in Fig. 7.7 acts like a low pass filter in which any spatial frequency in the mask that is less than NA/λ is imaged onto the wafer. In the absence of defocus, the diffraction orders pass through the projection lens unchanged. However, in the presence of defocus, the phase of each diffraction order changes as it passes through the projection lens. The change in the phase of the diffraction orders leads to a degradation of the image. If the image is degraded too much, there is a loss of contrast due to defocus.

In the presence of defocus the projection pupil function needs to be modified as shown in Eq. (7.28).

$$K(\xi,\psi) = \begin{cases} \exp\left(-i\dfrac{2\pi}{\lambda} z\sqrt{1-\xi^2-\psi^2}\right) & \sqrt{\xi^2-\psi^2} \le \text{NA} \\ 0 & \text{otherwise} \end{cases} \tag{7.28}$$

From Fig. 7.8, the discrete diffraction orders for a periodic structure are located at $m\lambda/P_x$ and $n\lambda/P_y$. The location of these discrete diffraction orders can be substituted into Eq. (7.28) for ξ and ψ resulting in Eq. (7.29). Equation 7.29 is the defocused projection pupil function for a periodic mask structure.

$$K(m,n) = \begin{cases} \exp\left(-i\dfrac{2\pi}{\lambda}z\sqrt{1-\left(\dfrac{m\lambda}{P_x}\right)^2-\left(\dfrac{n\lambda}{P_y}\right)^2}\right) & \sqrt{\left(\dfrac{m\lambda}{P_x}\right)^2-\left(\dfrac{n\lambda}{P_y}\right)^2} \le \text{NA} \\ 0 & \text{otherwise} \end{cases} \tag{7.29}$$

In Fig. 7.15, the phase of the projection pupil is plotted for various levels of defocus. As the defocus increases the amount of variation from the center of the projection pupil to the edge of the projection, pupil increases. This implies that the phase change increases between the diffraction orders passing through the center compared to the diffraction orders passing through the edge of the projection pupil. This phase

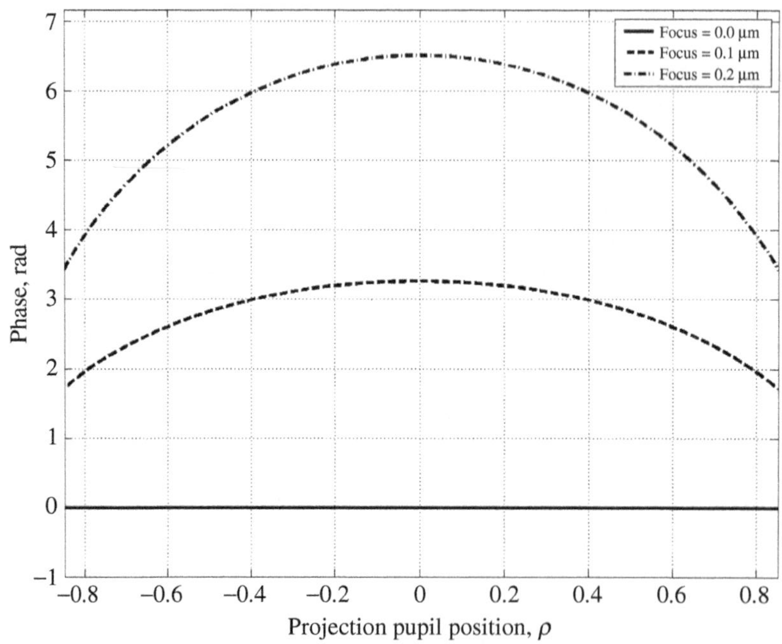

Figure 7.15 The phase of the projection pupil as a function of the radial position, ρ, in the projection pupil. As the defocus increases, the curvature in the phase increases.

change implies that the ray at the edge will lead or lag the ray at the center. This leading or lagging of the rays creates a loss of contrast in the image. The loss of contrast increases as the amount of leading or lagging increases.

The loss of contrast for a defocused image can be calculated exactly for each illumination point. The defocused image is calculated by substituting the defocused projection pupil function K in Eq. (7.29) into the equation for calculating the electric field at the wafer given in Eq. (7.25).

For off-axis illumination, the projection pupil function needs, K, to be modified as shown in Eq. (7.30). In Eq. (7.30), the areas of the projection pupil are shifted by α and β coordinates of the illumination pupil. In order to calculate the image of a defocused projection pupil with off-axis illumination, Eq. (7.30) is included in Eq. (7.21) resulting in Eq. (7.31).

$$K(m,n) = \begin{cases} \exp\left(-i\frac{2\pi}{\lambda}z\sqrt{1-\left(-\alpha+\frac{m\lambda}{P_x}\right)^2-\left(-\beta+\frac{n\lambda}{P_y}\right)^2}\right) \\ 0 \end{cases}$$

$$\times\sqrt{\left(-\alpha+\frac{m\lambda}{P_x}\right)^2-\left(-\beta+\frac{n\lambda}{P_y}\right)^2} \leq \text{NA} \quad \text{otherwise} \tag{7.30}$$

$$e(x,y) = \sum_{m=-\infty}^{\infty}\sum_{n=-\infty}^{\infty} K(m,n)F(m,n)\exp\left[ix\frac{2\pi}{P_x}\left(m+\frac{P_x\alpha}{\lambda}\right)\right]$$

$$\times\exp\left[iy\frac{2\pi}{P_y}\left(m+\frac{P_y\beta}{\lambda}\right)\right] \tag{7.31}$$

Using Fig. 7.15 and Eqs. (7.25) and (7.29), the intensity while defocused is calculated for a mask producing three diffraction orders, −1, 0, +1, when using on-axis illumination. This is commonly called a three-beam imaging problem, e.g., a binary mask (transmittance, $T = 0$) produces three diffraction orders. Using the binary mask example in Sec. 7.2.1, an opaque 100-nm line with a 250-nm pitch on a λ = 193 nm system with NA of 0.85 produces diffraction orders at $m\lambda/P_x = -0.772$, 0, and +0.772. At 50-nm defocus, $z = 50$ nm, the projection pupil causes the three orders to be delayed resulting in a phase change between the 0 and the ±1 orders. The phase of the three orders is calculated using Eq. (7.29). For this example, the −1, 0, +1 orders have phase 1.0346, 1.6278, 1.0346 rad., respectively. Using these phases, the electric field in Eq. (7.11) is modified by the phase from the defocus applied to the diffraction orders as shown in Eq. (7.32).

$$E(x,z=50\,\text{nm}) = e^{i1.6278}\frac{a}{P_x} + e^{i1.0346}\frac{2}{\pi}\sin\left(\frac{\pi a}{P_x}\right)\cos\left(x\frac{2\pi}{P_x}\right)$$
$$\times\, e^{i1.6278}\left[\frac{a}{P_x} + e^{-i0.5932}\frac{2}{\pi}\sin\left(\frac{\pi a}{P_x}\right)\cos\left(x\frac{2\pi}{P_x}\right)\right] \tag{7.32}$$

The intensity of the electric field in Eq. (7.32) is calculated using Eq. (7.13) resulting in Eq. (7.33). In Eq. (7.33), it is important to note that the phase difference between the 0 and ±1 orders produces a cross term in the intensity, the cos(–0.5932) term. This cross term causes a loss of intensity through focus. The intensity for this three-beam imaging example is plotted in Fig. 7.16 at three focal points, $z = 0$, 50, and 100 nm. For the image that is in focus ($z = 0$ nm), the contrast and NILS are large. The NILS at $z = 0$ nm is 1.22. However, as the image is defocused the contrast and NILS decrease. At $z = 50$ nm the NILS is

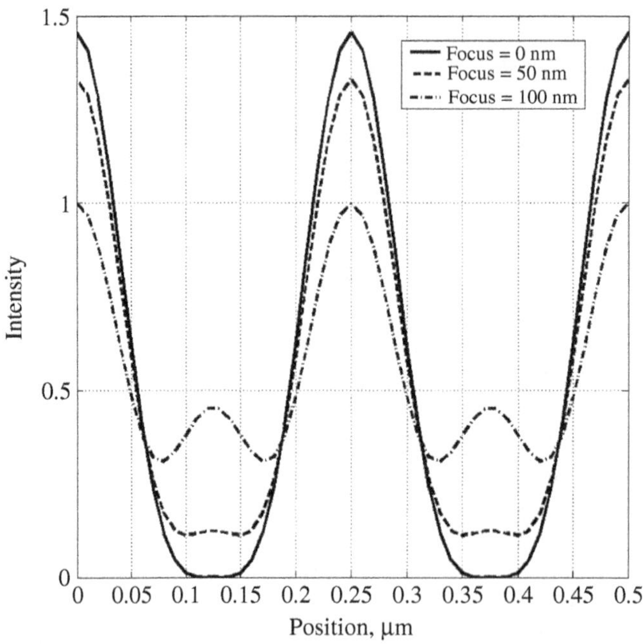

Figure 7.16 Aerial images for three diffraction orders (–1, 0, and +1) at three foci. As the defocus increases, the contrast decreases.

0.84, and at $z = 100$ nm the NILS is 0.37. At 100-nm focus, the image has degraded, and line may not print.

$$I(x, z = 50\,\text{nm}) = \left(\frac{a}{P_x}\right)^2 + \frac{4}{\pi}\cos(-0.5932)\sin\left(\frac{\pi a}{P_x}\right)\cos\left(x\frac{2\pi}{P_x}\right) + \frac{4}{\pi^2}\sin^2\left(\frac{\pi a}{P_x}\right)\cos^2\left(x\frac{2\pi}{P_x}\right) \tag{7.33}$$

The intensity while defocused can also be calculated for two-beam imaging. A two-beam image can be created in two ways, the first is using an alternating phase-shift mask with on-axis illumination, and the second is using an attenuating phase-shift mask with off-axis illumination. Using the alternating phase-shift mask approach, the image with two beams is calculated again by using Eqs. (7.25) and (7.29). An alternating phase-shift mask can be created by setting the transmittance to 1, phase to 180°, and setting $a = P_x/2$ in Eq. (7.6). With these conditions the zero order is 0 and the first order has a magnitude of $2/\pi$. A 125-nm alternating phase-shift line on a 250-nm pitch produces −1 and +1 diffraction orders at $m\lambda/P_x = -0.772$ and $+0.772$ on a $\lambda = 193$ nm system with NA of 0.85. At 50-nm defocus, $z = 50$ nm, the projection pupil causes the two orders to be delayed resulting in a phase change in the orders. The phase of the two orders is calculated using Eq. (7.29), and the −1 and +1 orders have a phase of 1.0346 rad. Using this phase, the electric field in Eq. (7.14) becomes Eq. (7.34) in which the phase from the defocus has been applied to the orders. In Eq. (7.34) for the two-beam imaging case, there is no cross term between the ±1 and the zero order.

$$E(x, z = 50\,\text{nm}) = e^{i1.0346}\frac{2}{\pi}\cos\left(x\frac{2\pi}{P_x}\right) \tag{7.34}$$

At 100-nm defocus, $z = 100$ nm, the −1 and +1 orders have a phase of 2.0693 rad, and Eq. (7.34) becomes Eq. (7.35). In Eq. (7.35) only the phase term changes. This is an important observation because the intensity is found by squaring the magnitude of the electric field and this phase term disappears as shown in Eq. (7.36). Equation (7.36) shows that an image created with two beams has the same intensity at all focal planes, i.e., the depth of focus is infinite. The intensity for this two-beam imaging example is plotted in Fig. 7.17 at three focal points, $z =$ 0, 50, and 100 nm. Therefore in order to improve the DOF, techniques that sample the projection pupil in only two locations have the largest DOF. In the following sections, resolution enhancement techniques that

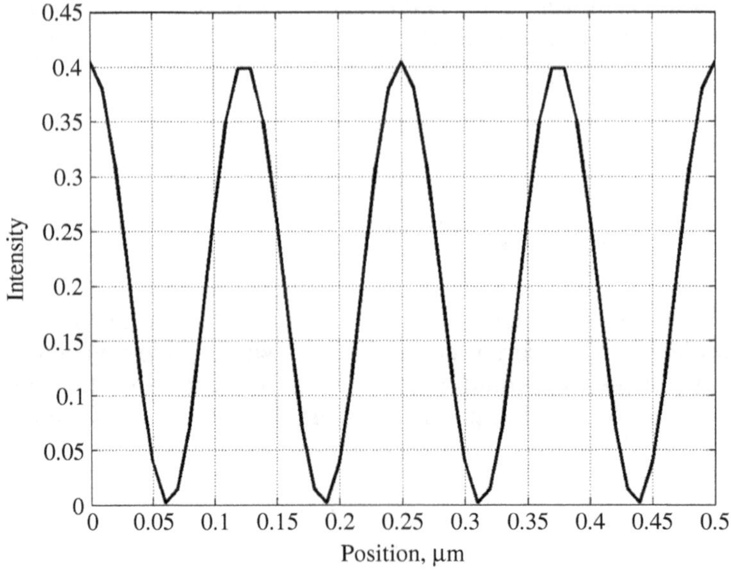

Figure 7.17 Aerial images for two diffraction orders (–1 and +1) at three foci. The images do not change as the defocus changes, i.e., the DOF is infinite.

use two locations of the projection pupil or that use the least amount of locations create an image with the largest DOF.

$$E(x, z = 100\,\text{nm}) = e^{i2.0693}\frac{2}{\pi}\cos\left(x\frac{2\pi}{P_x}\right) \tag{7.35}$$

$$I(x, z) = \frac{4}{\pi^2}\cos^2\left(x\frac{2\pi}{P_x}\right) \tag{7.36}$$

7.3 Scalar Imaging Theory

In Sec. 7.2, Fourier optics was discussed for imaging a mask with a single point in the illuminator. However, commercially available lithography systems do not support imaging with a single point in the illuminator. In these commercial systems, the illumination has a finite size. Therefore a method for imaging with a finite size illuminator is required. Scalar imaging theory allows one to calculate the image of the mask from a finite size illuminator, and scalar imaging theory is discussed in this section.

Scalar imaging theory ignores the vector effects of electromagnetic radiation. The mask is infinitely thin and is represented with an electric field transmission function f as discussed in Sec. 7.2. Furthermore, scalar imaging theory assumes that the electric and magnetic fields are not coupled, and the intensity can be calculated from the electric field. Scalar imaging theory is a good approximation when the features are large compared to the wavelength. Since lithography uses a reduction system, scalar imaging is a good approximation for most lithography applications. For hyper-NA (see Sec.7.7.1) and for masks with mask topography on the order of a wavelength (see Sec.7.5.2), scalar imaging theory begins to break down, and rigorous *electromagnetic field* (EMF) theory must be used for calculating the diffraction orders.[9] Scalar imaging theory is still used for calculating the image on the wafer from the EMF diffraction orders.

Imaging with a lithography system is analogous to microscopy. Methods for defining an image through microscopy have been available for many years. The first mathematical method for modeling imaging with a microscope was described in 1873 by Abbe and is often referred to as Abbe formulation or method.[4] Hopkins developed another modeling technique later in 1951 and this is often referred to as Hopkins formulation or method.[10] Each formulation, Abbe and Hopkins, has advantages and disadvantages which are described in this section.

Because of the advantages and disadvantages, each formulation is better suited to a specific type of problem. Abbe formulation integrates over the source last while Hopkins formulation integrates over the source first. In general, Abbe formulation is used in most of the commercially available lithography simulation packages—Prolith,[11] Solid-C,[12] and Panoramic.[13] Abbe formulation is also the best method for optimizing the illumination. Hopkins is the best method for optimizing the mask. Consequently, Hopkins formulation is used in most of the commercially available OPC tools. Hopkins formulation, however, was used in one of the first lithography simulation tools, SPLAT.[14]

Before describing Hopkins and Abbe imaging, a brief description of coherence is necessary.[15] The temporal coherence of the light in a lithography system describes the ability of the light to interfere with a delayed version of itself (shifted in time but not in space). The spatial coherence of the light describes the ability of the light to interfere with a spatially shifted version of itself (shifted in space but not in time). In the lithography system, the illumination source is temporally coherent and spatially incoherent. This means that the light can interfere with itself when it is delayed in time; however, the light cannot interfere with itself when it is shifted in space. A temporally coherent source has an infinitely long coherence length, which is set by bandwidth of the illumination source. Typically the bandwidth is very small, and essentially the laser is monochromatic. Although in this chapter it is assumed that the laser has only one wavelength, a polychromatic source can also be modeled by summing the image intensity from each wavelength weighted by the energy of each wavelength in the source. In the spatial domain, the illumination source is incoherent. This means that each point of the source has a random phase when compared to another point of the source. Because the phase is random, the electric field from each source point cannot interfere with another source point. Consequently, the intensity from each source point is added because the intensity represents the average energy from the random phase source.

The illumination optics in lithography systems are Köhler rather than critical.[16] In a Köhler illumination system, each point of the source is imaged as a plane wave incident onto the mask, i.e., the focus position of the illumination is at infinity. A critical illumination system is not used in commercially available lithography systems. In a critical illumination system, each point of the source is imaged to a point on the mask, i.e., the focus position of the illumination is on the mask plane. The Köhler illumination system and the rest of the optical lithography system are shown in Fig. 7.1.

7.3.1 Abbe formulation

The intensity on the wafer from the mask is calculated by dividing the source into a number of individual point sources. Since the source is incoherent, the total intensity on the wafer is calculated by summing the intensity resulting from each individual point source.[4] The intensity for an individual point source was given in Eq. (7.13) by calculating the electric field using Eq. (7.25). The total intensity is then calculated by summing the individual intensity from each point in the source weighted by illumination pupil intensity $J_s(\alpha, \beta)$ as shown in Eq. (7.37).

$$I(x,y) = \int_{\sqrt{\alpha^2+\beta^2} \le \sigma_{\max}\mathrm{NA}} \int J_s(\alpha,\beta) \left| \sum_m \sum_n e^{i\frac{2\pi x m}{P_x}} e^{i\frac{2\pi y n}{P_y}} F(m,n) K\left(\alpha + \frac{m\lambda}{P_x}, \beta + \frac{n\lambda}{P_y}\right) \right|^2 \tag{7.37}$$

Equation (7.37) can also be written as Eq. (7.38) by explicitly replacing the magnitude square with multiplication by the complex conjugate.

$$I(x,y) = \int_{\sqrt{\alpha^2+\beta^2} \le \sigma_{\max}\mathrm{NA}} \int J_s(\alpha,\beta) \sum_m \sum_n \sum_p \sum_q e^{ix\left[\frac{2\pi}{P_x}(m+p)\right]} e^{iy\left[\frac{2\pi}{P_y}(n+q)\right]} F(m,n) \times F^*(-p,-q) \cdot K\left(\alpha + \frac{m\lambda}{P_x}, \beta + \frac{n\lambda}{P_y}\right) K^*\left(\alpha - \frac{p\lambda}{P_x}, \beta - \frac{q\lambda}{P_y}\right) d\alpha d\beta \tag{7.38}$$

Equations (7.37) and (7.38) are mathematical expressions of imaging from the mask to wafer in an optical lithography system. It is important to note that the source is integrated over last, i.e., the intensity is calculated for each illumination point, and then the total intensity is calculated by integrating over the source multiplied by the intensity from each of these source points. Because the source is integrated over last, Abbe imaging is better suited for applications in which the illumination is optimized. When optimizing the illumination the quantity after the four summation signs can be precalculated, and then the source systematically varied to find an optimal illumination.

7.3.2 Hopkins formulation

In Eq. (7.38), the source is integrated over last.[10,15] However, because integration is a linear operation, Hopkins noticed that the integration over the source can be rearranged such that the source is integrated over first. Hopkins formulation for imaging is shown in Eq. (7.39). In Eq. (7.39), the quantity after the integral has been given the name, *transfer cross coefficients* (TCC). The TCC in Eq. (7.39) is explicitly described in Eq. (7.40) resulting in Eq. (7.41) for calculating the image in the optical lithography system. The TCC completely describes the illumination and projection optics of the optical lithography system. The TCC is independent of the mask. Since the TCC is independent of the mask, Hopkins formulation is better suited for applications in which the mask is optimized. In optical proximity correction, the mask is modified at many locations such that the mask better produces the design image on the wafer. Consequently, Hopkins imaging is used in most if not all the commercially available OPC software.

$$I(x,y)=\sum_m\sum_n\sum_p\sum_q e^{ix\left[\frac{2\pi}{P_x}(m+p)\right]}e^{iy\left[\frac{2\pi}{P_y}(n+q)\right]}F(m,n)F^*(-p,-q)\cdot\int_{\sqrt{\alpha^2+\beta^2}<\sigma\mathrm{NA}} \times\int J_s(\alpha,\beta)K\left(\alpha+\frac{m\lambda}{P_x},\beta+\frac{n\lambda}{P_y}\right)K^*\left(\alpha-\frac{p\lambda}{P_x},\beta-\frac{q\lambda}{P_y}\right)d\alpha d\beta \quad (7.39)$$

$$\mathrm{TCC}(m,n,p,q)=\int_{\sqrt{\alpha^2+\beta^2}<\sigma\mathrm{NA}}\int J_s(\alpha,\beta)K\left(\alpha+\frac{m\lambda}{P_x},\beta+\frac{n\lambda}{P_y}\right) \times K^*\left(\alpha-\frac{p\lambda}{P_x},\beta-\frac{q\lambda}{P_y}\right)d\alpha d\beta \quad (7.40)$$

$$I(x,y)=\sum_m\sum_n\sum_p\sum_q e^{ix\left[\frac{2\pi}{P_x}(m+p)\right]}e^{iy\left[\frac{2\pi}{P_y}(n+q)\right]}F(m,n)F^*(-p,-q)\mathrm{TCC}(m,n,p,q) \quad (7.41)$$

Hopkins formulation has one disadvantage. Hopkins formulation assumes that the TCC is independent of the source. When the diffraction orders are dependent on illumination, Hopkins formulation should not be used, and Abbe formulation should be used instead. Typically the diffraction orders are independent of the illumination because reduction is used in lithography. The maximum angle of incidence impinging on the mask is shown in Eq. (7.42). In Eq. (7.42), the reduction ratio of the lithography system is M, and the refractive index of the mask is n_{mask}. The refractive index of the mask is needed in Eq. (7.42) because the pattern of the mask faces the entrance pupil of the projection lens. The refractive index further decreases the angle impinging on the mask pattern. In producing the image on the wafer, the angles impinging on the mask are usually small (less than 10°). For these small angles, the diffraction orders are independent of the illumination. However, as the NA increases to above 1 in immersion optical lithography systems, the assumption that the diffraction order is independent of the illumination may no longer be valid. In Fig. 7.18, the magnitude of the first diffraction order as a function of illumination incident angle is plotted for a 60-nm phase-shifted line on a 120-nm pitch (wafer dimensions). From Fig. 7.18, the first order is independent of incidence angle until 18°. In most lithography systems, the angles impinging on the pattern are less than 15°, e.g., an NA of 0.85 on a 4X reduction system (M = 0.25) results in a maximum incident angle of 8.14° while an NA of 1.3 has a maximum angle of 12.5°.

$$\theta_{\text{illum max}} = \sin^{-1}\left(\frac{M\sigma_{\text{max}}\text{NA}}{n_{\text{mask}}}\right) \tag{7.42}$$

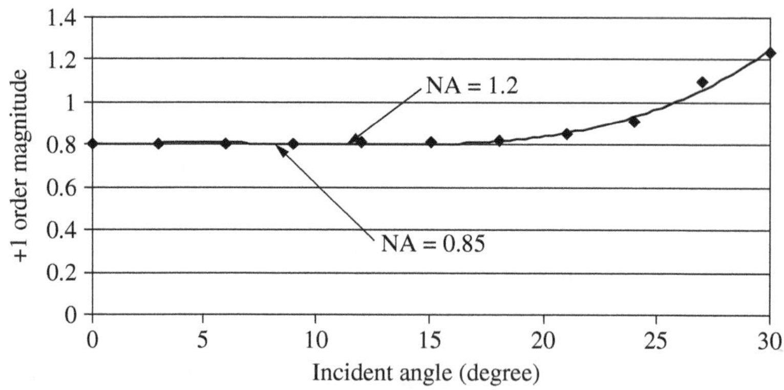

Figure 7.18 The magnitude of the +1 order as the mask as a function of incident angle of off-axis illumination. Note that the magnitude of the order is independent of the angle until 18°.

The TCC represents the complete image transfer function from the mask to the wafer. Since the TCC represents the system completely, the lithography community has borrowed ideas from the image processing community to increase the speed of calculations with the TCC. One technique used by OPC software is the sum of coherent systems, SOCS.[17–21] In the sum of coherent systems approach, the TCC is decomposed into a set of orthogonal kernels. Each of these orthogonal kernels acts a coherent system and represents a separate kernel for forming the image. A diagram for SOCS is shown in Fig. 7.19. In Fig. 7.19, the TCC is decomposed into a set of eigenvectors $\Phi_i(m, n)$ and eigenvalues λ_i by using *singular value decomposition* (SVD). The intensity is then calculated for each of these eigenvectors by using Eq. (7.25) in which $K(m, n)$ is replaced by $\Phi_i(m, n)$. This is shown in Eq. (7.43). The total intensity of the image is a summation of the intensities from each of these eigenvectors weighted by λ_i. This is described mathematically by Eq. (7.44). The SOCS approach utilizes the fact that SVD has the best energy compaction, i.e., the least amount of eigenvectors is needed to represent the TCC. In Fig. 7.20, the eigenvalues are plotted for three different illumination modes, quasar (quadrupole), annular, and conventional. It is important to note that the eigenvalues decay rapidly such that very few (less than 10) eigenvectors are needed to calculate the total intensity. Because very few eigenvectors are needed, the intensity of the mask pattern can be calculated rapidly. Consequently, the SOCS approach is used in many OPC software tools because the intensity must be calculated for millions of mask patterns. When the illumination is a single point source,

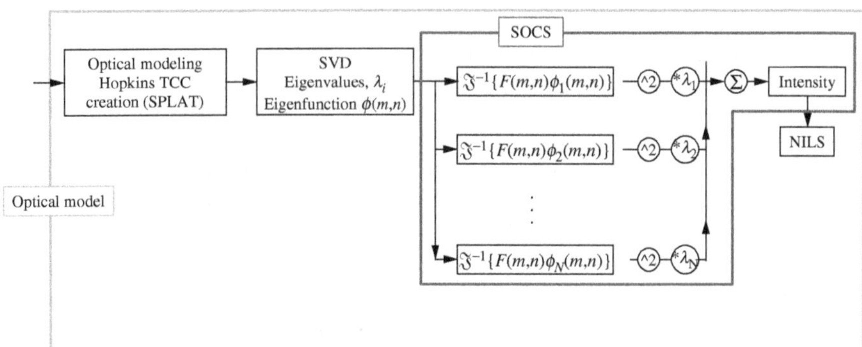

Figure 7.19 Block diagram describing the sum of coherent systems (SOCS). Eigenfunction kernels Φ are calculated by using SVD on the TCC.

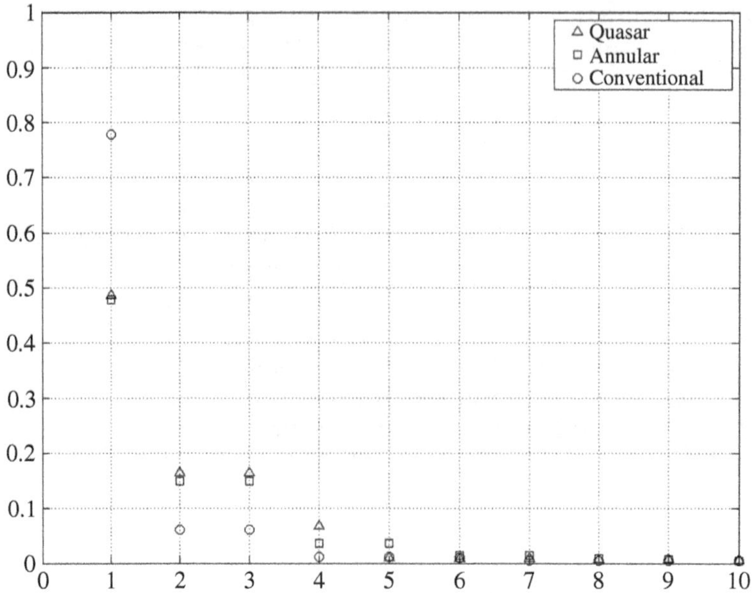

Figure 7.20 Eigenvalues for three different illuminations on an ArF scanner with NA = 0.75. The conventional illumination has $\sigma = 0.35$. The quasar and annular illumination have $\sigma_{in} = 0.5$ and $\sigma_{out} = 0.8$.

there is only one SOCS eigenvector kernel. This single kernel is the point spread function and the eigenvalue is 1.

$$e_i(x,y) = \sum_{m=-\infty}^{\infty} \sum_{n=-\infty}^{\infty} \Phi_i(m,n) F(m,n) e^{ix\left(\frac{2\pi}{P_x}m\right)} e^{iy\left(\frac{2\pi}{P_y}n\right)} \tag{7.43}$$

$$I(x,y) = \sum_i \lambda_i e_i(x,y) e_i^*(x,y) \tag{7.44}$$

7.4 Optical Proximity Correction

As the k_1 continues to decrease, the design pattern no longer matches the pattern printed on the wafer. This is due to the low pass filtering of the projection lens. Due to this low pass filtering, features in proximity to each other induce errors in the printing of the features, such as CD errors and corner rounding errors. In Fig. 7.21, a wafer *secondary electron microscopy* (SEM) is shown for a mask without OPC and with OPC. The SEM image shows that the lines are bridging without OPC. The bridging in the geometry is due to the proximity of the design inducing errors in the wafer printing. In order to fix these proximity effects, the mask needs to be changed by biasing the mask or by adding other subresolution features.

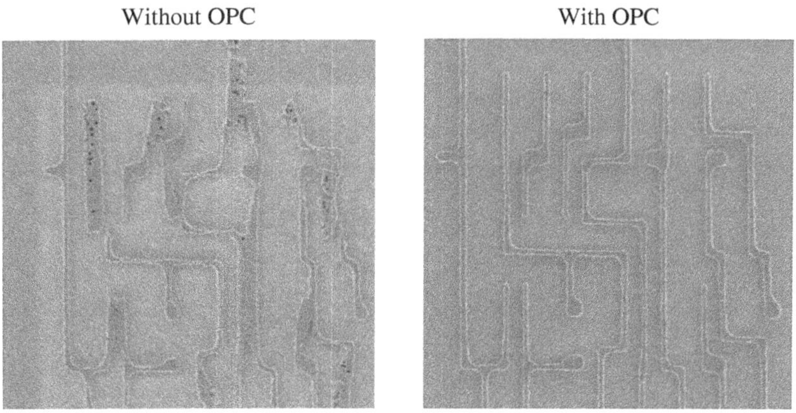

Figure 7.21 SEM images of a pattern with and without OPC. (Courtesy of J. Word, Mentor Graphics.)

In the early stages of OPC implementation, biasing and subresolution feature placement was accomplished through rule-based OPC. Rule-based OPC was done with existing DRC tools to correct proximity errors based on a set of rules. These types of rule-based corrections consisted of biasing the feature as a function of the separation (space width) from another feature or consisted of adding hammerheads or other end-of-line corrections. As the k_1 decreased further, rule-based OPC began to fail, or the rules became so complex the rules were impossible or slow to implement. Consequently, OPC evolved to use models based on the illumination optics, on the projection optics, and on the photoresist kinetics. Due to the complexity of the corrections needed, advanced processes use model-based OPC instead of rule-based OPC.

In Fig. 7.22 the proximity errors are shown for various k_1 factors. The desired pattern is a plus sign. As the k_1 factor decreases the printed pattern shape becomes more distorted from the desired plus sign pattern. The end of the line begins to pull back and the corners become more rounded. As the k_1 decreases, the printed image becomes more distorted. Consequently, at lower k_1, OPC is absolutely necessary and is a critical component of all RETs. In Sec. 7.4.1, modifications based on rule-based OPC are described, and these modifications include biasing, *end-of-line* (EOL) corrections (serifs and hammerheads), and assist features (AFs). In Sec. 7.4.2, model-based OPC is described.

OPC has been used in microphotography prior to the application of OPC to lithography. The first microlithography corrections involved placing serifs at the corners of features to prevent rounding in the image.[22] In the mid 1980s, techniques for correcting proximity errors in lithography were being developed and reported by Prof. Saleh at the

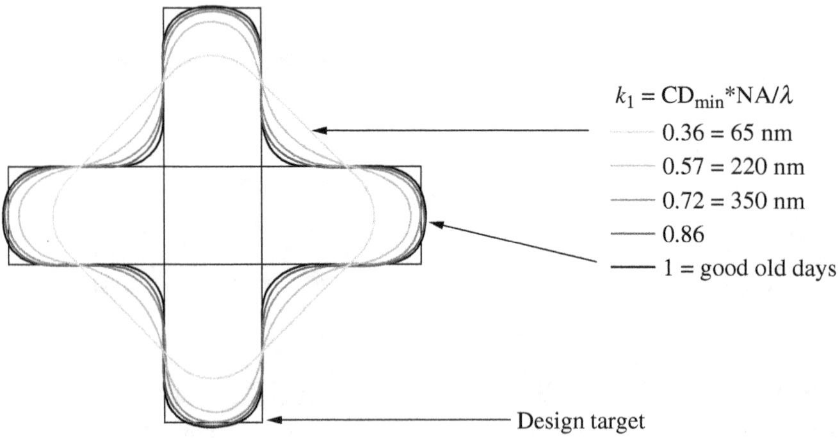

Figure 7.22 Printed images of a plus sign at decreasing k_1. At $k_1 = 0.36$ the printed image no longer represents the desired plus sign. (*Source:* Lucas, K., et al., *Proc. SPIE* 4691, 2002).

University of Wisconsin.[23–25] Some of the first commercial applications of OPC to optical lithography were initiated by SEMATECH. These commercial applications were developed by Stirniman[26,27] and by Cobb,[28] which are now available in commercially available software. The application of serifs to correct corner rounding in optical lithography was reported by Ito[29] and by Starikov.[30] The use of assist features to improve process window was developed by Chen.[31,32]

7.4.1 Rule-based OPC

In rule-based OPC, the printed image is corrected by applying bias, serifs, and hammerheads to the mask. As the name rule-based OPC implies the bias, serifs, and hammerheads are applied to the design based on a set of rules. The set of rules is encompassed in a rule table. This rule table is generated from measurements of experimental wafer data. The experimental data are collected by printing and measuring the results from an OPC test mask. On this OPC test mask, there are various test structures needed for generating the rule table. These test structures typically include measuring the CD as a function of pitch, linearity of the wafer CD as a function of mask CD, and end-of-line pullback. The wafer measurements from the test structures are then translated into a set of rules. The rules are then coded into the OPC language needed for the OPC software. This OPC language is called the rule table. Using this rule table, the design pattern is input into the rule-based OPC software, and a corrected mask pattern is created. Bias, serifs, and hammerheads are applied to correct the mask based on the rule table. A cartoon illustrating rule-based OPC is shown in Fig. 7.23.

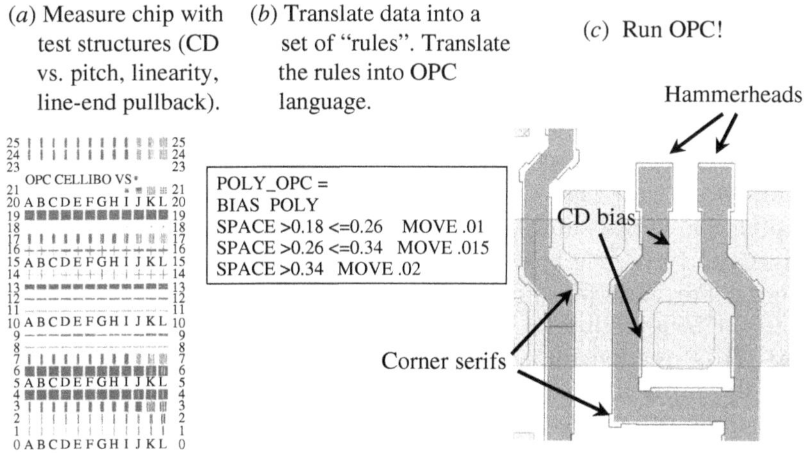

Figure 7.23 Block diagram showing the steps of rule-based OPC. (Courtesy of James Word, Mentor Graphics.)

Biasing. As the k_1 decreased, lithographers began to notice that the wafer CD changed as the pitch changed on the mask. The proximity of the mask surroundings began affecting wafer CD. Optical proximity correction was first conceived to eliminate the wafer CD variation as a function of pitch. Consequently, the first requirement of rule-based OPC is to ensure that the wafer CD printing does not change as the pitch varies. In Fig. 7.24, the printed wafer CD is plotted as a function of pitch. Figure 7.24 shows that the wafer CD changes as the pitch changes, i.e.,

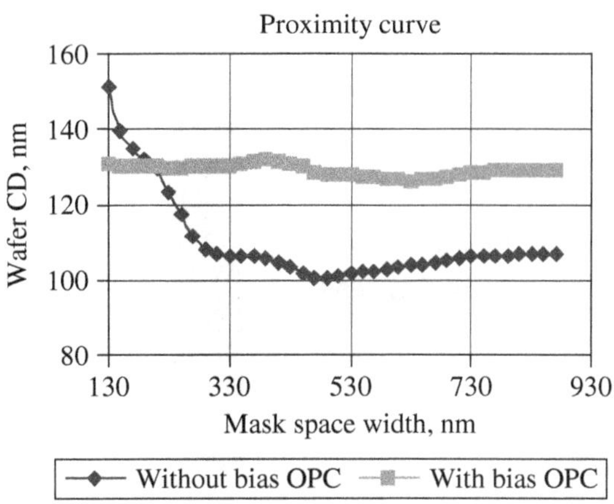

Figure 7.24 Proximity curve before applying bias OPC and after applying bias OPC. The proximity curve is flatter with bias OPC.

the wafer CD depends on the proximity. This curve is often called the wafer proximity curve. Ideally the wafer proximity curve should be flat. The goal of OPC is to flatten this proximity curve.

The wafer proximity curve is flattened by biasing the line as a function of pitch. For each pitch, the amount of biasing is different. Since space width is easier to determine than pitch, in rule-based OPC the bias is determined for each space width, and a rule table is created. In this rule table, a bias amount is specified for a range of space widths. The bias amount and the space width range are determined through experiments. For example, in Fig. 7.24, the dense lines are printing wider than the isolated lines. Consequently, positive bias (the linewidth is increased) is needed for the isolated lines, and negative bias is needed for the dense lines. The OPC program then applies the bias correction based on the space width in the rule table. After applying the bias from the OPC rule table, the proximity curve is flattened as shown in Fig. 7.24.

The bias changes the wafer CD. The bias, however, does not change the size of the process window. Bias OPC causes all the mask features to print at the same dose. Thus, bias OPC improves the overlapping process window of all the features on the mask.

End-of-line correction. As the k_1 decreases, the end of line on the printed wafer image prints smaller than the design pattern. This is called end-of-line shortening. For the gate, EOL shortening can cause a short, which destroys the functionality of the transistor. Therefore, EOL shortening must be corrected with OPC. In rule-based OPC, the EOL shortening is corrected by extending the EOL, by adding serifs, or by adding a hammerhead. In Fig. 7.25, both a serif and hammerhead

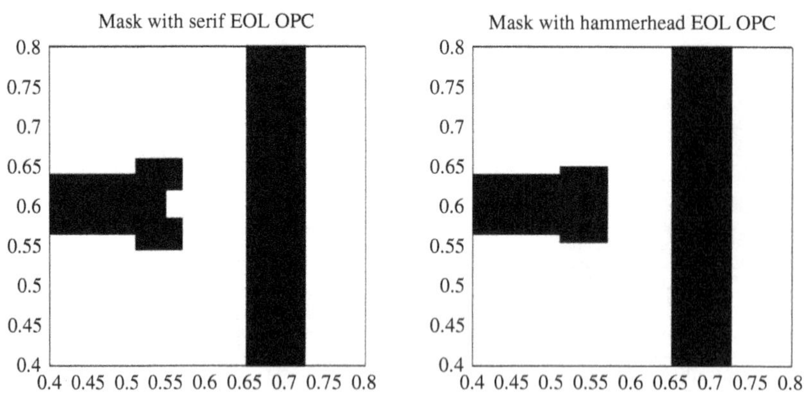

Figure 7.25 Two examples of rule-based EOL OPC serifs and hammerheads.

correction is applied to the end of the line. Typically, the choice of serif or hammerhead correction depends on mask manufacturing constraints and depends on the wafer printing results. At low k_1, a mask with a hammerhead correction is easier to manufacture than a mask with a serif correction. Also at low k_1, the wafer printing results with a hammerhead correction differs little from the results with a serif correction. Consequently, hammerhead corrections are widely used for EOL correction.

After adding a hammerhead to the mask, the EOL prints closer to the design compared to the mask without an EOL correction. In Fig. 7.26, the simulated wafer printing image is shown for an EOL with a hammerhead and for an uncorrected EOL. The uncorrected EOL is equivalent

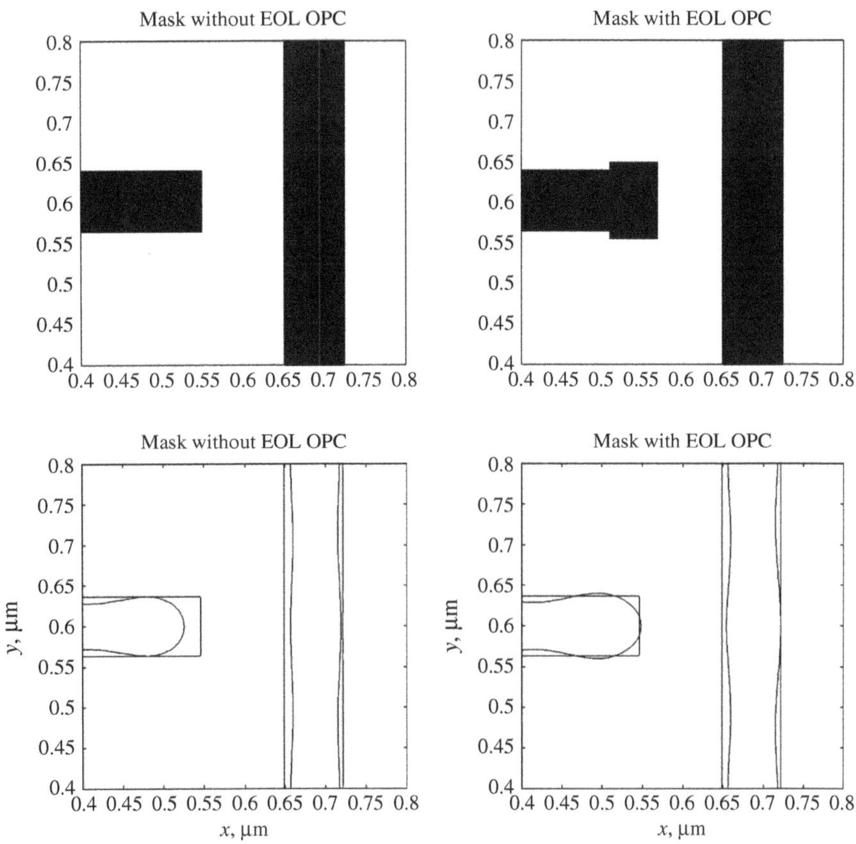

Figure 7.26 Printed images comparing the effect of hammerhead OPC. The mask without EOL OPC has 20 nm pullback while the mask with hammerhead OPC has no pullback.

to the original design. The wafer printing result without a hammerhead pulls back 20 nm from the original design. However, after adding a hammerhead, the wafer printing results does not pull back from the design.

End-of-line corrections are typically handled with model-based OPC. However, prior to model-based OPC, EOL corrections were handled similar to corrections made with rule-based bias OPC. In rule-based bias OPC, the space width is detected, and an appropriate bias is applied. Likewise in rule-based EOL OPC, the space width is detected, and an appropriate hammerhead is applied. An appropriate hammerhead is typically specified with three parameters. These three parameters are—hammerhead length (direction perpendicular to the line), hammerhead width (direction parallel to the line), and overlap of the hammerhead length with the line feature. In most rule-based OPC programs, these three parameters are varied depending on the space width. Similar to rule-based bias OPC, these three parameters are determined through experiments, and a rule table for the EOL correction is created based on the space width. The OPC program will then apply the EOL correction based on the rule table.

Assist features. In addition to biasing and end-of-line correction, assist features are another widely used OPC technique. Assist features have also been called *scattering bars* (SB) or *subresolution assist features* (SRAF). Assist features, like biasing, modify the printed CD; however, assist features also increase the density of the feature. In Fig. 7.27 the ILS as a function of pitch is shown. In Fig. 7.27, the dense pitches have the higher ILS. Therefore, the ILS of an isolated line can be improved by placing subresolution assist features around the isolated line. These assist features cause the isolated line to print similar to a dense line.

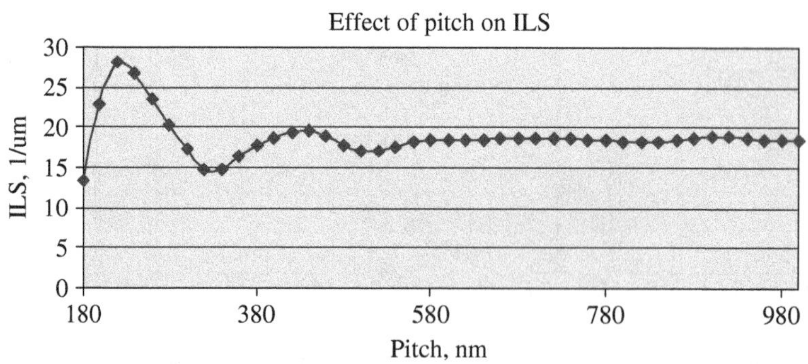

Figure 7.27 Image log slope through pitch of a 90 nm chrome line imaged with an ArF scanner NA = 0.75 illuminated with quasar $\sigma_{in} = 0.57$ and $\sigma_{out} = 0.87$.

If the assist feature was the same size as the main feature, the ILS of the main feature would be equivalent to the ILS of the dense line. The ILS of the dense line is the highest ILS possible. However, the assist feature cannot be the same size of the main feature because the assist features will print. Hence there is a trade-off between ILS and assist feature size. The size of the assist feature depends on the transmittance of the main feature, on the transmittance of the assist feature, and on the contrast of the photoresist. An assist feature that is half the size of the main feature does not print when the transmittance of the main feature equals the transmittance of the assist feature.

The placement of the assist feature is also critical. The assist features are typically placed in a location that maximizes the ILS of the main feature. An example of assist feature placement is shown in Fig. 7.28. The assist features have been applied to a resolution test structure. For the isolated feature two assist features have been placed on either side, while one assist feature has been placed in between the semidense features. These assist features cause the lines to appear optically dense, which improves the ILS of the main feature. In Fig. 7.29, the aerial image of an isolated line is plotted and compared to an isolated line with one SB per side. The NILS with one SB per side is 2.73 while the NILS without SB is 2.59. The NILS improves because the energy in the projection pupil is confined to the outer portion of the pupil.

In addition to improving the NILS by using the outer portion of the projection lens, the SB can also improve the DOF by forcing the energy in projection lens to be distributed onto a smaller ring. As discussed in Sec. 7.2.4, the DOF improves by distributing the energy into a smaller ring. In Fig. 7.29, the energy for the feature with SB is concentrated to the

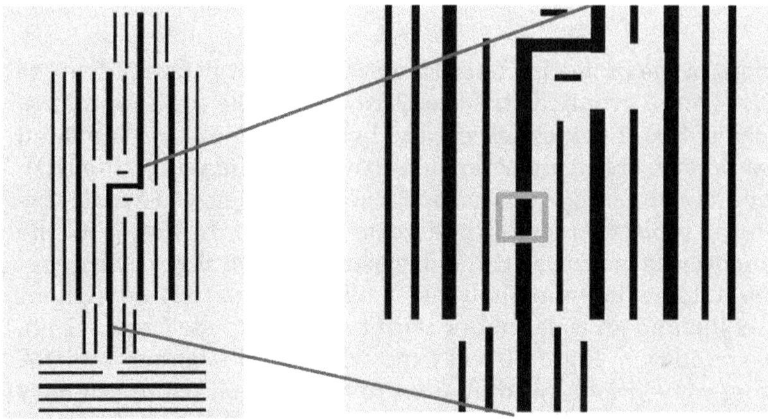

Figure 7.28 Example of a mask with assist features placed with rule-based OPC and with rule-based bias OPC.

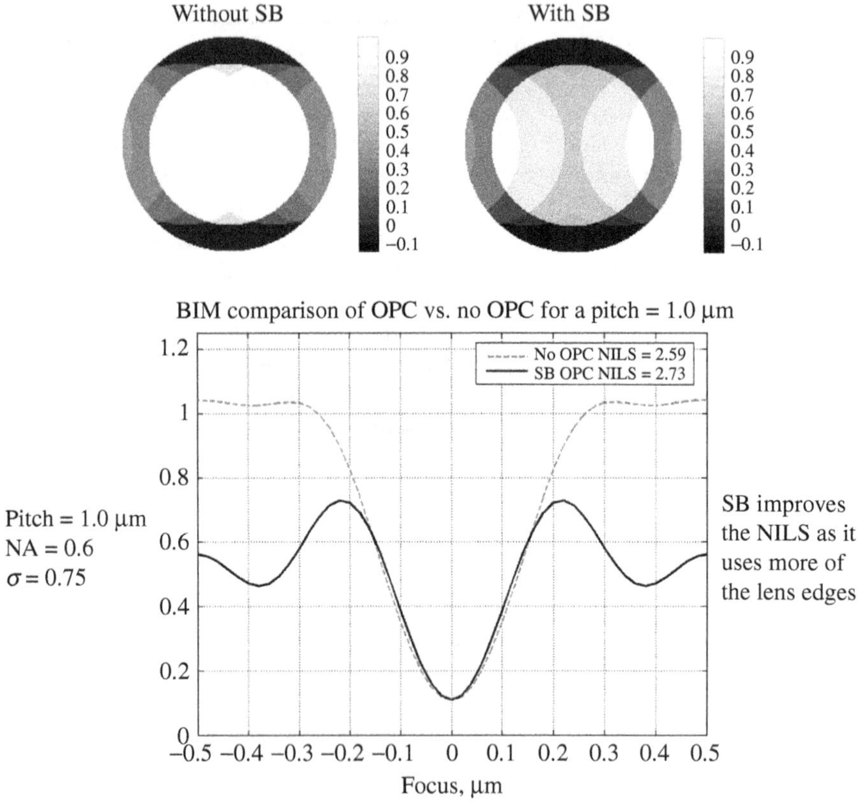

Figure 7.29 Projection pupil filling of a 180-nm line with and without SB. Notice that the pupil filling is confined to the outer portion of the projection lens with SB. The confinement of the light to the outer portion of the pupil improves the NILS of the image.

outer portions of the projection lens. However the energy for the feature without SB is more evenly distributed throughout the projection lens. As discussed in Sec. 7.2.4, by forcing the light to be more concentrated in the outer portion of the projection lens (two-beam imaging), the DOF will improve as compared to a situation where the energy is evenly distributed in the projection lens (three-beam imaging). In Fig. 7.30, the process window was measured through experiments for three features—a dense 150-nm line (300-nm pitch case 1-023), an isolated line without SB (case 1-140), and an isolated line with two SB per side (case 1-152). The process window in Fig. 7.30 was exposed on a KrF scanner with NA of 0.63 using conventional illumination ($\sigma = 0.75$) and using a binary mask. The maximum DOF for the isolated line without SB is 0.65 μm, and the maximum DOF with SB is 0.77 μm. The SB improves the maximum DOF by 0.12 μm.

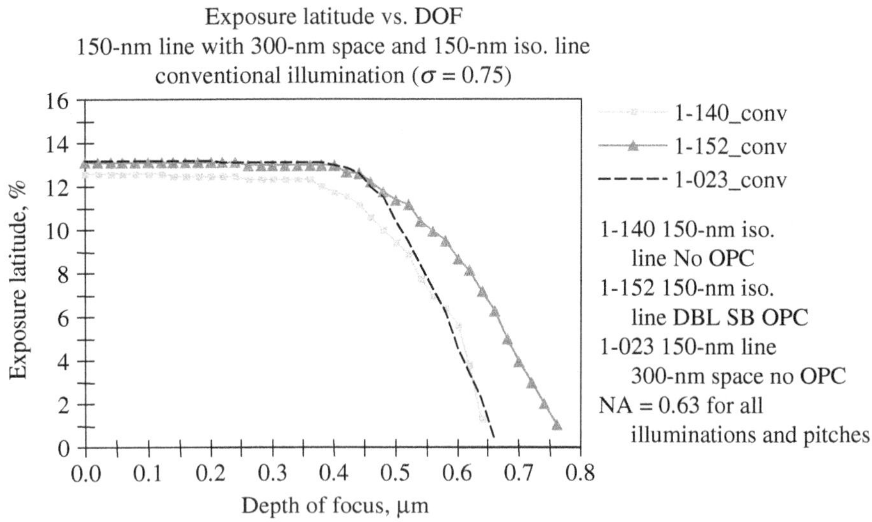

Figure 7.30 Process window of a 150-nm chrome isolated line without AF (1-140) and with AF (1-152) compared to the process window of semidense line (1-023). Conventional illumination ($\sigma = 0.75$) on a Kr Scanner NA = 0.63.

The benefit of using SB can be improved further by combining SB with off-axis illumination. In Fig. 7.31, the experimental process window was measured again for these three features. With quasar illumination, the maximum DOF of the dense line, case 1-023, is greater than 0.8 μm versus a maximum DOF of 0.65 μm with conventional

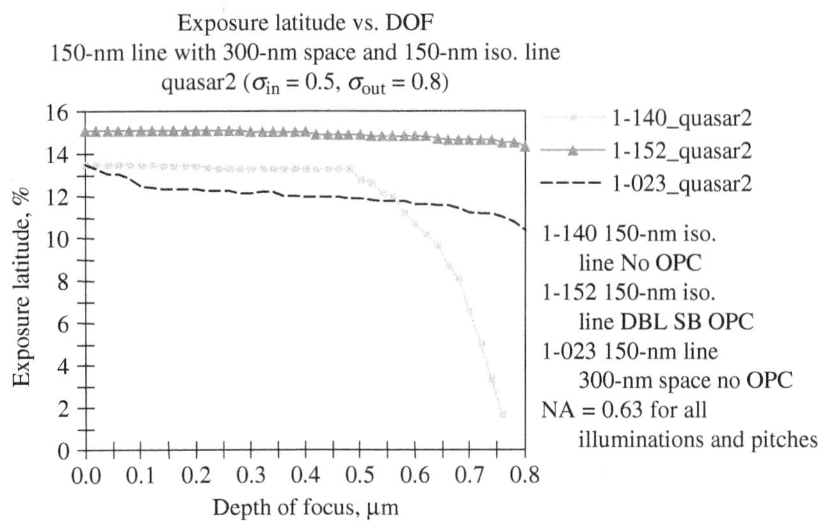

Figure 7.31 Process window of a 150-nm chrome isolated line without AF (1-140) and with AF (1-152) compared to the process window of semidense line (1-023). Quasar illumination ($\sigma_{in} = 0.5$ and $\sigma_{out} = 0.8$) on a Kr scanner NA = 0.63.

illumination. For the isolated line without SB, the maximum DOF is 0.77 μm when exposed with quasar. By placing two SB per side, the maximum DOF is greater than 0.8 μm as the process window is nearly isofocal. By using off-axis illumination and scattering bars, the energy in the projection lens can be forced to the outer portion of the lens. By forcing the energy to the outer portion of the lens the NILS and the EL are increased. By using off-axis illumination and scattering bars, the energy can also be forced into a smaller ring width. By forcing the energy to a smaller ring width, the DOF is improved.

7.4.2 Model-based OPC

At lower k_1, the complexity of doing rule-based OPC increases such that it is impossible to define a rule to correct proximity problems for every permutation in the design. Furthermore, creating a rule for complex two-dimensional design is especially difficult. Consequently, at lower k_1 (less than 0.4), model-based OPC has replaced rule-based OPC. In model-based OPC, both the effect of the scanner on the aerial image and the effect of the photoresist processing are modeled. This model is used to correct the mask; hence, the name model-based OPC.

In model-based OPC, three inputs are required. The first required input is the technology file. This file describes the areas to apply model-based OPC or describes the types of model-based OPCs. These types of model-based OPCs include linewidth bias corrections or EOL corrections. The second required input is the optical model. The optical model describes the scanner illumination and projection optics. In addition to describing the scanner optics, the optical model may include the effect of imaging into the thin-film photoresist or the effect of the mask topography. The third required input is the process model. The process typically includes a photoresist model; however, the process model may also include an etch model.

After specifying the required inputs, the design data, usually in gds II format, is loaded into the model-based OPC software. The model-based OPC software then breaks the design data into a number of fragmented edges. The OPC software then assigns a fragmentation point to each edge fragment. The fragmentation of an edge depends on the surroundings of the geometry. For example, if the edge is an end of line, more fragmentation points are assigned at the end of line. Or if there is another piece of geometry close to a critical feature, more fragmentation points are assigned to the critical feature and to the nearby geometry. The assignment of the fragmentation points is a very complex process that depends on the geometry and the optical model.

At each fragmentation point, the optical model and process model are applied, and the expected wafer position is measured at each fragmentation point. Typically the expected wafer position is the critical

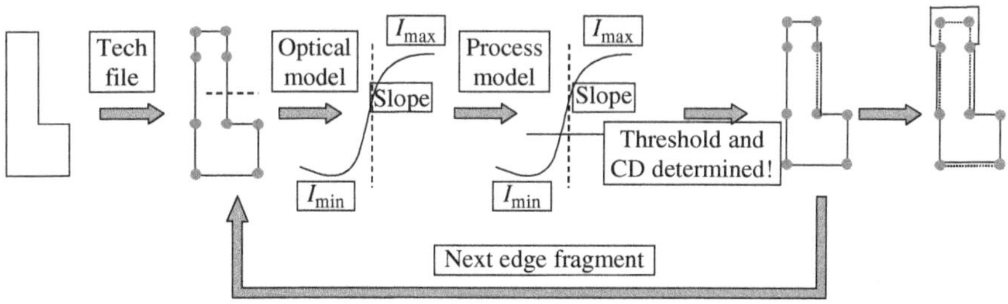

Figure 7.32 Block diagram showing the steps of model-based OPC. (Courtesy of Chi-Min Yuan, Freescale Semiconductor.)

dimension, of the wafer feature. If the CD does not match the design geometry, the fragmented edge is biased such that the CD moves to match the design geometry. The matching between the CD and the design is done through an iterative process. A cartoon of the model-based OPC process is shown in Fig. 7.32. In Fig. 7.32, the technology file assigns the fragmentation points. The fragmented geometry is then input into the optical model and into the process model. A threshold is then applied to output image, and the CD is measured. If the CD does not match the design, bias is applied. After applying the bias, the software corrects the next edge fragment. In Fig. 7.33, the output of *model-based OPC software* (MBOPC) is shown before and after MBOPC. Before MBOPC, the printed wafer linewidth has a 15-nm error from the original design. After MBOPC, the

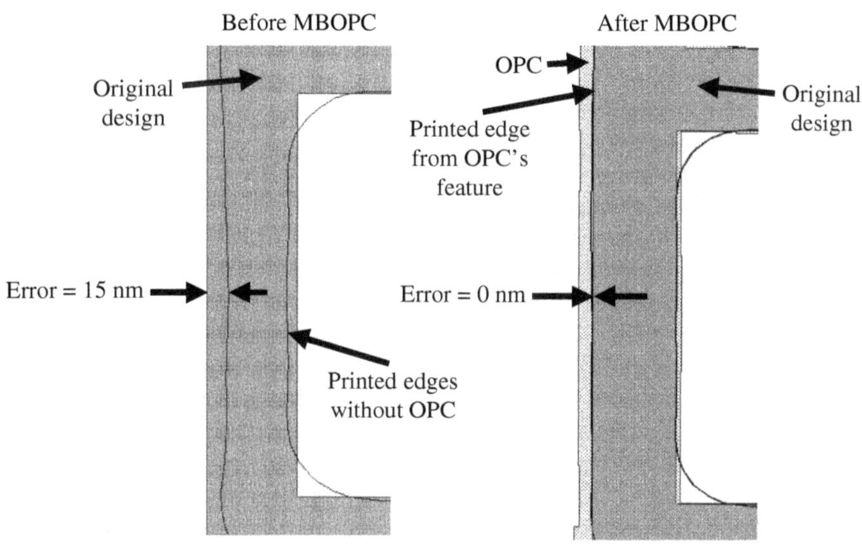

Figure 7.33 The printed image before and after MBOPC. The CD error without MBOPC is 15 nm while the CD error with MBOPC is 0 nm. (*Source:* Posnikov, S., et al., *Proc. SPIE* 5042, 2003).

edges have been biased such that there is no error between the printed wafer CD and the original design.

Most commercially available model-based OPC software uses a Hopkins type of formulation. In Sec. 7.3.2, Hopkins imaging was shown to create a transfer function, the TCC, that completely models the lithography scanner including the illumination and projection pupil. Since OPC changes the mask geometry many millions of times, Hopkins imaging is the best choice for OPC software because Hopkins formulation is independent of the mask geometry. In order to take the optical and process model of the OPC model into account, most commercially available MBOPC software use a sum of coherent systems approach. In SOCS, the SOCS kernel is orthogonal to the other SOCS kernels; consequently, the effect of each kernel on the printed wafer result can be calculated independent of the other kernels. Using SOCS, the MOPC software convolutes each kernel with the design mask creating the intensity from each SOCS kernel. The total intensity is the sum of the intensities from each SOCS kernel. A threshold is applied to the total intensity creating the printed wafer image. Depending on the printed wafer image, bias is applied to the design. The design after OPC is then fed back into the SOCS kernels until there is no error between the printed wafer image and the design. A cartoon of this SOCS-type MBOPC method is shown in Fig. 7.34. Since MBOPC is a slow process, all the commercially OPC software vendors have proprietary techniques to improve the speed without sacrificing accuracy.

The model-based OPC adds bias and EOL correction to mask. The bias and EOL corrections, however, do not change the size of the process window. Bias OPC causes all the mask features to print at the same dose. MBOPC improves the overlapping process window of all the features on the mask. The process window, however, is increased by combining MBOPC with assist features. The assist features improve the DOF while the MBOPC improves the overlapping process window.

The overlapping process window at low k_1 can be increased further by combining assist features, MBOPC, and off-axis illumination. Section "Assist features" showed that the process window with assist features and off-axis illumination is larger than the process window with assist features and conventional illumination. When using assist features and off-axis illumination, MBOPC is still needed to force the features to print at the same dose. Off-axis illumination, however, increases the process window of some pitches while it decreases the process window of other pitches. The decrease in the process window for some pitches can be eliminated by using assist features. Unfortunately, assist features cannot be applied to all pitches. For example, in semidense pitches there is no space for assist features. Consequently, some of the pitches may not have sufficient process window. These pitches with insufficient

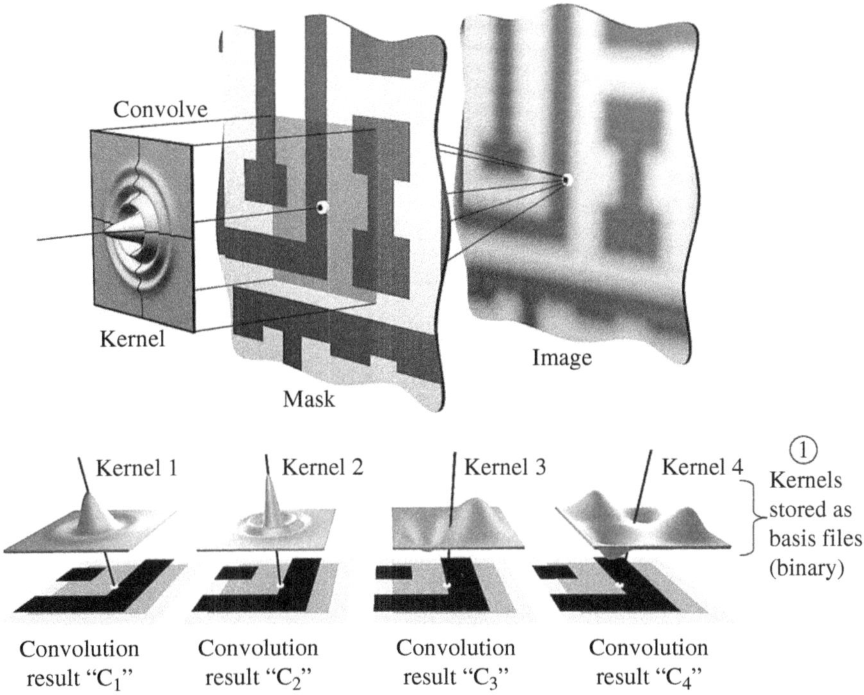

Figure 7.34 Cartoon demonstrating a SOCS method in MBOPC. The intensity is calculated for each kernel through convolution. The total intensity is the summation of the intensity from each kernel. (Courtesy of Synopsys.)

process window are called forbidden pitches.[33] In Fig. 7.35, the DOF at 8 percent EL is plotted for two types of illumination—quasar and annular. The target linewidth is an 85-nm binary line on an ArF scanner with NA of 0.85. The required DOF is 0.25 μm. For pitches less than 310 nm, an assist feature was not placed due to insufficient space. Figure 7.35 shows that there are two pitch regions where the DOF is less than 0.25 mm. These pitch regions are the forbidden pitches. The forbidden pitches with quasar illumination are from 250 to 290 nm and from 470 to 490 nm. The forbidden pitches with annular illumination are from 230 to 290 nm and from 450 to 490 nm. The forbidden pitch region is different for the quasar and annular illumination. The forbidden pitch region is highly dependent on the illumination. The forbidden pitch regions can be avoided by modifying the illumination at the expense of the process window, or by removing the forbidden pitches from the design. By removing the forbidden pitches from the design, the design is being created with manufacturing constraints in mind. Removing forbidden pitches is one concept used in *design for manufacture* (DFM).

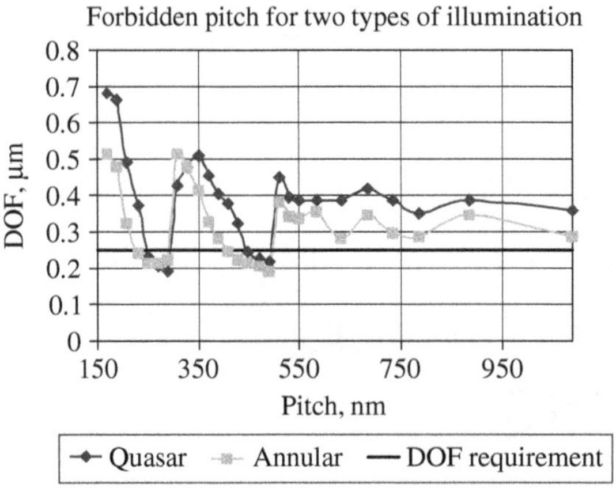

Figure 7.35 Forbidden pitch for an 85-nm line using binary mask on an ArF with NA = 0.85 using two illuminations—annular (σ_{in} = 0.7 and σ_{out} = 1) and quasar (σ_{in} = 0.7 and σ_{out} = 1).

The techniques described in these sections are used together with the RETs described in the next sections. For example, off-axis illumination improves the resolution of dense features, and assist features cause isolated features to print like dense features. The DOF is larger by combining assist features with off-axis illumination as compared to the DOF of using assist features or off-axis illumination individually. Model-based OPC is currently used on all layers regardless of the RET. For example, MBOPC is needed after assist features are placed or if off-axis illumination is used. In the next section on phase-shift mask, the concepts developed in this section and previous sections are required to increase the process window and to improve the overlapping process window.

7.5 Phase-Shift Mask

The resolution can be decreased and process window can be increased by using phase-shift masks. There are several types of *phase-shift masks*. Three types are discussed in this section. These types include attenuating PSM, alternating PSM, and *chromeless phase lithography* (CPL). Attenuating PSM is also called embedded PSM, halftone PSM, and weak PSM. Alternating PSM is also called Levenson PSM, complementary PSM, and strong PSM. CPL is also sometimes called *chromeless mask* (CLM).

All phase-shift mask techniques involve introducing a 180° phase shift in the mask. This 180° phase shift causes an increase in the peak-to-valley amplitude of the electric field mask transmission. The 180° phase also causes the electric field mask transmission to have positive and negative values. The increase in peak-to-valley amplitude causes an increase in the amplitude of the diffraction orders other than the zero order. The positive and negative values in the mask transmission cause the zero order to decrease. The increase in the high orders and the reduction of the zero order mean that two-beam imaging can be achieved by proper selection of illumination. As discussed in Sec. 7.2.4, two-beam imaging causes an increase in the process window.

In Fig. 7.36, cross sections of the mask are shown for the three types of phase-shift mask. In the alternating PSM, glass is either added or removed to create a 180° phase-shift mask. For the subtractive alternating PSM, a ray traveling through the π shifter is 180° out of phase compared to a ray traveling through the zero shifter. In the subtractive alternating PSM method (currently the most widely used), the 180° phase shift is accomplished by etching into the glass. In attenuating PSM, a ray traveling through the attenuating material has a 180° phase shift and is attenuated as compared to a ray traveling through glass. The attenuation of the ray is the key difference between attenuating PSM and alternating PSM. In alternating PSM, lines print at the phase edges. Unwanted phase edges cannot be avoided with alternating PSM. The unwanted phase edges produce unwanted lines on the wafer; consequently, a second exposure is required to remove the unwanted lines in alternating PSM. However, in CPL, two phase edges are placed close together removing the need for a second exposure. A CPL mask is more similar to an attenuating PSM than to an alternating PSM. A CPL mask is an attenuating PSM with 100 percent transmittance in the attenuating material.

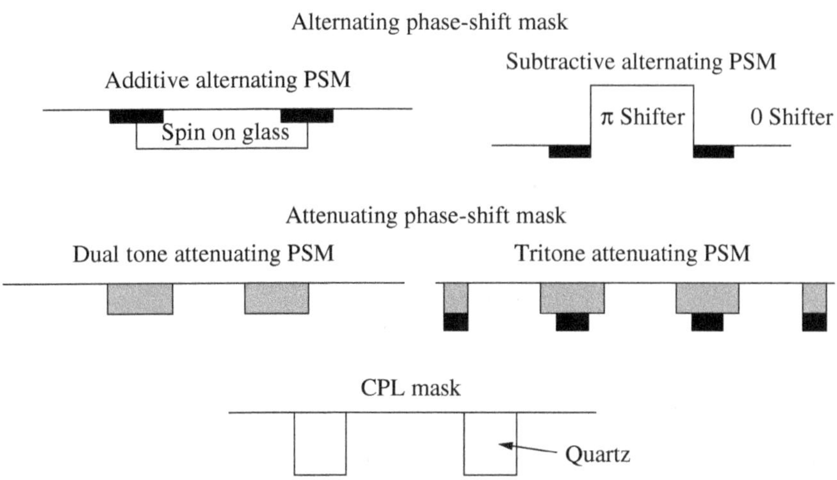

Figure 7.36 Cross sections of the mask for various phase-shift masks.

Various types of phase shifting in an object plane have been used to increase the contrast in the image plane. In optical lithography, Goodman[34] and Shibuya[35] have published ideas on using phase shift to improve contrast; however, neither Goodman nor Shibuya made masks and performed experiments. Levenson was the first to make and expose a phase-shift mask in the lithography community.[36] Hitachi Central Research Laboratory in Japan has also published numerous articles on phase-shifting masks. Lin[37] of IBM and later Terasawa[38] of Hitachi combined the idea of phase shifting with halftoning to create what is now called an attenuating PSM. The use of the chromeless phase edge to create a line was reported by Toh.[39] Later, Chen combined the idea of using chromeless phase edges with off-axis illumination.[40] Alternating PSM has been used widely to print the gate of a circuit[41] while attenuating PSM has been used widely to print the back-end processes especially the contact hole layers.[42–45]

7.5.1 Attenuating PSM

Attenuating PSM has a material layer that both attenuates a ray and phase shifts the ray by 180°. Attenuating PSM has been widely used in contact holes and back-end layers in many fabs. The attenuating PSM material is placed in the background for a dark-field mask (contact hole) or is placed in the main feature regions for a bright-field mask (back-end or gate layer). Typical transmittance of the attenuating PSM material is 0.06 (6 percent); however, research and development has also been done for higher transmittance materials (up to 35 percent). In Fig. 7.37, a cross section of a typical attenuating PSM is shown. A typical attenuating PSM has an absorber layer and a phase-shift layer because with two materials one can tune the transmittance such that there is always a 180° phase shift. The ray E_1 passing through the glass has a transmission of 1 while the ray passing through the attenuating material, E_2, has a phase of 180° and a transmittance of T resulting in an electric field transmission of $-\sqrt{T}$. In order to cause E_2 to have the desired transmittance T and to be 180° out phase with ray E_1, the thickness of the shifter, t_2, and the thickness of the absorber, t_3, must be chosen to meet criteria given in Eqs. (7.45) and (7.46).

$$\frac{|E_2|}{|E_1|} = \sqrt{T} \tag{7.45}$$

$$|\arg(E_2) - \arg(E_1)| = \pi \tag{7.46}$$

In Eq. (7.45) E_1 is the electric field of the ray passing through the glass, and E_2 is the electric field of the ray passing through the attenuating

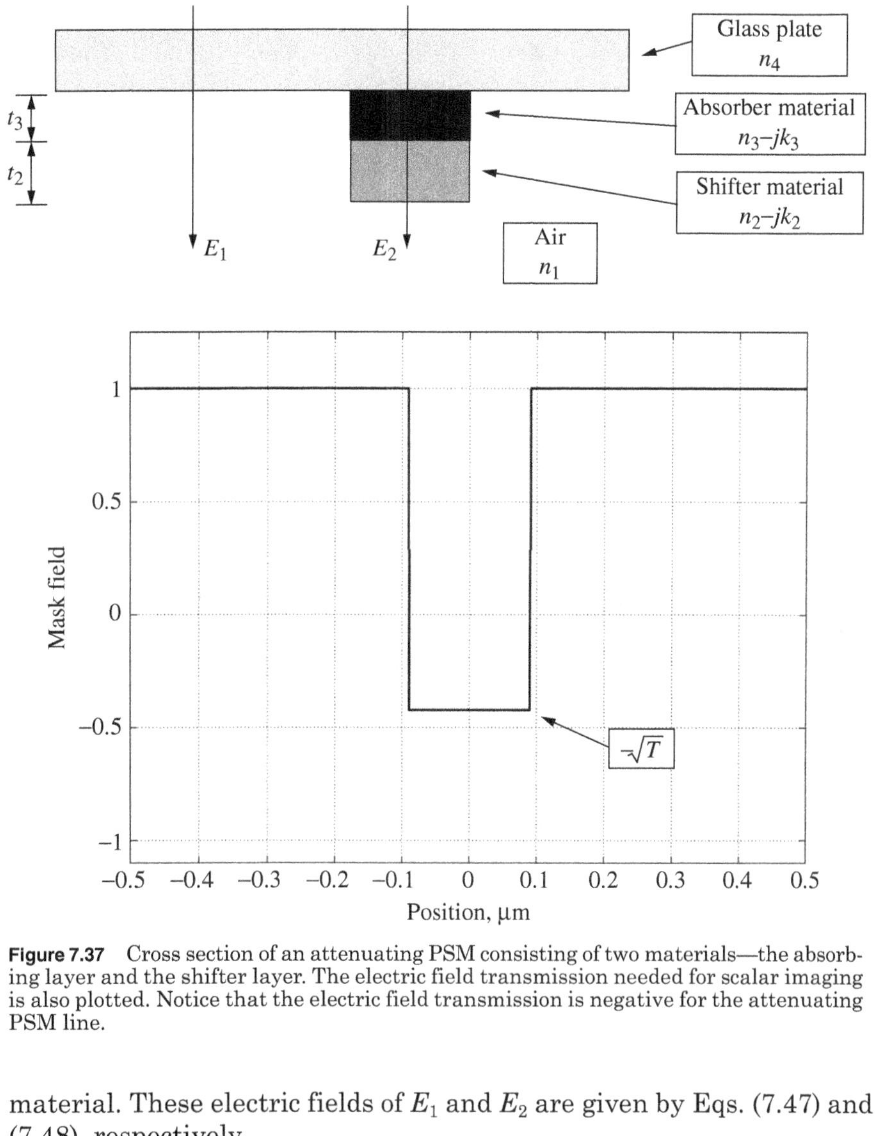

Figure 7.37 Cross section of an attenuating PSM consisting of two materials—the absorbing layer and the shifter layer. The electric field transmission needed for scalar imaging is also plotted. Notice that the electric field transmission is negative for the attenuating PSM line.

material. These electric fields of E_1 and E_2 are given by Eqs. (7.47) and (7.48), respectively.

$$E_1 = \tau_{14} e^{-jk_0 n_1 (t_2 + t_3)} \tag{7.47}$$

$$E_2 = \frac{\tau_{21}\tau_{32}\tau_{43} e^{-jt_2\gamma_2} e^{-jt_3\gamma_3}}{1 + \rho_{21}\rho_{32} e^{-j2t_2\gamma_2} + \rho_{32}\rho_{43} e^{-j2t_3\gamma_3} + \rho_{21}\rho_{43} e^{-j2t_2\gamma_2} e^{-j2t_3\gamma_3}} \tag{7.48}$$

where t_2 and t_3 are the thickness of the two materials to be solved and k_0 is the free space wavenumber, which is equal to $2\pi/\lambda_0$ where λ_0 is the free space wavelength. Also, in Eqs. (7.47) and (7.48), τ_{41}, τ_{43}, τ_{32}, and τ_{21} are the electric field transmission coefficients from material a into material $b(\tau_{ab})$, and ρ_{41}, ρ_{43}, ρ_{32}, and ρ_{21} are the electric field reflection coefficients. The transmission and reflection coefficients are given in the following equations. In Eq. (7.48), γ_2 and γ_3 are $\gamma_2 = k_0(n_2 - jk_2)$ and $\gamma_3 = k_0(n_3 - jk_3)$.

$$\rho_{41} = \frac{n_4 - n_1}{n_4 + n_1} \qquad \tau_{41} = 1 + \rho_{41}$$

$$\rho_{43} = \frac{n_4 - n_3 + jk_3}{n_4 + n_3 - jk_3} \qquad \tau_{43} = 1 + \rho_{43}$$

$$\rho_{32} = \frac{n_3 - jk_3 - n_2 + jk_2}{n_3 - jk_3 + n_2 - jk_2} \qquad \tau_{32} = 1 + \rho_{32}$$

$$\rho_{21} = \frac{n_2 - jk_2 - n_1}{n_2 - jk_2 + n_1} \qquad \tau_{21} = 1 + \rho_{21}$$

Attenuating PSM increases the DOF and EL by modifying the diffraction orders of the dense pitches. The process window is optimized when the 0 and ±1 orders have the same magnitude. The optimal magnitude of the 0 and ±1 orders depends on the amount of energy collected by the projection pupil.[46] The optimal magnitude also depends on the illumination. The diffraction orders for a binary mask and for a 6 percent attenuating PSM is shown in Fig. 7.38. In Fig. 7.38, the attenuating PSM causes the zero order to decrease and the higher orders to increase. Hence, the zero and higher orders are closer to balanced. The increase in the magnitude of the higher orders improves the NILS. The NILS for the binary mask is 2.468 while the NILS for the 6 percent attenuating PSM is 3.737. This increase in NILS leads to an increase in EL.

The increase of the higher orders with attenuating PSM also leads to an increase in the process window. Experiments for a 130-nm line printing were measured for a binary mask and for a 6 percent attenuating PSM using a KrF scanner with NA of 0.7 exposed with quasar illumination. Seven pitches were measured—310, 350, 390, 430, 510, 630 nm, and isolated. The process windows for each pitch are shown in Fig. 7.39 for the 6 percent attenuating PSM. Each of the pitches has over 0.7-μm DOF at 10 percent EL. From the individual process windows in Fig. 7.39, the overlapping process window was generated as shown in Fig. 7.40. In Fig. 7.40 for 6 percent attenuating PSM, the maximum EL is 9.5 percent, and the maximum DOF is 0.76 μm. Using the same scanner settings, the overlapping process window was measured for a binary mask and is also shown in Fig. 7.40. The maximum EL is 5.5 percent, and the maximum DOF is 0.46 μm for the binary mask. In comparing the binary

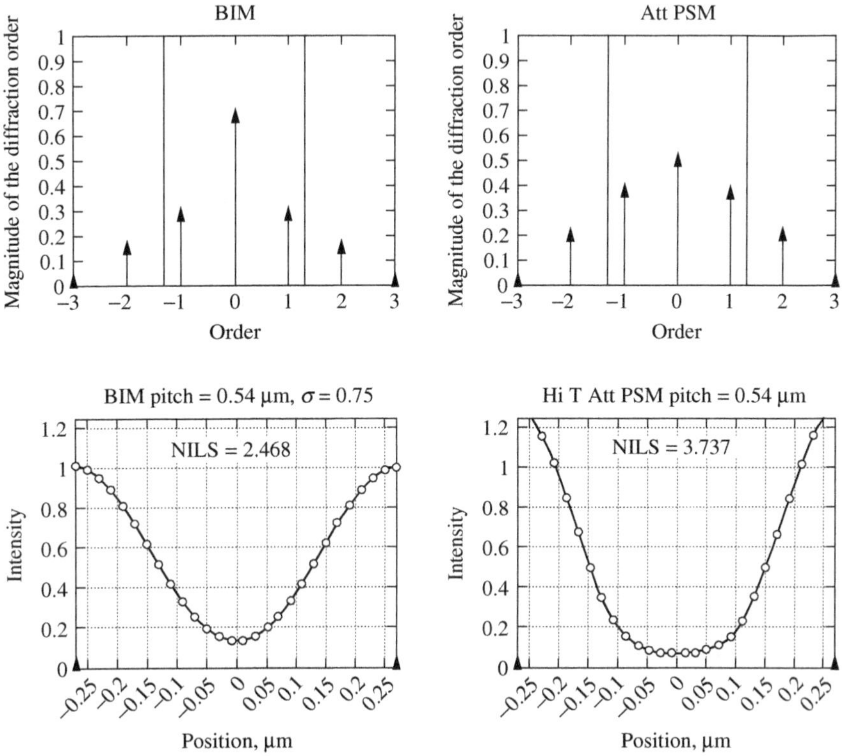

Figure 7.38 Comparison of the diffraction orders and of the intensity image for binary mask to the orders for 6 percent attenuating PSM. The attenuating PSM has less 0 order and more ±1; hence, the contrast and NILS are larger with 6 percent attenuating PSM. The simulations are for a 180-nm line on a 540-nm pitch imaged with a KrF scanner NA = 0.6 using conventional illumination $\sigma = 0.75$.

results to the 6 percent attenuating PSM, the 6 percent attenuating PSM significantly improves the process window as compared to the binary mask.

The mask transmittance of attenuating PSM is another parameter that can be optimized to improve the process window.[46] By increasing the transmittance, the magnitude of the diffraction orders are balanced resulting in the highest NILS. Each pitch on the mask has an optimal transmittance. In Fig. 7.41, the optimal transmittance of a 90-nm line for four pitches is plotted for an ArF system with NA of 0.85 exposed with quasar illumination (σ_{in} = 0.66 and σ_{out} = 0.96). For the dense pitch (180 nm), the optimal transmittance is 10 percent. As the pitch increases, the optimal transmittance increases. The optimal transmittance of the isolated pitch is 75 percent. Since the optimal transmittance

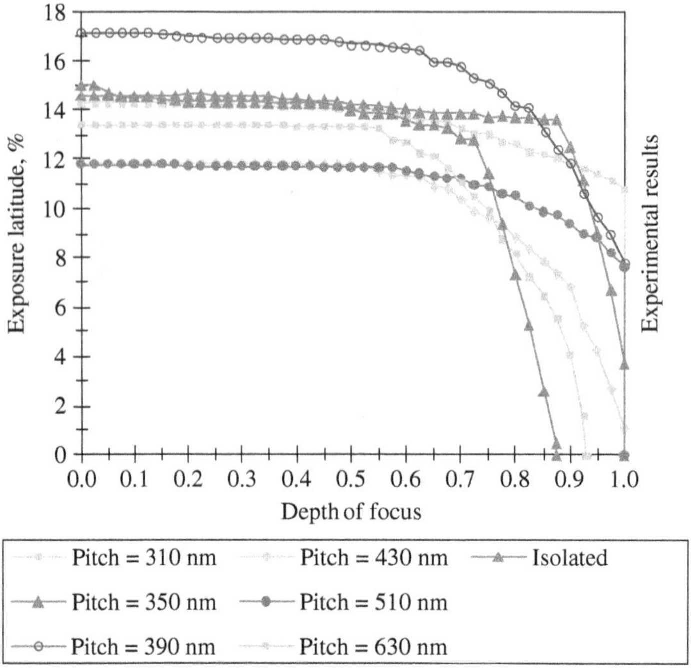

Figure 7.39 Experimental process window of a 130-nm line at various pitches for a 6 percent attenuating PSM. The exposures were done on a KrF scanner with NA = 0.7 using quasar illumination (σ_{in} = 0.5 and σ_{out} = 0.8).

depends on pitch, the transmittance should be selected such that the minimum NILS for the various pitches is maximized. For this example the minimum NILS is maximized for a transmittance of 14 percent.

In order to implement an attenuating PSM process, many resolution enhancement techniques should be combined to produce the largest

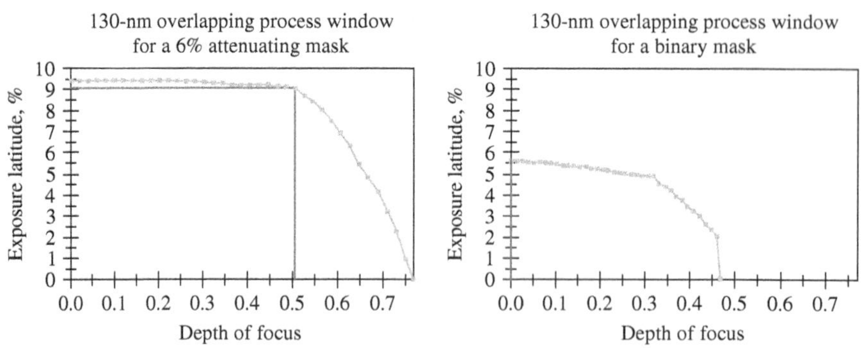

Figure 7.40 Overlapping process window through pitch for a 130-nm line for 6 percent attenuating PSM (left) and for binary mask (right). The exposures were done on a KrF scanner with NA = 0.7 using quasar illumination (σ_{in} = 0.5 and σ_{out} = 0.8).

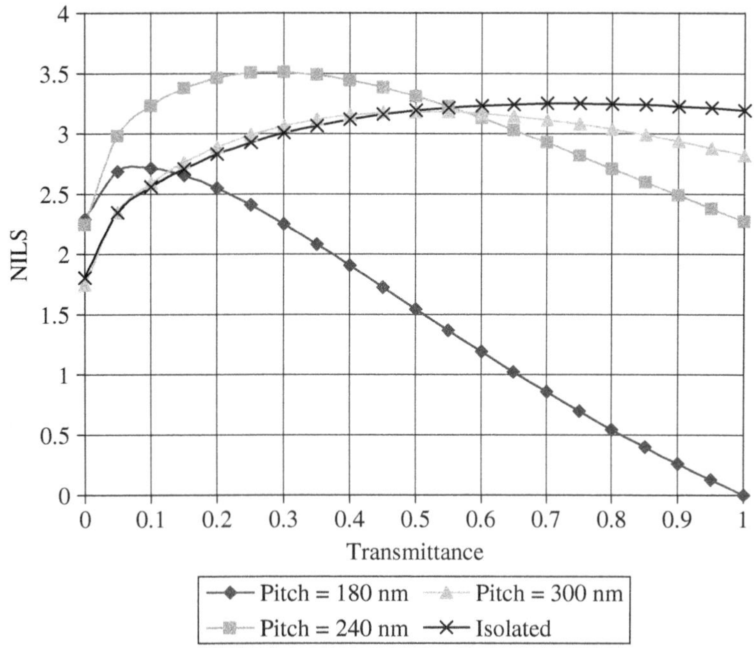

Figure 7.41 Optimal transmittance of an attenuating PSM for four pitches. The minimum NILS is highest for all four pitches when the transmittance is 14 percent. Simulations were done for a 90-nm line exposed on an ArF scanner with NA = 0.85 using quasar illumination ($\sigma_{in} = 0.66$ and $\sigma_{out} = 0.96$).

process window. First, the illumination is chosen to maximize the process window of the critical pitches. Then the transmittance is also chosen to maximize the process window of the critical pitches. After choosing the illumination and the transmittance, assist features are placed to maximize the process window of the isolated features. Finally model-based OPC must be performed to force all the features to print at the same dose. Combining all these resolution enhancement techniques results in the maximum process window at a low k_1 resolution.

7.5.2 Alternating PSM

Alternating PSM is a good technique for producing dark lines. Consequently, alternating PSM has been used to print the gate layer. The alternating PSM uses the same phase-shift principles for attenuating PSM by interfering two beams that are phase shifted by 180°. In alternating PSM, one beam is diffracted by a 0° opening in a dark-field mask. The other beam is diffracted by a 180° opening. The phase shift causes destructive interference, and a dark region in the image is then created between the 0° and 180° regions.

In alternating PSM, the phase-shift edge always produces a region of dark intensity. Consequently, in alternating PSM unwanted dark intensity areas are produced. If an alternating PSM is used with a single exposure, these dark intensity areas would produce unwanted lines, which would cause a short in the device. In order to remove these dark intensity areas, a second exposure is needed to trim these unwanted lines. Therefore, alternating PSM is commonly used with two exposures. The first exposure defines the small gate lines (gate exposure), and the second exposure trims unwanted lines and forms the interconnect areas (trim exposure). An example of alternating PSM exposure is shown in Fig. 7.42 for a gate pattern. In Fig. 7.42, the desired gate in red is shown over the active area in green. The first exposure mask, gate mask, has a chrome line (black) which separates a 0° area (white) from a 180° area (gray). The gate mask is a dark-field mask. The second exposure mask, trim mask, forms the interconnect. The trim mask has a large chrome area (black) to protect the gate defined with first exposure. The intensities from these two exposures are then added together to produce the intensity for the total exposure.

In alternating PSM, a 0° and 180° area must be placed around the geometry to create a dark region in the image. A dark region is only created in between 0° and 180° regions. The placing of 0° and 180° areas around chrome is called coloring. In some cases like a T intersection, it

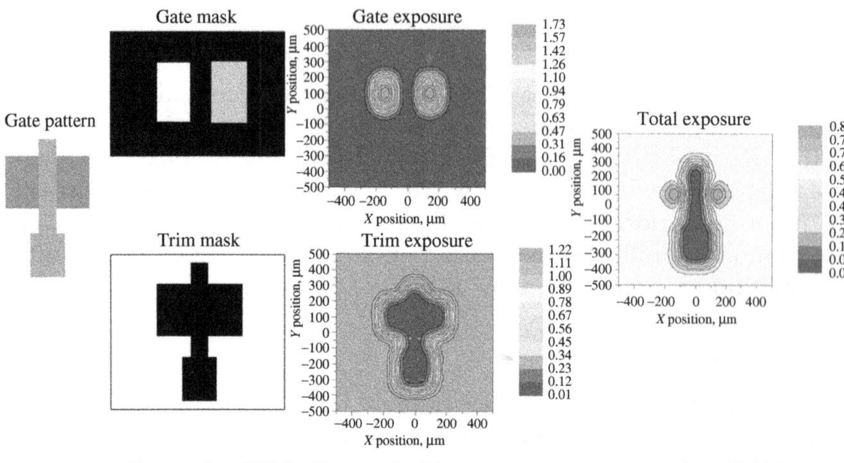

Figure 7.42 Alternating PSM relies on double exposure to pattern a clear field image of the gate pattern on the left (the gate is light gray and the active area is dark gray). The first mask is the gate mask and defines the gate over the active. The second exposure is the trim mask and defines the interconnect areas. The simulations are for a 100-nm gate on an ArF scanner. The gate exposure is simulated with NA = 0.75 using conventional illumination $\sigma = 0.3$. The trim exposure is simulated with NA = 0.93 using conventional illumination $\sigma = 0.75$.

is impossible to surround all the lines with 0° and 180° areas such that chrome always separates the 0° and 180° areas. In the T intersection example, it is possible to surround two of the chrome lines with 0° and 180° areas. However, one line will be surrounded by two 0° areas or by two 180° areas. The line surrounded by two 0° areas or by two 180° areas will create a line with much less contrast than a line surrounded by a 0° and 180° area. Consequently, not all designs can be phase shifted with alternating PSM, i.e., the design cannot be colored.

In order to understand the advantages of alternating PSM, the gate mask in Fig. 7.42 must be examined in detail. A cross section of the alternating PSM is shown in Fig. 7.43 along with the electric field transmission under the mask. The alternating PSM forms an electric field transmission with three transmissions— −1, 0, and +1. The −1 transmission is formed by the 100 percent transmitting 180° area in the gate mask. The chrome areas of the gate mask form the zero transmission. Finally, the +1 transmission is formed by the 100 percent transmitting 0° area in the gate mask. The −1 transmission is formed by etching into

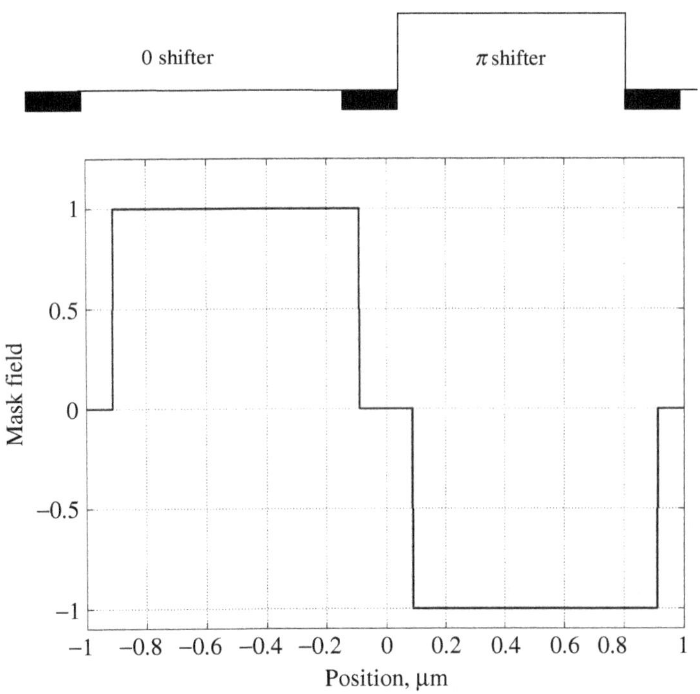

Figure 7.43 Cross section of an alternating PSM. The 180° area (π shifter) is formed by etching into the quartz blank. The electric field transmission needed for scalar imaging is also plotted. Notice that the electric field transmission is negative for the π shifter.

the quartz blank. This etch will cause a ray passing through the −1 area to be phase shifted by 180° with respect to the +1 area. The amount of etch needed can be calculated through geometrical optics and is given in Eq. (7.49).

$$t = \frac{\lambda}{2(n-1)} \tag{7.49}$$

In Eq. (7.49) the refractive index of the quartz blank is n and the wavelength is λ.

The phase shift required in Eq. (7.49) is derived through geometrical optics. However, the mask topography has an effect on the phase shift, which needs to be calculated through rigorous electromagnetic field simulation. Experimental printing and EMF simulation show that an alternating PSM prints with the 180° area being smaller than the 0° area. The 180° area printing smaller than the 0° area causes the line to shift toward the 180° area. The shift in the line position is unwanted and leads to overlay errors if not corrected.

The line shift is due to a nonzero zero order diffraction order. Ideally, the alternating PSM should not have a zero order. The zero order causes background light to add to the electric field. The addition of light to the electric field causes the electric field to increase at all positions. Since the 0° area has a positive electric field and the 180° area has a negative electric field, the increase in the electric field implies that the 0° area is brighter than the 180°. Consequently, the 0° area intensity is larger than the 180° area intensity, and the 0° area prints larger than the 180° area.

Scalar imaging theory predicts that the zero order is zero, and there is no shift in the line position. This does not agree with experiments. Experiments show that the line shifts toward the 180° area. Consequently, scalar imaging theory is no longer valid for calculating the image of the alternating PSM. The line shift can only be predicted with rigorous EMF theory. Rigorous EMF predicts that the zero order is nonzero as shown in Fig. 7.44. In Fig. 7.44, the magnitude of the diffraction orders is plotted for the alternating PSM shown in Fig. 7.43. In Fig. 7.44, the electric field is also plotted, and the figure shows that the nonzero zero order causes the electric field to shift upward. The nonzero zero order is caused by the electric field propagating through the 180° opening. The light propagating through the opening is propagating in a direction parallel to the etched edge. The light propagating through the 180° air opening is moving faster than the light in the quartz mask blank. Since the phase of the light must be continuous across the air and quartz interface, the light in the quartz at the interface will propagate

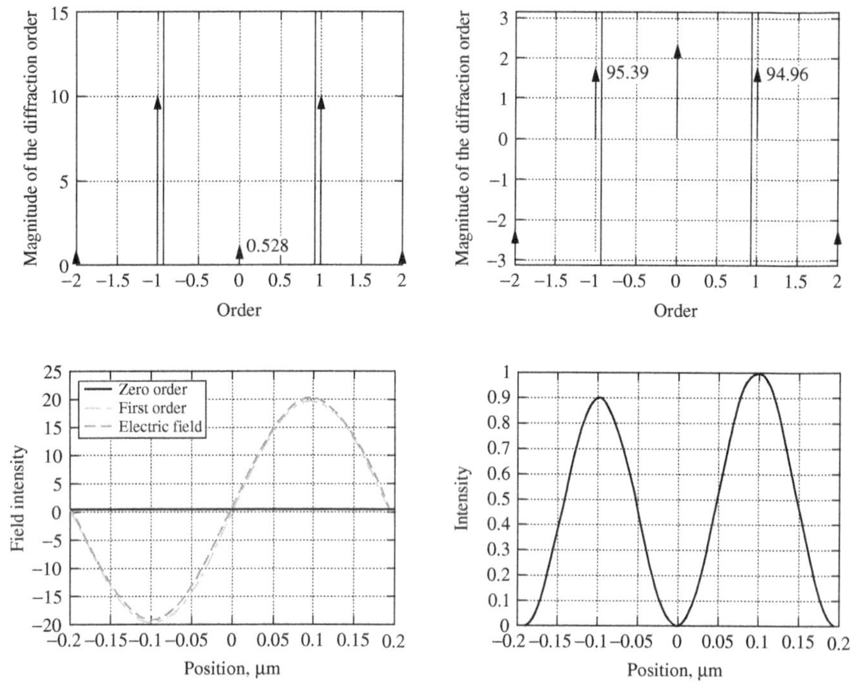

Figure 7.44 Diffraction orders of alternating PSM in Fig. 7.43 when using rigorous EMF simulation. The mask topography causes the magnitude of the zero order to be nonzero (top left). The nonzero zero order causes the 180° area to have less intensity than the 0° area (bottom right). Simulations for 130 nm chrome line exposed on KrF scanner NA = 0.57 using conventional illumination $\sigma = 0.35$.

at an angle away from the etched edges. The light in the quartz propagates with an angle that is equivalent to the critical angle. This angular propagation in the quartz causes light to be sucked out of the 180° air opening. Consequently, there is less light in the etched 180° opening compared to the unetched 0° opening.

Several methods have been proposed to correct the nonzero zero-order problem with alternating PSM. These methods include moving the line during OPC or correcting the shift through mask making. The mask making corrections include using an isotropic etch undercut[9] or using anisotropical etch of both the 0° and 180° areas.[47] In Fig. 7.45, the aerial image of an alternating PSM with an anisotropic etch is shown in order to compare its image to the aerial image of a mask with an anisotropic followed by an isotropic etch. In Fig. 7.45, the aerial image of the anisotropic etch shows that the peak intensity in the etched region (180°) is less than the peak intensity of the unetched region (0°). This reduction in the peak intensity will cause the 180° area to print smaller

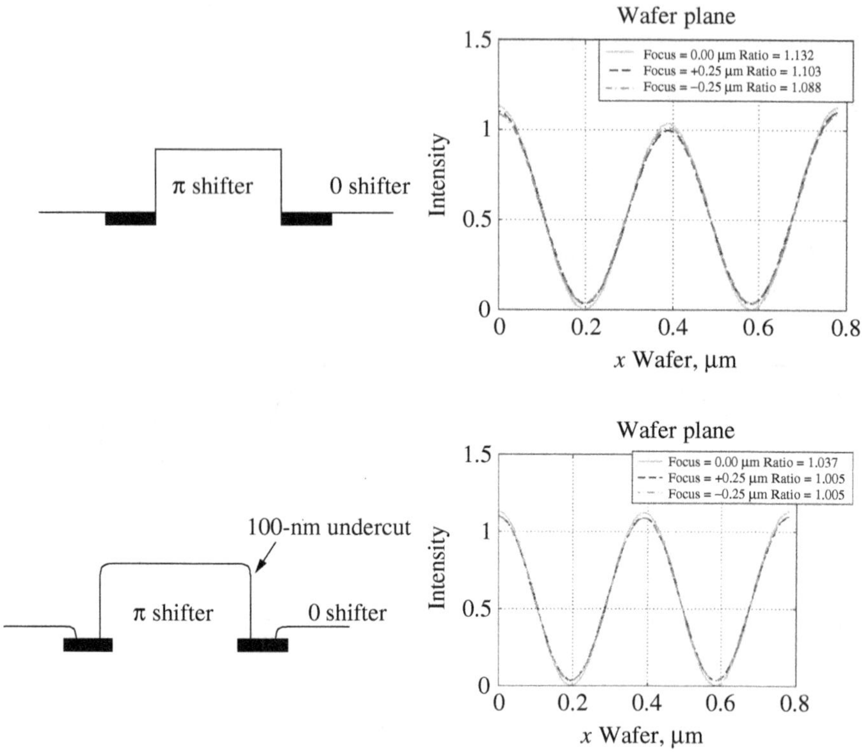

Figure 7.45 Cross sections of a mask without undercut (top) and with undercut isotropic etch (bottom). The simulation of the aerial image without undercut etch has the 180° area (π shifter) with lower intensity than the 0° area (zero shifter). The simulation of the aerial image with undercut etch has the 0° area and 180° area printing with the same intensity. Simulations for 130 nm chrome line exposed on KrF scanner NA = 0.57 using conventional illumination $\sigma = 0.35$.

causing the lines to shift toward the 180° region. The combination of the isotropic with anisotropic etch, however, causes the 0° and 180° regions to have the same peak intensity in the aerial images. The same peak intensities will cause the 0° and 180° regions to print with the same width, which removes the line shift. The combination of the isotropic with anisotropic etch removes the zero-order diffraction order created by the mask topography.

Unlike binary mask or attenuating PSMs, the electric field of an alternating PSM at the wafer has both positive and negative values. Thus the electric field is forced to be zero in between the 0° and 180° regions. This zero in the electric field produces an intensity with a large contrast or NILS as compared to binary or attenuating PSMs. Since the contrast with the alternating PSM is larger than binary or attenuating PSMs, the alternating PSM can produce a finer resolution line compared to binary or attenuating PSMs. In addition to finer resolution, the alternating PSM has an advantage over binary and attenuating PSM. The alternating PSM produces two diffraction orders while the binary and attenuating PSMs produce three diffraction orders. The diffraction orders are plotted for a binary and alternating PSM in Fig. 7.46. Since there are only two diffraction orders for the alternating PSM, the NILS of the alternating PSM is larger than the NILS with binary mask as shown in Fig. 7.46. The NILS for the alternating PSM is 3.96 while the NILS for the binary mask is 2.47. The higher NILS for the alternating PSM indicates that finer resolution and larger EL can be obtained with the alternating PSM.

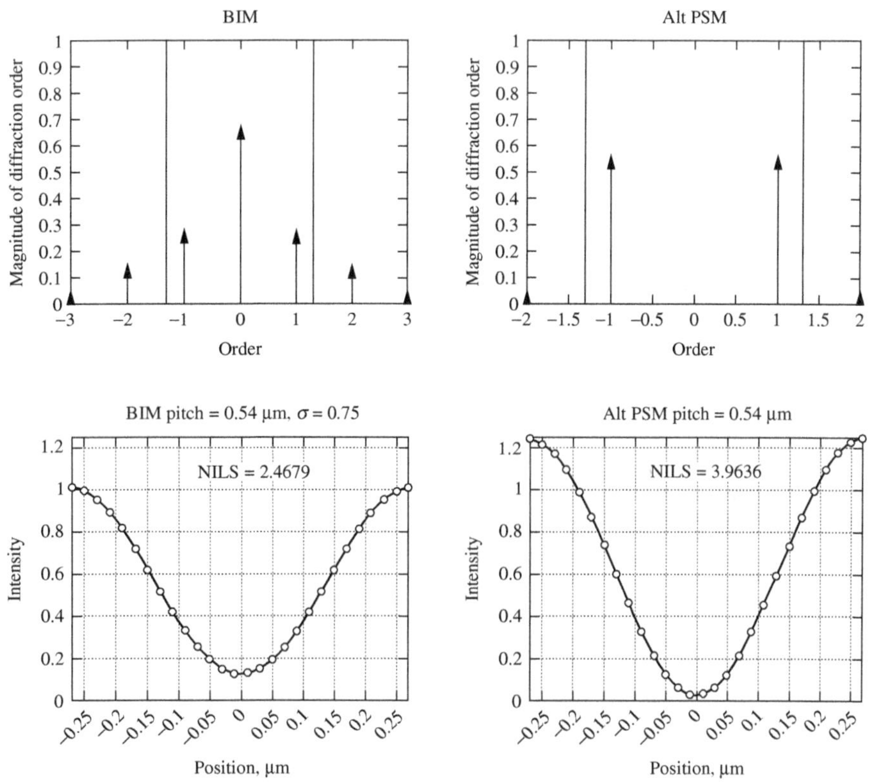

Figure 7.46 Comparison of the diffraction orders and of the intensity image for binary mask to the orders for alternating PSM (using scalar simulation). The alternating PSM has no 0 order and more ±1; hence, the contrast and NILS are larger with alternating PSM. The alternating PSM simulations are for a 180-nm line on a 540-nm pitch imaged with a KrF scanner NA = 0.6 using conventional illumination $\sigma = 0.35$. The binary simulations are using conventional illumination $\sigma = 0.75$.

In the alternating PSM, the two diffraction orders will produce an image with large depth of focus because the diffraction orders will be located on the same radius in the projection lens. In binary or attenuating PSM, the three diffraction orders will have less DOF in comparison to the alternating PSM. In a binary or attenuating PSM, three diffraction orders cannot occupy the same radius in the projection lens. In Fig. 7.47 the aerial images of 130-nm line for a binary mask and for an alternating PSM are shown through focus. In a the binary mask, the contrast for both the dense and isolated pitches decreases faster through focus compared to the contrast of the alternating PSM. Therefore, the alternating PSM will have a larger DOF compared to the binary mask.

In order to implement an alternating PSM process, many resolution enhancement techniques should be combined to produce the largest process window. First, the illumination is chosen to maximize the process window of the critical pitches. Since there is no zero order with the alternating PSM, the alternating PSM does not benefit from off-axis illumination.

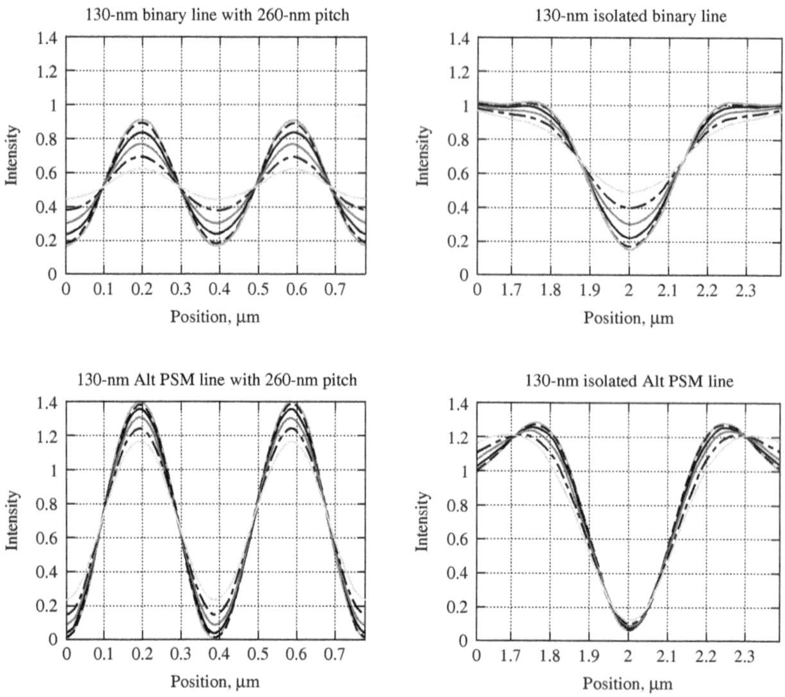

Figure 7.47 Simulations through focus of binary mask (top) versus alternating PSM (bottom) at two pitches—260 nm (left) and isolated (right). The alternating PSM simulations have more DOF than the binary simulations. Simulations are for a 130-nm line. The binary simulations are for NA = 0.7 using conventional illumination $\sigma = 0.75$. The alternating PSM simulations are for NA = 0.57 using conventional illumination $\sigma = 0.35$.

The best illumination for an alternating PSM is conventional illumination. The sigma of the conventional illumination is set to maximize the DOF. However, as the sigma decreases, 2D structures introduce ringing in the aerial image, which cannot be corrected with MBOPC. Consequently, there is a tradeoff between maximizing the DOF and minimizing 2D proximity effects. In addition to proper selection of the illumination, model-based OPC must be done to force all the features to print at the same dose. However, since an alternating PSM requires two exposures, MBOPC must be used on two masks to produce one image. This complicates the MBOPC process for an alternating PSM. Combining all these resolution enhancement techniques results in the maximum process window at a low k_1 resolution.

7.5.3 Chromeless phase lithography

In chromeless phase lithography, phase transitions are etched into the quartz mask similar to an alternating PSM. However, unlike an alternating PSM, the phase transitions are closer together in CPL. Because the phase transitions are closer in CPL, the zero-order diffraction order is not eliminated as in an alternating PSM. Consequently, CPL requires off-axis illumination to achieve similar resolution and DOF that can be obtained with an alternating PSM. Unlike an alternating PSM, CPL only requires a single exposure to form a pattern on a clear field mask while an alternating PSM requires two exposures to form the same pattern. Since CPL only requires one exposure, the CPL process can print more wafers in a fixed period of time as compared to an alternating PSM process. In Fig. 7.48, cross sections of the alternating PSM and the CPL mask are

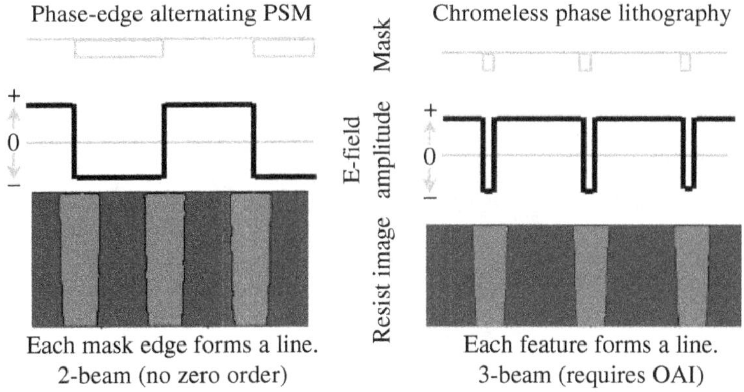

Figure 7.48 Cross section (top) of the alternating PSM (left) versus CPL (right). The electric field mask transmission (middle) and the resist images (bottom) are also plotted. The alternating PSM prints one line for etch phase edge while CPL prints one line for two phase edges.

shown along with the electric field of the mask. Each phase edge in the alternating PSM produces one line while the two phase edges in the CPL mask produce one line.

In Fig. 7.49, the diffraction orders for a CPL mask are plotted in comparison to the alternating PSM. The CPL mask produces three diffraction orders while the alternating PSM produces two diffraction orders. The aerial image at best focus is also plotted for the CPL and alternating PSM. The alternating PSM has better contrast and NILS than the CPL mask. By further optimization of the illumination, the CPL mask can produce an aerial image that is identical to the alternating PSM. It can be shown that the CPL and the alternating PSM are degenerate solutions at one pitch, i.e., the alternating PSM and CPL can produce the same aerial image. Although the CPL and alternating PSM are degenerate solutions at one pitch, each mask type has advantages and disadvantages. The CPL mask is a single exposure technique. In order for a CPL mask to produce the same aerial image as an alternating PSM, aggressive OAI must be used. Aggressive OAI, however, has forbidden pitches, i.e.,

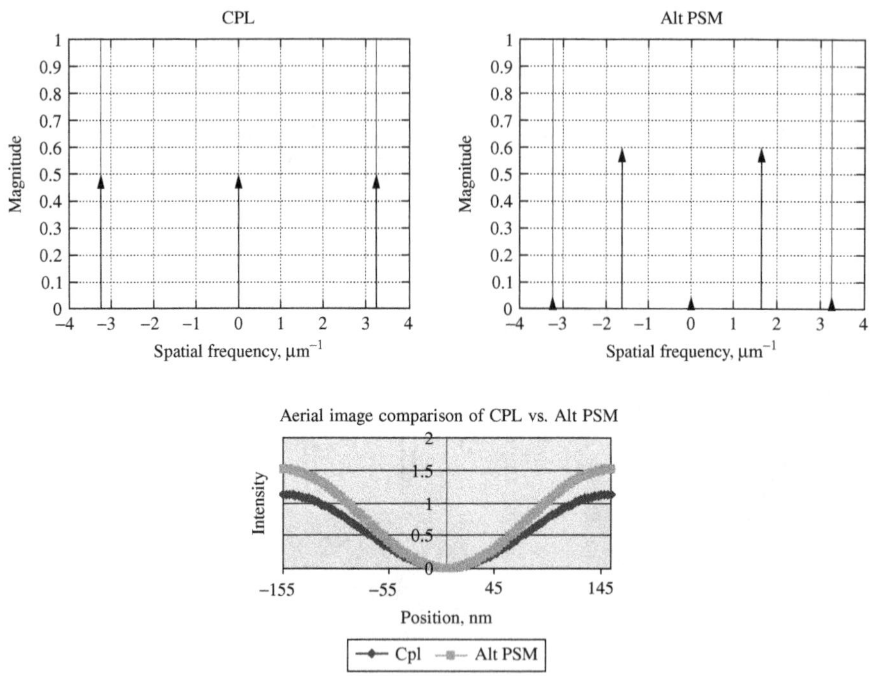

Figure 7.49 Diffraction orders of a CPL mask compared to an alternating PSM. The CPL and alternating PSM mask is for an 80-nm line on a 310-nm pitch. The aerial images are for a KrF system ($\lambda = 248$) with NA = 0.8. The alternating PSM uses conventional illumination $\sigma = 0.3$. The CPL mask uses quasar illumination with $\sigma_{in} = 0.68$ and $\sigma_{out} = 0.92$.

pitches with small or no process window. The alternating PSM is a double exposure technique. The alternating PSM, however, does not have a forbidden pitch. Alternating PSM is a double exposure technique without forbidden pitches while CPL is a single exposure technique with forbidden pitches. Consequently, each RET has advantages and disadvantages that must be carefully weighed.

Since the CPL mask produces three diffraction orders, the CPL requires OAI in order to improve the DOF. By using OAI, the diffraction orders are forced to sample the projection lens along a constant radius. Thus, the CPL with OAI behaves like an alternating PSM with on-axis illumination. In Fig. 7.50, the projection pupil is plotted through pitch when using OAI and a CPL mask. Using quasar OAI, most of the light collected by the projection lens occupies an area of the projection lens along the same radius. In addition to plotting the projection pupil, the aerial images at best focus are plotted for a 100-nm CPL line at three pitches. The NILS for all three pitches is well above two with the CPL mask.

The CPL mask is more similar to an attenuating PSM rather than an alternating PSM. The CPL mask is essentially an attenuating PSM with 100 percent transmission. Like an attenuating PSM, the CPL mask benefits from off-axis illumination and from assist features. For isolated lines, the DOF of a CPL (or an attenuating PSM) is improved by using AF as discussed in section "Assist features." In Fig. 7.51, the

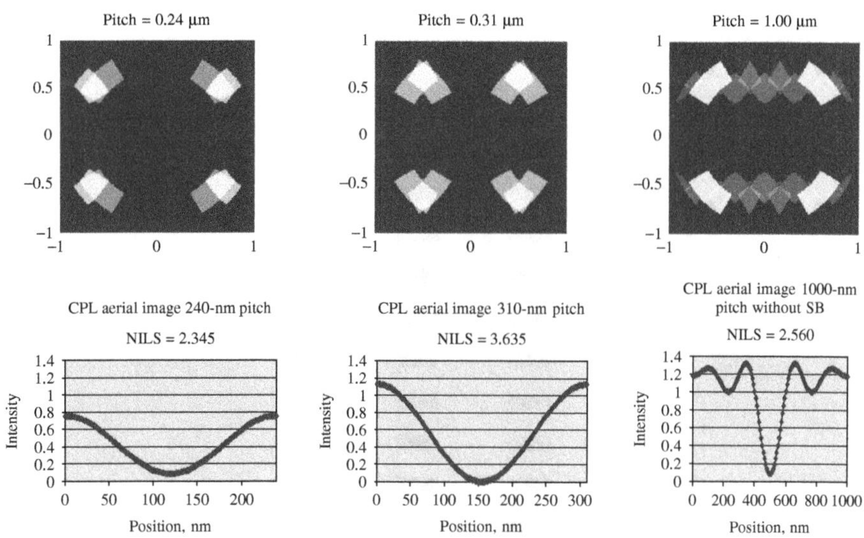

Figure 7.50 Sampling of the projection pupil by the diffraction orders. The projection pupils and aerial images are for a 100-nm CPL line with a KrF system ($\lambda = 248$ nm), NA = 0.8, and quasar illumination ($\sigma_{in} = 0.68$ and $\sigma_{out} = 0.92$).

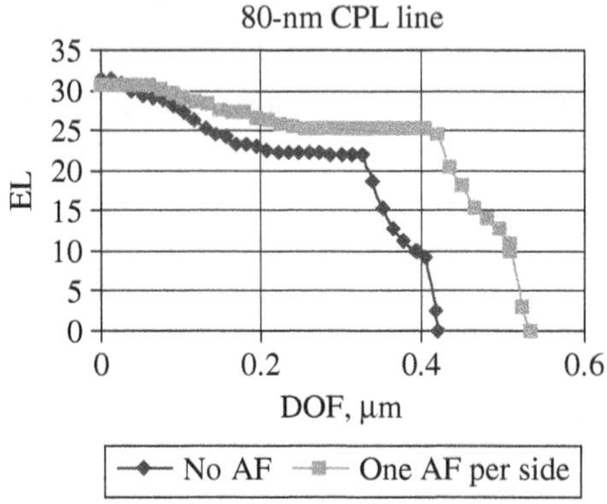

Figure 7.51 The process window of 80 nm CPL on an ArF system (λ = 193 nm), NA = 0.75, and quasar illumination (σ_{in} = 0.7 and σ_{out} = 0.9). The addition of one 40 nm chrome AF per side increases the maximum DOF by 115 nm.

process window is plotted for an 80-nm line with one chrome AF per side and without AF. By using the AF, the maximum DOF increases by 115 nm for an 80-nm chrome line created with a CPL mask.

In Fig. 7.41, it was shown that each pitch has an optimal transmittance. With CPL multiple transmittances can be made through a technique known as zebraing. An example of zebra CPL mask is shown in Fig. 7.52. In a zebra CPL mask, chrome is added to the lines to modify the transmittance. The chrome is added perpendicular to the line pitch. The pitch at which chrome is added perpendicular to the line pitch is called the zebra pitch P_{zebra}. The chrome must be added such that it does not cause modulation. The zebra pitch must be small enough that the projection lens does not collect diffraction orders created by the chrome zebra.

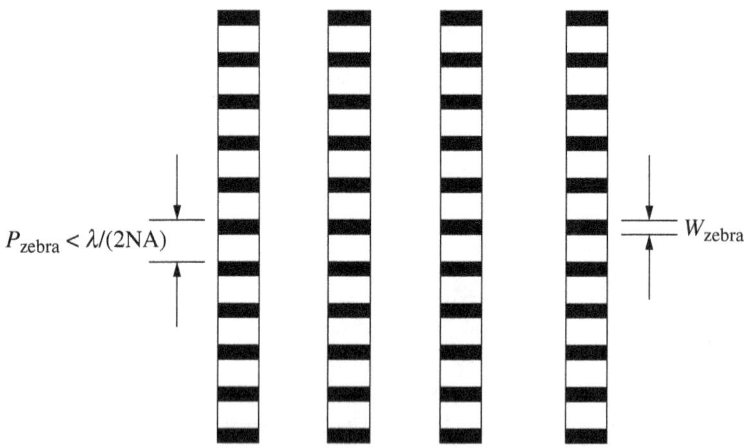

Figure 7.52 Using zebra chrome to modify the transmittance on a CPL mask.

Using Eq. (7.18), the zebra pitch can be calculated by setting n_{max} equal to 1 and solving for the pitch P. If the zebra pitch is less than this value, the zebra will not cause modulation. The minimum zebra pitch is shown in Eq. (7.50).

$$P_{zebra} < \frac{\lambda}{2NA} \tag{7.50}$$

Modifying the duty cycle of the chrome zebra can also modulate the transmittance of the mask feature. The transmittance of the feature is the duty cycle of the chrome zebra. The duty cycle is the ratio of the zebra width W_{zebra} to the zebra pitch P_{zebra}. Therefore, a CPL mask with zebraing is an attenuating PSM with variable transmittance. If the resolution of the mask writer is fine enough, any transmittance in Fig. 7.41 could be produced. However, since the mask writer has finite resolution, typically three transmittances are used—0, 50, and 100 percent.

Since a CPL mask is similar to an attenuating PSM, many of the resolution enhancement techniques that improve the process with an attenuating PSM improve the process with CPL. In order to implement a CPL process, many resolution enhancement techniques should be combined to produce the largest process window. First, the illumination is chosen to maximize the process window of the critical pitches. The selection of the OAI depends on the pitch and the orientation of the pattern. If the pattern has both features along the x- and y-axes ("Manhattan"), it is better to use an illuminator that is symmetric with respect to the x- and y-axes of the illuminator such as a quasar illuminator. If zebraing is to be used, the transmittance as a function of pitch is chosen to maximize the process window through pitch. After choosing the illumination and the transmittance, assist features are placed to maximize the process window of the isolated features. Finally model-based OPC must be performed to cause all the features to print at the same dose. Combining all these resolution enhancement techniques results in the maximum process window at a low k_1 resolution.

7.6 Optimization

In addition to lowering the resolution through phase-shift mask, the illumination intensity and the mask features can be optimized to further reduce the resolution and to improve the process window. In logic or ASIC designs, less aggressive off-axis illumination, such as annular illumination, is used with AF and/or phase-shift mask to achieve a lower k_1. In memory designs, more aggressive off-axis illumination is used to print $k_1 = 0.3$. This off-axis illumination is optimized through techniques discussed in this section. In addition to optimizing the illumination, the mask resolution techniques discussed in Secs. 7.4 and 7.5 can be combined with illumination optimization. Combining these techniques is called mask-source optimization and is also discussed at the end of this section.

Others have done various optimization techniques for optical lithography. Burkhardt described a method for optimizing the illumination for exposing a contact hole pattern.[48] In this method the illumination is optimized based on where the diffraction orders sample the projection lens. Gau also presented a method for optimizing the illumination for features at many pitches.[49] Rosenbluth invented the idea of combining optimization of the illumination with optimization of the mask, i.e., simultaneous mask and illumination optimization.[50]

7.6.1 Illumination (source) optimization

In this section two techniques for optimizing the illumination are presented. One technique involves examining the location of the diffraction orders in the projection lens. By optimizing the illumination, these diffraction orders can be forced to overlap in the projection lens. The diffraction orders are forced to overlap and occupy the same radius in the projection lens. As discussed in Sec. 7.2.3, when light is imaged on the same radius in the projection lens, the DOF increases. This technique of overlapping diffraction orders is sufficient when one wants to optimize the DOF for one pitch. However, this technique cannot increase the DOF for multiple pitches or for a two-dimensional pattern. Furthermore, it cannot optimize other parameters such as the pattern fidelity or *critical dimension uniformity* (CDU). When this simple overlapping diffraction order technique is insufficient, the technique proposed by Rosenbluth is capable of optimizing the process window, DOF and EL, for a two-dimensional pattern while simultaneously lowering the CDU. This Rosenbluth optimization technique is discussed in the end of this section.

By using the correct illumination, the diffraction orders are forced to overlap in the projection lens. As discussed in Sec. 7.2, forcing the orders to overlap in the projection lens maximizes the DOF. Using the theory presented in Secs. 7.2 and 7.3, a method for overlapping the diffraction orders is derived for one-dimensional lines at one pitch. For one pitch varying along the x direction, the diffraction orders change along the x direction and are centered at m/P, where m is the diffraction order number and P is the pitch along the x direction. Since the diffraction orders vary along the x direction only, the optimal illumination has poles along the x direction as well. In order to produce an image without image placement shift, the illumination must be symmetric with respect to the x- and y-axes. Therefore, two poles in the illumination, dipole illumination, are needed to produce an image without image shift. The location of the center sigma σ_c of the dipole illumination is calculated by forcing the diffraction orders to overlap in the projection lens.

The location of the diffraction orders in the projection lens is illustrated in Fig. 7.53 for dipole illumination centered at σ_c. In Fig. 7.53, the energy in the projection lens is plotted for the three orders that the lens can collect—−1, 0, and +1. The zero order from dipole illumination causes the energy in the projection lens to be centered at $\sigma_c\text{NA}/\lambda$. For the ±1 order, the projection lens collects only one pole in the dipole. The DOF is maximized when the energy from the zero order overlaps with the ±1 orders in the projection lens. This is accomplished when σ_c is chosen such that Eq. (7.51) is satisfied.

$$\sigma_c = \frac{\lambda}{2\text{NA}P} \tag{7.51}$$

The result in Eq. (7.51) is a simple rule that can be used for optimizing the illumination of a single pitch. Using Eq. (7.51), the optimal σ_c is at 0.63 for a pitch of 180 nm on an ArF scanner with NA = 0.85. Equation (7.51) can be compared to simulation to test its validity. In Fig. 7.54, Prolith simulations show the DOF as a function of σ_c for the 180-nm pitch case. The Prolith simulations indicate that the optimal σ_c is 0.63. This result is identical to Eq. (7.51) and indicates that Eq. (7.51) is capable of determining the optimal illumination.

Quadrupole illumination is a better choice of illumination when there are features with both x and y orientation. The equation for optimizing dipole illumination is easily modified to calculate the best σ_c for quadrupole illumination. For quadrupole illumination, the best σ_c is given in

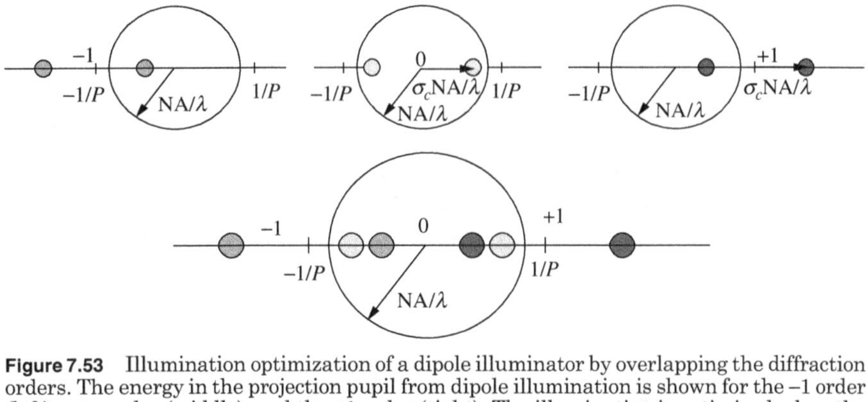

Figure 7.53 Illumination optimization of a dipole illuminator by overlapping the diffraction orders. The energy in the projection pupil from dipole illumination is shown for the −1 order (left), zero order (middle), and the +1 order (right). The illumination is optimized when the energy in the projection pupil (bottom) has the 0 order overlapping with the ±1 orders.

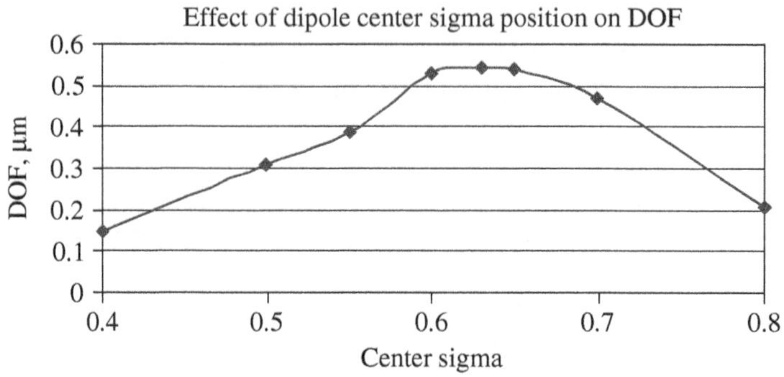

Figure 7.54 The DOF as a function σ_c for dipole illumination. The DOF is maximum for $\sigma_c = 0.63$. Result for a 90-nm binary line on a 180-nm pitch for $\lambda = 193$ nm and NA = 0.85.

Eq. (7.52). In Eq. (7.52) square root of 2 is added because the poles in the quadrupole illumination are along 45°. Like Eq. (7.51), Eq. (7.52) is valid for optimizing the illumination for lines at one pitch. Equation (7.52), however, can be used for lines at one pitch along both the x- and y-axes while Eq. (7.51) is valid only for lines along one axis (either x or y).

$$\sigma_c = \frac{\sqrt{2}\lambda}{2\mathrm{NA}\,P} \tag{7.52}$$

Equation (7.51) is a convenient equation for optimizing the illumination for periodic 1D lines with one pitch. However, 1D periodic lines with one pitch are rarely encountered in a design. Equation (7.51) cannot be used for optimizing multiple pitches or for optimizing a 2D pattern. A better method for optimizing multiple pitches or 2D patterns was developed by Alan Rosenbluth.[50] In the Rosenbluth method, the minimum NILS through focus is maximized at user selected fragmentation points while simultaneously forcing the intensity at these points to be the same. Maximizing the minimum NILS optimizes the EL while the intensity constraint improves the pattern fidelity or CDU. Thus the Rosenbluth method optimizes the overlapping process window at the fragmentation points.

The Rosenbluth method is best illustrated graphically. A common pattern in DRAM (dynamic random access memory) is the isolation layer. This layer looks like a brick wall and is commonly referred to as a brick wall. Using the Rosenbluth method the illumination is optimized for a brick wall pattern in Fig. 7.55. Using the Abbe imaging theory in Sec. 7.3.1, the aerial image of the brick wall is determined for each illumination point. In Fig. 7.56, the aerial image of the brick wall is shown for two illumination points. One point that is close to the center of the illuminator produces an inverted image, i.e., dark areas of the brick wall pattern are bright

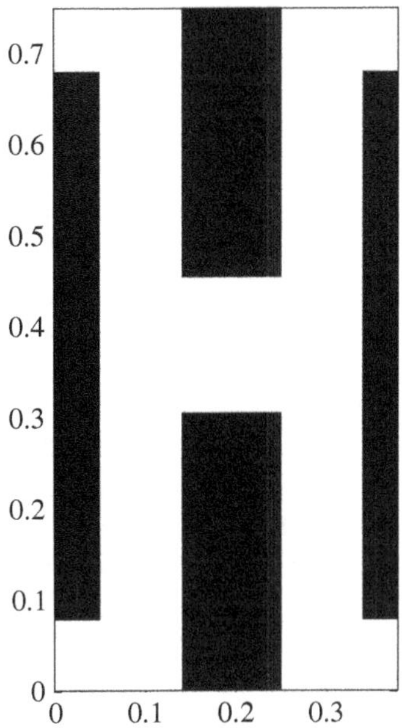

Figure 7.55 Brick wall pattern for illumination optimization. The brick wall pitch is 230 nm resulting in a k_1 half-pitch of 0.37. Optimization for $\lambda = 248$ nm and NA = 0.8.

and vice versa. This inverted image is highly undesirable; consequently, points in the center of the illuminator have a detrimental effect on the aerial image. Another point at the outer part of the illuminator produces an aerial image that resembles the brick wall design, i.e., dark areas in the brick wall are dark in the aerial image. Consequently, the point in the outer part of the illuminator is advantageous. The Rosenbluth method of illumination optimization calculates the effect each illumination point has

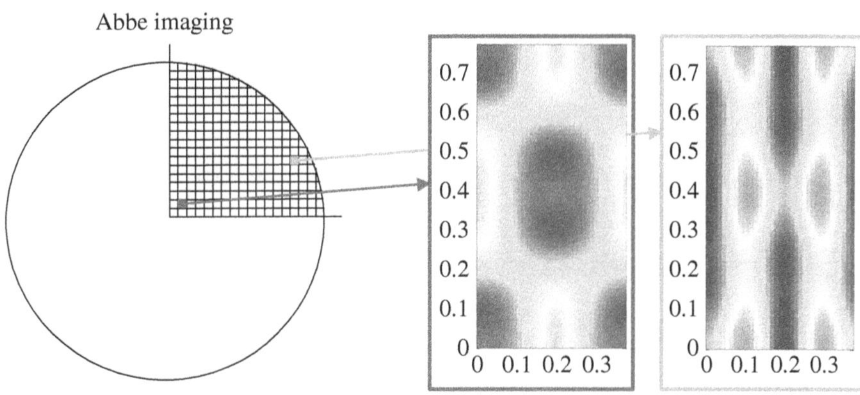

Figure 7.56 Effect of illumination points on the aerial image for the brick wall pattern in Fig. 7.55. The center illumination point reduces the contrast as the aerial image is inverted while the outer illumination point improves the contrast.

on the aerial image. The Rosenbluth method then optimizes the illumination by increasing the illumination intensity of the illumination points that maximize the minimum NILS.

In Fig. 7.57, points of the illumination are shown that increase the NILS at three fragmentation points. The fragmentation points along the edge of the brick wall benefit from dipole illumination. The NILS at the end of line, however, improves when illumination points along the y axis are used. It is important to note that the NILS scale is smaller for the EOL compared to the NILS scale of the two points along the edge. Since the NILS at the EOL is smaller, the EOL fragmentation point dominates the optimization because the Rosenbluth method maximizes the minimum NILS. Consequently, illumination points, which improve the EOL NILS, are weighted more by increasing the intensity of these illumination points. For the brick wall pattern in Fig. 7.55, the illumination, which maximizes the minimum, NILS at the three fragmentation points is a hexapole illuminator and is plotted in Fig. 7.58.

Using the hexapole illumination in Fig. 7.58, the process windows were simulated with Prolith. The overlapping process window at all three fragmentation points is plotted in Fig. 7.59 for the optimized hexapole illumination and is compared to the overlapping process window for annular illumination. The optimized hexapole illumination

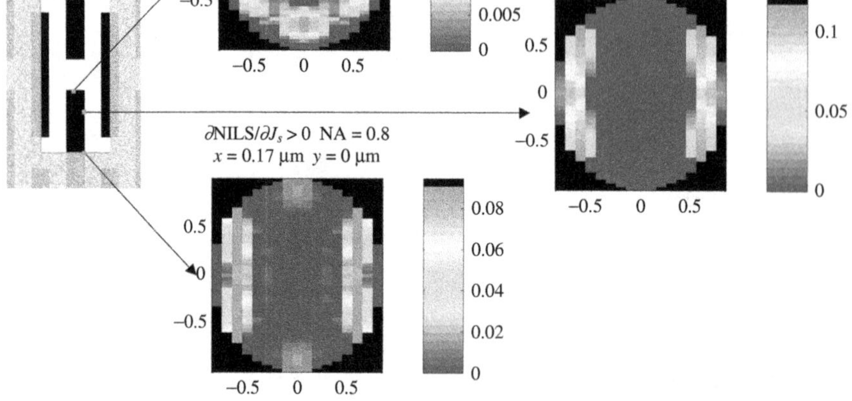

Figure 7.57 The effect of the illumination on the NILS at three fragmentation points. The EOL fragmentation point has the lowest scale; therefore, the EOL dominates the optimization.

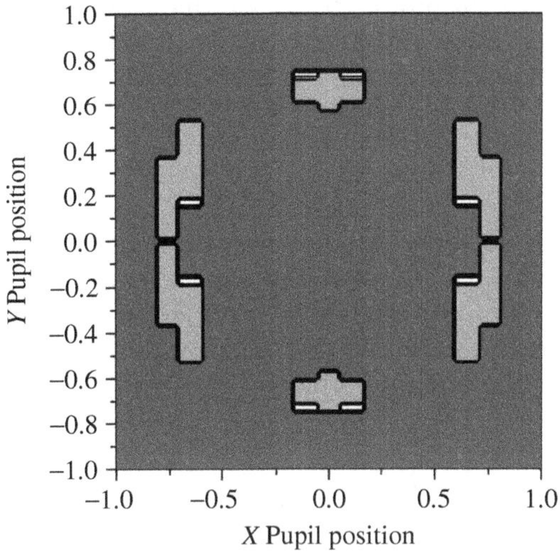

Figure 7.58 Optimized illuminator for the brick wall pattern in Fig. 7.57. The optimal illuminator resembles a hexapole illuminator.

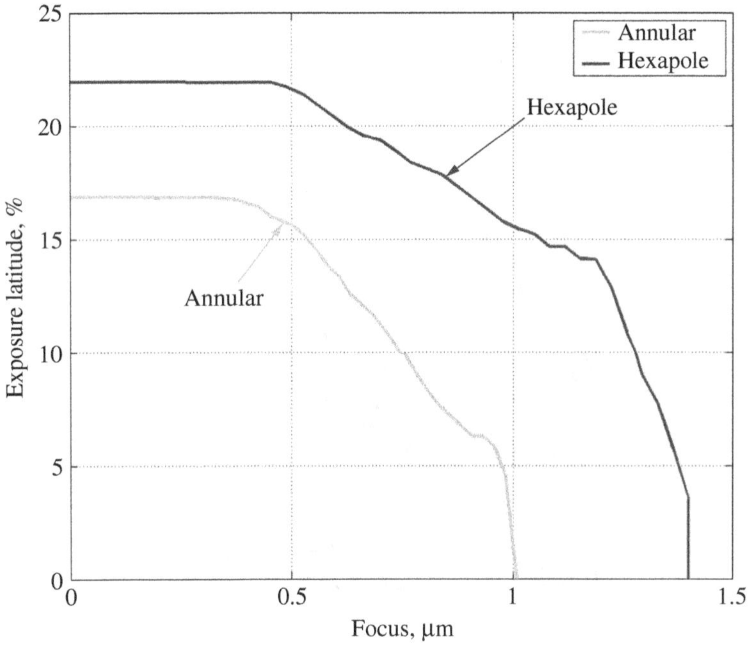

Figure 7.59 Overlapping process window for the brick wall pattern in Fig. 7.55. The optimized hexapole illumination in Fig. 7.58 produces an ED window with twice the area as the annular illumination ($\sigma_{in} = 0.56$ and $\sigma_{out} = 0.86$). Prolith simulations with $\lambda = 248$ nm NA = 0.8.

doubles the area under the process window curve in comparison to the area with annular illumination. This improvement in the process window indicates that illumination optimization is another RET capable of improving the process window for a low k_1 design.

7.6.2 Mask-source optimization

In order to achieve the lowest k_1 possible, the source and mask must be optimized simultaneously. Various mask RETs can be combined with source optimization in order to simultaneously optimize the source and mask. A less aggressive version of source and mask optimization combines illumination optimization with AF placement. In Fig. 7.60,

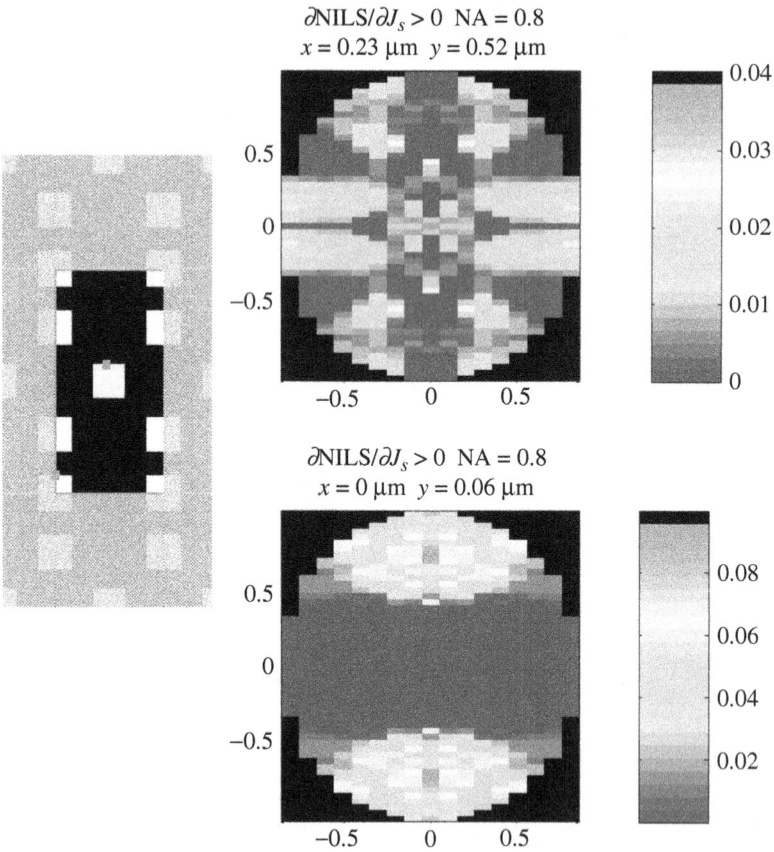

Figure 7.60 Locations in the illuminator that improve the NILS (right) at two fragmentation points for the contact hole pattern (left). Note that the locations in the illuminator to improve the NILS are different for the two fragmentation points. The optimization is for 120 nm contact with 80 nm minimum space. The cell size is 230 nm by 920 nm. The optimization is for a KrF scanner with NA = 0.8.

the illumination points that improve the NILS are plotted for two fragmentation points of a contact hole pattern. This contact hole pattern includes a dense string of contact holes along the y axis and includes an isolated contact hole. The dense contact hole prefers illumination points along the y axis while the isolated contact hole prefers illumination points in the center of the illuminator. These two points are in conflict because illumination points that improve the NILS of the dense contact hole reduce the NILS of the isolated contact, and vice versa.

In this contact hole pattern, there is space to place AF. In Fig. 7.61, AF have been placed around the isolated contact hole. In Fig. 7.61, the illumination points that improve the NILS are again plotted for the

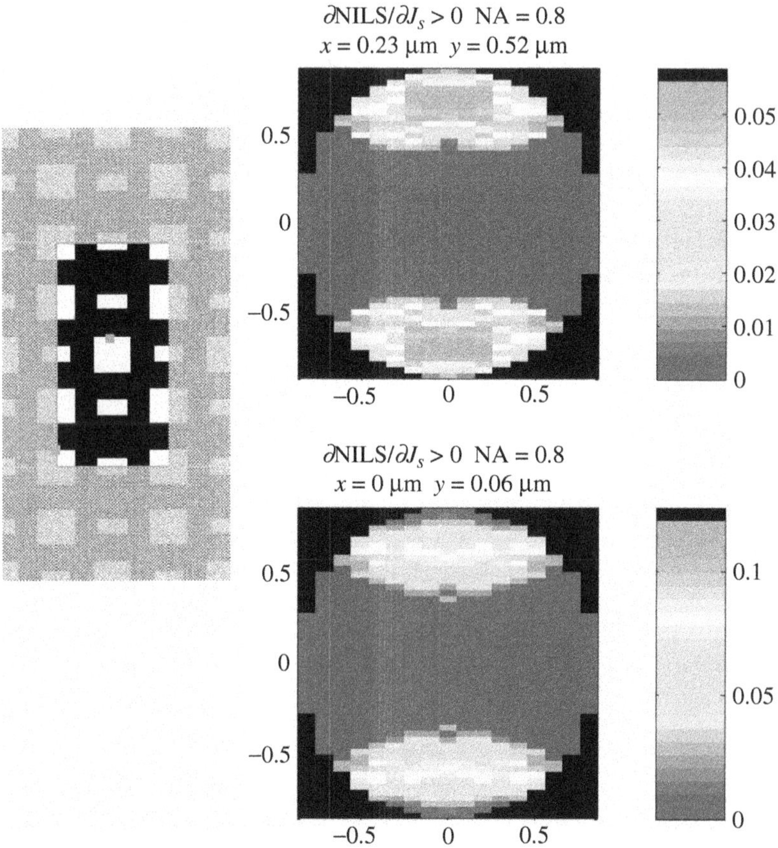

Figure 7.61 Locations in the illuminator that improve the NILS (right) at two fragmentation points for the contact hole pattern (left) after adding AF. Note that after adding AF the locations in the illuminator to improve the NILS are the same for the two fragmentation points.

Figure 7.62 The optimized illuminator for the contact hole pattern with AF in Fig. 7.61.

two fragmentation points. Figure 7.61 shows that illumination points, which improve the NILS of the dense contact after placing AF, also improve the NILS of the isolated contact. The AF have essentially increased the density of the contact hole pattern. By increasing the density of the isolated holes, the image of the isolated holes with AF is similar to the image of the dense holes. Consequently, illumination that improves the NILS of dense contacts also improves the NILS of isolated contacts with AF. The optimal illumination for the contact hole mask with AF is an octupole illuminator. The optimal illumination is plotted in Fig. 7.62 for the mask in Fig. 7.61.

In Fig. 7.63, the ED window for the isolated contact and for the dense contact is simulated for the mask in Fig. 7.61. The ED window is shown

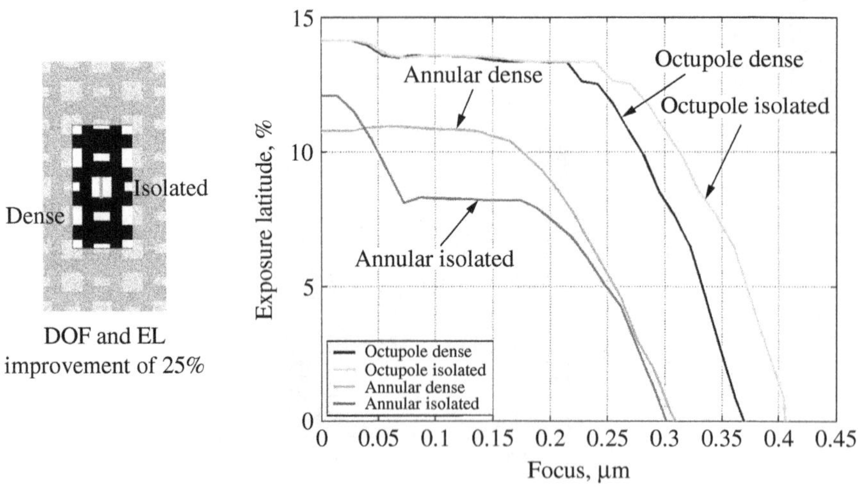

Figure 7.63 Comparison of the simulated process window for optimized octupole illumination in Fig. 7.62 to the process window for annular illumination ($\sigma_{in} = 0.5$ and $\sigma_{out} = 0.8$). The simulations use a KrF scanner NA = 0.8.

for the optimized octupole illumination and for annular illumination. The octupole illumination for the contact mask with AF has 25 percent more process window area as compared to the annular illumination for the mask with AF. Consequently by combining illumination optimization with AF, the process window is improved over a process with AF alone.

Most methods for placing AF rely on a rule-based placement of AF. At low k_1, biasing the mask requires a model-based method as rule-based mask biasing is no longer sufficient. Likewise, the placement of AF requires a model-based method rather than a rule-based method. A technique for placing the AF through a model is called *interference mapping lithography* (IML).[51] IML optimizes the placement of assist features around contact holes for a two-tone mask, such as a binary mask or an attenuating PSM. With IML, the optimal transmission and phase of the background can also be calculated for a two-tone mask. For a more aggressive k_1, IML optimizes the design of a CPL mask. IML optimization is performed by creating the interference map of the design. In the interference map, areas where the magnitude is greater than zero represent a location where light will interfere constructively. While areas where the magnitude is less than zero represent a location where light will interfere destructively. The aerial image of the contact hole can be improved by placing clear assist features (transmission of 100 percent) where constructive interference occurs in the interference map. The aerial image intensity can be increased further by placing 180° assist features (transmission of –100 percent) where destructive interference occurs in the interference map; hence creating a CPL mask.

In the previous example, the illumination was optimized after placing AF. A more powerful RET involves optimizing the illumination and optimizing the mask phase and transmission simultaneously. In a simultaneous optimization of the mask and illumination, the placement of structures with three transmittances is optimized through the Rosenbluth method. In the Rosenbluth method, the mask diffraction orders are optimized through Hopkins formulation.

Again using the contact hole pattern in Fig. 7.60, the illumination and the CPL mask were simultaneously optimized resulting in the CPL mask and the illumination shown in Fig. 7.64. In Fig. 7.64, the CPL mask is a diffractive optical element as the CPL mask does not resemble the design. In Fig. 7.65, the aerial image of the CPL mask is compared to the aerial image of the AF mask in Fig. 7.61. The CPL mask has a larger contrast than the AF mask. Consequently, the CPL mask has a larger process window compared to the mask with AF. In order to push the lithography system to a k_1 approaching 0.25, the source and mask must be optimized simultaneously.

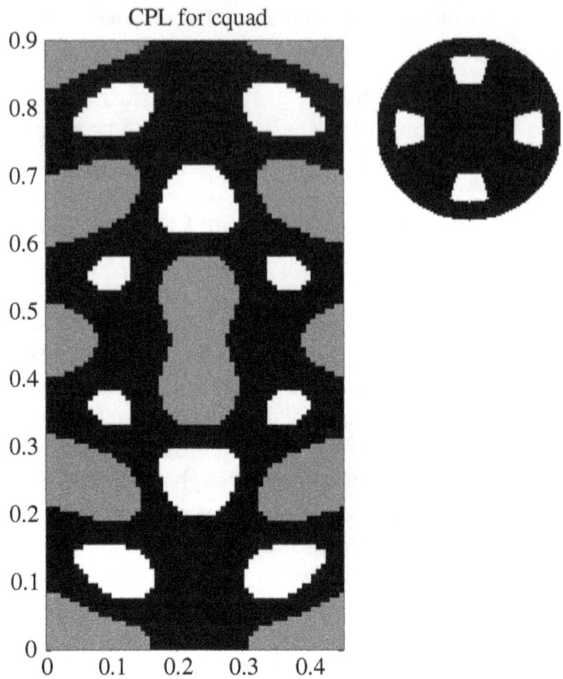

Figure 7.64 Optimize CPL mask and illumination for the contact hole mask in Fig. 7.60. In Fig. 7.64, the black areas have a transmittance of zero, the white areas a transmittance of +1, and the gray areas a transmittance of −1.

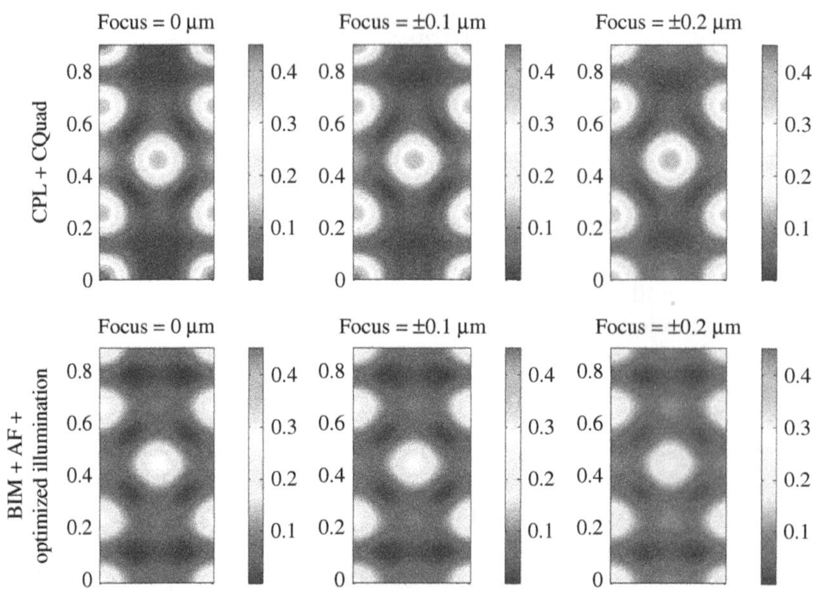

Figure 7.65 Simulated aerial images at three foci for the CPL optimized pattern in Fig. 7.64 and for the binary mask with AF in Fig. 7.61 with the optimized illumination in Fig. 7.62. Note the increase in contrast for the CPL mask.

7.7 Advanced Topics

In order to continue to decrease the minimum resolution, the wavelength can be decreased or the NA can be increased. Lithography tools are quickly approaching the limit on the maximum NA, an NA of 1. In order to increase the NA beyond 1, ambient material can be placed between the last element of the projection lens and the wafer. The maximum NA is limited to the index of refraction of the ambient material. Since the index of refraction of air is 1, the maximum NA of a lithography tool with an air ambient is 1. However, if the index of refraction of the ambient material is increased, the NA could be increased past 1. Currently, the lithography community is investigating the possibility of producing lithography tools with NA larger than 1 by immersing the wafer in a liquid ambient.[52–57] As the NA increases beyond 1, the effect of hyper-NA (NA > 1) on the RET must be considered; consequently, immersion lithography is described in this section.

When the NA increases, there are other factors, which impact the various resolution enhancement techniques described in the previous sections. At high NA, calculating the image in air is erroneous as the image must be calculated inside the resist thin -film.[58,59] The resist thin film refracts the rays being imaged onto the wafer. This refraction causes a shift in the best focus of the image. Furthermore, the polarization of the rays as the rays impinge on the thin film must be considered.[60] From Fresnel equations, one polarization state is transmitted differently into the resist as compared to the other polarization.[7] As well as changing the image in the resist, the polarization effects can be optimized such that polarization is used as an RET. Consequently, this section describes the effect of polarization on RET.

7.7.1 Immersion lithography

In microscopy, another technique for improving the resolution involves immersing the sample in a liquid, typically water or oil. By immersing the microscope lens, the NA of the microscope increases by the index of refraction of the immersion liquid, $\mathrm{NA} = n\mathrm{NA}_{\mathrm{air}}$. Consequently, an NA larger than 1, hyper-NA, is possible by using an immersion liquid with a large index of refraction n. Similar to microscopy, the NA of the lithography system is increased by immersing the gap between the last projection lens element and the wafer. Like the microscope, the NA of the lithography scanner increases by the index of refraction of the immersion liquid. In order to increase the NA as much as possible, the immersion liquid index of refraction must be maximized. However, the immersion liquid must not absorb the light; otherwise the dose would have to increase, which would reduce the throughput of the lithography system. At a wavelength of 193 nm, water has very little absorption, and the index of refraction is 1.43. Therefore, water is a likely immersion liquid at 193 nm.

Another way of understanding the improvement in resolution is to consider that an immersion system effectively reduces the wavelength by the index of refraction of the immersion liquid. For an ArF immersion scanner, the effective wavelength is 193/1.43 = 135 nm. Consequently, an immersion ArF system allows the continuation of the lithography roadmap at a wavelength of 135 nm.

By increasing the NA of the lithography system above one, a smaller feature is resolvable. For a system with an NA less than 1, there is a benefit in converting a dry lithography system into an immersion lithography system. On lithography systems with NA less than 1, the depth of focus increases if the projection optics is converted for immersion. The increase in DOF is understood by comparing the ray trace of a dry system to the ray trace of an immersion system. In Fig. 7.66, the ray trace of both a dry and an immersion system is shown. In the wet system, the angle impinging on the resist is smaller than the angle impinging on the dry system. Because the angle is smaller on an immersion system, the immersion system can tolerate more deviation in the separation between the bottom of the projection lens and the wafer.

Using Prolith simulation, the improvement in DOF is demonstrated in the immersion system over the dry system. In Fig. 7.67, the ED window is plotted for a dry system and an immersion system (wet). The maximum DOF for the immersion system is almost twice as large as the dry system. In addition to demonstrating the DOF improvement with simulation, experiments were done on an NA of 0.75 system. The experimental results of 90-nm dense line are shown in Fig. 7.68. In Fig. 7.68, the DOF of the

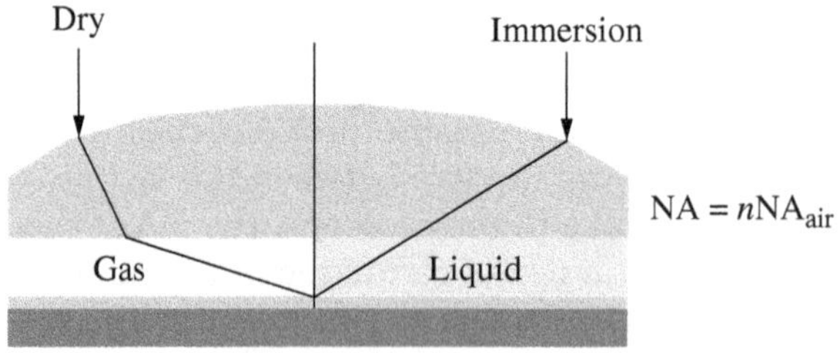

Figure 7.66 An immersion lithography system fills the interface between the last lens element and the wafer with liquid.

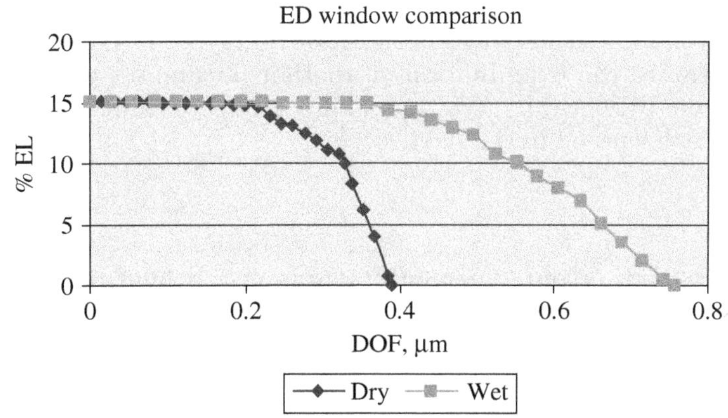

Figure 7.67 Simulations of a 65-nm chrome line on dry system and on an immersion system (wet). The maximum DOF on the immersion system is almost double the maximum DOF on the dry system. The simulations are for a binary mask with pitch = 200 nm, λ = 193 nm, NA = 0.85 using quasar illumination σ_{in} = 0.65 and σ_{out} = 0.95.

wet system is approximately 0.95 µm and is 1.7 times larger than the DOF of the dry system.

By using an immersion liquid, the NA of the lithography system can be increased above 1, or the DOF of a dry lithography system with NA less than 1 can be improved. There are many technical and engineering

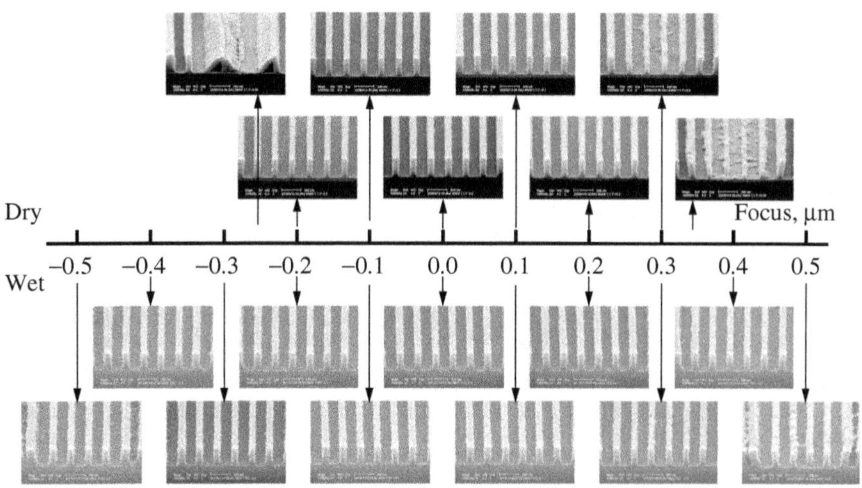

Figure 7.68 Immersion (wet) resist images and dry resist images. The DOF of the wet process has 1.7 times more DOF than the dry process. The images are for a 90-nm dense line on a binary mask. ArF scanner with NA = 0.75 and annular illumination σ_{out} = 0.89 and σ_{in} = 0.59 (Courtesy of ASML).

challenges to implement immersion in a lithography scanner. These challenges are beyond the scope of this book. However, hyper-NA lithography systems require the examination of another parameter previously neglected on lithography systems. This other parameter is polarization and is discussed in the next section.

7.7.2 Polarization

Prior to using resolution enhancement techniques, lithographers reduced the resolution by decreasing the wavelength and/or increasing the NA. However, simply using low wavelength or high NA did not improve the process window enough to implement finer resolution in a low k_1 process. In order to achieve sufficient process window at finer resolution, lithographers began to modify the mask as discussed in Sec. 7.5 or to optimize the illumination as discussed in Sec. 7.6.1. Together wavelength, NA, phase-shift mask, and off-axis illumination provided enough process window at these low k_1 resolutions. However, as the NA begins to increase past 1.0, the hyper-NA regime, there is another parameter that the lithographer can utilize to optimize the process window at low k_1. This other parameter is using the benefits of polarization.

In order to understand the benefits of polarization, the effect of polarized light as it propagates through the lithography system must be understood. From optics theory, the thin film on the top of the wafer preferentially allows certain polarized light to propagate into the thin film. In addition to the effect of polarized light on the wafer, the mask also preferentially allows certain polarized light to pass through. The effect of thin-film imaging is discussed in section "Wafer polarization effects" followed by the effect of mask polarization in section "Mask polarization."

Wafer polarization effects. In order to understand the effect of polarized light on lithography imaging, the light must be decomposed into an orthogonal basis in order to represent the polarization. The basis used in imaging theory is the *transverse electric* (TE) and the *transverse magnetic* (TM) *polarization*. TE polarized light is often called *s* polarized or perpendicular polarized light, and TM polarized light is often called *p* polarized or parallel polarized light. Transverse electric polarization means that the electric field is perpendicular to the plane of propagation, and transverse magnetic polarization means that the magnetic field is perpendicular to the plane of propagation. In Fig. 7.69, the polarization orientation of the electric field is plotted for TE and TM light. Since TE and TM form an orthogonal basis set, the

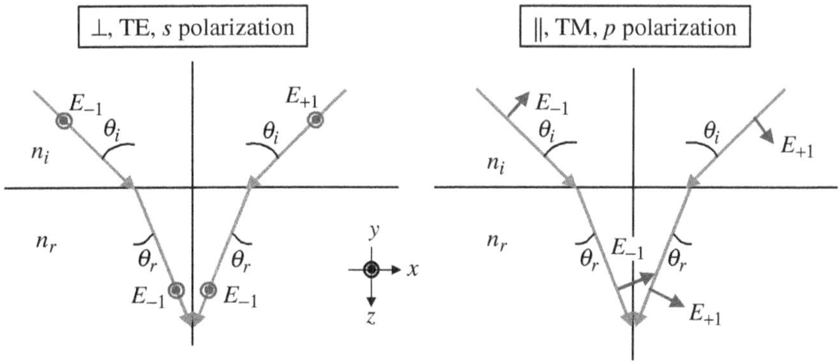

- TE: $I = E{\cdot}E^* = |E_y|^2$ amplitude of $I_{TE} = A^2$
- TM: $I = E{\cdot}E^* = |E_x|^2 + |E_z|^2$, $|E_z|^2 = 0$, $|E_x|^2 = A^2\cos^2\theta_r$, amplitude of $I_{TM} = A^2\cos^2\theta_r$
- $\cos\theta_r$ reduces the contrast of TM light. In order to reduce the effect of the $\cos\theta_r$ term, want to decrease θ_r (normal propagation, paraxial)

Figure 7.69 The polarization state for TE (left) and for TM (right) polarization. For TE, the electric field points out of the page while for TM, the electric field is in the plane of the page.

effect of TE light on imaging is independent of the effect of TM light on imaging. Since the TE and TM light are independent of each other, the total intensity is the summation of the intensity from TE light with the intensity from TM light. Unpolarized light consists of equal parts of TE and TM light. Consequently, the total image intensity created with unpolarized light is calculated by adding the intensity with TE to the intensity with TM.

In Fig. 7.69, two beams of light from the −1 order and the +1 order interfere to create an image in the photoresist. Since the photoresist refracts the incident light, the incident light is bent closer to the thin film normal with refraction angle of θ_r. When the −1 and +1 order with TE polarization interferes, all the electric field is pointing out of the page in the y axis. Since all the electric field is in the y axis, the TE light will interfere without any losses due to polarization. However, when the −1 and +1 order with TM polarization interferes, there is some light along the x and z axes. The −1 order with TM polarization has a component along the z axis that is in the opposite direction to the +1 order component along the z axis. Therefore, the TM electric field along the z axis stands cancelled and does not contribute to interference. Only the TM electric field along the x axis contributes to interference. The amount of TM light along the x axis is given by the cosine of θ_r. This implies that as θ_r increases the amount of TM interference decreases. Since θ_r

increases as the NA increases, high NA systems lose contrast due to the decease in TM interference.

The loss of contrast due to TM interference is plotted in Fig. 7.70 for a two-beam imaging case. For this case, the resist is assumed to have an index of refraction of 1.7. In Fig. 7.70, the TE contrast always has a contrast of 1 because all the TE light interferes. The TM contrast, however, decreases as NA increases. For an NA of 1.7, which is the index of the resist, the TM contrast would actually be zero, and TM polarization would not contribute to imaging. The two-beam imaging case represents the theoretical maximum contrast. Due to the finite size of the illumination, an actual lithography process is never truly two-beam imaging; consequently, the contrast of an actual lithography process is less than the contrast of the two-beam case. In realistic lithography systems, the TM light can also cause the image to invert at very large NA. The inverted image has a negative contrast, and the TM light is actually detrimental to contrast.

In order to increase the contrast in high NA lithography systems, the bad polarization, TM, must be reduced. The TM polarization is eliminated or reduced in lithography systems by using polarized optics. The optics of lithography systems can be polarized at many locations; however, two convenient places to polarize the optics are at the illumination pupil or at the projection pupil. Polarization of the optics in the projection pupil can cause distortion in the image. Polarization of the optics in the projection lens without distortion is difficult and costly to achieve. Hence, lithography systems polarize the light in the illumination optics. In the illumination optics various polarization states can be achieved,

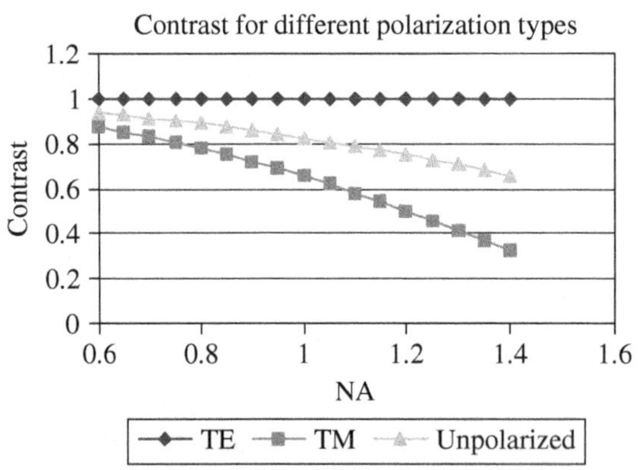

Figure 7.70 Contrast for three polarization states—TE, TM, and unpolarized—as a function of NA for an image formed by two beams, the ±1 diffraction orders. The TE polarization has the largest contrast. The contrast with TM polarization decreases as NA increases.

such as TE, TM, x, or y polarization. TM polarization is the detrimental polarization and should be eliminated. The obvious choice of illumination is TE polarized optics. TE polarization has the most benefit on imaging for a wide variety of lithography applications. However, in some lithography applications, such as having lines with only one orientation, x or y polarization has a larger benefit than TE polarization. For example, if one has all lines orientated along the y direction, y-polarized light is a better polarization state than TE polarization.

The effect of polarized illumination on imaging is best studied through an example. In Fig. 7.71, a line is imaged with unpolarized and TE, TM, x- and y-polarized annular illumination. The illumination and polarization orientation is plotted for each illumination. The unpolarized illumination has equal amounts of TE and TM polarized light. The TE polarized illumination is polarized along the azimuthal while TM polarized illumination is polarized along the radius. In Fig. 7.71, the TE polarized illumination has more contrast than the image with unpolarized light as there is about a 20 percent increase in contrast using the TE illumination. For TM polarization, the lines would not be resolved and only the line ends would print. The x- and y-polarized light are interesting cases as well. The x-polarized illumination increases the contrast between the lines but reduces the contrast at the end of line. Meanwhile the y-polarized illumination improves the contrast of the end of line but reduces the contrast between the lines. The intensity from unpolarized illumination is the summation of the intensity from TE and from TM illumination. Consequently, the unpolarized illumination produces an image with more contrast than the image with TM illumination;

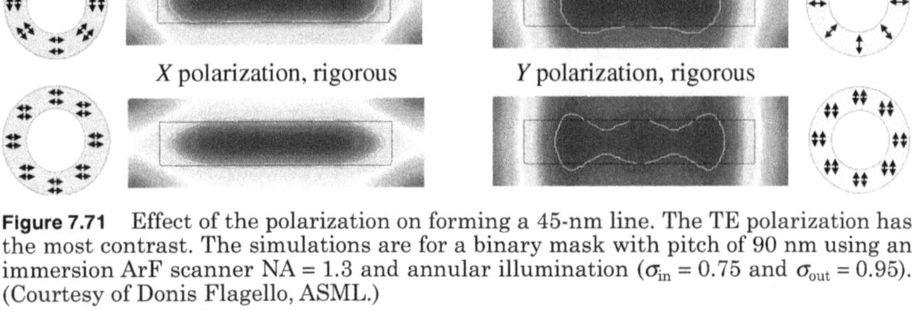

Figure 7.71 Effect of the polarization on forming a 45-nm line. The TE polarization has the most contrast. The simulations are for a binary mask with pitch of 90 nm using an immersion ArF scanner NA = 1.3 and annular illumination (σ_{in} = 0.75 and σ_{out} = 0.95). (Courtesy of Donis Flagello, ASML.)

however, the unpolarized illumination produces an image with less contrast than the image with TE illumination.

Mask polarization. If the optics is polarized in the illumination, the light impinging on the mask will be polarized. Therefore, the interaction of polarized illumination on the mask must be examined. As stated in the section on alternating PSM, the mask topography has an effect on the image formation, and must be studied with rigorous EMF simulation rather than with scalar mask simulation. The mask topography causes the diffraction orders to change. The change in the diffraction orders depends on the polarization state of the incident illumination.

In Fig. 7.72, the 0 and ±1 diffraction orders are plotted for TE and TM illumination for 110-nm thick binary mask. Figure 7.72 shows the fraction of TE polarization as a function of the wafer linewidth. The pitch is twice the linewidth. When the fraction of TE is positive, there is more TE light being diffracted compared to TM light. When the linewidth is between 15 and 70 nm, the fraction of TE is positive. This implies that the mask is actually reducing the amount of bad TM polarization. Since the mask is reducing the TM light, the contrast calculated with rigorous EMF simulation is actually larger than the contrast with scalar simulation. In Fig. 7.72, the fraction of TE is also larger for the ±1 order compared to the zero order. This also benefits the contrast because the ±1 order causes modulation in the image while zero order causes no modulation. Therefore, in this particular mask example, the mask is improving the contrast by increasing the TE light, and by increasing the ±1 order while reducing the zero order.

Figure 7.72 also shows the effect of incident illumination on the fraction of TE. For 0° incidence, the fraction of TE in the ±1 order has a large

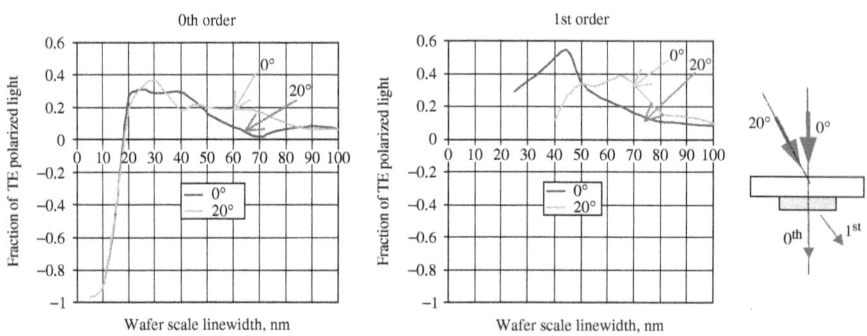

Figure 7.72 The increase in the TE light is plotted as function of wafer linewidth and incident illumination angle for the zero and first order. Fraction of TE = $(I_{TE} - I_{TM})/(I_{TE} + I_{TM})$. Positive fraction of TE implies more TE light than TM light (Courtesy of Douis Flagello, ASML).

difference compared to 20° incidence. This implies, the assumption that the diffraction orders are independent of angle used in scalar imaging is no longer valid. Since the incidence angle on the mask increases as NA increases, rigorous EMF simulation must be used in calculating the diffraction orders for hyper-NA lithography systems.

When the linewidth is less than 15 nm, there is more TM polarized light than TE polarized light. In this regime, the mask acts like a wire-grid polarizer. The mask begins to behave like a pair of sunglasses in which the polarization perpendicular to the lines is passed. As the TM polarization increases the contrast of the image in the resist decreases. Therefore, for an ArF scanner, a feature size below 15 nm will have reduced contrast. The polarization of the mask prevents extending 193-nm lithography systems beyond 15 nm.

7.8 Conclusions

In order to extend the optical lithography roadmap, resolution enhancement techniques that were once only research topics are now necessary for production. In the past the lithography roadmap has mainly relied on lowering wavelength or increasing NA. However, as k_1 continues to decrease many of the techniques discussed in this chapter must be combined. Combining these techniques requires an integrated approach in which the mask, illumination, and resist must be optimized together. The integration of these techniques requires heavy use of software. In the future, the lithography roadmap will require software that is as sophisticated as the hardware.

Currently, the resolution enhancement techniques optimize the printing of a fixed design. However, to achieve the finest resolution possible, many resolution enhancement techniques must be used. The RETs discussed in this chapter, however, can be economically expensive. Another possibility of achieving finer resolution while not increasing the cost is to modify the design such that lithography can print the design with less costly RETs. Modifying the design for less expensive manufacturing is called *design for manufacturing*. As they continue, RETs will eventually include optimizing the design for manufacturing. Hence, RETs and DFM will combine to maintain the intent of the design while reducing the cost of the RETs.

The techniques described in this chapter are limited to a minimum k_1 of 0.25.[61] However, by combining RETs and DFM with multiple exposures, it is possible to remove the restriction on k_1. The minimum pitch is always limited to $0.5\lambda/\text{NA}$ with a single exposure. However, by using two exposures, the k_1 limit of 0.25 can be broken. In two exposures, the first exposure prints every other line in a dense line pattern followed by a second exposure to print the lines not imaged with the first exposure.

In order to use multiple exposures in cost-effective way, the method of printing the design would have to be considered in the design phase. One possibility of designing for multiple exposures is to create a design with features placed on a regular grid. In order to follow the semiconductor roadmap, economics will dictate that RETs, DFM, and multiple exposures must be combined to achieve the lowest k_1 possible with the least amount of money.

References

1. Levinson, H., *Principles of Lithography*, SPIE-the International Society for Optical Engineering, Bellingham, Washington, ISBN 0819440450, 2001.
2. Wong, A., *Resolution Enhancement Techniques in Optical Lithography*, SPIE-the International Society for Optical Engineering, Bellingham, Washington, ISBN 0819439959, 2001.
3. Schellenberg, F. M., *Selected Papers on Resolution Enhancement Techniques in Optical Lithography*, SPIE Milestone Series, vol. MS 178, the International Society for Optical Engineering, Bellingham, Washington, 2004.
4. Abbe, E., "Beiträge zur Theorie des Midroskops und der mikroskopischen Wahrnehmung," *Archiv för Mikroskopische Anatomie*, Vol. 9, pp. 413–468, 1873.
5. Airy, G. B., "On the diffraction of an object-glass with circular aperture," *Trans. Cambridge Philos. Soc.*, Vol. 5(3), pp. 283–291, 1835.
6. Lord Rayleigh, "Investigations in optics, with special reference to the spectroscope 1-6," *The London, Edinburgh, and Dublin Philosophical Magazine and Journal of Science*, Vol. 8(49), pp. 261–274, 1879.
7. Hecht, E., *Optics*, 2d ed., Addison-Wesley Publishing, Reading, Mass., 1987.
8. Goodman, J. W., *Introduction to Fourier Optics*, 2d ed., McGraw-Hill, New York, 1996.
9. Wong, A., "Rigorous three-dimensional time-domain finite-difference electromagnetic simulation," Ph.D. Dissertation, University of California, Berkeley, 1994.
10. Hopkins, H. H., "The concept of partial coherence in optics," *Proc. Roy. Soc. A*217, p. 408, 1953.
11. Prolith is commercially available lithography simulation software produced by KLA-Tencor (www.kla-tencor.com).
12. Solid-C is commercially available lithography simulation software produced by Sigma-C (www.sigma-c.com).
13. Panoramic is commercially available lithography simulation software produced by Panoramic Technologies (www.panoramictech.com).
14. Lee, D., et al., *SPLAT v5.0 Users' Guide*, Memorandum No. UCB/ERL M95/13 University of California, Berkeley, 1995.
15. Goodman, J. W., *Statistical Optics*, John Wiley & Sons, New York, 1985.
16. Born, M. and Wolf, E., *Principles of Optics*, 6th ed., Pergamon Press, Oxford, 1980.
17. Gamo, H., "Matrix treatment of partial coherence," *Progress in Optics*, North-Holland, New York, vol. 3, p. 187, 1964.
18. Wolf, E., "New spectral representation of random sources and the partially coherent fields that they generate," *Opt. Commun.* 38(1), 1981, pp. 3–6.
19. von Bünau, R., "Depth of focus enhancement in optical lithography," Ph.D. Dissertation, Stanford University, Stanford, California, 1995.
20. Socha, R., "Propagation effects of partially coherent imaging in optical lithography and inspection," Ph.D. Dissertation, University of California, Berkeley, 1997.
21. Cobb, N., "Fast optical and proximity correction algorithms for integrated circuit manufacturing," Ph.D. Dissertation, University of California, Berkeley, 1998.
22. K. Eastman, *Techniques of microphotography: precision photography at extreme reductions*, Eastman Kodak, Rochester, N.Y., 1963.
23. Saleh, B., et al., "Reduction of errors of microphotographic reproductions by optimal corrections of original masks," *Opt.l Eng.*, Vol. 20(5), pp. 781–784, 1981.

24. Nashold, K. M., "Image construction through diffraction-limited high-contrast imaging systems: an iterative approach," *J. Op. Soc. Am. A*, Vol. 2(5), pp. 635–643, 1985.
25. Saleh, B., et al., "Image construction: optimum amplitude and phase masks in photolithography," *Appl. Opt.*, Vol. 24(10), pp. 1432–1437, 1985.
26. Stirniman, J. P., et al., "Fast proximity correction with zone sampling," *Proc. SPIE* 2197, 1994.
27. Stirniman, J. P., et al., "Quantifying proximity and related effects in advanced wafer processes," *Proc. SPIE* 2440, p. 252, 1995.
28. Cobb, N., et al., "Fast, low-complexity mask design," *Proc. SPIE* 2440, p. 313, 1995.
29. Ito, T., et al., Photo-projection image distortion correction for a 1μm pattern process, *Transactions of the Institute of Electronics, Information, and Communication Engineers (IEICE)*, Vol. J68-C(5), pp. 325–332, 1985.
30. Starikov, A., "Use of a single size square serif for variable print bias compensation in microlithography: method, design, and practice," *Proc. SPIE* 1088, 1989.
31. Chen, J. F., Matthews, J. A., "Mask for photolithography," *US Patent* 5,242,700; filed January 16, 1992, issued September 7, 1993.
32. Chen, J. F., et al., "Practical method for full-chip optical proximity correction," *Proc. SPIE* 3051, p. 790, 1997.
33 Socha, R., et al., "Forbidden pitches for 130nm lithography and below,"*Proc. SPIE* 4000, p. 1140, 2000.
34. Goodman, D., "Some partially coherent images," *Optical Sciences Center Newsletter*, Vol. 12(1), pp. 5–7, 1978.
35. Shibuya, M., "Projection master for use with transmitted illumination," Japanese Patent Publication Sho 57-62052, filed September 30, 1980, published April 14, 1982.
36. Levenson, M. D., et al., "Improving resolution in photolithography with a phase shifting mask," *IEEE Trans. Electron Dev.* ED-29(12), pp. 1828–36, 1982.
37 Lin, B. J., "Phase-shifting and other challenges in optical mask technology," *Proc. SPIE* 1496, pp. 54–79, 1990.
38. Terasawa, T., et al., "Image characteristics of multi-phase-shifting and halftone phase shifting masks," *Jpn. J. Appl. Phy.* Vol. 30(11B), pp. 2991–2997, 1991.
39. Toh, K., et al., "Chromeless phase-shifted masks: a new approach to phase-shifting masks," *10th Annual Symposium in Microlithography*, J. N. Wiley, ed., 1990.
40. Chen, J. F., et al., "Binary halftone chromeless PSM technology for quarter wavelength lithography," *Proc. SPIE* 4346, 2001.
41. Liu, H. Y., "The application of alternating phase-shifting masks to 140 nm gate patterning: line width control improvements and design optimization," *Proc. SPIE* 3236, p. 328, 1998.
42. Schellenberg, F. M, et al., "Optimization of real phase mask performance," *Proc. SPIE* 1604, p. 274, 1991.
43. Ma, Z. M, et al., "Preventing sidelobe printing in applying attenuated phase shift reticles," *Proc. SPIE* 3334, 1998.
44. Iwasaki, H., et al., "High transmittance rim-type attenuated phase shift masks for sub-0.2 μm hole patterns," *Proc. SPIE* 3412, 1998.
45. Socha, R., et al., "Design of 200 nm, 170 nm, 140 nm DUV contact sweeper high transmission attenuating phase shift mask: experimental results part 2," *Proc. SPIE* 3679, 1999.
46. Hsu, S., "RET integration of CPL technology for random logic," *Proc. SPIE* 5377, p. 510, 2004.
47. Petersen, J., et al., "Designing dual-trench alternating phase-shift masks for 140 nm and smaller features using 248 nm KrF and 193 nm ArF lithography," *Proc. SPIE* 3412, p. 503, 1998.
48. Burkhardt, M., et al., "Illuminator design for the printing of regular contact patterns," *Microelectron. Eng.* 41, p. 91, 1998.
49. Gau, T. S., et al., "The customize illumination aperture filter for low k_1 photolithography process," *Proc. SPIE* 4000, p. 271, 2000.
50. Rosenbluth, A., "Optimum mask and source patterns to print a given shape," *Proc. SPIE* 4346, p. 486, 2001.

51. Socha, R., et al., "Contact hole reticle optimization by using interference mapping lithography," *Proc. SPIE* 5377, p. 222, 2004.
52. Flagello, D. G., et al., "Optical lithography in the sub-50 nm regime," *Proc. SPIE* 5377, p. 21, 2004.
53. Lin, B. J., "Immersion lithography and its impact on semiconductor manufacturing," *Proc. SPIE* 5377, p. 46, 2004.
54. Owa, S., et al., "Feasibility of immersion lithography," *Proc. SPIE* 5377, p. 264, 2004.
55. Smith, B., et al., "Feasibility of immersion lithography," *Proc. SPIE* 5377, p. 273, 2004.
56. Streefkerk, B., et al., "Extending optical lithography with immersion," *Proc. SPIE* 5377, p. 285, 2004.
57. Honda, T., et al., "ArF immersion lithography: critical optical issues," *Proc. SPIE* 5377, p. 319, 2004.
58. Yeung, M., et al., "Extension of the Hopkin's theory of partially coherent imaging to include thin-film interference effects," *Proc. SPIE* 1927, pp. 452–462, 1993.
59. Flagello, D. G., "Theory of high-NA imaging in homogenous thin films," *J. Opt. Soc. Am.* A13(1), p. 53, 1996.
60. Adam, K., et al., "Polarization effects in immersion lithography," *Proc. SPIE* 5377, p. 329, 2004.
61. Brueck, S., "There are no fundamental limits to optical lithography," *International Trends in Applied Optics*, A. H. Guenther, ed., SPIE Press, San Diego, CA, 2002.

Chapter

8

NGL Mask Technology Introduction

Emily Fisch Gallagher, Michael Lercel, and Dhirendra P. Mathur

It is the nature of the next generation to be beyond what is well defined. This applies to adults who wonder what children will bring to the world. It also applies to lithographers and mask makers as they survey the available technologies, weighing the relative advantages and disadvantages of implementing each in semiconductor wafer manufacturing. The purpose of this chapter is to present all of the likely options for *next generation lithography* (NGL) masks. A section will be devoted to each major NGL mask technology—157 nm, *extreme ultraviolet lithography* (EUVL), x-ray, *electron projection lithography* (EPL), and *nanoimprint lithography* (NIL). The focus of this chapter is the NGL masks themselves, not the entire lithographic technology. All of the currently accepted NGL mask contenders are included, though others may emerge at some point in the future, and a few have never been seriously embraced.

Before delving into the details of the NGL masks, it is useful to review photolithographic and industry trends. Photolithography is the process used to create microscopic, two-dimensional patterns in a photosensitive resist material. These patterns are projected replicas of a master pattern on a photomask. Historically, the photomask has been a durable, thin, patterned layer of chromium or molybdenum silicide on a transparent quartz plate. However, the next generation lithography is not limited to optical exposure systems and the standard 6025 substrate (which is a solid square of dimensions: 6 in. × 6 in. × 0.25 in. or, in metric units, 152.4 mm × 152.4 mm × 6.3 mm). Progress in the field of lithography is marked by the ability to create smaller images. The incentive for this is largely economic. Decreasing the minimum image size by a factor of two increases not only the gate speeds, but also increases the number of circuits/area that can be built by a factor of four. Consequently, aggressive rates of lithographic improvement have been built into industry roadmaps. The International Technology Roadmap for Semiconductors (ITRS) is probably the most frequently quoted source; the general trend is that the minimum wafer image size is reduced by 70 percent every two years.[1] Understanding how lithography drives the improved resolution will help to put the incentive for developing NGL mask technologies into perspective.

Extending optical lithography remains the most familiar and least disruptive choice for improving resolution on wafers. The resolution of a feature formed with lithographic stepper can be defined by the minimum linewidth $W_{\min}$ that can be resolved. For illumination at a wavelength λ and numerical aperture NA of the projection lens, $W_{\min}$ is defined by

$$W_{\min} = k_1 \frac{\lambda}{\mathrm{NA}} \tag{8.1}$$

where k_1 is constant for a given mask type and lithographic process. $W_{\min}$ is considered equivalent to half-pitch and the terms are used interchangeably.

Clearly there are several ways to improve the resolution:

- Decrease the k_1 by optimizing illumination conditions, resists, and mask types. The various techniques used to decrease k_1 are known as *resolution enhancement techniques* (RETs).
- Increase NA with advanced lens designs.
- Reduce λ by changing the illumination source. Optical lithography for chip production has already progressed from 432-nm light (G-line of Hg source) to the current 193-nm light (ArF excimer laser).

Historically, technological step-function improvements in resolution have occurred with a wavelength change. The advent of immersion steppers has created another option for dramatic improvement by effectively increasing the lens NA by introducing a high-index fluid between the lens and the wafer. Figure 8.1 shows one option for using 193-nm and EUVL to meeting the ITRS roadmap challenges for resolution targets through time.

Immersion lithography does not drive dramatic changes in mask technology, and is not treated in this chapter. Instead, the third method of improving resolution, by reducing illumination λ, will define the NGL

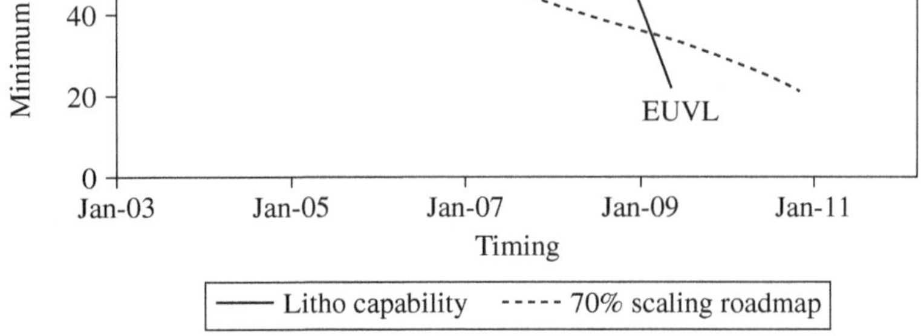

Figure 8.1 Minimum half-pitch as a function of introduction into manufacturing, assuming $k_1 = 0.4$. The dotted line indicates the target feature size through time. The solid line projects lithographic capability if only 193-nm lithography is used until EUVL becomes available. The shallow learning slope is a function of increasing k_1. The two step functions are driven by the NA increase from dry to immersion and by the wavelength change from 193 to 13.5 nm for EUVL.

mask technologies described here. While some lithographic details must be presented to explain the masks coherently, only the details relevant to mask making will be included. References will guide the reader seeking further lithographic explanation. This chapter does not rank the next generation lithography options in terms of the likelihood of their adoption into mainstream manufacturing. It does present all of the currently accepted mask options for the post-193-nm lithographic future.

8.1 157-nm Optical Lithography

The next shortest available excimer laser source after 193 nm is the fluorine laser that emits light at 157 nm. However, the more energetic 157-nm exposure wavelength introduces new challenges too. It creates transmission loss and radiation damage in standard mask materials including the pellicle. New materials must be inserted into many aspects of the mask creating a significant amount of development effort. An additional challenge is the fact that common molecular species such as oxygen, water, and carbon all absorb 157-nm radiation strongly. This affects both the environment that is appropriate for exposure and inspection of the mask and also what is defined as a defect on the mask surface. What was invisible to previous lithographic generations is now a printable defect. Consequently, actinic measurements must be preceded by a UV-ozone clean to remove the surface contamination, and exposure must be done in a carefully controlled gas environment.

However, there are some advantages as well. The 157-nm light source exists. Stepper architecture requires only evolutionary changes from existing 193 platforms. The technique is still based on optical projection, so that the RETs can be readily transferred. Maintaining the 4× magnification from wafer to mask is still possible, relaxing the mask feature tolerances and resolution requirements.[2]

8.1.1 The 157-nm mask blank

A mask blank is here defined as the ready-to-process combination of a transparent substrate and the desired absorbing films. Substrates for photomasks are typically glass or quartz. Materials appropriate for stepper lenses are usually candidates for the mask substrate since both must be transparent and must not darken after prolonged exposure to the illumination wavelength. However, the requirements of reticles are slightly different. Lifetime requirements are relaxed, but a larger area of high-quality material is required. It is also critical that the thermal expansion of the substrate be very low, or at least well-controlled through the range of temperatures a mask experiences. Additionally, the substrate must tolerate the mask process steps without degrading.

At 157 nm, fluoride crystals such as CaF_2 are the material of choice for the stepper lenses. However, CaF_2 does not work well as a substrate; it is expensive, has a thermal expansion coefficient 50 to 80 times greater than fused silica and is not compatible with all photomask processing steps. The standard 193-nm substrate material is quartz. Since the fundamental absorption edge of the Si-O-Si network is 9 eV, much higher than the F_2 laser light's 7.9 eV, it is clear that extrinsic defects are responsible for the absorption.[3] After considerable work reducing the OH concentration and introducing fluorine into the glass matrix, the quartz transmission at 157 nm was improved to a theoretical maximum of 88 percent.[4] The resultant material is called *modified fused silica* (MFS), and is the generally accepted mask substrate material.

An absorbing film or film stack will be patterned on the top of the MFS substrate to create a mask. A commonly used absorbing layer is chromium, usually with deposited oxygen and nitrogen to create an antireflective coating on the top surface. Thicknesses range between approximately 70 and 100 nm. Thinner films yield higher-resolution imaging; however, it important to maintain an optical density of ~>3 in the absorbing stack to be a sufficiently strong absorber of radiation. *Optical density* (OD), is defined by

$$\mathrm{OD} = \log_{10}\left(\frac{1}{T}\right) \tag{8.2}$$

where T is the transmittance, or the ratio of transmitted to incident light intensity. At 157 nm, chromium remains a good absorber of radiation, and the transmission does not appear to be changed after prolonged 157-nm laser irradiation.[5] So, the 193-nm lithography absorber workhorse remains suitable at 157 nm. Unfortunately the attenuating phase-shifting material used for 193-nm masks, MoSi, is not so easily transferred to 157 nm.

There are both optical and physical requirements for the attenuate PSM. Typically 5 to 30 percent of the incident radiation is transmitted, at the same time the phase of the light must be shifted by 180°. This combination of properties enhances image contrast without requiring complex data manipulation. There are many combinations of materials that can be engineered specifically for the correct phase and transmission at 157 nm. A bilayer stack has proved the most attractive solution with a metal-based attenuating layer and a phase-shifting oxide layer. Each can be tuned separately to achieve the desired attenuation and phase properties. There are several available film options, some of these are available from commercial mask blank suppliers.[6–8]

A hypothetical mask specification for a 157-nm mask blank consolidates all of the points discussed previously in tabular form, as shown in Table 8.1. There are area publications and SEMI specifications that include related definitions and measurement procedures.[9–13] The values included in the table are illustrative, specific mask applications may have different requirements.

8.1.2 157-nm mask fabrication

Mask processing does not change significantly from earlier optical technologies in most respects. There are some important deviations in the required cleans, pellicle material, and the films that were discussed in Sec. 8.1.1; however, the basic mask structures are very similar. The

TABLE 8.1 A Hypothetical Mask Specification for a 157-nm Mask Blank

Substrate	Target	Comment
Material	MFS	MFS = modified fused silica
Overall dimension	152.4 × 152.4 mm^2	
Critical area for measurements	142 mm × 142 mm	Unless otherwise noted
Flatness range (μm)	0.5 – 0.1	
transmission at 157 nm (if MFS)	>85 percent	Uniformity <±0.25 percent
Birefringence	<2 nm/cm	
Index homogeneity at 632.8 nm	<1.0 ppm	
Attenuator film		
Transmission at 157 nm	6.0	
Deviation from target within plate	±0.15 percent	
Deviation from target plate to plate	±0.30 percent	
Phase at 157 nm	180°	
Deviation within plate	±0.5°	
Deviation plate to plate	±1°	
Chrome film		
Optical density	>3	
Thickness within plate (nm)	±2	
Thickness plate to plate (nm)	±4	
Resist film		
Type	CAR	Compatible with mask process
Thickness within plate (nm)	Target ±2	No individual value exceeds range
Thickness plate to plate (nm)	Target ±3	No individual value exceeds range

chrome-on-glass (COG) mask is most identical once material modifications are made to the quartz substrate and chrome absorber.

The 157-nm alternating phase-shift mask introduces an additional change to required trench depth. A brief description of the alternating PSM mask in general will be useful. This type of mask was first proposed in the 1980s to extend the available stepper capabilities.[14] Alternating PSM masks use a chrome image bordered by alternating areas unetched (0°) quartz and quartz etched to the depth required to achieve a 180° phase shift when imaged on the wafer. Excellent resolution is achieved because as imaging light's phase goes from positive to negative, it passes through zero. The intensity (proportional to the square of the phase) is consequently forced to zero between peaks, making a very dark and sharp line on the wafer. The design is more complex and the costs are higher for alternating PSMs in general, but the resolution and process window enhancements are very attractive. for 157 nm the only adjustment to the mask relative to previous optical lithography technologies is that the trench depth must be tuned to provide a 180° phase shift at the new exposure wavelength:

$$\text{Depth} = \frac{\lambda}{2\Delta n} \tag{8.3}$$

λ is 157 nm and Δn is the difference between the index of refraction of air and that of the substrate. The mask fabrication *steps* for a 157-nm alternating PSM, shown in Fig. 8.2, are identical to those for a longer-wavelength optical mask. The materials, inspection tools, repair techniques, and cleaning requirements contain the differences.

Attenuated PSM masks replace the chrome absorber of binary masks with a material that attenuates the incident light while also shifting its phase by 180°. The mask layouts look similar, but the phase change between clear and opaque regions enhances resolution and improves process window. For 193-nm attenuated phase-shift masks and earlier lithographic generations, a single layer was used. A bilayer is required to obtain the desired attenuation (typically 6 to 20 percent) while creating a 180° phase shift. This bilayer complicates processing only slightly as can be seen in the mask process flow illustrated in Fig. 8.3. First, the resist is patterned using standard techniques, typically 50-keV electron beam exposure of the chemically amplified resist. After the resist is developed, it becomes the mask for the first *reactive ion etch* (RIE) that removes chromium at a faster rate than the resist. This step does not have to be especially selective to the underlying films since the exposed areas will soon be removed as well. The patterned chromium is then subject to a second RIE process, one that removes the attenuating film(s),

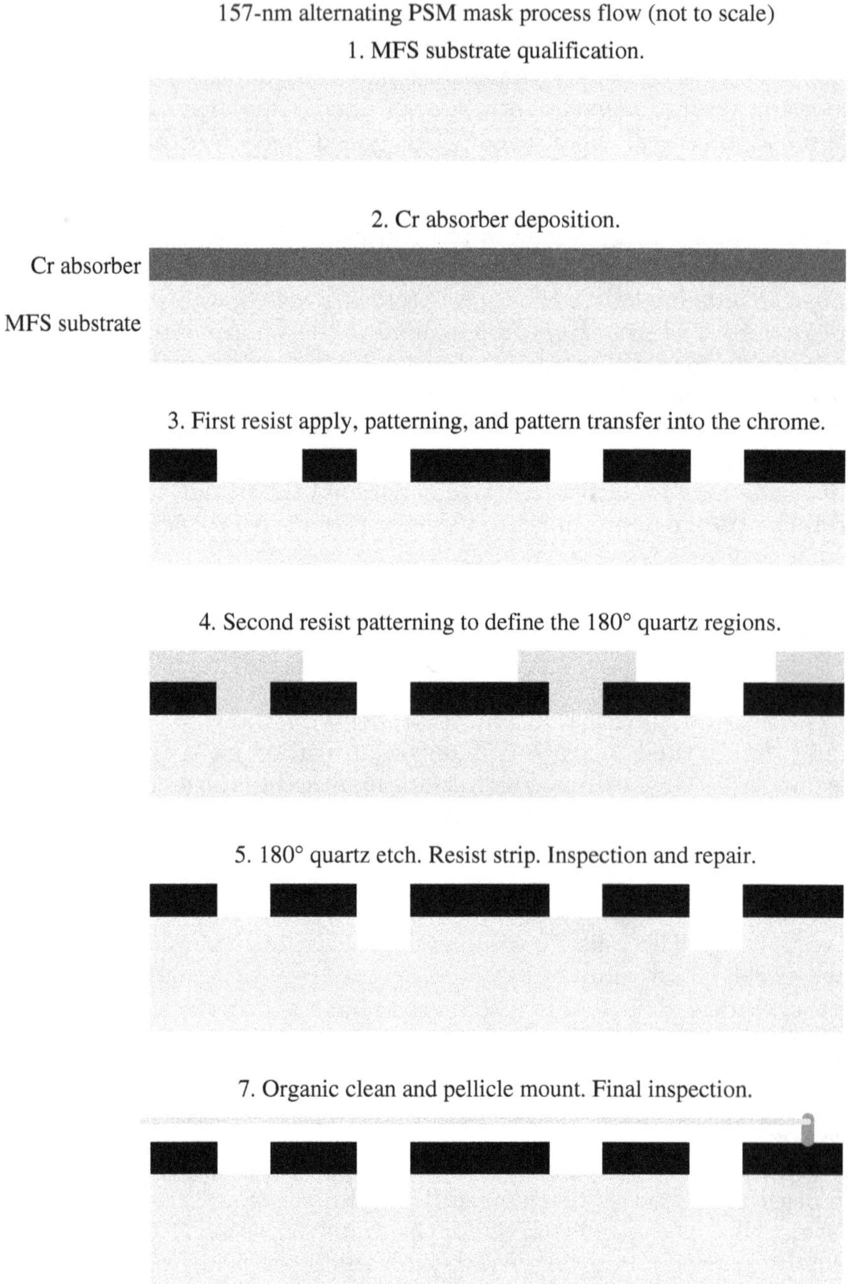

Figure 8.2 Schematic cross sections illustrate the mask fabrication steps required to build an alternating phase-shift mask.

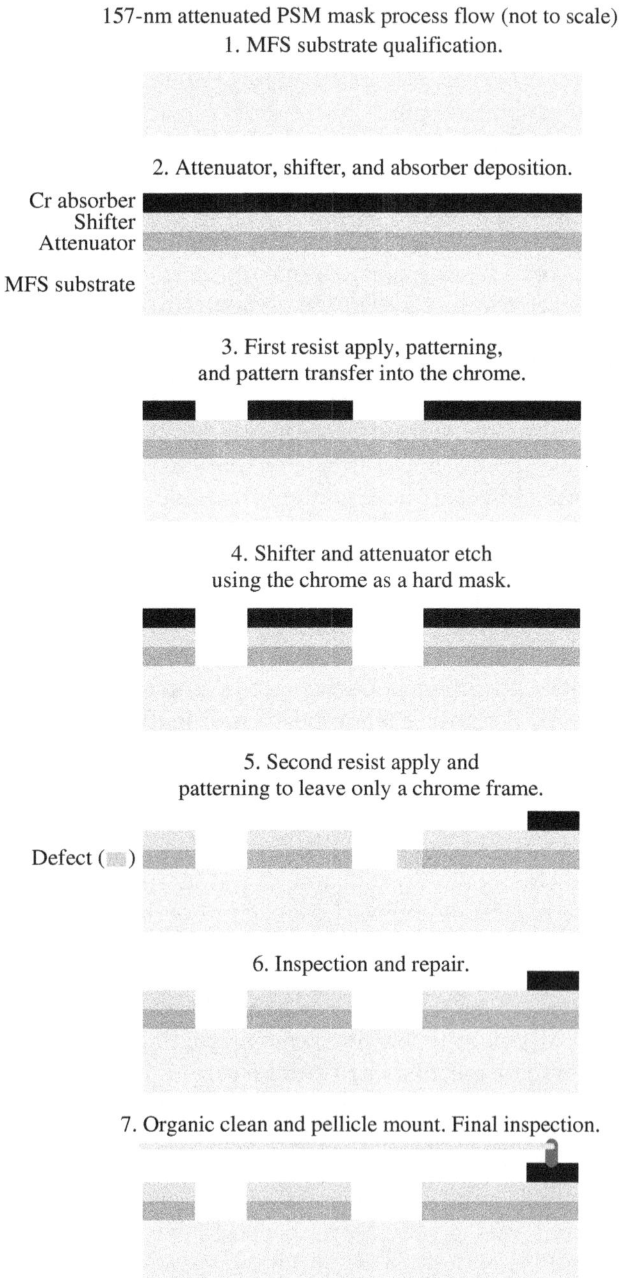

Figure 8.3 Schematic cross sections illustrate the mask fabrication steps required to build a 157-nm bilayer attenuated phase-shift mask.

but not the chromium. In some cases, a third RIE may be required to remove the attenuating layer with an etch process that is very selective to the underlying MFS. Now the substrate pattern has been transferred through to the quartz. The patterned photoresist must be stripped (if it has not already been removed–there is some flexibility in the location of this process step), the mask is cleaned, and a resist is again coated on the mask. This second patterning step is a low-resolution process to define the region just outside the patterned area that must maintain an opaque frame to reduce flare. A laser exposure in conjunction with wet chromium etch usually is a cost-effective way to remove the chromium from the chip areas. The RIE steps have been modified for the new materials associated with 157-nm masks, but no other process step has deviated from earlier optical fabrication flows.

It is generally assumed that a cleaning process can be developed to remove unwanted particles and surface films. Consequently, cleaning technology development often lags the more complex development areas like patterning and defect inspection. For 248- and 193-nm optical masks, conventional wet wafer chemistry has been successfully adapted for mask cleaning—sulfuric acid, peroxide, and ammonia solutions with and without megasonics energy. For 157-nm masks, a new clean is required to address surface organic contamination which has not been visible at longer wavelengths. Photomasks exposed to ambient air accumulate a film which is primarily hydrocarbon based and leads to transmission loss; as little as a monolayer on a modified fused silica substrate causes 1 to 2 percent of transmission loss.[15] This accumulation of partially opaque films may well introduce the largest change in the reticle manufacturing process flow. Fortunately UV light in the presence of oxygen is an effective "dry" clean of hydrocarbons. Light energy at wavelengths shorter than 175 nm breaks molecular oxygen bonds to create atomic oxygen (O) and ozone (O_3); both are highly reactive species that volatize components of the hydrocarbon film. The resultant gases can then be removed effectively with a gas purging system. Dry cleaning systems are commercially available, but do not remove particles so that more conventional wet cleaning methods are still required.

Dramatic changes must be made to all equipment that use the 157 nm exposure wavelength. This occurs in the mask house when actinic inspection, defect verification, and optical measurements are performed and also in the wafer fabricator when the mask is printed in the lithographic stepper. The OH-group that was problematic in the fused silica substrate creates similar problems when it is found between the exposure system and the mask. Environmental purging with, for example, dry N_2 is required. Gases such as H_2O and O_2 are severely restricted, typically to concentrations lower than 1 ppm each when optimal transmission is required. Additionally, only materials with very low outgassing can be used.[16] These additional environmental and cleaning requirements introduce significant cost and complexity to the mask fabrication and use of the 157-nm masks.

8.1.3 157-nm mask pellicle protection

Completed optical masks are fitted with a pellicle at the mask house to prevent particles from falling in the focal plane and printing during wafer exposure. A simple frame supports a thin, transparent film several millimeters above the patterned side of the reticle. The pellicle film must have adequate transmission at the actinic wavelength, not alter the stepper imaging of the mask, and be resistant to damage under laser irradiation. Fluorocarbon polymers less than 1 μm thick and supported by a frame several millimeters from the patterned surface have worked well at 248 and 193-nm exposures. At 157-nm exposure, the conventional polymer properties are not transparent enough and are easily damaged.[17] To date, no appropriate material has been developed.

An alternate strategy is to deploy inorganic pellicles made of a material that is known to be compatible with 157-nm radiation. An obvious candidate is the mask substrate material, modified fused silica. This type of pellicle is most commonly known as a *hard pellicle*. For this application, the fused silica must be polished much thinner than the mask blankwhile remaining flat and stress free. Gravitational forces will distort the pellicle unless it is sufficiently thick. This limits the minimum pellicle thickness to ~800 μm if the orientation is to be horizontal and 300 μm if the orientation is to be vertical. Even at these thicknesses, the pellicle becomes an optical element which must be considered in the exposure and final mask inspection systems. Extremely tight tolerances for the hard pellicle, difficult mounting and alignment procedures, and the need for antireflective coatings are all significant challenges with this option.[18]

The final option is to not use a pellicle at all. This would put the development burden on processing, handling, and storage. The industry is not eager to return to the days of unprotected reticles in manufacturing. Unfortunately, the other NGL options are unlikely to have a pellicle, so those vulnerable times will probably be revisited at some point.

Despite the various mask and stepper challenges associated with 157-nm lithography, articles proclaimed in the spring of 2002 that it would be ready ahead of other NGLs.[19] In 2003, the same news sources acknowledged that 193-nm immersion lithography is feasible.[20] Immersion lithography permits exposures at an effective NA that is greater than the theoretical limit of dry systems, 1. The extension improves resolution without changing λ. Introducing a fluid with $n > 1$ between the wafer and stepper lens is a much simpler technology tweak than the entirely new tool set and manufacturing methods required to support 157-nm lithography. While the technical hurdles associated with 157-nm development have been solved, the lithographic community may well invest their development efforts into an NGL that provides a larger resolution benefit.

8.2 Extreme Ultraviolet Lithography

The *extreme ultraviolet* (EUV) wavelength range used in lithography has photon energies around 100 eV, or wavelengths from 10 to 15 nm. In practice, 13.4-nm light is used in EUV steppers. This is at the short-wavelength limit of the extreme ultraviolet, or at the long wavelength limit of the x-ray spectrum. The proximity of EUV to x-ray provided a large initial advantage to the technology development because the multilayer interference coatings used for x-ray astronomy have very good EUV reflectivity at normal incidence. Simple optical projection lens designs are possible for the conventional 4× image size reduction.[21] Consequently, the mask tolerances are relaxed relative to 1× mask technologies, as was the case for 157-nm masks.

There are many practical difficulties associated with EUVL. The stepper optics and EUV illumination system must be entirely under vacuum to prevent absorption and scattering of the radiation. The multilayer interference mirrors are difficult to fabricate and reflect only a narrow band of radiation that varies strongly with angle of incidence. Limited reflectivity of the mirrors limits the total number of mirrors that can be used in the lens. The source power must be relatively high to compensate for the reflectivity losses in the lens and at the mask, yet high power sources for EUV radiation have been difficult to develop. Despite these challenges, enormous resources have been invested in this technology and feasibility of the technology was demonstrated by the year 2000.[22]

The 13.4-nm exposure light deployed for extreme ultraviolet lithography also introduces mask-specific changes. The strongly absorbed EUV light dictates that all optical elements, including the mask, must now be reflective. The reflective EUVL mask differs from its predecessors in many ways that will be clarified in the following sections.

8.2.1 EUVL mask blanks

The substrate and multilayer coating. Instead of a transparent quartz substrate, the EUVL mask is built on a low thermal expansion material. Substrates for EUVL need not be transparent to the EUV radiation. However, they must have an extremely low *coefficient of thermal expansion* (CTE) to minimize expansion and contraction of mask patterns due to energy absorption during writing and lithographic exposure. The best EUVL mask films reflect about 70 percent of the energy that impinges on them. The rest is absorbed and heats the substrate material. Since CTE changes with temperature, the temperature must also be carefully controlled. The upper limit of 30 ppb/K has been proposed at room temperature. Flatness must also be carefully controlled to minimize overlay errors in the exposure tool. The quality area is limited to <50 nm peak valley. *The high-order thickness variation* (HOTV) is limited to <50 nm over a spatial period equal to the side of a mask blank, 152 mm. This approximates the $\lambda/20$ rms requirement of optical photomasks at 633-nm wavelength. Flatness restrictions on local slope prevent local tilt from deflecting the incident light to an unwanted location. Surface roughness must also be minimized to enable high-quality films to be deposited on the substrate.

Defects must be limited on both front and back surfaces. Defects on the back lead to chucking errors and could become dislodged. Defects on the front side deform the multilayer (ML) deposition. Defects under the ML have a more dramatic effect than defects of the same size on the absorber.[23] A 3-nm bump on the substrate shifts the entire multilayer stack by $\sim\lambda/4$. Since the exposure light is reflected from the mask surface, this will introduce a phase shift of $\sim\lambda/2$ and would have a dramatic effect on printing. A standard exists that defines and summarized all of the EUVL substrate requirements, SEMI P37-1101.[24]

The substrate meeting the requirements is then coated with a mirror surface, for this application a multilayer or Bragg mirror that must have a high reflectivity at the exposure. The reflectivity of a single layer interface would be quite small. However, the multilayer introduces multiple reflections. It is important to tune the mask blank reflectivity both to match the band-pass wavelength of the projection optics of the stepper, and to accommodate the illumination incident angle of 5°. The multilayer of choice for EUVL masks is composed of alternating layers of Mo and Si. The characteristic reflectivity curve for 40 Si/Mo pairs at 5° is shown in Fig. 8.4. The peak reflectivity near the 13.4 nm is a key advantage of the Si/Mo choice; others are that the constituents are not health hazards, can be deposited under a range of conditions, and result in a stable structure with abrupt interfaces.[25]

The typical mirror stack consists of 81 layers, 40 Si/Mo bilayers, and a Si cap. There is no reflectivity advantage to additional bilayers since the reflectivity saturates at that point as shown in Fig. 8.5. The capping layer is typically silicon, but could be another material capable of protecting the multilayer from damage during processing and exposure to air. Alternate capping layers, such as a boride, nitride, carbide, or other material, can be added on top of the silicon capping layer to enhance the corrosion resistance without reducing the reflectivity of the multilayer.[26]

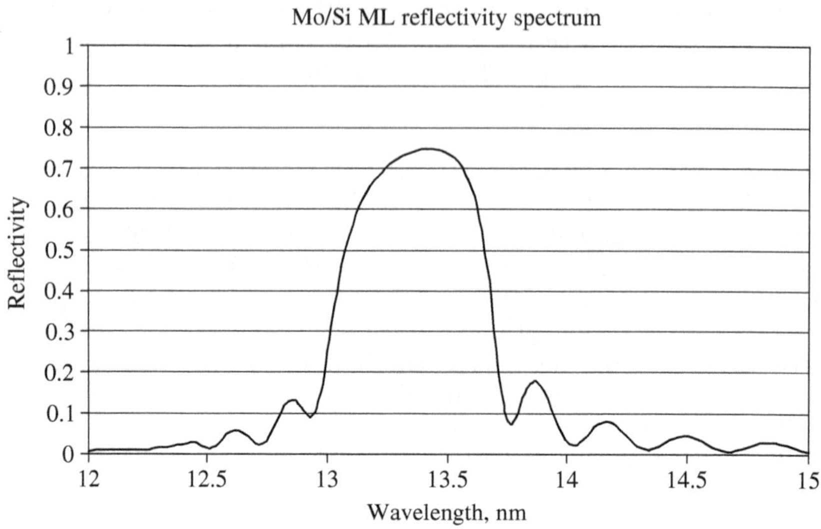

Figure 8.4 The characteristic reflectivity of 40 Si/Mo bilayers measured at 5° from normal. (Reprinted with permission from L. Kindt, IBM, unpublished data, 2005.)

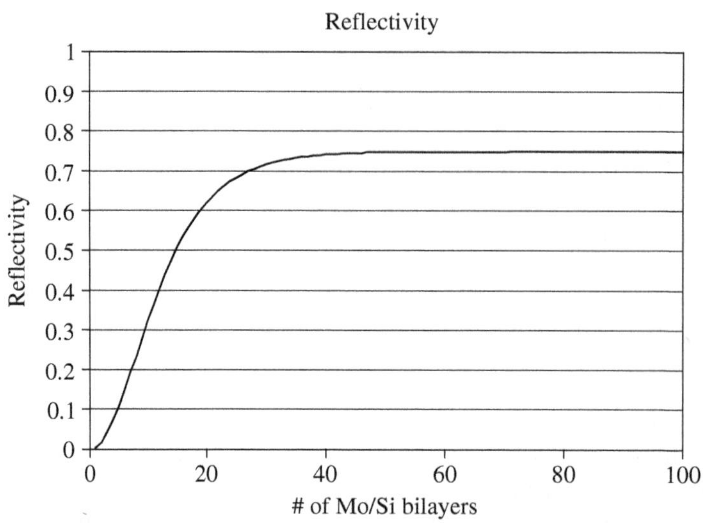

Figure 8.5 The characteristic reflectivity of Si/Mo multilayers measured as a function of the total number of bilayers measured at 13.4 nm and 5° from normal. (Reprinted with permission from L. Kindt, IBM, unpublished data, 2005.)

Alternative material pairs for the EUV Bragg mirror exist, beryllium and molybdenum is one example, although the wavelength of peak reflectivity may be different from 13.4 nm. An additional barrier layer, such as boron carbide, is sometimes added between the Si and Mo to prevent any mixing at the interfaces and thus increasing the life of the Bragg mirror.[27]

The multilayer reflectivity decreases above ~150°C. This is the result of heat-induced stress relief in the multilayer, changing layer spacing away from the optimal period for maximum reflectivity.[28] Consequently, it is important that all deposition and mask processing steps maintain low temperatures.

Mask blank repair was not required for optical masks because quartz is relatively inexpensive and because the defects were not considered repairable. The cost of EUVL mask blanks has driven efforts to reduce defects, for instance through wet cleaning, and to reduce the printability of existing defects. One method uses electron beam to heat the multilayer, causing local formation of $MoSi_2$. The silicide density is greater than that of the multilayer, so the change in surface height is reduced.[29] Another approach uses an ion beam to mill out the area in the vicinity of an embedded defect. Since the removed layers create an effective drop in reflectivity, the defect will have less contrast than would be the case on a full ML stack. The local area of reduced contrast creates an environment where the defect is less likely to print.[30] It may also be necessary to deposit a local cap to protect the now-exposed multilayer. Both repair methods would require actinic review to verify that the repair will print properly.

Additional materials. Once blank quality is confirmed, the buffer and absorber layers are deposited. Conventional EUVL masks require a buffer layer to protect the capping layer during mask processing.[31] The absorber layer is the top-most mask film required for a standard mask blank. Additional layers can be deposited anywhere within the film stack for different purposes, such as etch stops or conductive inspection/repair layers. The films must be carefully chosen. At a minimum, the mask films must be compatible with each other and with the EUVL stepper conditions, both radiation and environmental. The mask films must also demonstrate sufficient etch selectivity, low etch bias, and appropriate EUV light absorbing properties. They must be very thin to minimize the shadowing effect that occurs when the mask is exposed to EUV light at the nonnormal incidence angle (5°) that is required for reflective exposure of a mask. Many have studied the desirable properties in detail, and they are summarized in Table 8.2.[32–34] A tabular summary of desirable features of the EUVL mask absorber and buffer layers is shown in Table 8.2.

TABLE 8.2 A Tabular Summary of Desirable Features of EUVL Mask Absorber and Buffer Layers

	Absorber layer	Buffer layer
Overall description	Chosen primarily for ability to block EUVL light, but there are many other considerations	Chosen primarily to protect the Mo/Si multilayers during mask processing, but there are many other considerations
EUV properties	High absorption at 13.4 nm, good radiation stability	Good radiation stability
Physical properties	Uniform stress to avoid image placement changes after patterning; durable to chemical cleans during mask processing	Uniform stress (see absorber), capable of protecting ML during mask process, removable selective to the absorber and ML
Temperature	Must be deposited and etched at temperatures <150°C to avoid damaging ML	Must be deposited and etched at temperatures <150°C to avoid damaging ML
RIE	Initial resist pattern transfer into the absorber demands compatibility with resist and selectivity to the buffer material	Etched selective to the absorber which acts as a hard mask and must be stop on the ML cap
Inspection	High contrast to ML and low R at inspection wavelength	High reflectivity at inspection wavelength; contrast must be similar with and without the buffer layer
Repair	Material properties capable of being matched by repair deposition and removed by repair removal techniques	Must protect the silicon cap during repair (e-beam or FIB as required)

8.2.2 The standard subtractive-absorber binary EUVL mask

A binary EUVL mask consists of absorbing regions and reflective regions. The reflective regions are created by removing the absorber and corresponding portions of the buffer layer underneath. The absorbing regions are created by the portions of the absorber layer (and corresponding layers underneath) have been preserved. Together the absorbing regions and reflective regions collectively define the desired mask pattern. This *subtractive-absorber mask* type is the most common. The fabrication steps for such a mask build are shown in Fig. 8.6 and described in more detail in this section.

The mask pattern is formed by coating the absorbing stack with a resist layer and patterning it using standard mask processes. A dry etch transfers the pattern through the top absorber layer. The buffer

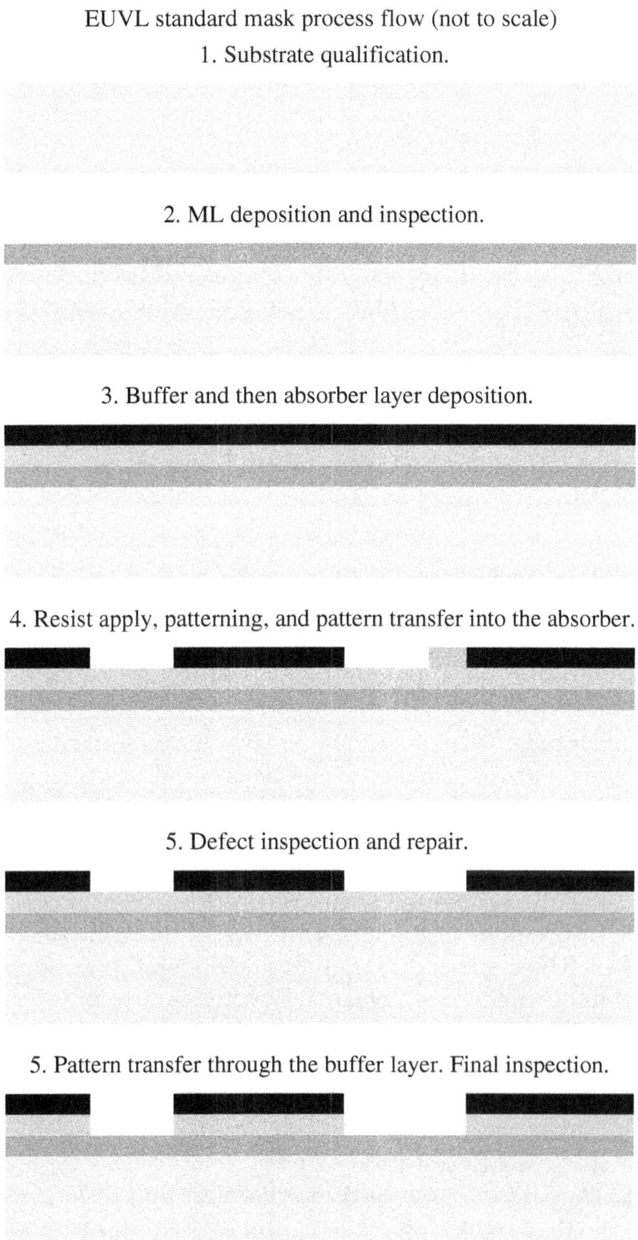

Figure 8.6 Schematic cross sections illustrate the mask fabrication steps required to build an EUVL mask.

layer material and thickness is chosen to protect the mirror from damage of all kinds—chemicals, repair processes, and particles. Inspection and repair are performed while the buffer layer is still intact. The final design is then transferred through the buffer layer to expose the reflective ML surface. The patterned absorber acts as the hard mask for the buffer layer etch. The mask is cleaned and a final inspection is performed to confirm mask quality. The material properties of the absorber stack and exposed multilayer are characterized, and the final defect inspection is performed. The resulting mask is composed of reflective regions where the ML surface has been exposed and absorbing regions where the absorbing stack remains.

The mask type described in this section is a binary. Light is completely absorbed if it impinges on the absorbing stack, or it is completely reflected if it impinges on the multilayer. However, there are other ways to fabricate the binary mask, and other types of reflective masks that can be designed for EUVL.

8.2.3 The PSM alternating EUVL mask

The alternating PSM mask is well known in its optical manifestation, and has already been reviewed in Sec. 8.1.2 for 157-nm applications. Since the EUVL mask is reflective instead of transmissive, 0° and 180° regions are now formed in the reflective ML regions instead of in the clear quartz. A step *underneath* the absorber feature may be used to create the desired phase effect.[35,36] This step can be created by etching into the low thermal expansion substrate before ML deposition. Another option is to deposit a thin film on the substrate and pattern it to leave additive steps. The deposited layer has several advantages. The thickness, or step, can be carefully controlled with known thin-film deposition practices. It is also possible to inspect and repair the thin film before depositing the ML. The step heights that give a 180° shift between incident and reflected light are defined by the following equation:

$$\text{Depth} = \frac{(2m+1)\lambda}{4\cos\theta} \tag{8.4}$$

where m is an integer, λ is the exposure wavelength, and θ is the angle of illumination. Using the 13.4-nm exposure wavelength and 5° angle of incidence, the possible depths are 3.37, 10.10, and 16.84 nm. From a film deposition perspective, thicker steps are better because the films are more uniform and depth can be more easily controlled. However, the manufacturability consideration must be balanced against mask printability. It was found that the deeper step heights (depths) associated with higher-order solutions to Eq. (8.4) introduce an image imbalance which shifts the image position, particularly at tight pitches.[37]

8.2.4 The subtractive-ML binary mask

Instead of creating absorbing features on top of the ML as was described in preceding sections, the absorbing features can be etched into the ML itself. This can be accomplished with a simple process that is shown schematically in Fig. 8.7.[38–40] The starting mask blank is very similar to the conventional, subtractive-absorber, EUVL masks. Instead of the

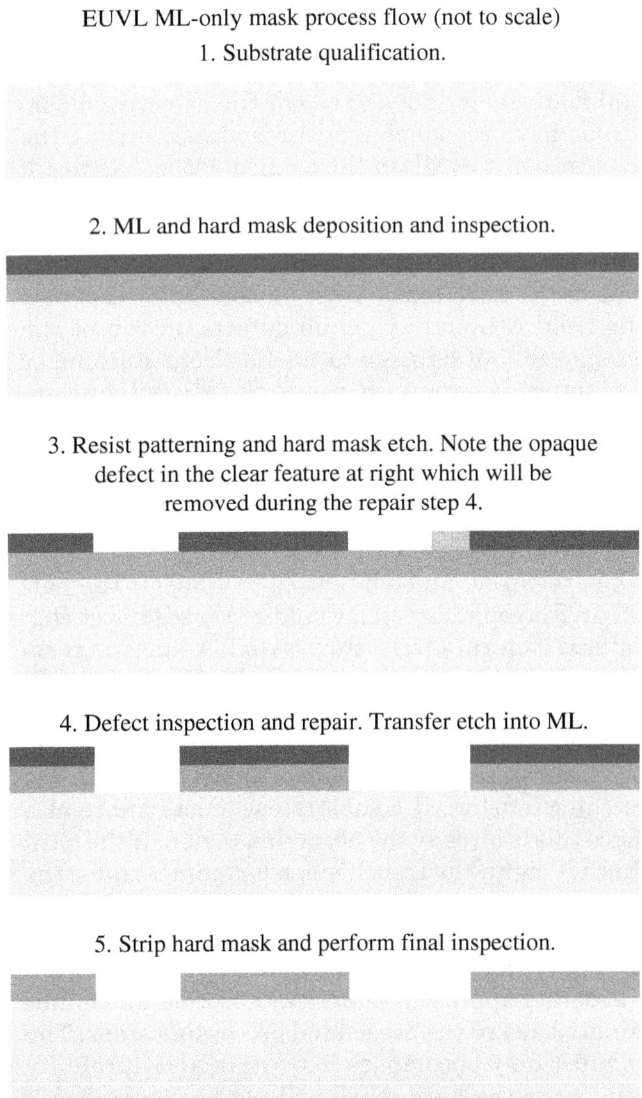

Figure 8.7 Schematic cross sections illustrate the mask fabrication steps required to build a subtractive-ML EUVL mask.

buffer and absorber layer stack, a hard mask is deposited directly on the multilayer stack. The choice of an appropriate hard mask film is important. There are two options—it can be a sacrificial layer, which will not be present on the final mask, or it can be a film that is left in place. If the hard mask is removed, the EUVL properties of the material, such as absorption and radiation durability, no longer matter. Still, there are several factors to be considered. The RIE chemistry used to pattern the hard mask must also have excellent selectivity to the underlying ML during the mask open etch so that the capping layer is not damaged should a repair be needed. Since this region of the ML will eventually be removed, surface quality is only critical where patterning errors have occurred and additional material is added to repair the defective mask. The hard mask must also have reasonable etch resistance during the transfer of the pattern through the ML in the opened areas. Finally, if necessary, there must be a selective process to strip the hard mask while leaving the ML unchanged. For this sacrificial hard mask, a thin chromium layer satisfies all of these requirements.

Inspection and repair of the hard mask must be done after it is patterned, but before the final ML etch. Opaque defects on top of the exposed ML must be removed, but damage to the ML from staining or localized heating, for example, is permissible since the ML will be completely removed in these areas. Clear defects created by the absence of hard mask material in regions must be patched using a process that does not damage the surface of the ML, such as electron beam assisted deposition. If the hard mask is sacrificial, these repairs must also be completely removed during the hard mask strip.

Once the hard mask is repaired, an etch is used to transfer the pattern into the ML stack. An isotropic, dry etch would work, but a wet etch could be used if the lateral etch rate were suppressed. A fluorine reactive ion etch chemistry is capable of etching both the Mo and the Si layers of the ML, along with offering an adequate etch rate and vertical sidewalls. The multilayer etch can be either timed or end pointed on the ultralow expansion (ULE) substrate. An adequate overetch of the ML can be allowed, since overetching into the ULE substrate will not compromise the mask properties since this is already the absorbing region. If the hard mask is transparent to EUV light, the trench mask is complete after the final ML etch and a subsequent clean. If the hard mask is sacrificial, it must be stripped after the ML etch, but before the final clean.

The subtractive-ML approach has many advantages over the conventional subtractive-absorber approach. The tricky absorber and buffer layer choices are eliminated as are the associated processing steps. The hard mask properties must only encompass inspection and durability during mask processing since the hard mask will not be a component of the final mask. Repairs are also simplified. In the nonreflective regions, the defects need only be removed if they are reflective. Repairs in the reflective ML regions can be minimized through inspection and repair of the hard mask before ML etch.

8.2.5 The damascene EUVL mask

The damascene process, used commonly in wafer semiconductor fabrication, has also been proposed for masks.[41] The damascene process flow differs from those already discussed and is shown in Fig. 8.8. The low thermal expansion substrate with multilayers is the same except that the silicon capping layer is much thicker, 60 to 100 nm. Standard pattern transfer now defines and etches the regions that are to be opaque. After the photoresist is stripped, inspection and repair is performed. If there is an extension into the etched region, this can be removed and any staining or overetch will not be problematic under what will be the absorber. If the defect is in the unetched region that will be reflective, the repair is more difficult since index-matched silicon deposition is required. Next a metal with good fill properties, such as tungsten (W), is deposited over the entire mask surface. *Chemical mechanical polishing* (CMP) is used to planarize the metal so that the silicon is again exposed in reflective regions, and the metal remains only in the absorber regions. An additional capping layer may be deposited for optimal reflectivity and durability.

The damascene mask type has been built and printed.[42] The advantage of this mask concept is that the surface of the mask is planar and consequently easier to inspect and clean and less likely to trap particles. However, the absence of standard CMP equipment and processes for the standard 6025 square mask blanks format pose a significant hurdle to wider acceptance.

8.2.6 EUVL mask pellicle protection

There is no pellicle solution for EUVL masks because all materials are too absorptive at 13.4 nm. Additionally, the pellicle would be traversed twice during exposure making an already unsolved engineering problem even harder. Attention has focused on removable pellicles that would not be in place during exposure and active cleaning within the exposure chamber. One popular concept called thermophoretic protection exploits a small temperature gradient between the mask and its surroundings to move molecules away from the mask.[43] Establishing effective mask handling, inspection, and protection protocols are among the largest EUVL mask challenges.

Good mask prototypes exist for imaging experiments. However, defect-free manufacturing will require much more focus on defects—how to inspect and repair them during mask processing and how to keep them away from completed masks.

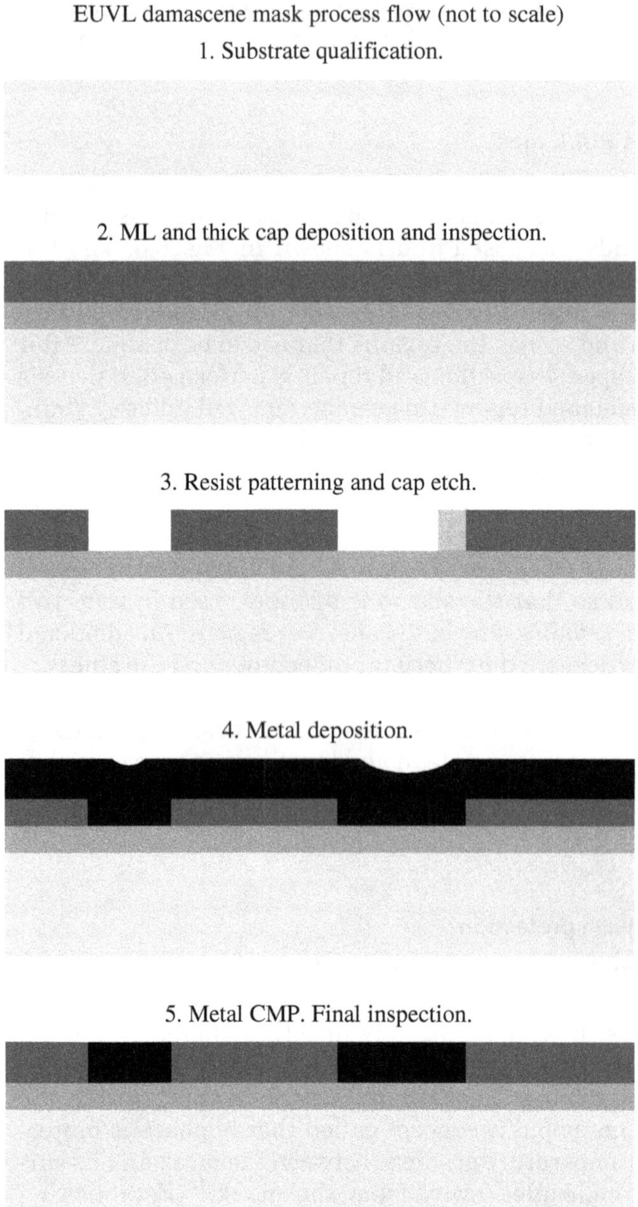

Figure 8.8 Schematic cross sections illustrate the mask fabrication steps required to build a damascene EUVL mask.

8.3 Membrane Masks Introduction

Masks for x-ray, electron, and ion beam lithography require a structure significantly different from that of conventional photomasks. The high energy of the exposing radiation interacts with all elements so that there are no truly transparent materials. All of these lithography methods use exposing radiation energies >1 keV which lie far above all valence and conduction bands. The choice of the correct materials can minimize, but never remove, the interaction of the radiation with the mask substrate. Proximity x-ray lithography typically uses ~1 nm wavelength radiation (~1200 eV energy). Electron beam projection techniques typically use 100-keV incident electrons, and electron beam proximity printing techniques operate at lower energy, but still typically 1 to 2 keV. Although the materials' choice is difficult for these exposure techniques, *proximity x-ray lithography* (PXL), EPL, *low energy electron proximity lithography* (LEEPL), and *ion projection lithography* (IPL) provide the advantage that the wavelength of the exposing energy is far smaller than the features that are being printed.

Thin membranes provide the only semitransparent structure for these lithography technologies. Two forms of membrane masks are commonly used. Stencil masks consist of a thin membrane with clear openings patterned into the membrane. Stencil mask clear regions are therefore effectively completely transparent, but a thin membrane is still required to maintain reasonable aspect ratios of the clear openings. Because the stencil mask membrane is discontinuous, certain geometries (such as donuts) cannot be patterned without the use of two complementary exposures. The other category of masks, continuous membrane masks, consists of a thin freestanding film with a pattern of absorber or scatterer material deposited on top of the membrane. Continuous membrane masks are therefore limited in transparency by the absorption of the thin film, but allow for all patterns to be formed without requiring separate, complementary exposures.

Of the NGL mask formats shown in Fig. 8.9, all are membrane masks except for EUVL (bottom left). The EPL masks on the right are either a continuous membrane with a thin scatterer (top right) or a stencil (bottom right). The membrane masks are all based on silicon wafer starting substrates.

Several common challenges exist for all membrane mask fabrication. The membrane must be kept thin to allow sufficient transparency, but a thin membrane is much less rigid than a typical optical lithography masks. Distortion caused by mask fixturing and stress relaxation must be tightly controlled. The membranes are typically defined in a much thicker substrate material (silicon wafers are common). This requires a challenging patterning and etching step that is significantly different from any existing photomask process.

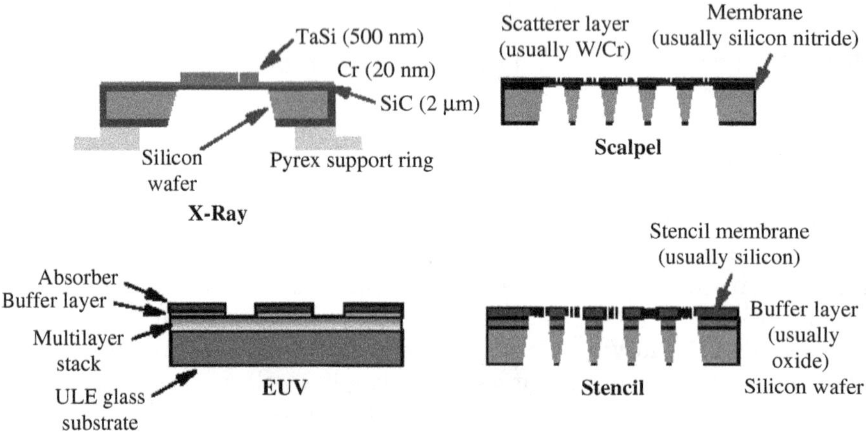

Figure 8.9 NGL mask formats except for EUVL (bottom left). The EPL masks on the right are either a continuous membrane with a thin scatterer (top right) or a stencil (bottom right).

Key requirements as defined by the ITRS roadmap for NGL masks are listed in Table 8.3. The patterning requirements are for 4× reduction. HP refers to half-pitch. All dimensions are in nanometers.[44,45]

Membrane masks are typically formed by depositing the membrane layer on top of a thick substrate. A deep etch process is used to remove the substrate material from underneath the membrane region. Typical photomask processing methods adapted to the new materials and patterning requirements are then used to define the absorber region.

The process flow for fabricating a membrane mask is shown schematically in Fig. 8.10. The membrane regions are typically defined in the substrate before the fine device patterns are formed on the front side. The schematic is for a stencil mask, but x-ray and *scattering with angular limitation projection electron lithography* (SCALPEL) mask process flows are similar, with the stencil etch replaced by that of a thin scatterer layer.

TABLE 8.3 Key Requirements as Defined by the ITRS Roadmap for NGL Masks

	100 nm	70 nm	50 nm	45 nm HP	32 nm HP	22 nm HP
Mask minimum image size	260	180	120	70	50	36
Image placement	21	15	12	8	5.5	4
CD uniformity	10	7	5	2.5	1.8	1.2
Defect size	80	55	40	35	25	15

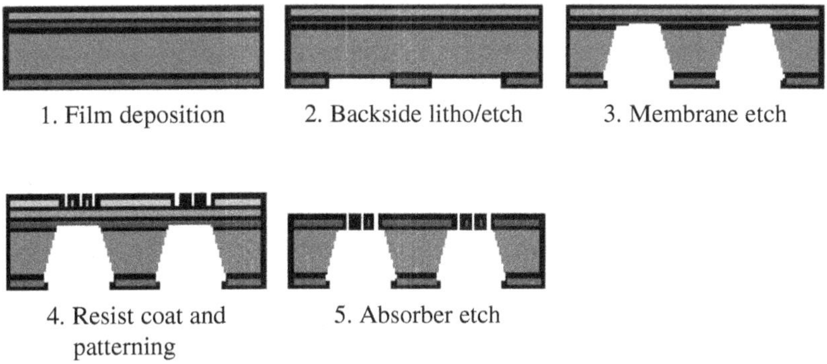

Figure 8.10 The process flow for fabricating a membrane mask.

8.3.1 X-ray lithography masks

Proximity x-ray lithography has been explored as a possible next generation lithography since the early 1980s. The fundamental concept is very simple–a beam of x-rays illuminates a membrane mask held in close proximity to a photoresist coated wafer.[46] A pattern of an absorber on the membrane defines opaque regions and the membrane itself is the clear field. The mask defines a complete field that is then stepped from field-to-field to complete the wafer exposure. Therefore the membrane region is a single and complete stepper field (typically 26 mm × 32 mm but can be larger or smaller). Because the mask is in close (~10 to 50 μm) proximity to the wafer, the pattern on the mask is the same size as that printed on the wafer. Typically this is referred to as a 1× mask referring to the lack of pattern reduction.

The x-ray mask structure is shown in Fig. 8.11, detailing typical thicknesses of layers. The membrane is ~2 μm thick SiC with an absorber ~500 nm thick of Ta-, W-, or Au-based material. The silicon wafer used to define the membranes is attached to a thick glass support ring for rigidity.

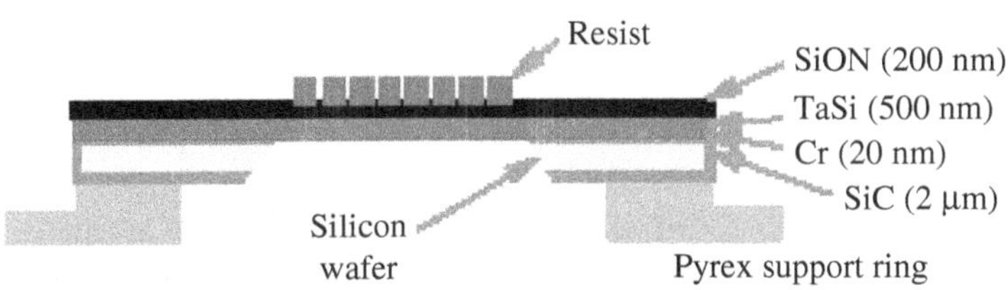

Figure 8.11 The x-ray mask structure.

PXL mask structure. The membrane thickness and material are chosen to provide sufficient transparency to the x-ray radiation while providing the maximum stiffness and durability possible. Soft x-rays in the ~1 nm absorption region are typically absorbed least by low atomic number elements, and absorbed most by high atomic number transition metal elements. Membrane material choices therefore usually include carbon or silicon. High stiffness (large Young's modulus) and good intrinsic stress control are also required, as well as durability to most common cleaning and process chemicals. Early membranes were fabricated from silicon or silicon nitride because these materials are readily available in silicon process lines. However, silicon membranes were not stiff enough, and silicon nitride stress control was insufficient to meet distortion targets for <250 nm node lithography. Therefore methods of depositing silicon carbide[47] and diamond[48] cleanly and reproducibly were developed as both of these materials provide high stiffness and excellent durability to most process chemicals. The membrane material must also be resistant to x-ray radiation damage as the mask may be used for 10,000 wafers or more with >100 fields/wafer. During the life of the x-ray mask, the total dose delivered may be >80 kJ/cm^2.[46] The method of deposition of the membrane can have a large impact on the radiation durability. Many films deposited through chemical vapor deposition incorporate residual amounts of hydrogen or oxygen which is easily liberated through high doses of ionizing radiation. The change in film composition causes stress changes and membrane relaxation which then result in unacceptable image placement distortions.

Stress-induced distortion can be calculated from finite element analysis of the stress fields and the pattern. Fundamentally the distortion is a simple application of Hooke's law

$$\Delta x = \frac{\sigma X}{E} \tag{8.5}$$

where σ is the stress, X is the width of characteristic pattern, and E is Young's modulus. From this simple relationship, increase in Young's modulus can linearly decrease the required stress to meet the same displacement criteria.

The x-ray field is approximately 26 mm × 32 mm to match other lithographic field sizes but can be larger or smaller. The membrane size should ideally match the field size to minimize patterning-induced distortions. However, the mask blanks are typically fabricated in standard sizes and stored until patterning.

The membrane is typically ~2 μm thick to provide sufficient mechanical strength and stiffness, but still allow reasonable x-ray transmission. For 2-μm thick SiC, the transmission is ~53 percent at 1-nm x-ray wavelength.[49]

TABLE 8.4 The Stiffness, or Young's Modulus, of Membrane Materials

Material	Silicon	SiC	Diamond
Young's modulus	160 GPa	~450 GPa	~900 GPa

The stiffness, or Young's modulus, of membrane materials is listed in Table 8.4 and the attenuation length of various films used in x-ray lithography at a wavelength of 1 nm is listed Table 8.5. Ta forms the bulk of many x-ray mask absorber films.

The absorber material that defines the lithographic pattern must also have controllable stress, good radiation durability, and durability to cleaning chemicals. However, it must be very highly absorbing as well as easy to pattern. The transmission in percent decreases with increasing layer thickness

$$T = e^{-d/\Lambda} \tag{8.6}$$

where d is the layer thickness and Λ is the attenuation length. Table 8.5 lists Λ for various films.

Early x-ray masks were fabricated with gold absorbers formed by electroplating gold into patterns defined in photoresists. However, most recent x-ray mask fabrication has been through reactive ion etch patterning of thin films consisting of tantalum or tungsten compounds. TaSi, Ta_4B, TaGe, and W-Ti are common x-ray absorbers.[47,50] Tantalum and tungsten have high x-ray absorption but are usually alloyed to form nanocrystalline or amorphous materials with better stress control. Often the materials are annealed after deposition to produce the correct stress values and stress gradients. These materials are easy to etch because tungsten fluorides (such as WF_6) are volatile and both tantalum chlorides and fluorides are volatile. The typical thickness of these materials is 200 to 500 nm to provide sufficient x-ray absorption. The absorber material is therefore chosen for its lithographic and patterning properties.

The substrate supporting the membrane must be rigid and durable. A 100-mm silicon wafer (~0.5 mm thick) is often used because of its compatibility with wafer process equipment. However, the silicon wafer

TABLE 8.5 The Attenuation Length of Various Films Used in X-Ray Lithography at a Wavelength of 1 nm

Material	Ta	Silicon	SiC	Diamond
Attenuation length	0.25 μm	4.9 μm	3.2 μm	3.8 μm

is still not rigid enough and distortions of the wafer shape can cause pattern placement errors. The wafer is usually bonded onto a thick Pyrex ring (7 mm thick) for purposes of shape control and mask handling.[46] The only requirement is to keep the silicon wafer from distorting so anodic bonding of the Pyrex and wafer are common as this is durable and does not suffer from degradation due to cleaning chemicals or x-ray radiation. The wafer is bonded to the ring before any patterning steps that are image placement sensitive, but the bonding ring is incompatible with standard wafer process tools unless fixturing changes are made to allow the ring to be handled by the tool.

PXL mask patterning. The completed x-ray mask blank then consists of the absorber (for example, TaSi) on the membrane (SiC) defined in a silicon wafer and bonded to a thick support ring. The absorber is then patterned using processes similar to those used for advanced photomasks. The x-ray mask blank is coated with a photoresist that is sensitive to e-beams and patterned with a mask writer. After baking and develop of the photoresist, the absorber is etched using a reactive ion etch. The mask is then inspected for image size, image placement, and defects.

Because the mask is not reduced by 4× during the wafer exposure, the mask writer for x-ray lithography must be able to define smaller features, more precisely placed than one used for a 4× exposure technique (such as optical lithography). The mask writer resolution can be achieved quite easily through research electron beam patterning tools that use single focused beams; however, these single beam tools typically suffer low throughput. Most commercial mask writers use shaped beams that vastly increase the throughput but at a cost of lower resolution and worst image placement. Good progress has been made on engineering of an electron beam writer that could simultaneously meet the image size and image placement specifications required for the 65-nm node.[50] X-ray masks require little lithographic proximity correction which reduces the number of shapes relative to the number on an advanced optical lithography mask with complex *optical proximity correction* (OPC).

Patterning on a membrane adds complication in photoresist processing. Most modern photoresists are chemically amplified referring to the activation of a photoacid generator by the incident radiation that, through a bake step, causes an amplified response in change in dissolution. For accurate lithography by any technique, good photoresist thickness control and good bake temperature control are always required. Because the membrane is much thinner than the surrounding substrate, mechanical deformation can lead to small changes in resist thickness at the edge, and thermal gradients existing during the bake step can introduce image size variations at the membrane edge.[51,52]

The reactive ion etch patterning step is similar to that used in photomask processing, but differs because of the absorber material and aspect ratio. The absorber thickness is fixed by x-ray attenuation requirements; therefore, the aspect ratio of the etched features increases with decreasing feature size. At 90-nm node dimensions, the aspect ratio is approximately 5:1 (height to width). At the 45-nm node, this ratio increases to nearly 10:1. For accurate profile and *critical dimension* (CD) control at the high aspect ratios, a thin hard mask is often used. The photoresist must only provide a barrier to pattern the thin hard mask, and then the more robust hard mask (typically SiON or CrN) provides the etch barrier for the thicker absorber film in a process very similar to some advanced gate stack patterning schemes in *complementary metal-oxide-semiconductor* (CMOS) processing. The hard mask must also have low stress to prevent image placement distortion.

An SEM micrograph of a pattern on an x-ray mask is viewed from a 30° tilt angle, as shown in Fig. 8.12. In this example, a wire terminates on a contact pad.

Aspect ratios increase with smaller device dimensions, as shown in Fig. 8.13. The x-ray absorber thickness is fixed by x-ray attenuation requirements and does not scale with smaller lithography nodes.

One of the significant challenges in patterning membrane masks is meeting the image placement requirements and this is amplified when it is a 1× mask. As described in this section, the materials are selected for good stress control but some pattern distortion occurs as the films are removed during patterning. The pattern is not always isotropic or of uniform density so the distortions vary across the field. If the pattern density and the film stress are known, the resulting distortion can be predicted with reasonable accuracy. Such methods of pattern-specific

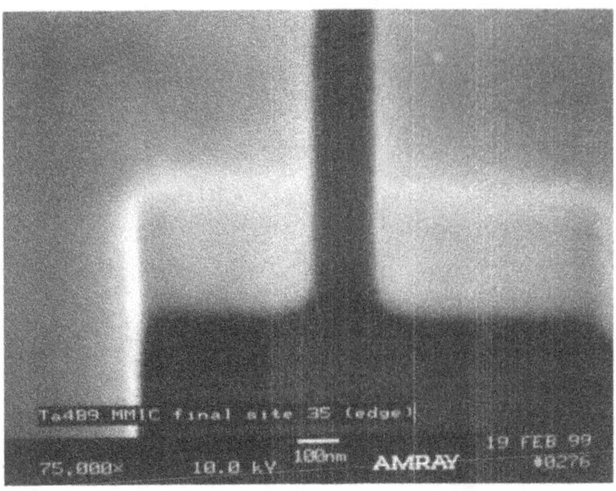

Figure 8.12 An SEM micrograph of a pattern on an x-ray mask viewed from a 30° tilt angle.

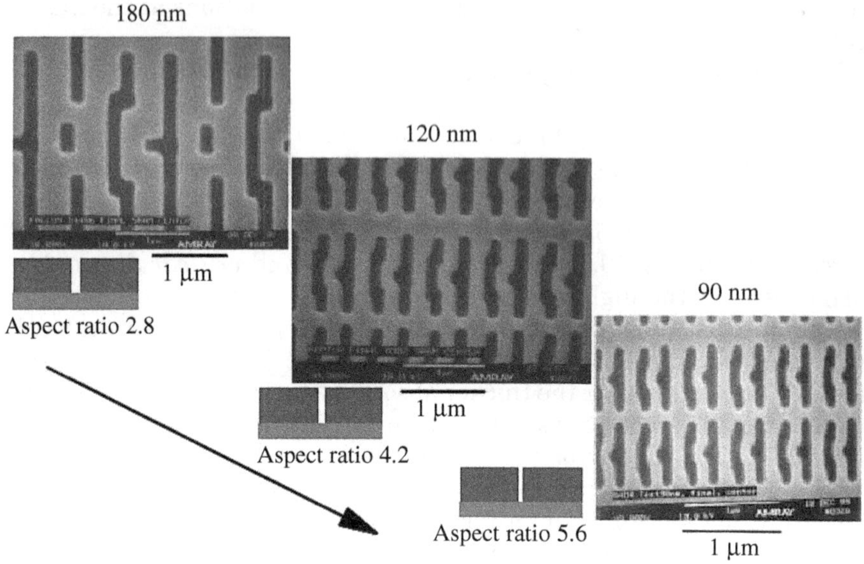

Figure 8.13 Aspect ratio increases with smaller device dimensions.

emulation have been implemented with some success.[53] Correction techniques are only useful when sufficient process control exists in film formation and patterning as random errors must be minimized. Another distortion source is generated by x-ray lithography point sources used without a collimator. The resultant pattern must also be corrected for the point source illumination. The edges of the pattern will have a magnification distortion dependent on the field size and the stand-off distance of the mask to the wafer.

Defect inspection usually demands electron beam techniques to achieve the required resolution at 1× mask dimensions. The presence of the high atomic number absorber on the low atomic number membrane provides ideal contrast for electron beam inspection and techniques from wafer fabrication, and defect control can be applied. However, such systems are often slow and are inspected by comparing one field to another, an inspection method commonly called *die:die*. When the inspection compares a die to the original database pattern it is called *die:database* and provides verification that no systematic errors are introduced during patterning. In single-die mask designs, die:database pattern inspection is the only option. Unfortunately, methods for predicting what the database pattern should look like on the mask must use e-beam rendering algorithms that are less mature than optical ones. Defect repair typically requires focused charged beams. The 1× x-ray mask dimensions actually drove the development of many of the techniques now applied to photomask repair, such as replacing ion beam with e-beam. The materials used are generally different for x-ray mask applications. For example, repair of clear defects usually uses gold or another high atomic number element for absorption.

PXL mask fabrication line. In principle, an x-ray mask fabricator looks similar to one used for photomasks. A few specialized tools are required for the blank fabrication, including membrane etch and absorber deposition, but these could be located in a blank fabrication facility. However, most of the patterning tools cannot be shared with photomask tools because different mask form factors are used, and the specifications for 1× process tools are unique. Additionally, x-ray masks must be repeatedly cleaned to remove contamination from particles or resist outgassing. Optical reticles typically have a pellicle to protect the patterns from contamination. The close proximity of the x-ray mask to the wafer during use prevents the use of traditional pellicles. Protective covers could be deployed to keep particulates from damaging the patterned absorber, but these covers must be kept defect free. This is similar to the reticle protection problem that exists for EUVL masks.[46]

Phase-shift x-ray masks. X-ray masks can be built as phase-shift masks. An additional film is applied to the back of the membrane with a thickness chosen to give 180° phase shift at the appropriate x-ray wavelength. For example, approximately 1 μm of Si_3N_4 can provide the 180° phase shift and still allow 60 percent transmission. Simulations predict that printing with the bright peak enhanced by the phase shift is then possible.[54] These masks require the additional backside phase shifter with its own characteristic etch properties. Of course, patterning this additional backside film must not cause image placement distortions.

8.3.2 Electron and ion beam lithography masks

Charged particle lithography fits into several categories. Electron projection lithography includes two related techniques—EPL and scattering with angular limitation projection electron lithography. There are only minor differences between the two system implementations. Ion projection lithography is fundamentally similar except that protons or helium ions are used instead of electrons.[55] Proximity printing is also possible in an implementation much like x-ray lithography, this is called low energy electron proximity lithography. All of these charged particle masks are membrane masks whose requirements are fundamentally the same as those used for x-ray lithography. Thin, free-standing membranes must be fabricated in a supporting substrate. The requirements for electron absorption are different, and the membrane size varies by the specifics of the lithography system design.

EPL is a system marketed by Nikon that projects small square subfields in a step and repeat sequence.[56] The subfield size is dictated by the optics design and is typically 250 μm at the wafer (1 mm at the mask with 4× reduction). The mask can be either a continuous membrane mask with scatterer layer or a stencil mask. The stencil mask is more common. SCALPEL is a technique very similar to EPL, also using 100- keV electrons.[57] The mask may also be either stencil or continuous membrane but the latter is more common. Scanned stripes of 12 mm × 1 mm (at the mask) are typically used but the mask requirements are fundamentally the same as for EPL.

Photographs of electron projection masks illustrate the subfield membranes, as shown in Fig. 8.14. During exposure the subfield position on the wafer is corrected by the electron optics to remove the struts. At left are 12 mm × 1 mm membranes for a step and stripe exposure system, such as SCALPEL. At right are 1 mm × 1 mm membranes for a step and repeat exposure system, such as the Nikon EPL system.

Electron beam projection uses high- energy electrons (100 keV) which have a very large penetration depth in most materials so complete absorption (equivalent to high optical density) is not possible. The electrons can be scattered out of the electron optical path and filtered out with a downstream aperture. The EPL and SCALPEL masks are designed to cause this amount of scattering because absorbers hundreds of microns thick are not practical for the required patterning dimensions.

In Fig. 8.15, a schematic of EPL or SCALPEL system shows the mask (stencil here) scattering electrons that hit the membrane or absorber layer. These electrons are deflected out of the electron optical path, and

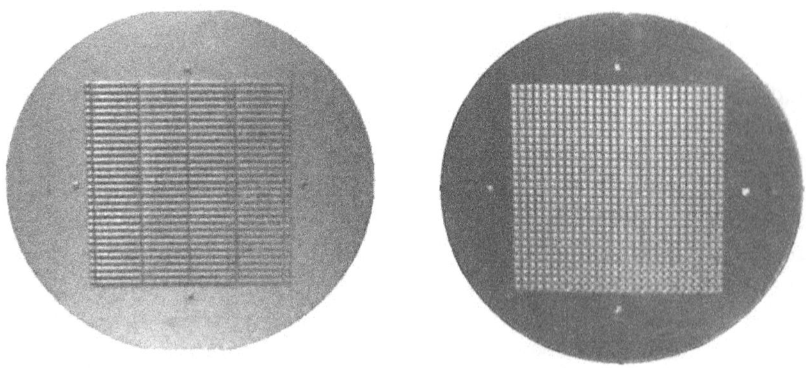

Figure 8.14 Photographs of electron projection masks.

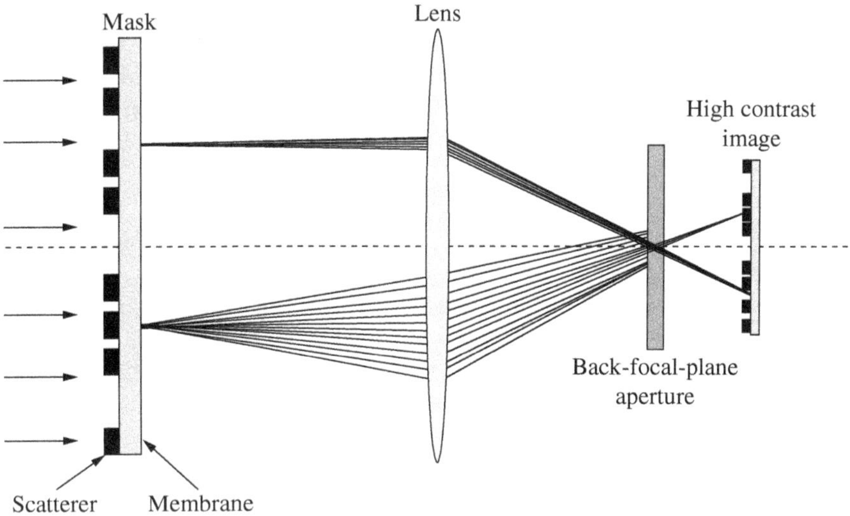

Figure 8.15 A schematic of EPL or SCALPEL system.

although some may transit the back plane aperture, the overall contrast is low. Electrons that are not deflected by the membrane are focused by the electron optics and pass through the back plane aperture to form a high-contrast image on the wafer.

A schematic of stencil and continuous-membrane scattering masks is shown in Fig. 8.16. At left, the stencil mask allows 100 percent transmission of electrons through the open slots. Electrons that strike the silicon membrane (~2 μm thick) on average lose about 1.2 percent of their total energy and are scattered far from the electron optical path. At right, the continuous membrane mask deflects a small number of electrons because it has a thin membrane (~150 mm). The patterned scatterer

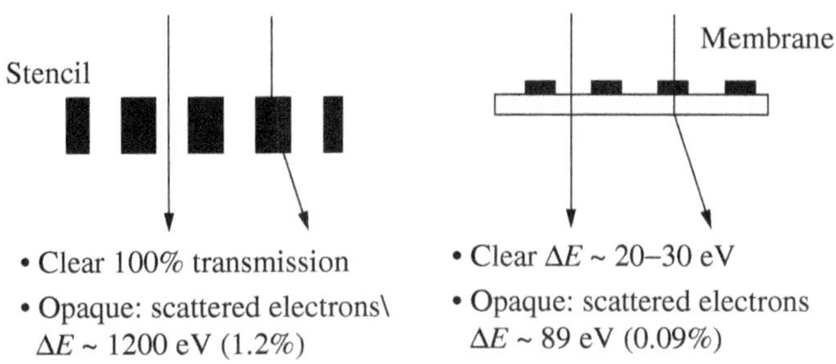

Figure 8.16 A schematic of stencil and continuous-membrane scattering masks.

layer causes the electrons to lose approximately four times more energy on average than those scattered by the membrane alone.

Electron range in materials is statistical; the distance depends on each electron's interaction with the material. A common measure of the range is the Grun range

$$R_g = \frac{0.04E_{\mathrm{el}}^{1.75}}{\rho} \tag{8.7}$$

where ρ is the density of the material and E_{el} is the electron energy (in keV). The Grun range is an approximate description of the electron penetration distance that provides a guide to density and electron energy scaling. The maximum scattering angle can be represented approximately for silicon as

$$\Theta_m = \tan^{-1}\left(\frac{140t_m^{0.75}}{E_{\mathrm{el}}^{1.75}}\right) \tag{8.8}$$

where Θ_m is in degrees and t_m is the thickness of the scatterer layer.[58] Under typical EPL conditions (100 keV, 1 μm membrane), the maximum scattering angle is about 2°. The amount of scattering is dependent on the materials properties including the atomic composition and the ability of the membrane to support plasmons.

The approximate electron range (Grun range) in silicon is plotted versus incident beam energy in Fig. 8.17. At 100 keV, the electron range is ~54 μm.

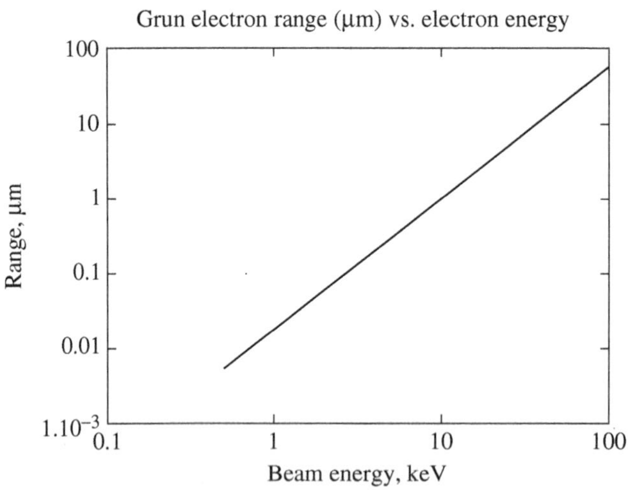

Figure 8.17 Approximate electron range (Grun range) in silicon versus incident beam energy (at 100 keV, the electron range ~54 μm).

Electron beam lenses are difficult to fabricate with large field sizes and low aberrations so EPL and SCALPEL systems print only small subfields (~250 μm at the wafer) at one time. These subfields are stitched together to make the full field. Stitching adds complexity at the subfield boundaries but has the advantage of requiring smaller membrane than those used for x-ray lithography. It is easier to control image placement distortions on smaller membranes, and struts can be included between the membranes for mechanical rigidity. Ion beam lenses have smaller aberrations, and therefore full fields can be printed. LEEPL does not involve an imaging lens so a single large membrane is required.

Membrane requirements. Stencil masks are typically one to two μm thick and can provide sufficient electron scattering without additional absorbers. Adding completely transparent clear openings in the membrane as is done on stencil masks creates excellent contrast. However, stencil masks do not allow arbitrary geometries; complementary mask pairs are required to print all geometries. The stencil membrane stress control is very critical. Removal of the membrane to form the pattern will result in image placement errors caused by stress relief. Therefore, the membrane should have uniform and low film stress and be stiff (high Young's modulus). The membrane openings should be easily formed by reactive ion etching. Silicon is a common stencil mask material because the stress can be low when controlled by doping with ion implantation and because mature silicon etch processes exist.[59,60] Diamond or diamond-like carbon deposited by CVD is another choice because of high stiffness and the existence of etch processes.

The scatterer on continuous membrane masks must only scatter the high-energy electrons from the optical path. Thin layers of high atomic number elements are sufficient, for example, <100 nm of tungsten. Because the membrane is very thin, stress control in the scatterer is critical. Either an alloy with good stress control (such as TaSi) or a bilayer tuned for stress control (tungsten with chromium) is used. The membrane requirements are similar in concept to those of proximity x-ray masks. The membrane should have low electron scattering, be stiff and resilient to patterning-induced distortion, not suffer from radiation damage, and have sufficient mechanical strength to prevent breakage. Silicon nitride is a good example and it can be deposited with low defects in conventional CMOS processing tools. Diamond and diamond-like carbon are also excellent choices but stress and defect control is more difficult than for silicon nitride deposition.[48,58]

Mask substrate fabrication. The basic process flow for continuous membrane masks is very similar to that of x-ray lithography. A mask blank is formed by defining membrane regions in a substrate, and then a thin scatterer layer is patterned by conventional mask writing and etching techniques. The structure of the mask usually consists of small membranes separated by narrow support struts. Achieving high packing density of the small membranes requires narrow struts. Although wet etches of silicon are rapid and inexpensive, etching of a (100) silicon wafer results in tapered struts that are too wide for electron projection lithography. Therefore deep silicon etches often used in MEMS fabrication have been applied to EPL mask fabrication.[61] These reactive ion etches must achieve precise sidewall control while etching ~725 μm of silicon (with the membrane packing density, this is approximately 40 percent of the silicon wafer mass) in a reasonable time. The Bosch process, named for the inventing company, is the most common technique.[62] Alternating steps of nearly isotropic high-rate etching (usually with SF_6) and deposition of thin fluorocarbon films (usually with C_4F_8) can achieve silicon etch rates >5 μm/min. An etch stop is usually required because the thin membrane can be damaged by the high rate etch. When the starting substrate is a *silicon-on-insulator* (SOI) wafer, the buried oxide provides the etch stop.

The fabrication process flow for electron projection mask blanks is shown schematically in Fig. 8.18. The unpatterned membranes are formed by removing the supporting silicon wafer in the membrane regions. The blanks can be prefabricated and then patterned as needed through the resist coat, pattern writing, and etching steps that resemble a conventional photomask process flow. The stencil mask process etches slots through the supporting membrane. The continuous membrane process defines patterns in the scatterer layer.

The membranes are defined in the substrate either before or after the fine patterns are defined. The membrane flow etches the membranes first, allowing the substrates to be fabricated in bulk and stored. This membrane flow has the advantage of discovering any membrane definition damage before the expensive mask writing step. However, the disadvantage to this order is that etching the membranes can result in image placement distortions from wafer shape distortion. The wafer flow reverses the steps so that the fine patterns are exposed and etched on the bulk wafer. This wafer flow eliminates the membrane effects discussed in the x-ray mask section.[63]

Mask patterning. The mask writing step for EPL and IPL uses electron beam writing tools that are also used in optical mask fabrication. The mask chuck must be modified to handle the different mask format. The silicon wafer is quite flexible relative to a 6025 optical mask, consequently the mounting of the mask is critical to good image placement control. The shape of the mask must be identical during the mask writing and the wafer exposure steps.[64,65] Electrostatic chucking is most

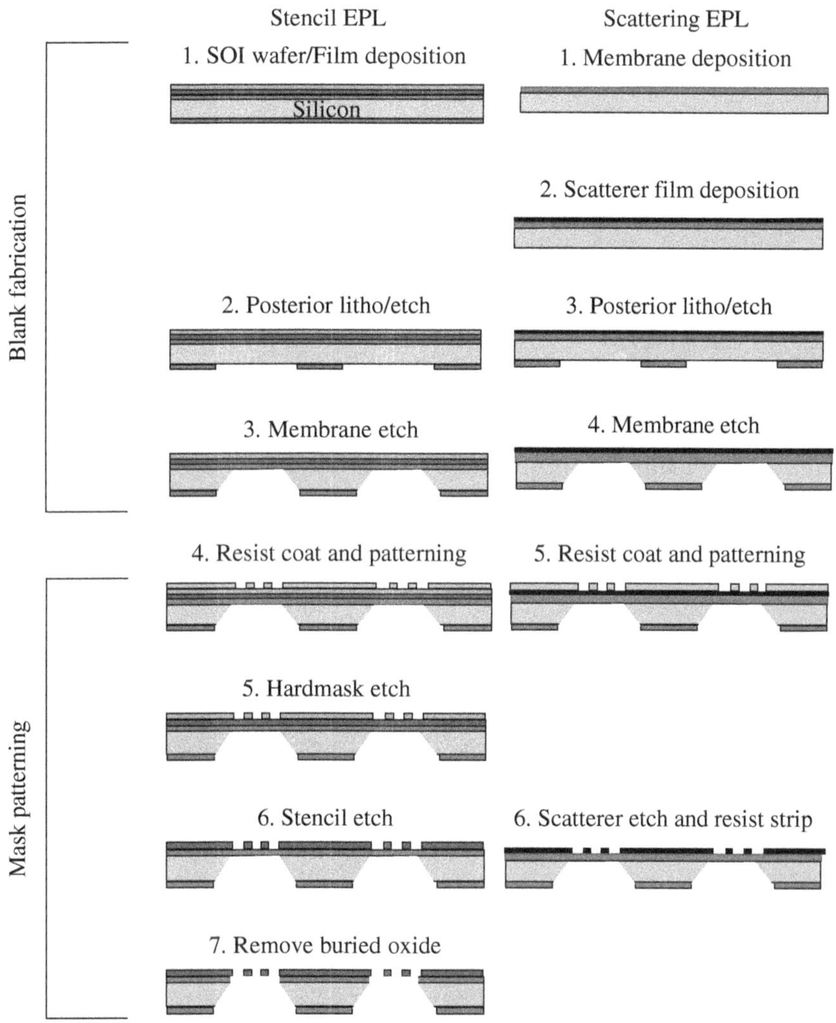

Figure 8.18 The fabrication process flow for electron projection mask blanks.

often used to hold the wafer format. The shape and location of the chucking pads and the applied force must be nearly identical on both writer and exposure tools if image placement is to be maintained.[66]

Stencil masks may also have to be split into two complementary pairs to create any arbitrary pattern. The pattern split is a complex data processing operation that identifies shapes for each complementary pair based on several criteria. Shapes incompatible with stencil masks (such as donuts and long freestanding structures) must be removed. Roughly equal pattern density and distribution between the two pairs is needed for optimal matching of the complementary masks. After the patterns are fractured, corrections for proximity effects and field-to-field stitching must be applied.[67]

Schematic illustrations of patterns fractured into complementary pairs are shown in Fig. 8.19. A donut (top left) and leaf pattern (bottom left) can be formed by the superposition of the two complementary pairs (right). Only one of many possible pattern fractures is shown. The small triangles on the line ends indicate stitching patterns that are sometimes applied to improve overlap of the complementary pairs when exposed on the wafer.

Chemically amplified resists are most often used for the mask patterning step. The thin membrane may cause resist thickness and thermal gradients during process which may affect CD control.[52,68,69] The reactive ion etch step does not have to define high aspect ratio patterns for continuous membrane masks because the scatterer layer is thin, but high selectivity to the membrane is required. Removal of any of the membrane thickness will cause transmission variations and wafer level CD control variation.

Stencil mask patterns have high aspect ratios because the membrane is ~2 μm thick and the mask minimum image size is ≤250 nm. The membrane thickness is determined by stiffness needs and electron scattering requirements that do not scale with decreasing image size. Therefore, the maximum aspect ratio increases with decreasing feature size. This is the same phenomenon that occurs in proximity x-ray masks. At the lithographic nodes of interest, high aspect ratio silicon (or diamond) stencil etches are required to transfer the resist pattern into the

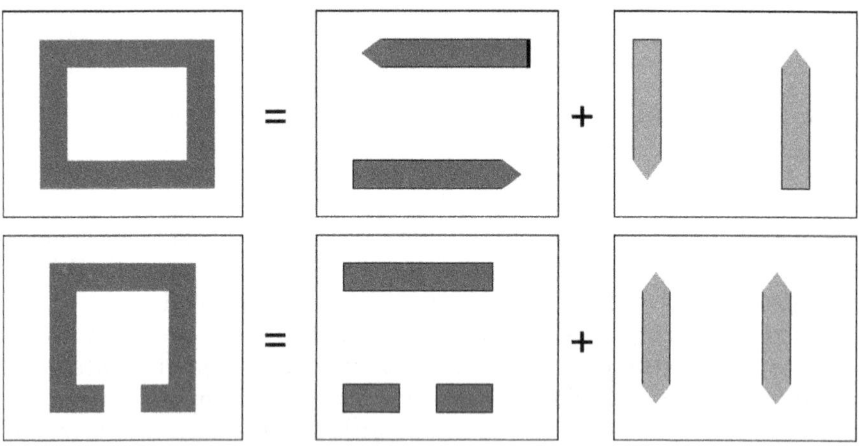

Figure 8.19 Schematic illustrations of patterns fractured into complementary pairs.

stencil membrane. Etch bias and sidewall angle must be carefully controlled across all image sizes and pattern polarities. Many CMOS silicon etches exist for both high aspect ratio features—contact holes, DRAM trenches, and gate stacks. However, these processes are optimized for one application. The Bosch multistep silicon etch is popular for silicon stencil pattern because it can perform well for both contact and line patterns.

For continuous-membrane masks, the thin scatterer layer is etched to define the patterns. The scatterer layer is typically <50 nm thick so aspect ratios are low even at the 45-nm node and beyond. However, high etch selectivity is needed so that none of the membrane is removed. Any change in membrane thickness will result in dose and image size variations.

Scanning electron microscopy (SEM) micrographs of patterns on completed electron projection masks at a tilted viewing angle are shown in Fig. 8.20. At left is a silicon stencil mask. At right is a continuous membrane mask.

As was the case for the PXL mask absorber, the residual stress in the membrane causes pattern deformations as the stencil layer or the scatterer film is etched. For stencil masks in particular, the removal of the entire membrane causes relatively large pattern distortions for small residual stress. Target values for stencil stress are <10 MPa (tensile). However, the projection capability of EPL or SCALPEL allows for small corrections to known distortions. For example, the membrane position is easily corrected by changing the stepping deflection. Rotation and magnification can be corrected with the electron optics. Distortion targets on the mask can be relaxed if the membrane distortions are both known and transferred to the scanner.[70] The scanner can only correct for distortions across the full field. Local pattern density differences cause uncorrectable within-membrane distortions. The distortions

Stencil mask Continuous membrane mask

Figure 8.20 SEM micrographs of patterns on completed electron projection masks.

depend on both the difference in pattern density and the orientation of the characteristic patterns (more distortion in the direction of lower effective stiffness).[71,72] These effects could be simulated and then compensated during the mask writer step.[73]

Other mask formats. Requiring two complementary mask pairs to create arbitrary patterns for EPL stencil masks has fueled the exploration of ultrathin continuous membrane masks with increased electron transmission. The membrane thickness must be reduced to ~40 nm or less while still providing sufficient stiffness and physical robustness. The ultrathin masks can be formed of C/Si_3N_4/C trilayer stacks. The tensile silicon nitride can balance the compressive stress of the diamond-like carbon. This type of membrane has transmitted 43 percent transmission of zero-loss electrons, ~13 times more than traditional SCALPEL masks.[74] For membranes <40 nm thick, zero-loss electron transmission of >50 percent is possible, though sufficient thickness control to maintain transmission uniformity becomes increasingly difficult as the membranes thin.[75]

Schematics of alternative EPL mask formats using thin membranes to eliminate the need for complementary mask pairs are shown in Fig. 8.21. At left is a diamond-like carbon mask with thick absorber and thin planar membrane. At right is an ultrathin membrane mask using a Si_3N_4/C membrane for structural rigidity of the silicon stencils.

LEEPL masks are similar to EPL stencil masks, both have patterns etched into the silicon membrane. However, LEEPL is a proximity technique and so the patterns are not reduced from the mask to the wafer, and remain 1×. Stencil aspect ratios are therefore higher, although not

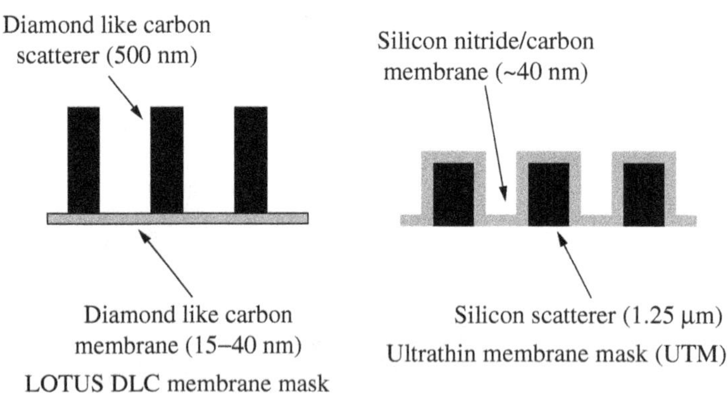

Figure 8.21 Schematics of alternative EPL mask formats using thin membranes to eliminate the need for complementary mask pairs.

always 4× because the membrane can be somewhat thinner because the exposure electrons have lower energies.[76] Because a full field exposure is used for LEEPL, pattern distortion requirements are very critical.

Defect control. Electron beam and ion beam masks can be inspected in reflection with regular optical mask inspection tools; however, these optical tools may not detect transmission defects such as defects in the bottom of stencil trenches or backside particles on continuous membrane masks.[77] Ideally electron beam inspection will be used since it is closer to an actinic inspection, although the electron energy will probably be different. Electron beam inspection has been used to successfully find buried defects in EPL masks.[78,79] The disadvantage of electron beam inspection is that it is typically much slower than optical methods, and inspection of the pattern to the original database image is more difficult. However, electron beam inspection methods are being developed for improved wafer level inspection and this technology could be easily applied to EPL mask inspection.

Electron beam masks can be repaired with standard focused ion beams with some process modification to account for the different materials. Continuous membrane masks are relatively straightforward; the repair process either removes an unwanted scatterer to repair opaque defects or adds an additional scatterer to repair clear defects. The resulting repairs must match the transmission of the standard film. Stencil masks are more challenging to repair because of the high aspect ratio of the stencil slots. Opaque defects are removed with a chemically assisted FIB process designed to etch the silicon membrane. Clear defects must be meticulously bridged by gradually building material over the defect. Electron beam repair creates less damage to clear regions and more control to opaque regions. However, the electron-assisted processes are generally slower than ion-assisted processes.[80]

The lack of transmissive materials prevents an effective pellicle solution for electron or ion lithography masks. So particulate control during the mask lifetime is also critical and mask cleaning in the fab will be required. Various methods of cleaning, including laser ablation of defects, have been explored.[81]

8.3.3 Summary

Membrane masks for x-ray or charged particle lithography have fundamentally different structures from optical masks because of the requirements of the exposing radiation. However, the methodology of patterning and producing defect-free masks from a starting blank for advanced CMOS nodes is the same as optical mask patterning. Much of the mask processing difference occurs in the blank fabrication. Once the blank is fabricated, then mask size, clamping, and etch processes need be adapted to the new format.

8.4 Templates for Imprint Lithography

The previous sections have described technologies that print smaller wafer features by shrinking the exposure wavelength, whether optically or through the use of charged particles. An alternative path to smaller wafer dimensions shifts the focus to the template (mask) and the materials used during the printing process. Imprint lithography is simply that— wafer resist is physically imprinted with a template pattern to create a 1× replica. When this is done on the nanometer scale, it is called nanoimprint lithography. This technology is not new; it has been used for over a century by artists in the form of silkscreen. Briefly, a stencil is cut from a thin film, fixed to a fine screen of silk, and ink is forced through the screen onto a piece of paper. Since the ink passes only through the cut regions, those areas are only inked and a negative image of the stencil is formed. The techniques proposed for use as an NGL retain the same basic components—a template, a chemical medium for pattern transfer, and a substrate. The materials and processes have become more complex, and the printed images can achieve sub-50-nm feature dimensions.[82] Indications are that imprint lithography resolution is only limited by the resolution of the template fabrication process. The main advantages of imprint lithography are that no source development is required and that small features can be achieved over a large area at low cost.

Imprint lithography is a general term that encompasses a family of processes and their associated templates. Since the template clearly corresponds to the mask of other NGL technologies, it will be the focus of this section. However, it is important to understand the imprint methods. The three most common imprint options are shown in Fig. 8.22.

Soft lithography covers a number of techniques that print and mold elastomer stamps based on a master template. The template can be made of silicon, metal, resist, or quartz. It is formed into the inverse of the desired pattern using conventional methods. The casting occurs when the master is coated with the elastomer polydimethylsiloxane (PDMS). PDMS is desirable because it is nontoxic, chemically inert, and moderately hard; however, other materials could be used. After the PDMS has solidified, it is removed and the result is a soft template.[83] Since PDMS is soft, it is not well-suited for devices requiring precise image placement, very small image sizes, or physical durability. However, the soft mask can assist in the formation of self-assembled monolayers that do not require any heat or light curing. This additive technique is called *microcontact printing* (μCP) and has been preferred for biological applications.[84]

Another general class of imprint lithography is nanoimprint lithography. NIL uses a solid template or mold made from, for example, silicon

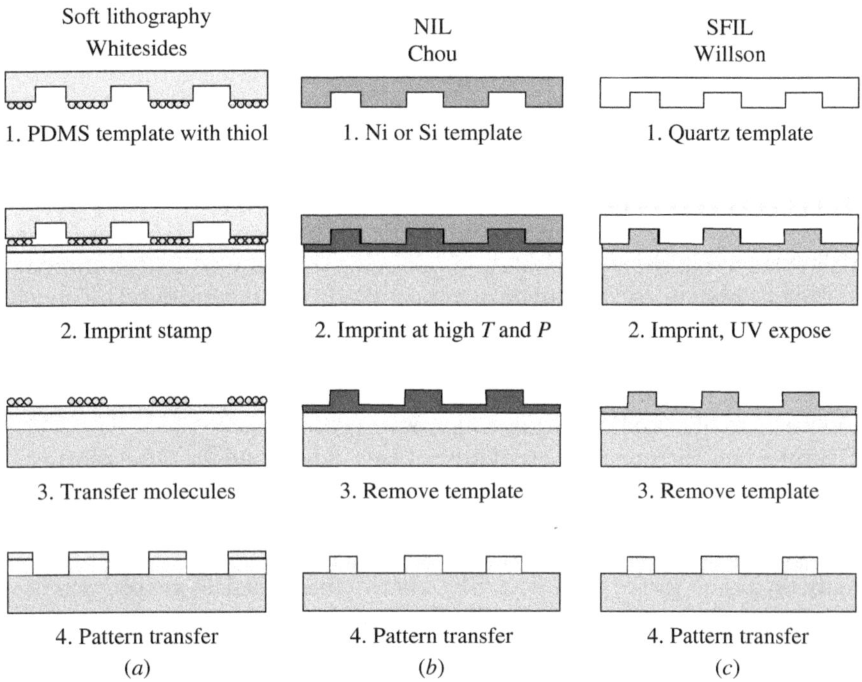

Figure 8.22 Three nanoimprint lithography process flows are shown: (*a*) soft lithography; (*b*) NIL; (*c*) SFIL (Reprinted with permission from Resnick, D. J., et al., "Imprint lithography: lab curiosity or the real nGL?," *Proc. SPIE* 5037, p. 12, 2003)

or silicon dioxide, with nanostructures on its surface. The wafer is coated with a resist that can be a thermal plastic, curable polymer, or other active material. The template is placed in physical contact with the wafer resist, and uniform pressure is applied to deform the resist into the negative of the template. If the resist is a thermal plastic resist, it is heated above its glass transition temperature during imprint and cooled before separation from the mold. This type of NIL is sometimes called heat embossed NIL, (HE NIL). If the resist is UV curable, it is cured while still in contact with the mold, before separation. This variation is called UV NIL.[85] A blanket RIE etch removes a layer of the wafer resist, eliminating the thinner stamp indentations completely.[85] The resultant wafer resist pattern is the inverse of the template. Features as small as 10 nm have been imaged using NIL.[86]

The final imprint lithography type is very similar to UV NIL, but specifies that the template is stepped repeatedly across the wafer. It is *step and flash imprint lithography* (SFIL).[87] The term *step and stamp imprint lithography* (SSIL) has also been used to describe the same technique. The distinguishing features are that a transparent rigid template is used and that the pattern transfer material is a low-viscosity, photocurable organic etch barrier. The etch barrier is spun onto the wafer, on top of the organic transfer layer. The template is placed in contact with the wafer, deforming the transfer layer and trapping the etch barrier in the template relief. UV light cross-links the polymer through the backside of the template. The template is then removed and stepped across the wafer to the next location so that the process can be repeated to cover the desired wafer area. After a brief barrier-layer etch, a low-aspect-ratio hard mask is formed. A second etch transfers this pattern through the transfer layer which can have high aspect ratios if desired. SFIL is well suited to lithographic applications. The photopolymer flows easily through the complex template designs. An additional advantage is that wafers are unlikely to be deformed by the minimal pressure and temperature gradients required. A final advantage is that the use of UV light during the cure facilitates the use of standard wafer alignment processes.[88]

Despite the different uses of templates described in this section, there are similarities. The templates are solid. The template features must be of the same dimensions as the desired wafer dimensions, although the aspect ratios can change during transfer. The template size dictates the wafer field size. SFIL is the imprint lithography variation most likely to be used as an NGL for wafer applications, so a compatible template is described in the following sections. Accordingly, a transparent substrate process flow will be used to discuss imprint templates in some detail.

8.4.1 Template fabrication

The templates for imprint lithography are 1×. Consequently, image tolerances will be correspondingly tighter for the template than for 4× masks at the same technology node. However, the membrane challenges of the x-ray and LEEPL technologies are avoided. Templates can be fabricated using a process very similar to optical alternating PSMs. Differences typically favor the ease of the template fabrication:

Since the pattern is physically transferred, no optical proximity correction tricks are required to create the desired wafer image. This simplifies the data submission process, reduces the data volumes and shortens write times.

Imprint systems use field sizes that are smaller than conventional steppers. This shortens the pattern generation time, decreases data volumes further, and improves process yields. The current sizes are small enough to allow multidie templates so that die:die inspections are possible and redundancy can be invoked.

No pellicle can be used. This is generally a technology disadvantage, but simplifies the template manufacturing process.

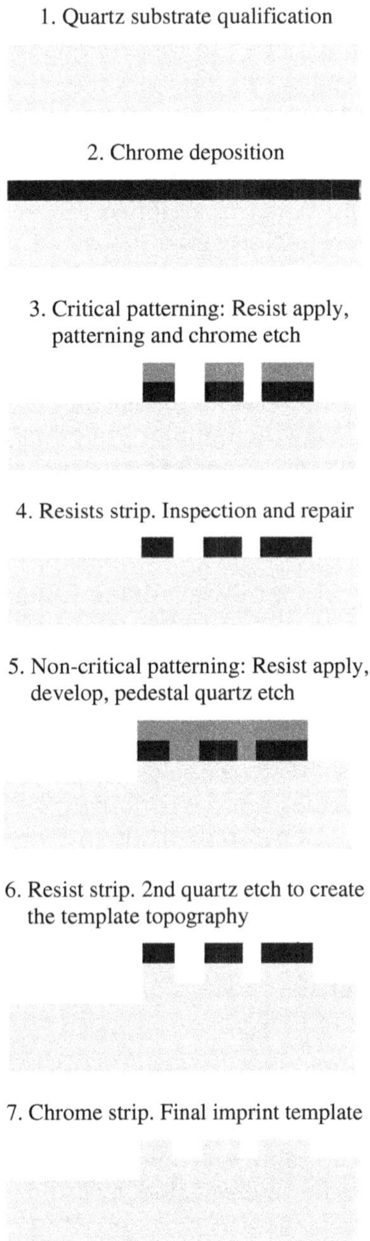

Figure 8.23 Schematic cross sections illustrate the mask fabrication steps required to build an imprint lithography template.

The process steps required to build a template are shown in Fig. 8.23. A standard COG blank is patterned to define both the fine template features and the pedestal region. This first patterning step is the one that defines the template's CD character and resolution and is typically an e-beam mask write. A second layer of resist blocks the template field during the pedestal etch. This is not a critical patterning step, and standard laser processes are sufficient. The resist is stripped and a second quartz etc transfers the fine chrome pattern into the quartz and, parasitically, deepens the pedestal or moat region. Finally the chrome is stripped. If the stepper requires a template smaller than the mask blank size, it can be diced using a variety of methods such as laser or precision saw.[89]

The template's role is twofold. It primarily imparts its topography to a wafer transfer layer. It is also imperative that the template releases the transfer layer after cure instead of sticking to it. A surface treatment is required to facilitate this release. Creating a lower surface energy on the template than on the transfer layer mediates the release of the template. The release properties must be maintained, even after thousands of imprints in a manufacturing process. Fortunately, available films appear to provide the required physical characteristics and durability.[90]

8.4.2 Remaining challenges in fabricating templates

There are many challenges associated with the SFIL technology that are beyond the scope of templates. Still, some issues are particular to the formation and will require significant progress before SFIL manufacturing can be a reality:

1. *Pattern generation*: 1× technologies always place an imaging burden on the mask patterning process. An additional challenge with the templates is the fidelity of the pattern transfer process. Any written defect will be transferred to the wafer

2. *Metrology and inspection*: Electron beams must be used for both measurements and defect inspection to achieve the required resolution. Since the fused silica template is an insulator, electrons will generate a static charge that can destroy the template imaging. It is possible to address this issue from the environmental side, by building e-beam systems with excellent charge control. A quicker path is to change the template itself. Incorporating a conductive, transparent film such as *indium-tin-oxide* (ITO) that can become part of the final template is one solution. Another option is to coat the template with a thin conductive layer, such as Cu or chrome, which would be removed after the measurements and inspection. This is less attractive

than the ITO solution because no final inspection is possible after the conductive film is removed.

3. *Repair*: Because the entire template is quartz, any defects are likely to be difficult to find and difficult to repair. Conventional defect terminology does not apply to templates. The defects are primarily pits where the quartz is supposed to be full-height, or quartz remaining where it should have been removed. The first class of defects, pits, are unlikely to cause issues if they are very small. Larger pits could be eliminated with resists and chrome inspection and repair steps. The unwanted full-or partial-height quartz is more problematic. Nanomachining techniques can scrape off large defects, but tool control is not yet sufficient for 45-nm node 1× applications.
4. *Template handling and protection*: The physical nature of imprint lithography precludes the use of a pellicle to protect the template. This moves the additional inspection burden to the wafer fabricator and is likely to increase the transfer of templates between the template and wafer fabrication sites.

8.4.3 Imprint lithography summary

The renewed interest in imprint technology for wafer applications has generated a wealth of publications on the solutions to the technical challenges. It is a technology that touches many disciplines and therefore draws on a wealth of intellectual and financial interest. It is important to note that there is growing support and development by commercial tool suppliers.[91] While imprint lithography is a relatively new player in NGL circles, it is generating considerable interest.

8.5 NGL Mask Technology Summary

The introduction to this chapter promised that no NGL technology winner would be declared. That promise will be kept. In the end, there are many factors that will determine which NGL is used. Understanding the future wafer requirements is critical. There may be different solutions for different pattern types (contacts versus lines) or for different designs (logic versus memory). Another consideration will be the relative advancement of the technologies. If one is ready for manufacturing first, it will clearly have an advantage. Finally the costs must be weighed. There are many possible scenarios. If the exposure source for one technology is astronomically expensive, industry focus may shift to an NGL that eliminates that concern entirely (imprint lithography). If mask costs rise too high for low-volume logic applications, maskless lithography may eliminate the need to order any mask at all. It is also possible that the pace of feature shrinks will slow dramatically to offset costs and another NGL will emerge. Ultimately the semiconductor market, the technology costs, and the wafer application will play a key role in selecting an NGL. These considerations are likely to dominate the NGL technological challenges discussed in this chapter.

References

1. *The International Technology Roadmap for Semiconductors*, 2004 Edition.
2. Hector, S., "Mask magnification: 8X over 4X? not yet," *Microlithography World*, 13(4) November, 2004.
3. Ikuta,Y., et al., "Defect formation and structural alternation in modified SiO_2 glasses by irradiation with F_2 laser of ArF excimer laser," *J. Vac. Sci. Technol.* B18, p. 6, 2000.
4. Smith, C. M, and L. A., Moore, *Proc. SPIE* 3676, p. 834 1999.
5. Bates, A. K. et al., "Review of technology for 157-nm lithography," *IBM J. Res. Dev.* 45, p. 5, 2001.
6. Becker, H., et al., "Development of a new PSM film system for 157-nm extensible to high-transmission 193 nm lithography," *Proc. SPIE* 5256, p. 204, 2003.
7. Nozawa, O., et al., "Development of attenuating PSM shifter for F_2 and high transmission ArF lithography," *Proc. SPIE* 5130, p. 39, 2003.
8. Carcia, P. F., et al., "Thin films for phase-shift masks," *Vacuum and Thin Film*, HIS Publishing Group, Libertyville, IL, p. 14, 1999.
9. Walton, R., "Photoblanks for advanced lithography based on Corning high-purity fused silica (HPFS)," *Proc. SPIE* 5040, p. 1062, 2003.
10. SEMI P1-1101, Specification for hard surface photomask substrates.
11. SEMI P2-0298, Specification for chrome thin films for hard surface photomasks.
12. SEMI P3-0298, Specification for photoresist/e-beam resist for hard surface photoplates.
13. SEMI P29-0997, Guideline for description of characteristics specific to halftone/attenuated phase shift masks and mask blanks.
14. Levenson, M. D., N. S.,Viswanathan, and R. A. Simpson, "Improving resolution in photolithography with a phase-shifting mask," *IEEE Trans. Electron Devices* ED-29(12), p. 1812, 1982.
15. Bloomstein, T. M., et al., "Laser cleaning of optical elements in 157 nm lithography," *Proc. SPIE* 4000, p. 1537, 2000.
16. Okoroanyanwy, U., et al., "Experimental investigation of fabrication process-, transportation-, storage-, and handling-induced contamination of 157 nm reticles and vacuum-UV cleaning," *Proc. SPIE* 5377, p. 272, 2004.
17. Rothschild, M., et al., *J. Vac. Sci. Technol.* B 17, p. 3262, 1999.
18. Bruls, R., et al., "Hard pellicle investigation for 157-nm lithography: impact on overlay" *Proc. SPIE* 5504, p. 1, 2004.
19. "157-nm lithography will be ready when needed, concludes Sematech meeting," *Silicon Strategies*, May 15, 2002, 8:48 AM EST.
20. Clark, P., "Industry scrambles to embrace immersion lithography," *Silicon Strategies*, Dec. 22, 2003, 10:42 AM EDT.
21. Jewell, T. E., "Optical system design issues in development of projection camera for EUV lithography," *Proc. SPIE* 2437, p. 340, 1995.
22. Tichenor, D. A., et al., "EUV engineering test stand," *Proc. SPIE* 3997, p. 48, 2000.
23. Lin, Y., and J. Bokor, "Minimum critical defects in extreme-ultraviolet lithography masks," *J. Vac. Sci. Technol.* B15, p. 2467, 1997.
24. SEMI P37-1101, Specification for extreme ultraviolet lithography mask substrates, http://www.semi.org.
25. Barbee, T. W. Jr., S. Mrowka, and M. C. Hettrick, "Molybdenum-silicon multilayer mirrors for the extreme ultraviolet," *Appl. Opt.* 24 pp. 883–886, 1985.

26. Montcalm, C., et al., "Passivating overcoat bilayer for multilayer reflective coatings for extreme ultraviolet lithography," *US Patent* 5,958,605, 1999.
27. Bajt, S., et al., "Improved reflectance and stability of Mo/Si multilayers," *SPIE* 4506, p. 65, 2001.
28. Mirkarimi, P. B., and C. Montcalm, "Advances in the reduction and compensation of film stress in high-reflectance multilayer coatings for extreme ultraviolet lithography," *Proc. SPIE* 3331, p. 122, 1998.
29. Mirkarimi, P. B., et al., "Advances in multilayer defect planarization in masks for extreme ultraviolet lithography," *Proc. SPIE* 4688, 2002.
30. Barty, A. et al., "Method for repairing Mo/Si multilayer thin film phase defects in reticles for extreme ultraviolet lithography," *J. Appl. Phys.* 91, p. 81, 2002.
31. Kleineberg, U., et al., "Bufferlayer and caplayer engineering of Mo/Si EUVL multilayer mirrors," *Proc SPIE* 4506, p. 113, 2001.
32. Yan, P-y, et al., "EUV mask absorber characterization and selection," *SPIE* 4066, p. 116, 2000.
33. Hector, S. D., EUVL masks: requirements and potential solutions," *SPIE* 4688, p. 134, 2002.
34. Racette, K., et al., "Correlation of absorber film properties with patterned film characteristics on square format extreme ultraviolet lithography (EUVL) masks," *First International EUVL Symposium*, 2002.
35. Kubaic G. D., and D. R. Kania, eds., "OSA trends in optics photonics vol. 4," *Extreme Ultraviolet Lithography,* Optical Society of America, Washington, DC, 1996.
36. Sugawara, S., A. Chiba, and I. Nishiyama, et al., "Phase shift mask in EUV lithography," *Proc of SPIE* 5037, p. 850, 2003.
37. Yan, P-y, "EUVL alternating phase shift mask imaging evaluation," *SPIE* 4889, p. 1099, 2002.
38. Fisch, E. E., et al., patent application US24009408A1, 2004.
39. Schwarzl, S., "Method for fabricating a lithographic reflection mask in particular for the patterning of a semiconductor wafer, and a reflection mask," US Patent Application # 2002-0192571, 2002.
40. Tennant, D. M. et al., "Reflective mask technologies and imaging results in soft x-ray projection lithography," *J. Vac. Sci. Technol.*, B9, p. 6, 1991.
41. Troccolo, P., "A method of fabricating reflective masks," *US Patent* 5,795,684, 1998.
42. Yan, P-y, "EUV mask patterning approaches," *SPIE* 3676, p. 309, 1999.
43. Klebanoff, L. E., et al., "Verification studies of thermophoretic protection of EUV masks," *Proc. SPIE* 4688, p. 182, 2002.
44. Semiconductor Industry Association, *International Technology Roadmap for Semiconductors*, International Sematech, Austin, Tex., 2004.
45. Semiconductor Industry Association, *International Technology Roadmap for Semiconductors*, International Sematech, Austin, Tex., 2001.
46. Silverman, J. P., "Proximity X-Ray Lithography," *Sematech NGL Workshop*, 1998.
47. Brooks, C. J. et al., "Advanced refractory-metal and process technology for the fabrication of x-ray masks," *Proc. SPIE* 3676, p. 14, 1998.
48. Brooks, C. J. et al., "Characteristics of CVD diamond films for x-ray mask substrates," *J. Vac. Sci. Technol.* B17, 3144–3148, 1999.
49. www.cxro.lbl.gov
50. Tsuboi, S. et al., "Recent progress in 1X x-ray mask technology: Feasibility study using ASET-NIST format TaXN x-ray masks with 100 nm rule 4 GBit dynamic random access memory test patterns," *J. Vac. Sci. Technol.* B19, p. 2416, 2001.
51. Lercel, M. J. et al., "Etching processes and characteristics for the fabrication of refractory x-ray masks," *J. Vac. Sci. Technol.* B16, p. 3577, 1998.
52. Lercel, M., and C. Magg, "Image size control in next generation lithography masks," *Proc. SPIE* 3997, p. 266, 2000.
53. Puisto, D., M. Lawliss, and M. Sturans, "Overlay enhancement with product-specific emulation in electron beam lithography tools," *J. Vac. Sci. Technol.* B12, p. 3436, 1994.
54. Taylor, J. W. et al., "Modeling for sub-50 nm x-ray applications with phase masks," *Proc. SPIE* 5374, p. 311, 2004.

55. IPL Project Group, "White paper on ion projection lithography," *Sematech NGL Workshop*, Colorado Springs, 1998.
56. Liddle, J. A., et al., "SCALPEL: a projection electron-beam approach to sub-optical lithography," *Sematech NGL Workshop*, 1998.
57. Suzuki, K., "EPL technology development," *Proc. SPIE* 4754, p. 775, 2002.
58. Yamashita, H., et al., "Mask electron modeling for Coulomb interaction by mask-scattered electrons in electron-beam projection lithography," *J. Vac. Sci. Technol.* B16, p. 3227, 1998.
59. Degen, A., et al., "Stress engineering of SOI silicon stencil masks boron doping concentration," *Proc. SPIE* 3997, p. 395, 2000.
60. Eguchi, H., et al., "Low-stress stencil masks using a doping method," *Proc. SPIE* 5256, p. 871, 2003.
61. Dauksher, W. J., et al., "Modeling and development of a deep silicon etch process for 200 mm election projection lithography mask," *J. Vac. Sci. Technol.* B19, p. 2921, 2001.
62. Laermer, T., and T. Schilp, *US Patent* 5,498,312, 1996.
63. Lercel, M. et al., "Fabrication of masks for electron-beam projection lithography," *J. Vac. Sci. Technol.* B18, p. 3210, 2000.
64. Dicks, G. et al., "Mask blank fabrication, pattern transfer, and mounting distortion simulations for the 8-in. format SCALPEL Mask," *Microelectron. Eng.* 5758, 433, 2000.
65. Chen, C-f. et al., "Simulating the response of electron beam projection lithography masks under standardized mounting techniques," *J. Vac. Sci. Technol.* B19, p. 2646, 2001.
66. Takikawa, T., et al., "200-mm EPL stencil mask fabrication and metrology at DNP - IP and CD accuracy within subfield," *Proc. SPIE* 5374, p. 727, 2004.
67. Yamashita, H., K. Takeuchi, and H. Masaoka, "Mask split algorithm for stencil mask in electron projection lithography," *J. Vac. Sci. Technol.* B19, p. 2478, 2001.
68. Magg, C., and M. Lercel, "Chemically amplified resists for electron beam projection lithography mask fabrication," *Proc. SPIE* 3997, p. 276, 2000.
69. K. Nordquist, et al., "Inter- and intramembrane resist critical dimension uniformity across a SCALPEL mask," *J. Vac. Sci. Technol.* B18, pp. 3242–3247, 2000.
70. Ocola, L., et al., "Pattern placement correction methodology for 200 mm SCALPEL masks," *J. Vac. Sci. Technol.* B19, p. 2659, 2001.
71. Lercel, M., et al., "Patterning-induced image placement distortions on electron beam projection lithography membrane masks," *J. Vac. Sci. Technol.* B19, p. 2671, 2001.
72. Magg, C., et al., "An evaluation and comparison of the pattern transfer induced image placement distortions on e-beam projection lithography masks," *Proc. SPIE.* 4343, p. 374, 2001.
73. Reu, P. L., et al., "Simulating the effects of pattern density gradients on electron-beam projection lithography pattern transfer distortions," *J. Vac. Sci. Technol.* B19, p. 2652, 2001.
74. Wood, O. R., et al., "Ultrathin membrane masks for electron projection lithography," *J. Vac. Sci. Technol.* B22, p. 3072, 2004.
75. Yamashita, H., et al., "Lithographic performance of diamond-like carbon membrane mask in electron projection lithography," *J. Vac. Sci. Technol.* B22, p. 3067, 2004.
76. Nozue, H., A. Yoshid, and A. Endo, "Recent progress of LEEPL mask technology," *Proc. SPIE* 5374, p. 1043, 2004.
77. Pettibone, D. W., "UV inspection of EUV and SCALPEL reticles," *Proc. SPIE* 4186, p. 241 , 2001.
78. Yamamoto, J., et al., "EPL stencil mask defect inspection system using a transmission electron beam," *Proc. SPIE* 5037, p. 531, 2003.
79. Lercel, M. "Analysis of critical parameters for EPL mask fabrication," *Proc. SPIE* 4688, 2002.
80. Koops, H. W. P., et al., "Electron beam mask repair with induced reactions," *Proc. SPIE* 5148, p. 90, 2003.
81. Kawata, K., et al., "Stencil reticle development for electron beam projection system," *J. Vac. Sci. Technol.* B17, p. 2864, 1999.

82. Li, M., et al., "Four-inch photocurable nanoimprint lithography using NX-2000 nanoimprinter," *Proc. SPIE* 5374, p. 209, 2004.
83. Xia, Y., and G. M. Whitesides, "Soft lithography," *Angew. Chem. Int. Ed.*, 37(5), p. 550, 1998.
84. Michel, B., et al., "Printing meets lithography: soft approaches to high-resolution patterning," *IBM J. Res. Dev.* 45, p. 697, 2001.
85. Chou, S. Y., P. R. Krauss, and P. J. Renstrom, "Nanoimprint lithography," *J. Vac. Sci. Technol.* B14, p. 4129, 1996.
86. Chou, S., *DARPA Adv. Lithography Rev.*, Feb. 2004.
87. Colburn, M., et al., "Step and flash imprint lithography: a new approach to high resolution patterning," *Proc. SPIE* 3676(I), p. 379, 1999.
88. Hand, A., "One on one: a closer look at nanoimprinting," *Semicond. Int.*, 2004.
89. Bailey, T. C., et al., "Template fabrication schemes for step and flash imprint lithography," *Microelectron. Eng.* 61-62, p. 461, 2001.
90. Bailey, T. C., et al., "Step and flash imprint lithography: a low-pressure, room temperature nanoimprint patterning process," *Alternative Lithography. Unleashing the Potentials of Nanotechnology*, C. S. Torres, ed., Elsevier, Amsterdam, The Netherlands, 2002.
91. Hand, A., "Nanoimprint suppliers gathering support," *Semicond. Int.*, 2005.

Index

ABOUT THE EDITORS

Benjamin G. Eynon, Jr., is senior director of RAPID Marketing with KLA-Tencor Corporation. He has over 18 years of experience in the semiconductor industry in technology development, technology and manufacturing group management, team organization and leadership, and international professional conferences committee membership, and is a reviewer of peer-reviewed technology journal articles. He holds multiple U.S. patents and has over 25 professional publications.

Banqiu Wu, Ph.D., is a senior staff scientist with Photronics, Inc.. Prior to taking his present position, he was a research scientist at the University of Alabama and an associate professor at the Harbin Institute of Technology in China. Dr. Wu has nearly 20 years of research and industrial experience covering photomask fabrication, plasma etching, plasma enhanced chemical vapor deposition, plasma synthesis, and advanced material processing. He has published over 50 articles in several refereed journals and conference proceedings, and holds multiple patents.

Printed in the USA
CPSIA information can be obtained
at www.ICGtesting.com
CBHW080934250924
14027CB00008B/2

9 780071 445634